Smart Grid Fundamentals

Smart Grid Fundamentals

Energy Generation, Transmission and Distribution

Radian Belu

CRC Press
Taylor & Francis Group
Boca Raton London New York

CRC Press is an imprint of the
Taylor & Francis Group, an **informa** business

First edition published 2022
by CRC Press
6000 Broken Sound Parkway NW, Suite 300, Boca Raton, FL 33487-2742

and by CRC Press
4 Park Square, Milton Park, Abingdon, Oxon, OX14 4RN

ISBN: 978-1-4822-5667-3 (hbk)
ISBN: 978-1-032-19194-2 (pbk)
ISBN: 978-0-429-17480-3 (ebk)

DOI: 10.1201/9780429174803

Typeset in Times LT Std
by KnowledgeWorks Global Ltd.

Dedication

---

*To my wife Paulina Belu, my best friend and partner in life, for
her continuous encouragement, patience and support.*

*To my children Alexandru, Mirela and Maria-Ruxandra, and to my grandchildren
Stefan-Ovidiu and Ana-Victoria, my greatest joy in life and hope for the future.*

To all my teachers, professors and mentors.

*This book is also dedicated to memory of my parents Grigore
and Gheorghita Belu, my best teachers and mentors.*

Contents

Preface

Another book on smart grids published? Too much seems to have been published recently on smart grids and intelligent power systems. Over the last two decades, journals and whole conferences are devoted to the topic. Well-funded research programs and government-regulated subsidies stimulate and encourage investment in the new power systems. Smart grid, in several ways, represents a compilation of concepts, technologies and operating practices intended to bring the electric grid into the 21st century. All elements of smart grid include important engineering, economic and policy issues. The smart grid has offered unprecedented opportunities and alternatives for utilities, research and academic communities and end-user participations to the electric grid management, operation, control and market, enhancing the reliability, resilience, sustainability and capabilities for industry and customer choices in energy systems. The smart grid has made possible to set up microgrids that could be operated as standalone (island) operation modes in critical operating conditions or grid-connected in normal operation. Such small installations can enhance the reliability of regional electric power systems when the larger grid is faced with major contingencies. In addition, the smart grid allows microgrids to optimize the use of volatile and intermittent renewable energy resources and enhance the sustainability of regional power systems. Power system state estimation plays key role in the energy management systems (EMS) of providing the best estimates of the electrical variables in the grid that are further used in functions such as contingency analysis, automatic generation control, dispatch and others. The purpose of this textbook is to provide a comprehensive overview of smart grids, their role in the development of electricity systems, as well as issues and problems related to smart grid evolution, operation, management, control, protection, entities and components. This is accomplished by defining smart grids, highlighting the major drivers for deployment, outlining the range of actors and issues that need to be engaged and tackled and a vision for electricity system structure and development. The goal and aim of this book are to provide the engineers, students and interested readers with the essential knowledge of the power and energy systems, smart grid fundamentals, concepts and features, as well as the main energy technologies, including how they work and operate, characteristics and they are evaluated and selected for specific applications. The book is divided in 11 chapters, covering energy and environmental issues, basic of power systems and introduction to renewable energy, distributed generation and energy storage, smart grid challenges, benefits and drivers and smart power distribution. Last four chapters are focusing on smart grid communication, transmission, power flow analysis, smart grid design tools, energy management and microgrids. Students or readers using this book must have only fundamental knowledge of mathematics, physics and basic of power engineering as usually expected for most of the students enrolled in engineering programs. Likewise, this book assumes no specific knowledge of smart grids, or in-depth power and energy engineering; it guides the reader through basic understanding form topic to topic and inside each topic. This book originates from courses that the author taught in the areas of energy and power engineering, renewable energy systems, distributed generation, industrial energy systems, smart grids, microgrids and energy management, as well as from the research projects that the author was involved in the last 25 years. The book can be used as required or recommended textbook for an introduction course in smart grids that can be delivered in a 15-week semester or a 10-week quarter. The instructor can easy customize the materials, chapters and topics included in agreement with specific course outline. The author is fully indebted to the students, colleagues and co-professionals for their feedback and suggestions over the years, and last but not least to the editor technical staff for support and help. A rich, comprehensive and up-to-date literature is also included for professionals, engineers, students and interested readers in smart grid topics.

MATLAB® is a registered trademark of The Math Works, Inc. For product information, please contact:

The Math Works, Inc.
3 Apple Hill Drive
Natick, MA 01760-2098
Tel: 508-647-7000
Fax: 508-647-7001
E-mail: info@mathworks.com
Web: http://www.mathworks.com

About the Author

Radian Belu, PhD, is an associate professor in the Electrical Engineering Department at Southern University and A&M College, Baton Rouge, Louisiana. He has a PhD in power engineering and a PhD in physics. Before joining Southern University, Dr. Belu held faculty and research positions at universities and research institutes in Romania, Canada and the United States. He also worked for several years in the industry as project manager, senior engineer and consultant. His research focuses on energy conversion, renewable energy, microgrids, power electronics, climate and extreme event impacts on power systems. He has taught and developed courses in power engineering, renewable energy, smart grids, control, electric machines and environmental physics. His research interests include power systems, renewable energy systems, smart microgrids, power electronics and electric machines, energy management and engineering education. Dr. Belu has published 4 books, 15 book chapters and over 200 papers in referred journals and in conference proceedings. He has been a PI or Co-PI for various research projects in the United States and abroad.

1 Energy, Environment, Power System History, Evolution and Trends

1.1 ENERGY AND ENVIRONMENT, STATUS AND FUTURE POWER SYSTEMS

There are strong correlations between living standards, as measured by the gross domestic product per capita and the energy consumption per capita. It is quite natural that all developed or developing countries try to increase energy consumption per capita in order to maintain or to improve the living standard. The net effect is a significant increase of global energy demands, together with significant impacts on our environment due to the ways that the energy is generated and used. Electricity is the most versatile energy form and it can be accessed by over 5 billion people around the world through a series of tried-and-tested technologies and infrastructures. Modern power systems are based on centralized generation power plants, supplying the end-users or consumers, via long-established, unidirectional transmission and distribution systems. Power systems have served us well, in many cases for more than a hundred years, but times are changing, the power systems are facing new challenges. Societies and communities are demanding better quality, more robust and cleaner energy supplies to reduce pollutant emissions and so the demands for electricity are rising. This means more electricity must be generated from a larger variety of energy sources. Wind, solar, biofuel, and geothermal plants will all be needed, as well as coal, gas and nuclear energy, with significant consequences for the power system. Modern power systems or better said *electric energy systems* are made up of distinct sections, generation, transmission and distribution. The generation section includes the main components of power plants, electric generators and prime-movers, e.g. steam or hydro turbines. The energy sources used to generate electricity in the today power systems are fossil fuels, nuclear and atomic energy, hydropower. In thermal power plants, the heat is generated by burning fossil fuels, such as coal, oil or natural gas, or through a nuclear reaction, then converted into mechanical energy by the thermal turbines, and eventually into electricity. In hydroelectric power plants, water flows through hydro turbines convert the water kinetic energy into rotating mechanical energy. The turbines, either thermal or hydro are the prime-movers of the electrical generators that convert the mechanical energy into electric energy, distributed through power transmission and distribution networks to the consumers. Energy sources are loosely divided into three categories: *fossil fuels*, *nuclear fuels* and *renewable energy resources*. Fossil fuels include oil, coal and natural gas, while renewable energy sources include hydropower, wind, solar, geothermal, marine, hydrogen, biomass and biofuels. All these energy resources can be further classified as primary and secondary energy resources, while the secondary energy resources include all renewable energy sources, with the exception of hydropower. Over 95% of the electricity is generated by using primary energy sources. The chapter starts with a brief discussion of the energy sources and related energy issues. The chapter also introduces the reader to fundamental power and energy concepts, a brief history and evolution of power systems. The chapter concludes with a brief discussion of present and future energy demands and environmental implications of energy generation and use, followed by a discussion of the issues and the needs that are leading to the transition to smart grids. Several examples and end-of-chapter problems and questions are also included to enforce the discussed concepts and issues.

DOI: 10.1201/9780429174803-1

Energy conversion is concerned mainly with converting direct or indirect forms of solar radiant energy to electrical, mechanical or chemical energy. Energy cannot be created or destroyed, but is only transformed from one form to another. The sole exception is the atomic energy, which is derived from a reduction in mass of the nuclear fuels. Thus, strictly speaking, energy engineering is simply the engineering of transformation of energy between its different forms. We can distinguish between *primary energy forms*, found naturally in the environment, and *secondary energy forms* into which the former ones can be converted to enable easier transfer, uses and storage. Notice that there are energy forms that need less processing, being closer to the form in which they are used, the final state in which they are used for some purpose. An example of a primary energy form is oil which is found in nature, and a secondary form is electricity which is not found in nature but which is obtained from oil. The useful energy form may be work that is needed to perform a certain mechanical task. Energy is an essential component of our daily lives and a vital source of economic development and national welfare. During industrial revolution technologies have been discovered to convert heat into electricity, the most versatile and convenient energy form. Electricity enabled astonishing scientific and technological advances, transforming our civilization and way of life. However, it comes with unprecedented fossil fuel uses and adverse environmental impacts, making us dependent of a complex energy infrastructure for transportation, communication, heating, lighting, manufacturing and goods' distribution. Energy conversion is a multidisciplinary and disparate engineering subject, requiring deep understanding of physical and engineering principles. Energy issues tend to be open and controversial ones, needed to be address with open minds, being a rewarding and intellectually stimulating exercise. Throughout our history, the energy harnessing in its various forms presented great challenges and stimulated scientific and technological discoveries.

Modern energy technologies are the results of decades of advances and discoveries in science, technology and gradual equipment and devices design improvements. To meet the world needs of economic growth, there is a dramatic increase into the energy demands. Unlike developed countries, developing countries are struggling to meet such increasing energy demands and energy supply challenges. Increased energy needs put considerable stresses on the earth's resources and have had increasingly adverse environmental impacts. We are at cross-roads when energy uses must be critically assessed to determine more sustainable, environmentally friendly, and efficient approaches for energy generation, distribution and uses. The 21st-century economies face a two-fold energy challenge: meeting the needs of billions of people, still lack access to energy services while participating in a global transition to clean, low-carbon energy systems, both demanding urgent attention. The access to reliable, affordable and socially acceptable energy services is a pre-requisite for alleviating poverty and meeting development goals. However, pollutant emissions from developing countries are growing rapidly, significantly contributing to environmental problems, putting the people health and prosperity at risks. Energy use is marked by four trends: (1) rising consumption and transition from traditional energy sources to commercial energy forms, e.g. electricity or fossil fuels, (2) steady improvements into the efficiency of energy technologies, (3) fuel diversification and de-carbonization trends for electricity generation and (4) improved pollution control and lower emissions. These trends have largely been positive. However, the technology improvement rate has not been sufficient to keep pace with the rapid energy demand growth negative consequences. The challenge is not so much to change course as it is to accelerate progress, toward increased energy efficiency and lower-carbon energy sources. This has many concurrent benefits for developing countries in terms of public health, broader access to basic energy services and future economic growth. Moreover, to the extent that sustainable energy policies promote the development of indigenous renewable-energy industries, having additional benefits of creating new economic opportunities, reducing exposure to volatility of energy markets and conserving resources for internal investments, by reducing overall energy costs. In this context, renewable energy is becoming more relevant part of the energy solutions, included in national policies, with goals to be a significant part of generated energy in the near future.

Electricity, not a primary energy source is by far the most versatile, convenient and efficient mode, in which primary energy of fossil fuels, nuclear energy or energy of water is transferred, converted and used. For almost any equipment, the electricity generated in power stations and transferred to the equipment is a much cheaper than any other energy forms converted locally. Power systems are based on centralized generation plants that supply end-users via long-established, unidirectional power transmission and distribution systems. Electrical energy infrastructure, the power grid, after several decades of evolution and development are experiencing tremendous changes due to factors, such as: aging assets and infrastructure, increasing energy demands, limited generation and transmission capacity, needs for increased energy efficiency, environmental concerns, integration of renewable energy sources or increased concerns about system vulnerability to natural hazards and human attacks. The operation of today power grids is based on four levels resulting from its structure, as shown in Figure 1.1:

1. *Power Generation*: vast majority of electricity is generated by large power units located in strategic points with respect of power grid, usually near primary energy sources.
2. *Transmission System*: the grid section transferring electricity from large power plant to large consumption centres or other sub-transmission and power distribution systems. It is the backbone of the whole power system, containing sophisticated equipment and has a highly centralized management.
3. *Power Distribution Networks*: the interface between transmission and the end users (the customers), are connected to the transmission networks through substations, via power transformers and, for economic and practical operation reasons are usually operated in radial structures.
4. *End-users or consumers*: mostly passive consumers characterized by *non-controllable loads* and do not contribute to the power system management and operation.

Our power systems have served us well, well over one hundred years, but times are changing. There are increased demands for cleaner energy supplies to mitigate electricity generation environmental impacts, while the electricity uses keep increasing, more electricity must be generated from a larger variety of energy sources. Wind, solar, biofuel and geothermal energy sources are needed, as well as coal, natural gas or nuclear energy, with significant consequences for the power system. Such mixtures of renewable energy generation systems are introducing variations in the grid power quality. Weather patterns affect the availability of wind and solar energy, and the emergence of distributed power generation (e.g. rooftop solar panels) complicate the matters further, requiring residential networks to receive and deliver power. The existing electricity infrastructure is unable to manage such complexity, and needs to change. It needs to be equipped with advanced and intelligent communication and information technologies to monitor, analyze and manage supplies and electricity demands. However, the current power grid has not seen, with notable few exceptions, e.g. the use of computer and information technologies or electronic protection significant major changes. Traditionally, the operational objectives of the conventional electric utilities and power grids have been to provide reliable energy at minimum cost, stated as: (1) to supply quality power at constant voltage and frequency; (2) to minimize adverse effects on people and the environment; (3) to maintain adequate system security and reliability and (4) minimize energy losses. However, technological advances in small generators, renewable energy, power electronics, distributed

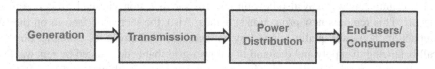

FIGURE 1.1   Typical power system structure.

generation, intelligent control, monitoring, protection, management and energy storage have provided new opportunities at the power distribution level, as well as in generation and transmission. These advances resulted in economic incentives to decentralize, change and expand the existing power systems.

The electricity grids are facing three looming and critical challenges: *its organization and structure, its technical ability to meet future, e.g. 25-year electricity needs* and *its ability to increase efficiency without diminishing, reliability, resilience, service quality and security*. Starting from 1995, the amortization and depreciation rates are exceeding the utility construction expenditures. Since this crossover point in 1995, utility construction expenditures have lagged behind asset depreciation, resulting in a mode of operation of the system that is analogous to harvesting more rapidly than planting replacement seeds. Because of these diminished "shock absorbers", the electric grid is becoming increasingly stressed, and whether the necessary carrying capacity or safety margin to support anticipated demand will exist is in question. In addition, energy, communications, transportation and financial infrastructures are becoming increasingly interconnected, thus posing new challenges for their secure, reliable, resilient and efficient operation. All of these infrastructures are complex networks, geographically dispersed, nonlinear and interacting both among themselves and with their human owners, operators and users. Energy, environmental issues, supply security, higher demands for increased power quality, resilience and faster service restoration have caused a marked increase in electricity production from renewable energy sources, local, distributed generation or extended use of energy storage since the beginning of the 21st century. The concept of sustainable development and concerns for the future generations are challenging us to develop new technologies for energy production, and new patterns of the energy uses. Their rapid emergence can make the understanding and therefore the perception of these new technologies difficult.

Our over a century-old power grid is the largest interconnected system on Earth, so massively complex and inextricably linked to human involvement and endeavour that it has alternately and appropriately been called an engineering ecosystem. It consists of more over 10,000 electric generating units with more than 1,000 GW of generating capacity connected to more than 300,000 miles of transmission lines. In many ways, the present power grid works exceptionally well for what it was designed to do, e.g. keeping the electricity costs down or providing quality services. Because electricity has to be used the moment it is generated, the power grid represents the ultimate in just-in-time product delivery. Everything must work almost perfectly at all times, and it does. The major evolution drivers for power systems are the needs to meet rising electricity demands, better service quality and lowering the costs, while reducing pollutant emissions to avoid irreversible changes to the environment. All this must be achieved without compromising the quality, reliability and supply security on which our economies are increasingly dependent. However, the even as demands have skyrocketed, there has been chronic underinvestment in getting electricity where it needed, through transmission and distribution, which are further limiting grid efficiency and reliability. While hundreds of thousands of high-voltage transmission lines course throughout the United States, only about 1,000 additional miles of interstate transmission have been built since 2000. As a result, system constraints worsen at a time when outages and power quality issues are estimated to cost the US businesses more than $100 billion on average each year. Nowadays, the advent of more and more distributed generation is changing this top-down paradigm. Wind-farms, biomass plants and large-scale photovoltaics (PV) are feeding electricity into the MV levels, while smaller PV systems as well as combined heat and power plants (CHP) and other small-scale electricity sources are generating electricity on the low-voltage (LV) level. This change in generation has several consequences. For example, power may also flow into the reverse direction, caused by the feed-in at lower voltage levels. This requires new protection schemes. Also, the increased feed-in on the LV level has an impact on the power quality, caused by voltage band violations, for example. Furthermore, the overall balancing of supply and demand in a power grid that is dominated by renewable energy sources is unequally more complicated. The reason for this is a combination of two factors. First, the electricity generated by many of the renewable energy sources directly depends on the weather conditions, causing intermittent feed-in. Second, as the grid itself cannot store electricity, measures have to be taken for bridging the times with little renewable energy feed-in. In the short term, conventional power plants need to compensate the fluctuating feed-in from renewable energy sources. However, in the long run these plants shall be replaced by renewable sources completely. To achieve this, new and cheap storing technologies are required and additional flexibility on the demand side has to be exploited to be able to shift loads from low generation to high generation times.

The electric industry is poised to transform the electric power systems, from a highly centralized, producer-controlled network to one that is less centralized and more consumer-interactive. The move to a smarter grid promises to change the entire power industry business model and the relationship with all stakeholders, involving and affecting utilities, regulators, energy service providers, technology vendors and electricity users of all types. The power grids are undergoing changes towards more localized energy production and storage supported by distributed energy generation, larger energy storage and increased use of computing, IT and communication. On a worldwide basis, the smart grid (SG) developments are a consistent answer to the problems of an efficient and sustainable electricity delivery through power distribution and consumer participation. A smarter grid makes such transformation possible, bringing the concepts and technologies that enabled information technology and computing to the electric grid. More importantly, it enables the best grid modernization ideas to achieve full potential, opening the door to new applications with far-reaching impacts: providing the capacity to safely integrate more renewable energy sources, electric vehicles and distributed generators into the network; delivering power more efficiently and reliably through demand response and comprehensive control and monitoring capabilities; using smart grid reconfiguration to prevent or to restore outages (self-healing capabilities); enabling consumers to have greater control over their consumption and to actively participate in the electricity market. SGs are a combination of information, communication, control and monitoring technologies and new energy technologies. For example, smart metering is going to provide fine-grained measurement and automatic remote reading of the consumption and production amounts. It enables flexible tariffing and dynamic load optimization, ultimately aiming at cost and consumption reduction. The related security requirements are mainly authenticity, integrity and privacy of metering data. Even more challenging is grid automation, which is critical for the safety and availability of the grid. The overall situation calls for an integrated security architecture that not only addresses all relevant security threats but also satisfies functional, safety, performance, process integration and economic side conditions.

There are several smart grid definitions and concepts, appearing critical to gather the available knowledge from both industry, academia and research laboratories in one handbook. In general sense the smart grid is an electricity network that can intelligently integrate the actions of all users connected to it, generators, consumers and those that do both, to efficiently deliver sustainable, economic and secure electricity supplies. It employs innovative products and services together with intelligent monitoring, control, communication and self-healing technologies to: (a) facilitating the connection and operation of generators of all sizes and technologies; (b) allow consumers to play a part in optimizing the system operation; (c) providing consumers with greater information and choice of supply; (d) significantly reduce the environmental impact of the whole electricity supply system and (e) deliver enhanced levels of reliability and security of supply. In summary, a smart grid for electricity is an intelligent, auto-balancing and self-monitoring power grid, accepting power generated from any source (coal, sun, natural gas or wind) by a facility of any size or location, centralized or distributed. It then uses advanced sensors, state-of-the-art applications and distributed computing to transfer that electricity effectively, with minimal human intervention, to customers for their uses. Distributed generation is rightly receiving an increased attention and it becomes an integral and critical part of new energy systems, providing consumers and energy providers with safe, affordable, clean, reliable, flexible and readily-accessible energy services. However, the conventional design of the network control with a centralized structure is not in line with the paradigm of the unbundled electricity generation systems and decentralized control, as highlighted by how the active networks are efficiently linking the small- and medium-scale power sources with consumer demands, allowing decisions to be made on how best to operate them in real time. It also looks at the level of control required: power flow assessment, voltage control and protection require cost-competitive technologies and modern communication systems with more sensors and actuators than presently used, certainly in relation to the distribution systems. The high-level requirements and trends towards decentralized generation systems are suggesting decentralized communication architectures. While data management is an important SG aspect, there also are needed ways to manage SG functions. To manage active networks, the grid computing needs universal access to computing resources. An intelligent grid infrastructure gives more flexibility concerning demand and supply, providing new instruments for optimal and cost-effective grid operation at the same time. A smart grid uses information and communications technology to gather and act on information, such as information about the suppliers and consumers behaviours, in an automated mode to improve efficiency, reliability, economics and sustainability of the electricity generation and distribution, in which electricity generation from renewable energy resources of small to medium powers are presented. The basic concepts necessary to understand the SG operating characteristics are also introduced, and the constraints and problems of distributed and renewable energy integration in the electrical networks are also discussed in this chapter.

1.2 FUELS AND SOURCES OF ENERGY

The vast majority of energy, including electricity consumed around the world is provided by fossil fuels, such as: coal, oil and natural gas. At the same time, all fossil fuels are not the same in terms of their availability and their environmental impacts. Fossil fuels are coming from layers of prehistoric carbonaceous materials, compressed over millions of years to form solid, liquid or gas high energy-dense concentrations. These can be relatively easy extracted, transported and combusted to meet human energy needs. Fossil fuels are varying in terms of energy density, the energy content per unit of mass or volume, current consumption rates in different countries or world regions, remaining availability of resources and emissions per unit of energy released during combustion. Fossil fuels are the dominant primary energy source worldwide, counting for over 85% of all world energy production. Because fossil fuels are so important for meeting our energy needs, sustainable energy consumption must consider high-efficient fuel combustion technologies, energy efficiency and conservation, distributed generation and the move toward the use of renewable and alternative energy sources. About 84% of global electricity is generated by fossil fuels, 3% from nuclear plants and the remaining 13% comes from renewable sources, such as hydropower, wind, solar, biofuels, geothermal, wave and tidal power. The US energy generation follows a similar pattern. About 41% of the US energy comes from oil, of which a significant part is still imported. However, there is strong tendency in declining the imports, mainly due to the present availability of shale oil (and natural gas) is decreasing this dependence, the United States is almost energy independent and is becoming one of the large exporters. The global per capita energy consumption is highest in the United States. With nearly 4.5% of world's population (313 million out of 7 billion), the United States consumes nearly 28% of global energy, reflecting a very high standard of living. In comparison, China, now the world's second largest economy, with nearly 19% of the world's population, consumes nearly half the total US energy. The world has enormous reserves of coal, and at the present consumption rate it is expected to last about 300 years. The natural gas reserves are expected to last around 150 years. However, the recent availability of large amounts of shale oil and gas is creating an economic boom in some countries, particularly in the United States, as mentioned previously. Uranium (U-235) has a very low reserve and is expected to become exhausted in about 50 years or so. Unfortunately, fossil fuel burning generates gases (SO_2, CO, NO_X, HC and CO_2) that cause environmental pollution.

From the three major fossil fuels used today, coal can be used directly only in stationary applications, mostly in industrial applications, and on limited scale in some of the domestic applications.

Natural gas and oil are more flexible in the ways that are combusted, being used extensively for transportation and also for stationary applications. Natural gas requires less pre-processing before combustion, and has less adverse environmental impacts than coal or oil. One important characteristic of the fossil fuels is the energy density, a measure of the amount of energy per unit of mass or volume available in that resource. It is important parameters because fossil fuels with lower energy density must be provide to energy conversion process in greater quantities, increasing the overall cost and the environmental impact, than the higher energy density ones. Coal energy density is expressed in GJ/tone or millions of BTU/ton, the one of oil in GJ/barrel or BTU/barrel, while the energy density of natural gas in kJ/m^3 or BTU/ft^3 when the gas is at atmospheric pressure and its value can be considered constant. However, in practice natural gas is compressed in order to save space during transportation and storage. Typical values of coal energy density are between 15 and 30 GJ/tone, about 5.2 GJ/barrel for crude oil and 36.5 MJ/m^3 for natural gas. Among all tree cola is the most difficult to extract, transport and use. The cost of these fossil fuels varies largely from country to country, year to year and region to region with price fluctuations not dictate only by the market. However, when the energy prices rise disproportionately, they tend to choke off the economic growth, leading to reduction in demands, so the energy prices to fall again. On long run, high energy costs motivated research and applications of high energy-efficient technologies, easing the pressure on energy demand and allowing the cost growth to slow down or reverse.

Nuclear fuels are heavy materials, used for energy generation, mostly uranium (U), but plutonium (Pu) is also used. Nuclear energy term has different meanings to different people. For an energy engineer nuclear energy is the controlled release of the nuclear fuel energy for electricity generation. Nuclear energy can be obtained either through *fission* (heavy atoms are splitting in two or more smaller atoms) or fusion (joining together of to light atoms into a larger one). Energy release during such processes is transferred to a working fluid, and then the thermal energy is converted into electricity in a similar conversion process as the ones in a conventional thermal power plant. The energy releases during such processes are much larger than the chemical energy releases during the combustion processes. There are no or very little pollutant emissions during construction and operation of a nuclear power plant, as well as in all other related processes. However, the major problem with nuclear power is that the waste from nuclear plants remains radioactive for thousands of years, and we do not know how to safely dispose of such waste. It is quite possible that in the future such waste can cause considerable environmental damages. Managing the radioactive materials requires unique techniques, reactor shielding, careful moving and transportation of by-products and waste, power plant control and monitoring, etc. Data for uranium production are not freely available for political and security reasons. The largest uranium producers are Canada, Australia, Niger, Namibia, Russia, United States and former USSR states. Consumption of nuclear energy is mostly distributed in North America and Europe with about 80%, and the rest mostly in Asia.

Hydroelectric power plants harness the energy of the Earth hydrologic cycle, while converting it into electrical energy. Water from ocean, lakes, rivers, plants, etc. absorbs the solar energy evaporates into the atmosphere forming clouds, eventually return back to ground as precipitations, and through the ocean and lakes through water streams. The motion of water through ocean and lakes is due to kinetic energy, which can be harnessed by hydro-power plants and converted into electrical energy. If water is stored at high elevations, usually in a reservoir it poses potential energy proportional to that elevation. When this water is allowed to flow from high to low elevation, the potential energy is converted into kinetic energy, and converted through a hydro-power turbine into electricity. The common types of hydroelectric power plants are impoundment hydroelectric (involving a large water reservoir formed by a dam), diversion hydroelectric (some water of a river with strong current is diverted through hydropower turbines) and pumped storage hydroelectric systems, used as a form of energy storage. Amount of electricity generation depends on the water head behind the dam, reservoir capacity, flow rate, topography, and efficiency of the hydroelectric power plant components. The electricity generation system description is beyond the scope of this book, interested readers are directed to the book references or elsewhere in the literature.

It is commonly accepted that the earth's fossil energy resources are limited, and sometimes in the future their production comes beyond their peaks. At the same time there is strong opposition against the nuclear power in many parts of the world. In this scenario renewable energy resources have to contribute more and more to the world's ever rising energy needs. The major renewable energy resources are the sun, with some forms also attributed to the moon and the earth. Notable for their contribution to the current energy demands are water, wind, solar energy and biomass. Renewable energy is becoming increasingly used in electricity generation due to the technological advances in wind turbine, photovoltaic systems, energy storage, power electronics and control technologies. Renewable energy sources also provide cleaner energy, with fewer pollutants. For this reason, many policy experts and scientists advocate renewable energy sources over traditional fossil fuels. The difficulty is to achieve the technology, infrastructure and political support to make this transition. According to NREL reports electricity supply and demand can be balanced in each region with nearly 80% of electricity from renewable energy sources, including nearly 50% from variable renewable energy generation, according to power generation projections. However, as renewable energy electricity generation increases, additional transmission infrastructure is required to deliver power from cost-effective renewable energy resources to load centres, enable reserve sharing over greater distances and smooth output profiles of variable resources by enabling greater geospatial diversity. Many of the system flexibility resources and options can benefit from such transmission infrastructure enhancements to enable the transfer of power and sharing of reserves over large areas to accommodate the wind and solar energy variability in combination with variability in electricity demand.

1.2.1 RENEWABLE ENERGY SOURCES

Renewable energy is defined as energy coming from naturally replenished energy resources on a human timescale, such as: sunlight, wind, rain, tides, waves and geothermal heat. It can replace conventional fossil fuels in three areas: electricity generation (including off-grid energy service), heating and cooling and engine fuels. In all forms, renewable energies derive directly from the Sun, or from heat generated deep within the Earth. Included here is the electricity and heat generated from solar, wind, marine, hydropower, biomass, geothermal resources, biofuels and hydrogen derived from renewable energy sources. Vast majority of renewable energy sources derive from solar radiation. Direct solar energy refers to solar thermal energy conversion and PV, while indirect solar energy includes wind and wave energy, hydropower and biofuels. Other types of renewable energy are the tides (due to Moon attraction), and geothermal power. The magnitude of renewable energy sources is huge and may easily supply all our present and future energy needs. Non-solar renewable energies are those that do not depend direct on solar radiation. One type of renewable energy, not derived from natural resources is the conversion the wastes into energy, often producing methane that can be used to generate electricity or to fuel engines. Major advantages of the renewable energy include: sustainable, virtually inexhaustible and non-polluting, free fuels, ideal for off-grid and distributed generation, while the main drawbacks are: variability and intermittence, lower energy density, generation sites are often located far from populated areas, higher initial investment, higher maintenance costs and some, either limited, adverse environmental impacts. It is out of question that in the future large energy proportions will be from renewable energy sources. Renewable energy can be regarded as one of the fundamental premises for building a sustainable society. A truly sustainable energy source is meaning that the energy production and uses are endured by the biosphere, the biodiversity, environment and the social stability of the human culture.

Renewable energy resources exist over wide geographical areas, in contrast to conventional energy resources, concentrated often in a very limited number of countries. Rapid deployment of renewable energy and energy efficiency are resulting in significant energy security, climate change mitigation and economic benefits. There is also strong public support for promoting renewable energy. At the national level, many nations around the world already have renewable energy contributing significantly to their energy supply mixture. Renewable energy markets are projected to continue to grow strongly in the coming decades and beyond. For example, solar energy, the most important renewable energy sources, is plentiful, having highest availability compared to other energy sources. The amount of solar energy supplied to the Earth in one day is sufficient to supply the world total energy needs for an entire year. Solar energy is clean and almost free of emissions, it does not produce pollutants or harmful by-products. Solar energy conversion into electrical energy has many application fields. Residential, vehicular, space, aircraft and naval applications are the main solar energy applications. Sunlight has been used as an energy source by ancient civilizations to ignite fires and burn enemy warships using "burning mirrors". Until the 18th century, solar power was used for heating and lighting purposes. During the 1800s, Europeans started to build solar-heated greenhouses and conservatories. The main renewable energy sources give rise to a multitude of very different energy flows and carriers due to various energy conversion processes occurring in nature. In this respect, wind energy, hydropower, ocean current energy, bio-fuels all representing conversion of the solar energy. The energy flows available on Earth resulting from renewable energy sources vary tremendously, in terms of energy density or with regard to spatio-temporal variations. Major renewable energy sources include: solar radiation, wind energy, hydropower (in all forms), photo-synthetically fixed energy and geothermal energy. Appropriate techniques permit the exploitation and conversion of different renewable energy carriers into secondary or final energy, energy carriers or useful energy, respectively. Currently, there are tremendous variations in terms of utilization methods, technology and given perspectives. However, not all options are possible for every site or location. The most promising renewable energy options include:

- Solar heat provision by active systems (i.e., solar-thermal collector systems);
- Solar thermal electric provision (i.e., solar tower plans);
- Photovoltaic energy conversion systems;
- Power generation by wind energy conversion systems;
- Power generation by hydropower to provide electric energy;
- Utilization of ambient air and shallow geothermal energy for heat provision;
- Utilization of deep geothermal energy resources for heat and/or power provision;
- Wave and tidal energy for electrical energy provision and
- Utilization of photo-synthetically fixed energy for heat, power and transportation fuels.

Solar radiation represents the electromagnetic energy emitted by the Sun, while the terrestrial solar energy is the portion of the total solar radiation reaching the Earth's surface. The terms insolation and irradiance are used interchangeably to define solar radiation incident on the earth per unit of area per unit of time as measured in W/m^2 or $Wh/m^2/day$. The solar radiation amount, reaching the Earth's surface at any given location and time depends on many factors, including time of day, season, latitude, surface albedo, the atmosphere translucence and the weather conditions. The primary causes of winds (horizontal atmospheric motions) are the solar radiation uneven Earth and atmosphere heating and Earth's rotation. The atmosphere reflects about 43% off the incident solar radiation back into space, absorbs about 17% of it in the lower troposphere and transmits the remaining 40% to the surface of the earth, where much of it is then reradiated into the atmosphere. The Sun short wavelength radiation (0.15–4 mm) passes readily through the atmosphere, while the longer wavelengths (5–20 mm) are absorbed by the atmospheric water vapours. Thus, Earth radiation is primarily responsible for the warmth of the atmosphere near the Earth's surface. Heat is also transferred from the earth's surface to the atmosphere by conduction and convection. On average, the total amount of energy radiated to space from the earth and its atmosphere must be equivalent to the total solar radiation amount absorbed, or the temperature of the earth and its atmosphere would steadily increase or decrease. The more nearly perpendicular the sun's rays strike the earth, the more solar radiation is transferred through the atmosphere. During the year, tropical regions receive significantly more solar energy than the polar areas. Winds and ocean currents level out this thermal imbalance, preventing the tropical regions from getting progressively hotter and the Polar Regions from getting progressively colder. In addition, Earth's surface inhomogeneity, land, water, desert, forest, rocks, sands, etc. leads to differences in solar radiation absorption and reflection back to the atmosphere, creating large differences in atmospheric temperatures, densities and pressures, which in turn, are shaping the local wind regimes. It may be worth noting that the power densities per-area by kinetic wind energy and solar radiation are the same order of magnitude, when considering exploitable values in a given are. At 20 m/s the wind exerts on a vertical plane 1.04 kW/m^2, while the solar radiant flux density on a horizontal surface on June 21st at noon, 50 N latitude is 1.05 kW/m^2. However, the efficiency conversion is about 40% for wind energy conversion systems and about 15% for photovoltaic generators.

Renewable energy is generally defined as energy coming from naturally replenished energy resources on a human timescale, such as: sunlight, wind, rain, tides, waves and geothermal heat. Renewable energy can replace conventional fossil fuels in three areas: electricity generation (including off-grid energy service), heating and cooling and engine fuels. In all forms, renewable energies derive directly from the Sun, or from heat generated deep within the earth. Included in this definition is electricity and heat generated from solar, wind, ocean, hydropower, biomass, geothermal resources, biofuels and hydrogen derived from renewable resources. Vast majority of renewable energy sources derive from solar radiation. Direct solar energy refers to solar thermal energy conversion and PV, while indirect solar energy includes wind and wave powers, hydropower and biofuels. Other types of renewable energy are the tides (mainly due to Moon attraction) and geothermal power. The magnitude of renewable energy sources is huge and may easily supply all our present and future energy needs. Non-solar renewable energies are those that do not depend direct on solar radiation. One type of renewable energy, not derived from natural resources is the conversion the wastes into energy, often by producing methane that can be used to generate electricity or to fuel engines. It is very important to highlight the difference between primary energy sources and energy storage medium. Major advantages of the renewable energy include: sustainable, virtually inexhaustible and non-polluting, fuel is free, ideal for off-grid and distributed generation, while the main drawbacks are: variability and intermittence, lower energy density, generation sites are often located far from populated areas, large initial investment, high maintenance costs and sometimes adverse environmental impacts. It is out of question that in the future large proportions of energy are from renewable energy sources. Renewable energy can be regarded as one of the fundamental premises for building a sustainable global society. A truly sustainable energy source is not only be renewable, but is also sustainable, meaning that energy production and use is endured by the biosphere, the biodiversity and the social stability of the human culture.

1.2.2 Future Energy Demands and Environment

Coal and crude oil were not relevant to the energy production until the end of 19th century. Firewood and wind and hydropower techniques provide most a fluid of the energy used. Windmills and watermills were the common features of the landscape for most of the medieval age. Coal become the single most important energy source, after the discovery by James Watt in 1769 of steam engine, followed of the explosive spread of its usage in industry and transportation. The steam engine, and later the internal combustion engine, replaced mechanical wind and water installations at the end of nineteenth and begging of twenty centuries. During the first part of the last century oil took as it was needed to provide energy for increased motorized transportation. During these periods firewood lost its importance as an energy source, and large hydro-electric power systems replaced the watermills. Around and after Second World War the energy demand increased dramatically. For example, the energy consumption in the United States increased from 32 quads in 1950s over 100 quads in 2010s. One quad is equal to 10^{15} Btu. Similar trends are in all developed countries and in many developing nations. Much of the energy used today is in the form of electricity. Figure 1.2 shows the share of energy sources used to generate electricity in the United States. The majority of

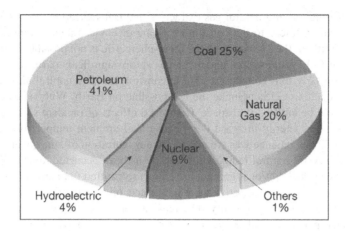

FIGURE 1.2 The US energy mix for electricity generation.

electricity comes from fossil fuels, hydropower, nuclear energy, while a tiny fraction (˜ 1%) comes from other sources. Global energy consumption in the last half century has increased rapidly and is expected to continue to grow in near future, with significant differences between the last 50 years. The past increase was stimulated by relatively "cheap" fossil fuels and increased rates of industrialization in developed countries yet while energy consumption continues to increase, while additional factors are making the picture for the future more complex. These include the very rapid increase in energy use in China and India (countries representing about a third of the world's population), the expected depletion of oil and coal resources sometimes in the future and the effect of human activities on environment. On the positive side, the renewable energies of wind, biofuels, fuel cells, wave energy, solar thermal and PV are finally showing maturity and the ultimate promise of cost competitiveness.

Our current living standard could not be sustained without energy. The provision of energy or the related energy services (e.g. heated living spaces, information and transportation) involves a huge variety of environmental impacts which are increasingly less tolerated by the society of the 21st century, making "the energy problem" a major topic in engineering education, research and policy of almost all countries. This attitude it is not expected to change within the near future. On the contrary, in view of the increasing knowledge and recognition of the effects associated with energy utilization in the broadest sense of term, increased complexity has to be expected. In 2010, the worldwide consumption of fossil primary energy carriers and hydropower account to approximately 450 EJ. Roughly 28% of this overall energy consumption accounts for Europe and Eurasia, about 27% for North America, 5% for Central and South America, roughly 5% for the Middle East, 3% for Africa and 32% for Asia and the Pacific region. North America, Europe and Eurasia as well as Asia and the Pacific region consume about 90% of the currently used primary energy derived from the fossil energy carriers and hydropower. This energy consumption has increased over 2.5 in the last four decades. All energy sources have some environmental impacts, however the impacts vary considerable across the energy spectrum. The energy sources, such as coal or oil, in general fossil fuels have higher harmful effects on the environment compared to the renewable energy ones, considered more benign, having very limited harmful impacts. Furthermore, fossil fuel environmental impacts have gotten worse, due to the population growth, increased energy use per capita, as a result of the rise in the worldwide loving standards. Living standards and the energy use per capita are strongly correlated either higher per capita wealth not invariably requires higher per capita energy use. Besides the pollutant emissions, during electricity production, pollutants are released into environment during the operation of fossil fuel-fired heating, with various environmental effects. In addition, the exploitation of fossil fuels is also associated with effects damaging the environment, during the well drilling, oil and natural gas extraction and transportation, processing in refineries or transportation from refinery to the consumers. Some of environmental impacts may less dramatic than others. Without the protection of the atmosphere, life is not possible. Minor atmospheric constituents, such as CO_2, water vapour and methane capture significant parts of the incoming solar radiation, acting similarly to a greenhouse. These constituents have natural and artificial origins. Existing natural greenhouse effects make the life possible on Earth. Without greenhouse effects, Earth radiation would be emitted into space. Combined effects of incident solar radiation and the retention of this heating energy increase the mean global ambient temperature to about +15°C. Over millennia, a delicate balance was created in the concentration of atmospheric gases and constituents. However, several natural temperature variations have occurred, during Earth history, as evidenced by the ice ages. Additional greenhouse gases are emitted to the atmosphere as a result of energy generation and other human activities. Their effects on the climate are not fully understood and accepted by the scientific community. The reasons for climate change are controversial and debatable. As a fact, part of global temperature rise of 0.6°C during the last century is linked to natural fluctuations, while the rest is believed to be of anthropogenic origin. Detailed prediction of the anthropogenic greenhouse effect consequences is not possible. Climatic models are only giving an estimate of what would happen if the current emissions remain the same or increases.

1.3   BRIEF HISTORY OF POWER SYSTEMS

Power engineering, power system analysis and theory are very broad subjects to be fully covered, even at the most basic level in a single textbook. Power and electrical engineers are concerned with every step and aspects in the process of generation, transmission, distribution and utilization of electricity. Power industry is probably the largest and the most complex industry in the history of mankind. Electrical and power engineers who work in this industry encounter challenging problems in designing and shaping future power systems to deliver increasing amounts of electrical energy at highest quality standards, in safe, clean, efficient and economical manner. Power system industry and utilities significantly contribute to the welfare, life standard, progress and technological advances of humanity. The growth and demand of electrical energy in the world as a whole and in each country in the last half a century was phenomenal, with over 50 times as much as the growth and demand rates in all other energy forms used during the same period. Today the installed electrical power capacity in the US is estimated at about 3 kW, with similar values in the E.U. and developed countries. There is also significant increasing in the installed capacity per capita in China, Brazil or India and in most of the developing countries. The first electric utility in North America was started in 1882 in New York City by Thomas Edison. In 1878 Thomas Edison began work on electric light and formulated the concept of centrally generated power with distributed lighting serving a surrounding area. Historically, Edison Electric Illuminating Company of New York inaugurated the first commercial power plant at Pearl Street station in 1882, with six engine-dynamo units (DC generators) and a capacity of four 250-HP boilers. It uses a 110-V DC underground distribution with copper cables insulated with jute wrapping. In 1882, the first water wheel-driven was installed in Appleton, Wisconsin. The introduction of DC motor by Sprague Electric, the growth of incandescent lighting and the development of three-wire 220-V DC systems, which allowed load to increase somewhat promoted the expansion of the Edison's DC systems. However, the voltage problems and transmission over long distances remain in the DC systems. These limitations of the maximum distance and load were overcome by the development of William Stanley of a practical transformer. With the transformer, the ability to transmit power at high voltage and lower line voltage drops made AC more attractive than DC. The first practical AC distribution was installed by W. Stanley at Great Barrington, Massachusetts, in 1885 for Westinghouse, who acquired the American rights to the transformer form its British inventors. First single-phase AC line of 4 kV was operated in United States in 1889, between Oregon City and Portland, over 20 km distance. The first three-phase line in Germany became operational in 1891, transmitting power 179 km at 12 kV. Southern California

Edison Company established the first three-phase 1.3 kV system in 1893. Nikola Tesla invention of the induction motor in 1988 helped replace DC motors and hastened the advance in use of AC systems. In 1884, City of Timisoara, Romania was the first city of Europe with electric street lighting installed. From these modest beginnings, the power grid has grown to cover the entire continents, providing almost everyone with reliable electricity. The grid evolved as a centralized unidirectional system of electric power transmission, distribution and demand-driven control. In the 20th century local grids grew over time, and were eventually interconnected for economic and reliability reasons. By the 1960s, the electric grids of all developed countries had become very large, mature and highly interconnected, with thousands of generation power stations delivering power to major load centres via high-capacity transmission lines, branched and divided to provide power to smaller industrial and domestic users over the entire supply area. The topology of the 1960s grid was a result of the strong economies of scale: large coal-, gas- and oil-fired power stations in the 1 GW (1,000 MW) to 3 GW scale are still found to be cost-effective, due to efficiency-boosting features that can be cost effectively added only when the stations become very large. Present day electric was designed to operate as a vertical structure consisting of generation, transmission and distribution, supported with controls and protection and monitoring devices and systems to maintain system reliability, stability and efficiency.

The US electrical system initially consisted of isolated local areas with two electric power components; generation and distribution. The grid was developed to interconnect isolated areas, enabling electric power transmission and creating a network of formerly isolated areas. Instead of having small power systems, or local areas of power generation and distribution, the system was expanded into three components; generation, transmission and power distribution, allowing electric power suppliers to take advantage of economies of scale. Power quality was poor before the transmission component was added, simply because of the nature of electricity supply and demand. For example, if a local power generator could not meet demand, the drag on limited generation capacity resulted in sagging electrical frequencies and voltages. During the early days of the power system, most of the power load was for lighting. When demand exceeded supply, the resultant, sags merely caused lights to dim. From a practical perspective, this was a benign problem, but as electricity started to be used for devices such as motors, radios and other communications devices, power quality became important. It became necessary for electrical voltage and frequency to be maintained at standard levels. Consistent power quantity, or reliability, followed much the same path as power quality. Before the transmission component was added to the electrical power system with the development of the grid, there was nowhere to get supplemental power when demand exceeded supply. These facts resulted in a power system that could not reliably furnish electric power. This was especially problematic when there was zero supply, as was the case when local generation facilities went off-line, usually because of equipment failure. Transmission, the third power component, allowed electricity to be transported, or transmitted, between formerly isolated areas, resulting in improved quality, supply security and reliability. Local power generators began to establish transmission interconnections, or power grids, by building transmission lines with neighbouring power suppliers to improve quality and reliability. Local grids developed into regional electric grids which, collectively, make up the national grid as we know it today. The national grid, with its interconnecting transmission capabilities, allows customers to receive power from adjacent power producers, resulting in a more reliable system than was initially available. The simplified diagram of the electricity distribution from generation stations to consumers, Figure 1.3, is showing the three main electric grid components, generation, transmission and distribution of the power grid with common voltage levels.

The electric power system evolves in a complex technical system with main function to deliver electricity between generation, consumption and storage. A power system is an interconnected network with components converting nonelectrical energy continuously into the electrical form and transporting the electrical energy from generating sources to the loads/users. The interconnection of solated power systems brought both benefits and risks. The most important benefit was the electricity generation could be shared among power distribution networks. Since power plants have

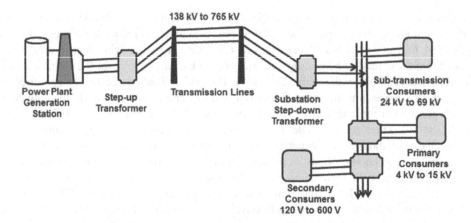

FIGURE 1.3 Electric grid structure, the most common voltage levels are included.

significant scale economies, allowing becoming cheaper to produce electricity. Another benefit was improved power grid reliability, since the failure of a local generator could be offset by another generator farther away, with consumers even noticing that there was a problem. Such was and is the case, the vast majority of time. However, the fact that localized distribution was now highly interdependent exposed utilities to the disruptive event risks miles away. It is inevitable that an electrical grid built on such a huge scale in a patchwork manner over 100 years has reliability issues. Several cascading failures during the past 40 years have highlighted the need to understand the complex phenomena associated with power systems and the development of emergency controls and restoration. Furthermore, the US electrical grid has been plagued by ever more and ever worse blackouts over the past 15 years. In an average year, outages total 92 minutes per year in the Midwest and 214 minutes per year in the Northeast. Japan, by contrast, averages only 4 minutes of interrupted service each year. In addition to the mechanical failures, overloading a line can create power-supply instabilities such as phase or voltage fluctuations. For an AC power grid to remain stable, the frequency and phase of all power-generation units must remain synchronous within narrow limits. A generator that drops 2 Hz below 60 Hz will rapidly build up enough heat in its bearings to destroy itself. As a result, circuit breakers trip a generator out of the system when the frequency varies too much. However, much smaller frequency changes can indicate instability in the grid: in the Eastern Interconnect, a 30 mHz drop in frequency reduces power delivered by 1 GW. Moreover, power outages and power quality disturbances cost the US economy more than $80 billion annually.

Power systems and communication infrastructure are close interrelated, either may not be apparent at first, but they are twin topics of electrical engineering. Both, electric grids and communication infrastructure is the same field with different emphasizing, and both are transmitting electric power. However, the communication infrastructures are trying to minimize power and maximize the information content, while the power grids are trying to maximize the power and to minimize the information content. It is of great scientific and practical interests to see and learn what happens when these two fields of electrical engineering are physically coming together in specific technologies, e.g. power line carrier, in which communication attempts to became similar to power, following the same conductive paths or wireless power transmission, in which power seeks to propagate through space, similar to wireless communication. The electricity or power grids are facing at least three looming challenges: *the optimal organization and structure, the technical ability to meet the future 25-year and 50-year period electricity needs* and *the ability to increase its efficiency without diminishing its reliability and security*. Starting at the very end of the 20th century, the amortization or depreciation rate exceeded utility construction expenditures for the first time. Since this crossover point in 1995, utility construction expenditures have lagged behind asset depreciation, resulting in a mode of operation of the system that is analogous to harvesting more rapidly than planting replacement seeds. Because of these diminished *shock absorbers*, the electric grid is becoming increasingly stressed, and whether the necessary carrying capacity or safety margin to support anticipated demand will exist is in question. However, system operators are facing new challenges including penetration of distributed energy resources (DER) in the legacy system, rapid technological changes, rapid technological changes and different types of market players and end users. In addition, energy, telecommunications, transportation and financial infrastructures are becoming increasingly interconnected, thus posing new challenges for their secure, reliable and efficient operation. All of these infrastructures are complex networks, geographically dispersed, nonlinear and interacting both among themselves and with their human owners, operators and users. A sustainable energy system involves key components, such as: increasing use of renewable energy resources, increased energy efficiency and use of electricity in the transportation sector. As an effect of this some key questions for the electric power system are; integration of intermittent electricity generation, e.g. from wind power, and connections of electrical vehicles to the electrical distribution system and different solutions for energy storage. From 1882 through 1972, the electric utility industry grew at a remarkable pace, based on outstanding technological and scientific advances and engineering creativity. In the United States, only the electric energy sales have grown to well over 400 times during this period. A growth rate was 50 times as much as the growth rate in all energy forms used during the same period. During the early begins, the electricity was generated in steam-powered and water-powered turbine plants. Today, steam turbine accounts for about 85% of the US electricity generation, while hydro-turbine for about 7%. Gas turbines are often used to meet the peak load demands. It is also expected that renewable energy generation, mainly wind and solar energy to grow considerably in the near future. Steam turbines are fuelled primarily by cola, gas, oil and uranium. In 1957, nuclear units of 90-MW steam turbine were installed at the Shipping-port Atomic Power Station, near Pittsburg. The United States is the world's largest supplier of commercial nuclear power, and in 2013 generated 33% of the world's nuclear electricity. After 1990s, fuel of choice for new power plants in the United States was natural gas due to its availability, low cost, higher efficiency, lower pollutant emissions, shorter construction and commission times, safety and lack of controversy associated with this type of power plants. In the last three decades was an increasing trend to utilize renewable energy sources for electricity generation, due to their abundance, almost zero emissions during the operation and technological advances making them economical viable options. Renewable energy sources include among others conventional hydro-power plants, solar-thermal conversion systems, wind energy system, PV, biomass energy or ocean energy. In 2012 approximately 12% of the US electricity was generated by renewable energy sources, while in Germany in 2014 achieved about 31% of electricity generated from renewable energy sources, mainly wind energy.

1.4 TRANSITION TO SMART GRIDS

The current energy scenario is becoming increasingly complex, and the need for an intelligent power grid is ever more urgent. However, it is introducing several challenges and issues to be solved and managed, such as the interconnection of different sub-grids, the requirement of maintaining power quality above stricter limits, the improvement of grid stability and resilience, the integration of clean energy generation and the use of high-efficient power units and energy storage systems (ESSs). Currently, the power grid is changing continuously, being adapted to fulfil minimum requirements in terms of nominal power, stability and reliability. The increasing demand for energy by industry and housing sector requires the commissioning of new power plants or the purchase of energy from neighbouring regions or countries. This necessitates the implementation of a highly meshed decentralized energy distribution grid, which allows the interconnection of power systems with different natures and from different regions. Multiple interconnections between different grids are currently required but this leads to challenges, such as the stability and reliability of the overall energy generation and distribution system. Demand for electrical energy has grown in such a way that it is pushing the existing power transmission and distribution systems to their limits. In today environment, new challenges have appeared to electrical and electronics power system engineers. In order to ensure static and dynamic power system stability, new technologies have been adopted or are being developed. In summary, the first century electricity generation powered incandescent lamps in our homes, motors in our factories, facilities and buildings, electronics and appliances. Now, in the second electricity century, the large-scale electrification of air, sea and land vehicles are undoubtedly and significantly transforming the way electricity is generated, transmitted, dispatched and consumed. The power grid modernization is intending to accommodate more complex power flows, to serve reliability needs and to meet future projected uses. It is leading to the incorporation of electronic intelligence capabilities for power control purposes and operations monitoring. To cope with these new demands, a more flexible, reconfigurable and information-intensive cyber and physical infrastructure, at least at the distribution level, but more likely throughout the entire system is necessary, expected and required. While the extent and final shape of this paradigm shift in design and operation of modern power systems is still being contemplated and debated, fundamental expectations for the next generation of power systems have already emerged, and they point to a more customer-centric perspective. In addition to expecting digital-grade *perfect power*, consumers expect the next generation *smart grid* to keep them informed, empowered and secure. Addressing these needs requires that new design approaches build intrinsically on enabling technologies, including power electronics, communication, distributed decision making, distributed generation (including renewable energy sources) and integration of local energy storage units in stationary and mobile platforms. The smart grid upgrades are successfully addressing three major weaknesses of the conventional power grids, as discussed here. Limited flow control and monitoring, by addressing the current power grid lack of granular-level directing, switching, controlling or monitoring capabilities, together with advanced remote repairing capabilities. Centralized generation, by allowing, encouraging and accommodating distributed generation at all grid levels and power generation by the end-users. Low utilization, as the current power grids are inducing wastes of the energy resources, as a result of low efficiency of many grid components, lack of scalable energy storage capabilities or the waste of the generated power of the fluctuating renewable energy systems. The smart grid conceptually consists of three layers, described here. First is *the infrastructure layer* that includes the physical enhancements and changes that are needed to be installed in the conventional power grids. Second are the advanced *communication layers* consisting of hardware, software and protocols for data and information flows needed for monitoring and control of power flows. Third are the *application layers* consisting of the applications and energy management systems needed to be installed once the other two layers are in place in the future electricity grids.

Current electric grid was designed to meet the needs of a different industry as one that is in full development and changes in the 21st century, with outdated monitoring, metering, control and operation technologies that is incapable of meeting today's industry requirements. As we already mentioned in previous chapter subsections, the existing grid was essentially designed as one-way conduit to transmit was energy amounts generated primarily at a limited number of large power stations to major load centres, based on the premise that customer load is given, requiring only that generation to be adjusted to meet the load requirements. Since very beginning of 21st century, there has been a constant and increased interest in the smart grid (SG). However, there are many who do not like the usage of the term, *smart grid*, but would have preferred to use *smarter grid*, because they feel that the existing grid is already smart. What are needed are *a more efficient or advanced power grid* in term of advanced communication and information technology and other advanced technology usage, and improved efficiencies. Recognition of that the information technology and computing can offer significant opportunities to modernize the operation and management of power system and electrical networks has coincided with the understanding the adverse environmental impacts of electricity sectors and effective de-carbonization of electricity industry at a realistic cost can be realized only if it is monitored and controlled effectively. In addition, a number of other reasons have now coincided with the stimulated SG interests. In many parts of the world, including the USA and the EU countries, the power system expanded rapidly from 1950s and transmission and distribution equipment and networks that were installed are now beyond the design life and in need of replacement and modernization. The replacement and modernization capital costs are very high and it is also even questionable the required manufacturing capabilities and skilled staff are now available. These needs to refurbish the transmission and distribution networks are a great opportunity to innovate with new designs and operating practices. The move toward the smart grid is fuelled by several needs. For example, there is a need for improved grid reliability while dealing with an aging infrastructure, and there is a need for environmental compliance and energy conservation. In addition, there is a need for improved operational efficiencies and customer service. The context of smart grid include: bulk power generation, transmission, distribution, customer operations, operation provider, customer and market, involving electric and secure communication flows within the system. Under federal law of Energy Independence and Security Act of 2007, the National Institute of Standards and Technology (NIST) has been given the key role of coordinating development of a smart grid framework for standards. Standardized architectural concepts, data models, and protocols are essential to achieve interoperability, reliability, security and evolvability. New measurement methods and models are needed to sense, monitor, control and optimize the grid's new operational paradigm. NIST launched a three-phase plan for the development and promotion of smart grid interoperability standards:

1. Engage stakeholders in a participatory public process to identify applicable standards, grasping in currently available standards and priorities for new standardization activities;
2. Establish a formal private-public partnership to drive long-term progresses and
3. Develop and implement a framework for testing and certification.

Smart grid, the vision of modernized and efficient electricity distribution enabled by the latest information and communication technologies (ICT), has been identified by governments and policy makers around the world as a way of addressing global warming and energy independence. Standards play a key role in the development, deployment and operation of smart grids worldwide. They are a proven tool to safeguarding interoperability, enabling the different components of a grid to exchange information and to mutually understand the information exchanged. Smart grid is a rather new concept that includes aspects of energy generation, transmission and distribution and aims for a more reliable service, higher efficiency, more security, two-way utility-user communications and promotion of green energy, among other goals. There is no a smart grid definition universally accepted and standardized. However, in broad sense the smart grid is referring to as: *a smart grid for electricity is an intelligent, auto-balancing and self-monitoring power grid.* It accepts power generated from any source (coal, sun, natural gas, biomass, hydropower or wind) by a facility of any size or location, centralized or distributed generation units. It then uses advanced sensors, state-of-the-art applications and distributed computing to transport that electricity efficiently and cost-effectively, with minimal human intervention, to customers for their consumption. In fact, the smart grids are distributed systems consisting of a very large number of components, agents, stakeholders or actors. The actors or agents are varying in their use of the power grid: some actors are producing energy, some consume energy and some do both. The actors or agents are behaving largely autonomously but have to communicate and work together to balance energy supply and demand. Keeping in mind that fundamental concepts such as reliability may be defined in terms of the quality of accessible power and information, and loads can be viewed as active participants in the electric network, with the ability to autonomously modify their behaviour as needed to improve the security, reliability and availability of the electrical network. This point of view is also introducing the reliability as an optimization process within a cyber-physical system in which loads that can be scheduled, accessible local stored energy and renewable energy sources provide the operational flexibility to mitigate problems before the onset of widespread blackouts and brownouts. These increased requirements suggest that the distribution power grid of the future may benefit from overlaid AC and DC links. The building blocks of such hybrid AC and DC power distribution are the microgrids and nanogrids, made up of a multiport power electronic interface that allows for the seamless integration of renewable energy sources, intelligent management of the energy storage units (stationary or mobile) and scheduling of the local appliances and equipment, an interface with the AC feeder and a DC linkage with neighbouring nanogrids and/or microgrids to form a DC loop (network) among these smart (intelligent) elements and components. An important advantage of this system topology is that it enables critical loads to be fed from multiple DC buses with automatic bus selection to ensure uninterrupted operation. In addition to the physical contact that allows for bidirectional flow of power between the AC grid and other consumers (i.e. nanogrids or microgrids), a reliable, robust and well-secured communication network is essential and critical. A system controller will ensure bidirectional interaction (cyber and physical) between the utility-side network and the customer-side nanogrids and microgrids. This unit, which may be physically centralized or distributed, oversees the local dispatch of energy in such a way as to optimize system reliability, quality and the real-time trading of electricity. This central unit is referred to as a microgrid controller; it may also contain its own centralized renewable energy source and energy storage unit.

In order to achieve the SG goals mentioned below and in particular the improvement in reliability, security and efficiency, it is essential to have well-developed digital technologies, e.g. computing, communication, smart monitoring and metering. Among the significant challenges facing development of a smart grid are the cost of implementing it, and the new standards that regulatory bodies have to enact. Interoperability standards certainly will allow the operation of highly interconnected systems that include distributed generation plants. Another big difficulty that the implementation of Smart Grid and distribution automation faces is the huge variety of technologies produced by multiple vendors. There is neither a single set of requirements nor a single technology path to a smart grid. There is, however, an emerging understanding that smart grids will include:

- Rely to the greatest extent possible on digital information and control, rather than on analogue technologies or constant human intervention.
- Increase the amount of generated electricity ultimately delivered to customers through new materials and a wide variety of grid optimization tools.
- Accommodate: distributed generation, including variable generation from renewable energy sources, demand side management; advanced building control technologies; "smart" consumer appliances; centralized and distributed energy storage and enabling new applications such as widespread charging of electric vehicles.
- Provide all customers with appropriate levels of information and control options.
- Rest on standards that accommodate both regulated and competitive entities to deliver a wide variety of products and services to all parts of the economy.

Clearly, the advanced and smart metering system is one of the SG major elements, but certainly is not the only one. Many elements and concepts are included in the overall SG fields. Those pertaining to distribution systems and in particular to the automation of distribution systems (or distribution automation) will be considered in later chapter of this book. According to the US Department of Energy (US DOE Modern Grid Initiative Report), a modern smart grid must satisfy the following requirements:

1. Motivate consumers to actively participate in operations of the grid;
2. Self-healing capabilities;
3. Resistant to attacks;
4. Provide a higher quality power that will save money wasted during outages;
5. Accommodate all generation and energy storage options;
6. Enable deregulated and broad electricity markets to flourish,

7. Higher efficiency and
8. Enabling higher penetration of intermittent power generation sources.

Costs of deploying the Smart Grid remains an issue, and study estimates vary. While some DOE programs have supported grid modernization, Congress has not explicitly appropriated funding for deployment of the smart grid. While concerns such as cybersecurity and privacy exist, most electric utilities appear to view Smart Grid systems positively. Costs could be reduced and system resiliency improved by further integration of automated switches and sensors, even considering the cost of a more cyber-secure environment. However, with the potentially high costs of a formal transition, some see the deployment of the Smart Grid continuing much the same as it has, with a gradual modernization of the system as older components are replaced. Notice also that estimating costs for smart grid systems can be difficult given that the digital technologies and electronic equipment are constantly evolving with tendency of price decreasing, while digital, monitoring, metering infrastructure and communication systems must be designed or augmented for cybersecurity and to work with new IT, computing and communication infrastructure. On the other hand, most of the overall SG benefits in terms money are well surpassing the total smart grid implementation cost estimates.

1.5 CHAPTER SUMMARY

The book is written as primer textbook and handbook for addressing the fundamentals and essential concepts of smart grids. It provides the working definition the concepts and functions, the design criteria and the tools and techniques and technology needed for building smart grid. The book is needed to provide a working guideline in the design, analysis and development of smart grids. It incorporates all the essential factors of smart grid appropriate for enabling the performance and capability of the power system. There are no comparable books which provide information on the *how to* design and analysis. The first chapter of this book gives a good summary of the structure and evolution of the power systems or electric grids. It starts with brief description of the energy and environment dual impacts and the ways that such problems and issues can be mitigated. This chapter gives a brief but comprehensive review of the world's energy resources and climate change problems due to fossil fuel burning, along with possible solutions or mitigation methods. The chapter also gives a brief history of the power grids, and how and why it evolves in the modern grid structure. A section of the chapter is devoted to the transition to the smart grids, their benefits, implementation and development challenges. The electrical grid comprises all of the power plants generating electricity, together with the transmission and distribution lines and systems that bring power to end-use customers. The modernization of the power grid to accommodate today's more complex power flows, higher power quality requirements, serve reliability and resilience needs and meet future projected uses is leading to the incorporation of electronic intelligence and computing capabilities for power control purposes, management and operations monitoring. Balancing supply and demand more accurately and making better use of renewable power sources are essential if the ambitious pollutant reduction targets that have been set in many countries are to be achieved. A smart grid offers significant opportunities for utilities and consumers to wisely manage the energy consumption by the usage of advanced metering and monitoring infrastructure and dual-way and real time communication. It also provides opportunities to wisely manage the fuel resources by potentially reducing the national need for additional generation sources, better integrating renewable and non-renewable generation sources into the grid operations, reducing outages and cascading problems and enabling consumers to better manage their energy consumption. Smart grid is not a single technology but a combine of several latest and diverse technologies in order to achieve a more sustainable electricity generation and uses.

1.6   QUESTIONS

1. Is the quality of life better with electricity or without? Explain.
2. Describe how a specific energy-related process can affect the environment.
3. List the main components of a today power system.
4. What is a primary fuel?
5. What is the difference between primary and secondary fuels?
6. What are the major fossil fuels? Briefly describe their characteristics.
7. Briefly describe the main components of a power system.
8. Briefly define the primary and secondary energy forms.
9. List the major renewable energy sources with potential grid applications.
10. What are the main challenges and issues of today power systems?
11. Briefly describe the major drivers of the power system modernization and change.
12. Which are the main smart grid features?
13. What are the three major weakness of the conventional electricity grid.
14. What are the major benefits of smart grids?
15. List and briefly describe the three smart grid layers.
16. What are the main challenges that the power grid is facing?
17. List in your own words the main benefits of smart grids.
18. Briefly describe the issues related to the wind energy grid integration
19. List the major renewable energy sources
20. What is the structure of present-day grid? What are its major advantages, issues and risks?
21. What are major drivers of smart grids?
22. What are the most important technologies in the future smart grid implementations? Briefly describe these technologies.
23. What the major technologies applicable to future smart grid?
24. List the main barriers of the smart grid deployment.

2 Power System Structure and Components

2.1 INTRODUCTION, OVERVIEW OF ELECTRIC UTILITY

The electric power system is one of the most complex technical systems, with its main functions to deliver and supply electricity from generation to the costumers (end-users) or in some cases to the energy storage systems. A power system is an interconnected electric network with components converting nonelectrical energy to electrical energy and transporting the generated electrical energy from generating units to the load centers and end-users. Power industry is the largest and most complex industry in the history of humanity. Electrical and power engineers who work in this industry encounter challenging problems in designing and shaping future power systems to deliver increasing amounts of electrical energy at highest quality standards, in a safe, clean, efficient and economical manner. Power system industries and utilities significantly contribute to the welfare, life standard, progress and technological advances of humanity. The growth and demand of electrical energy in the world as a whole and in each country in the last half century was phenomenal, with over 50 times as much than the growth and demand rates in all other energy forms used during the same period. Power and energy industries significantly contribute to the welfare, life standard, progress and technological advances of humanity. Today, the installed electrical power capacity per capita in the United States is estimated to be about 3 kW, with similar values in EU and developed countries. There is also significant increase in the installed capacity per capita in China, Brazil and India and in most of the developing countries. The electricity generation, transmission and distribution is the business of utility companies, performed through complex networks of interconnected generators, transformers, transmission lines, control, monitoring, measurement and protection equipment, developed over a century. A sustainable energy system involves key components of: increasing use of renewable energy resources, increased energy efficiency and use of electricity in the transportation sector. As an effect of this key questions for the electric power system are: integration of intermittent electricity generation sources and different solutions for energy storage.

Most of the modern power systems are three-phase, from generation, transmission to the users because is an efficient and economical way to perform its functionalities. Only very near to the users, the power is changed to single-phase power. However, industrial and commercial users are still using three-phase systems. In modern power systems, the generator voltages are transformed into high-voltages for efficient and economic transmission at long distances, and near the industrial and residential consumer locations the voltages are lowered form the transmission levels to the levels required by the consumers. The energy resources used to generate electricity in most of the power plants are fossil fuels (e.g. coal, natural gas or oil), nuclear or hydropower. In the last three decades, there has been an increasing trend to include renewable (alternative) energy sources for electricity energy generation, especially in the distribution section. Nuclear power and hydroelectric power are nonpolluting energy sources, whereas the last one is also a renewable energy source. Currently, in the United States, hydropower accounts for about 8% and nuclear power 20% of the electricity generation. The generated electricity is transmitted by a complex network composed of transmission lines, transformers, control and protective equipment. Transmission lines are used to transfer electrical energy from power plants to load centers. Transformers are used to step-up the voltage at power plants to very high values (200 kV to 1200 kV), in order to reduce currents and

DOI: 10.1201/9780429174803-2

losses and reducing the size of transmission wires and implicit reducing overall cost of the transmission system.

Power systems are undergoing significant changes in terms of how they operate, how the electricity is generated and transferred to the users and how the consumers interact and participate with the power systems. Future electricity distribution and generation with the extended uses of the distributed energy resources (DERs) and renewable energy sources (RESs) require the creation of a new utility grid architecture and structure. Economic, technological and environmental incentives and issues are changing the face of the electricity generation and transmission. Centralized energy generation structures, needing very complex and bulk power networks to produce and transfer energy to the consumers, have high investment, operation and maintenance costs, while the overall system efficiency is low due to the large losses in these networks and systems. At the same time, the existing electricity and energy infrastructures are becoming older, and in addition, there are new energy supply security and environmental issues with the construction of new network components (e.g. power plants or high-voltage transmission lines). Other critical issues of the power distribution are maintaining the required power quality and supply stability, while the customers have increased power quality and supply stability demands, due to the extended use of sensitive or critical loads, which may also significantly affect the power quality. For such reasons the centralized power generation is giving way a part of generation to smaller and distributed energy supplies. Distributed generation (DG) is loosely defined as a small-scale electricity generation, often located at the consumption points. Usually, the connection is to the power distribution or on the meter customer side. Distributed generation has become more present into the power systems for reasons such as: an alternative to the construction of large power plants, constraints on the construction of new transmission lines, higher power quality and supply stability demands. It is also economically attractive as the cost of small-scale generation keeps decreasing with technology advances, changing the economic and regulatory environment and the electricity market liberalization. Future electricity distribution networks with large DG and RES penetration, energy storage units, electric vehicles and customers with smart meters and controllable loads are requiring the creation of new grid architecture – the smart grid. But realizing the full smart grid (SG) benefits goes beyond near-term steps, e.g. smart meters or improving delivery efficiency.

Technologies are coming into the marketplace at different rates, and progress and development paces, too. The pressure of improving the overall power system efficiency, power quality, energy supply security, stability and environmental impacts has forced the energy industries to answer to these issues. There are also significant increases in the electricity and energy demands, while the DG and RES seems one way to cope with these demands and grid issues. DER and RES integration into the existing energy networks can result in many benefits, in addition to the reduced grid losses and environmental impacts, such as relieved transmission and distribution (T&D) congestion, peak demand shavings, voltage support, reduced price fluctuations and the deferred investments to upgrade existing systems. DERs have and are expected to have even more impacts in the future on the energy market. The energy storage systems (e.g. batteries, fuel cell stacks, flywheels or thermal energy storage) are included to harness the excess of produced electricity during the off-peak and low-demand periods, for the use during the peak periods or when needed, reducing the needs for the high-cost peak-load generators. The energy storage units are also providing the dispatchability of the renewable energy sources, having no dispatchability by their own, e.g. PV arrays or wind turbines. Modern power systems generates and supplies electricity through a complex process and system, consisting of electricity generation in large power plants, usually located close to the primary energy sources (e.g. coal mines, water reservoir), far away from the large consumer centers, delivering the electricity by a large passive but complex distribution infrastructure, involving high-voltage (HV), medium-voltage (MV) and low-voltage (LV) electric networks. Power distribution operates mostly radially, in which the power is flowing in one direction, from HV levels down to customers, along the distribution feeders. Nowadays, the technological advancements, environmental policies and the expansion of the electricity markets are promoting significant changes into the electricity industry. New technologies allow the electricity to be generated in smaller size units or in DG units, located in the MV and LV grid sections, closer to the users. Moreover, the increasing RES use in order to reduce the environmental impacts and diversify the supply leads to a new electricity supply schemes. In this new paradigm, the power production is not exclusive to the generation end, but is shifted to the MV and LV networks, with part of the energy supplied by the centralized generation and another part produced by the DG units, closer to the customers. Large-scale DG integration is a main trend into modern power systems. These generators are of considerable smaller size compared to the traditional generation units.

This chapter gives a conceptual introduction to the electric power system as a main part of the energy system. It will go through a brief description fundamental modeling of the AC system and the power flow, main components of a power system, power system operation and planning including aspects of reliability and market. We are trying here to conceptually describe the technical fundamental characteristics and performances of the electric power system, with main function to deliver and transfer electricity between generation, consumption and/or storage. A sustainable energy system involves key components of: increasing use of renewable energy resources, increased energy efficiency and use of electricity in the transportation sector. As an effect of this, some key questions for the electric power system are: integration of intermittent electricity generation, e.g. from wind power, and connections of electrical vehicles to the electrical distribution system and different solutions for energy storage.

2.2 STRUCTURE OF THE MODERN POWER SYSTEMS

A typical power system structure is shown in Figure 2.1. In this complex power system structure, electro-mechanical systems play a key role, and essential system components are the three-phase synchronous generators or alternators. The electric generator converts nonelectrical energy provided by the prime mover, usually steam or hydro turbines, into electrical energy. The mechanical power source, commonly known as prime mover, may be hydraulic turbines or steam turbines. The turbine's function is to run the electrical generators by converting the steam thermal energy or water kinetic energy into rotating mechanical energy. In thermal power plants, fossil fuels or nuclear reactions are used to produce high temperature steam, eventually passed through the turbine blades causing turbine to rotate. A hydro-electric plant consists of dam, holding water upstream at high elevations with respect to the turbine. When electricity is needed, the water flows through the hydro turbine blades through penstocks rotating the generator. Since the generator is mounted on the turbine shaft, it rotates with the turbine, generating electricity. At load level, an energy bulk is consumed by electrical motors, mostly induction type.

The generators used in power plants are synchronous machines (alternators). Synchronous generators have two synchronized rotating electromagnetic fields – one produced by the rotor driven at synchronous speed and excited by a DC circuit, and the other one is produced in the stator windings by the three-phase armature currents. The excitation system maintains the generator voltage and

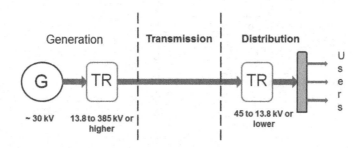

FIGURE 2.1   A basic structure of a simplified power system.

controls the reactive power flow. Due to their structure and construction, AC generators can generate high power and voltage, typically 30 kV. However, induction or DC generators can be found in standalone and low-power distributed generation. Synchronous machines have magnetic field circuits mounted on the rotor and are firmly connected to the turbine shaft. The alternator stator has windings wrapped around its core in a three-phase configuration. Insulation requirements and other practical design issues limit the generated voltage to some low values, up to 30 kV. The generator voltages (5 kV to 22 kV) are not high enough for efficient power transmission, being stepped-up by transmission transformers. The devices connecting generators to transmission subsystem and from transmission subsystem to distribution subsystem are the transformers. The transformers transfer the electricity with very high efficiency from one level of the voltage to another one, suitable for specific applications. Their main functions are stepping up the lower generation voltage to the higher transmission voltage and stepping down the higher transmission voltage to the lower distribution voltage. The main advantage of having higher voltage in transmission system is to reduce the losses in the grid. Since transformers operate at constant power, when the voltage is higher, then the current has a lower value. Therefore, the losses, a function of the current square, will be lower at a higher voltage. However, when the electrical energy is delivered to the load centers, the voltage is stepped down for safer distribution and usage requirements. When the electrical power reaches customers' facilities, it is further stepped down to the required levels depending on the various standards worldwide. Figure 2.2 shows the most common voltage ranges in the main sections of a power system. The electricity in an electric power system may undergo four or five transformations between generation and consumers.

The electric grid is a vast physical and human network connecting thousands of electricity generators to millions of consumers, consisting of public and private enterprises operating within a web of government institutions: federal, regional, state and municipal. The electric power system is a complex technical system with main function to deliver electricity between generation, consumption and storage. A power system serves one important function and that is to supply customers with electricity as economically and as reliably as possible. It is an interconnected network with components converting nonelectrical energy continuously into the electrical form and transporting the electrical energy from generating sources to the loads/users. Electric power systems are real-time energy delivery systems, meaning that power is generated, transported and supplied

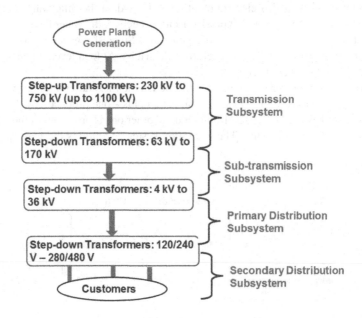

FIGURE 2.2 Electricity infrastructure from generation to power distribution.

the moment we are turning on the light switch. Electric power systems are not storage systems like water systems and gas systems. Instead, generators produce the energy as the demand calls for it. The system starts with generation, by which electrical energy is produced in power plants and then transformed in power stations to high-voltage electrical energy that is more suitable for efficient long-distance transportation. The power plants transform other energy sources in the process of producing electrical energy. For example, heat, mechanical, hydraulic, chemical, solar, wind, geothermal, nuclear and other energy sources are used in the electrical energy production. High-voltage (HV) power lines in the transmission portion of the electric power system efficiently transport electrical energy over long distances to the users. Finally, substations transform this HV electrical energy into lower-voltage (LV) energy that is transmitted over distribution power lines that is more suitable for the distribution of electrical energy to its destination, where it is again transformed for residential, commercial and industrial consumption. Power systems can be divided into four sub-systems:

1. *Generation* – Generating and/or sources of electrical energy.
2. *Transmission* – Transporting electrical energy from its sources to load centers with high voltages (115 kV and above) to reduce transmission losses.
3. *Distribution* – Distributing electrical energy from substations (44 kV to 12 kV or lower) to end-users.
4. *Utilization subsystem* – Using electrical energy in residential, commercial and industrial facilities.

Planning of the power systems or the power distribution networks is an essential task, intended to enable that the required electricity demands can be met based on various forecast loading horizons, meeting the supply security and reliability. There are three categories of planning involved in the power systems: the long-term, the network and construction planning. Long-term planning is to determine the most optimum network arrangements and the associated investment with consideration on future developments. Stage-by-stage development must be in line with the forecasted load growth, so that electricity demands can be timely met. The construction planning or design is the actual design and engineering work when the required circuits and substations have been planned and adopted. The planning of electric power distribution in buildings and infrastructure facilities is subject to constant transformation. The search for an assignment-compliant, dependable solution should fulfill those usual requirements placed on cost optimization, efficiency and time needs. At the same time, technical development innovations and findings from the practical world are constantly seeping into the planning process. Opportunities for improving the functioning and reliability of the grid arise from technological developments in sensing, communications, control and power electronics. These technologies can enhance efficiency and reliability, increase capacity utilization, enable more rapid response to remediate contingencies and increase flexibility in controlling power flows on transmission lines. If properly deployed and accompanied by appropriate policies, it can deal effectively with power distribution challenges, facilitate the integration of large volumes of renewable and distributed generation, provide greater visibility of the instantaneous state of the grid and make possible the engagement of demand as a resource. Increasingly greater demands are placed on modern building energy systems. At earlier planning stages, the demands for high level of safety and flexibility throughout the entire life cycle, low pollution levels, integration of renewable energies and low costs must be taken into account to exploit the full potential of economic efficiencies and fulfilling technical demands. A special challenge is the coordination of the individual installations. The main building installations are heating, ventilation, air conditioning and refrigeration, fire protection, safety, building control and monitoring systems and electric power distribution. With innovative planning, the requirements are not simply broken down to the individual installations, but have to be coordinated. Electricity networks that are developed on a more holistic basis reflecting system-wide planning and more objective cost-benefit analyses can enable more efficient, timely and cost-effective investments. Improved efficiency and transparency can help to build general acceptance towards new investments, helping also to foster local acceptance. Transparent and consultative network infrastructure planning processes can build local community understanding, allowing proponents to draw on specific local knowledge to support appropriate power system developments. Benefit-cost analysis needs to take account of all applicable costs to the greatest extent possible, including the environmental and distribution costs. Unaccounted costs can become a significant driver for local resistance during the siting process.

Modern power systems are made up of four major distinct components or sub-systems: generation, transmission, distribution and end-users or loads. The generation subsystem includes power plant generation units, e.g. turbines and generators. The energy resources used to generate electricity in most of the power plants are fossil fuels (e.g. coal, natural gas, or oil), nuclear or hydropower. In the last three decades, there has been an increasing trend to include renewable energy sources for electricity energy generation, especially in the distribution section. Nuclear power and hydropower are nonpolluting energy sources, while the last one is also a renewable energy source. Currently, in the United States, hydropower is accounting for about 8%, while the nuclear energy for 20% of the electricity generation. The generated electricity is transmitted by a complex network composed of transformers, transmission lines, control and protective equipment. The transmission systems consist of a set of lines, substations and equipment designed to connect large generation stations and consumption centers, with power consumption mainly being carried out in cities and industrial areas. Transmission system lines span over long distances and transport large quantities of energy, therefore operating at high-voltage levels (e.g. 400 kV, 720 kV). Transmission lines are used to transfer electrical energy from power plants to load centers. Transformers are used to step-up the voltage at power plants to very high values, up to 1200 kV, in order to reduce currents and losses, and reducing the size of transmission wires and implicitly reducing overall cost of the transmission system. The transmission lines are carrying power to load centers, where appropriate, steps it down to lower voltages up to 35 kV at bulk power substations. Sub-transmission systems are an intermediary link between the transmission and the distribution system. The lines that compose the sub-transmission system cover shorter distances than those in the transmission system; for that reason they operate at lower voltage levels (e.g. 132 kV, 66 kV, and 45 kV). Initial voltage reductions are required due to the difference in voltage level with respect to the transmission system. Large loads (such as large industrial plants and other high consumption facilities) can be directly connected to the sub-transmission systems. At load centers, the transmission line voltages are reduced by step-down transformers to lower values (4.5 kV to 35 kV) for power distribution networks. Large industrial customers can be directly supplied form these substations. This power system component is known as the transmission and sub-transmission systems or sections. Distribution of the electricity to the commercial and residential users takes place through power distribution systems consisting of substations where step-down transformers lower the voltages to a range of 2.4 kV to 69 kV. The distribution substation, a major distribution component, is the interconnection element between the distribution system and the upstream power delivery system. At the substation, the step-down (HV-MV) transformer reduces the sub-transmission voltage level to an appropriate value for primary distribution lines. Electricity is carried by main feeders to specific areas where there are lateral feeders to step it down to consumer levels. Different protection, switching and measurement equipment is installed at the substation to ensure a safe operation. The primary distribution lines spread across the consumption area served by the substation; these primary distributions lines are also known as feeders. At customer sites, the voltage is further reduced to values such as 120 V, 208 V, 280/277 V, etc. as required by users. Power systems are extensively monitored, controlled and protected. These complex transmission and distribution networks encompass large areas or regions. Each power system has several levels of protection to minimize or avoid the effects of any damaged or not properly operating system component on the system ability to provide safe reliable electricity to all customers. Any power system serves one important function and that is to supply the customers with electricity as economically and as reliably as possible. In summary the main functions of four sub-systems of the power systems are:

1. Generation subsystem – Units designed to generate electrical energy.
2. Transmission and sub-transmission subsystems – Subsystems designed to transfer electrical energy from its sources/generators to load centers with high voltages (115 kV and above) to reduce losses.
3. Power Distribution – Subsystems that are distributing electrical energy from substations (in the range 44 kV to about 12 kV) to end-users or customers.
4. Consumers, end-users or utilization subsystems

In the complex power system structure (e.g. Figures 2.1 and 2.2), electro-mechanical systems play a key role. An essential power system component is the three-phase AC synchronous generators or alternators, converting nonelectrical energy provided by the prime mover, usually a turbine, to electrical energy. The turbines rotate electrical generators by converting the steam thermal energy or water kinetic energy into rotating mechanical energy. In thermal power plants, fossil fuels or nuclear reactions are used to produce needed high temperature steam. Typical hydroelectric plants consist of dam, holding water upstream at high elevations with respect to the turbine. When electricity is needed, the water flows through the hydro turbine blades through penstocks rotating the generator. At load level, electricity is used by electrical motors, equipment and devices. Synchronous generators have two synchronized rotating electromagnetic fields, one produced by the rotor driven at synchronous speed and excited by a DC circuit, and another produced in the stator windings by the three-phase armature currents. The excitation system maintains the generator voltage and controls the reactive power flow. Due to their structure and construction, AC generators can generate high power and voltage, typically 30 kV. However, induction or DC generators can be found in standalone and low-power generation units. Synchronous machines have magnetic field circuits mounted on the rotor and is firmly connected to the turbine shaft. The alternator stator has windings wrapped around its core in a three-phase configuration. Insulation requirements and other practical design issues limit the generated voltage to some low values, up to 30 kV. The generator voltages are not high enough for efficient power transmission, being stepped-up by transmission transformers. Transformers are also connecting transmission subsystem to power distribution subsystems. The transformers transfer the electricity with very high efficiency from one level of the voltage to another, suitable for specific applications. Their main functions are stepping up the lower generation voltage to the higher transmission voltage and stepping down the higher transmission voltage to the lower distribution voltage. The main advantage of having higher voltage in transmission system is to reduce the losses in the electric grid, losses being proportional to square of the current. However, when the electrical energy is delivered to the load centers, the voltage is stepped down for safer distribution and usages. When the electrical power reaches customers' facilities, it is further stepped-down to the required levels depending on the various standards worldwide. The electricity in an electric power system may undergo four or five transformations between generation and consumers. A power system is predominantly in steady state operation or in a state that, with sufficient accuracy, could be regarded as steady state. In a power system, there are always small load changes, switching actions and other transients occurring so that in a strict mathematical sense most of the variables are varying with the time. However, these variations are most of the time so small that an algebraic, i.e. not time varying model of the power system, is justified.

The electricity is delivered from the generation station to the loads (end-users) through transmission lines and transformers. The bulk of the electricity produced in the power plants is transmitted to the load centers over long-distance high-voltage transmission lines, operating at very high voltages, 220 kV–1200 kV. The lines that distribute the electrical power within an area are called MV distribution lines. There are several other categories such as sub-transmission and HV distribution lines or power distribution networks, which are discussed later. The transmission lines are high-voltage conductors (wires) mounted on tall towers to prevent them from coming in contact with humans, trees, animals, buildings, equipment or ground. High voltage towers are usually made from galvanized steel, 25 m–45 m in height for strength and durability for the harsh environment that they may operate in. The higher the wire voltage, the higher is the tower. Since the steel is electrically conductive material, the high voltage wires are not mounted directly on the towers. Instead, the insulators, made of nonconductive materials mounted on the towers, are used to hold the conductors away from the tower structure. Insulators, in various shapes and designs, are strong enough to withstand the static and dynamic forces exerted by the conductors during windstorm, freezing rains, earthquakes or animal impacts. Before discussing the power system components, operation and structure, it is useful to define the most common power system terminology. *Power capacity* of a power system element is given by its apparent power, expressed usually in kVA or often in MVA. A *conductor ampacity* (A) is the highest current level that it can carry, with it being function of the conductor's cross-sectional area. The energy used by loads is the electric power over time, given in *kilowatt-hour* (kWh, MWh or GWh). Most of the generation and transmission circuits, as well as a part of the power distribution, are three-phase circuits, usually analyzed by using single-line diagrams and in per-unit notations, as discussed later in this chapter. A load connected to the power system is usually either absorbing or using both active and reactive power. Single-line diagram is a useful power system analysis tool and concept.

2.3 THREE-PHASE SYSTEMS

The generation, transmission and distribution of electric power is accomplished by means of three-phase circuits. An AC generator designed to develop a single sinusoidal voltage for each rotation of the shaft (rotor) is referred to as a single-phase AC generator. If the number of coils on the rotor is increased in a specified manner, the result is a poly-phase AC generator, which develops more than one AC phase voltage per rotation of the rotor. As we mentioned, at the generating station, three sinusoidal voltages are generated having the same amplitude but displaced in phase by 120°, so called a balanced source. If the generated voltages reach their peak values in the sequential order *abc*, the generator is said to have a positive phase sequence. If the phase order is *acb*, the generator is said to have a negative phase sequence, as shown in Figure 2.3. In a three-phase system, the instantaneous power delivered to the external loads is constant rather than pulsating as it is in a single-phase circuit. Also, three-phase electric motors, having constant torque, start and run much better than single-phase electric motors. This feature of three-phase power systems, coupled with the inherent efficiency and cost compared to single-phase (less wire for the same delivered power),

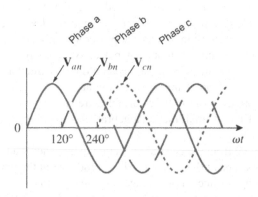

FIGURE 2.3 The representation of a three-phase voltage system.

accounts for its universal use. In general, three-phase electric systems are preferred over single-phase electric systems for reasons, such as:

1. Thinner conductors can be used to transmit the same kVA at the same voltage, which reduces the amount of copper required (typically about 25% less).
2. The lighter lines are easier to install, supporting structures can be less massive and farther apart.
3. Three-phase equipment and motors are running smoothly, having preferred running and starting characteristics compared to single-phase systems because of a more even power flows.
4. In general, the larger motors are three-phase because they are essentially self-starting and do not require a special design or additional starting circuitry.

Three-phase sinusoidal voltages and currents are generated with the same magnitude but are displaced in phase by 120°, by what is called a balanced source or generator, consisting of three identical coils a, b and c separated by an angle of 120° from each other. The generator is turned by the prime-mover, and three identical voltages (see Figure 2.3), V_{an}, V_{bn} and V_{cn}, separated by 120° phase angles are generated:

$$V_{an} = V_m \sin(\omega t) = V_m \langle 0^o$$
$$V_{bn} = V_m \sin(\omega t - 120^o) = V_m \langle -120^o \qquad (2.1)$$
$$V_{cn} = V_m \sin(\omega t - 240^o) = V_m \sin(\omega t + 120^o) = V_m \langle -240^o$$

Where, V_m is the peak value or the amplitude of the generated voltage. The sum of the three waveform voltages, by using trigonometric identities is:

$$V = V_{an} + V_{bn} + V_{cn} =$$
$$V_m[\sin(\omega t) + \sin(\omega t - 120^o) + \sin(\omega t - 240^o)] = 0 \qquad (2.2)$$

Three-phase systems may be labeled either by 1, 2, 3 or a, b, c or sometimes, by using the three natural color, Red, Yellow and Blue to represent them. The phase sequence is quite important for transmission, distribution and use of electrical power. If the generated voltages reach their peak values in the sequential order abc, the generator is said to have a positive phase sequence, shown in Figure 2.4a. If the phase order is acb, the generator is said to have a negative phase sequence, as shown in Figure 2.4b.

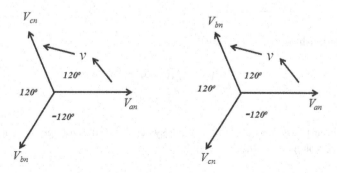

FIGURE 2.4 Positive (a) and negative (b) voltage sequences.

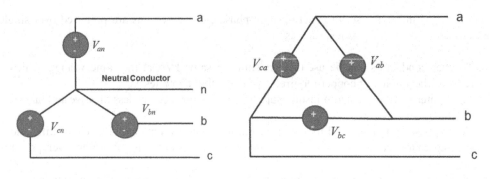

a) Wye (Y) connected sources   b) Delta connected sources

FIGURE 2.5   Wye (a) and delta (b) connected electric sources.

The three single-phase voltages can be connected to form three-phase systems in two ways: (1) star or wye (Y) connections (circuits), or (2) delta (Δ) connections (circuits), as are shown in the diagrams of Figure 2.5. In the Y-connection, one terminal of each generator coil is connected to a common point or neutral n and the other three terminals represent the three-phase supply. In a balanced three-phase system, knowledge of one of the phases gives the other two phases directly. However, this is not the case for an unbalanced supply. In a star-connected supply, the line current (current in the line) is equal to the phase current (current in a phase). However, the line voltage is not equal to the phase voltage. In a three-phase system, the instantaneous power delivered to the external loads is constant rather than pulsating as it is in a single-phase circuit. Also, three-phase motors, having constant torque, start and run much better than single-phase motors. This feature of three-phase power, coupled with the inherent efficiency of its transmission compared to single-phase (less wire for the same delivered power), accounts for its universal use. A power system has Y-connected generators and usually includes both Δ- and Y-connected loads. Generators are rarely Δ-connected, because if the voltages are not perfectly balanced, there is a net voltage and a circulating current around the Δ loop. The phase voltages are lower in the Y-connected generator, and thus less insulation is required. In Y-connected circuits, the phase voltage is the voltage between any line (phase) and the neutral point, represented by V_{an}, V_{bn} and V_{cn}, while the voltage between any two lines is called the line or line-to-line voltage, represented by V_{ab}, V_{bc} and V_{ca}, respectively. For a balanced system, each phase voltage has the same magnitude, and we define:

$$|V_{an}| = |V_{bn}| = |V_{cn}| = V_P \tag{2.3}$$

Here, V_P denotes the effective magnitude of the phase voltage. We can show that

$$V_{ab} = V_{an} - V_{bn} = V_P(1 - 1\angle -120^o) = \sqrt{3}V_P\angle 30^o \tag{2.4}$$

For similar relationships, we can obtain

$$V_{bc} = \sqrt{3}V_P\angle -90^o$$
$$V_{ca} = \sqrt{3}V_P\angle 150^o \tag{2.5}$$

In a balance three-phase Y-connected voltage system, the line voltage V_L whose magnitude is related to the phase voltage magnitude through:

$$V_L = \sqrt{3}V_P \tag{2.6}$$

Example 2.1 **A three-phase generator is Y-connected, as shown in Figure 2.5a. The magnitude of each phase voltage is 220 V RMS. For *abc* phase sequence, write the three-phase voltage equations, and calculate the line voltage magnitude.**

SOLUTION

The expressions of the phase voltages are

$$V_{an} = 220 \langle 0° \text{ V}$$
$$V_{bn} = 220 \langle -120° \text{ V}$$
$$V_{cn} = 220 \langle 120° \text{ V}$$

While the magnitude for the

$$V_{LL} = \sqrt{3} \times V_P = \sqrt{3} \times 220 = 380.6 \text{ V}$$

For a balanced system, the angles between the phases are 120° and the magnitudes are all equal. Thus, the line voltages would be 30° leading the nearest phase voltage. Calculations show that the magnitude of the line voltage is √3 times the phase voltage. The current, I_L (the effective value of the line current), is the same as the phase current, I_P (the effective value of the phase current), for the wye connected circuits, thus

$$I_L = I_P \tag{2.7}$$

In delta connection, as shown in Figure 2.5b, the line and the phase voltages have the same magnitude:

$$|V_L| = |V_P| \tag{2.8}$$

Similarly, in the case of a delta connected supply, the current in the line is √3 times the current in the delta. In a manner similar as for the Y-connected sources, we can easily prove:

$$I_{ab} = \sqrt{3} I_P \langle 0°$$
$$I_{bc} = \sqrt{3} I_P \langle -90° \tag{2.9}$$
$$I_{ca} = \sqrt{3} I_P \langle 150°$$

A balanced three-phase current system in a delta connection yields a corresponding set of balanced line currents related as:

$$I_L = \sqrt{3} I_P \tag{2.10}$$

Where the I_L is denoting the magnitude of the three equally line currents of the system.

2.3.1 BALANCED LOADS

A load on a three-phase supply usually consists of three impedances and one of the ways in which these can be connected are wye(Y) or star connection or delta (Δ), as shown in Figure 2.6. A balanced load would have the impedances of the three phases equal in magnitude and in phase. Although the three phases would have the phase angles differing by 120° in a balanced supply, the current in each phase would also have phase angles differing by 120° with balanced currents. Thus,

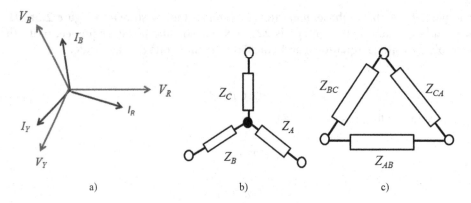

a)   b)   c)

FIGURE 2.6 (a) Phasor diagram; (b) Y-connected load; (c) delta connection.

if the current is lagging (or leading) the corresponding voltage by a particular angle in one phase, then it would lag (or lead) by the same angle in the other two phases as well (Figure 2.6a). For the same load, star-connected impedance and the delta-connected impedance will not have the same value. However, in both cases, each of the three phases will have the same impedance as shown in Figures 2.6b and 2.6c.

In a Y-connected load, the line currents are taken from the supply bare, therefore equal to the phase currents of the load. The star point (junction) of the load is connected to the generator star point, and the neutral current, in phasor notation, is expressed as:

$$I_N = I_A + I_B + I_C \tag{2.11}$$

With the neutral connected to the three-phase loads, the phase voltages across is corresponding phase of the load are:

$$V_A = Z_A I_A; \quad V_B = Z_B I_B; \quad V_C = Z_C I_C \tag{2.12}$$

For a balanced set of supply voltages and loads, the sum of the load voltages and currents is zero, as well as the neutral current. If the neutral current is zero, then the neutral connection between the load and the source is not needed. If the load impedances are not balanced, with the neutral conductor connected, a neutral current is flowing. However, if the neutral is not connected, then the star point neutral departs from the supply neutral (so-called "*floating neutral*"). The following equations then apply:

$$V_{AB} = Z_A I_A - Z_B I_B$$
$$V_{BC} = Z_B I_B - Z_C I_C \tag{2.13}$$
$$I_A + I_B + I_C = 0$$

These equations are sufficient to solve for the load phase currents (the source line currents) and hence the load voltages. In the delta-connected loads, the phase voltages are equal to the source line voltages for a balanced case. The relationships between supply line currents and the load phase currents are:

$$I_A = I_{AB} - I_{CA}$$
$$I_B = I_{BC} - I_{AB} \tag{2.14}$$
$$I_C = I_{CA} - I_{BC}$$

A distinct advantage of a consistent set of notation adopted in a three-phase circuit analysis is the symmetry of the expressions, resulting in an additional way to check their consistency and correctitude. By using Ohm's Law, the phase currents are given by:

$$I_{AB} = \frac{V_{AB}}{Z_{AB}}; \quad I_{BC} = \frac{V_{BC}}{Z_{BC}}; \quad I_{CA} = \frac{V_{CA}}{Z_{CA}} \tag{2.15}$$

Example 2.2 A Y-connected balanced three-phase load consisting of three impedances of

$$Z_L = 44 \langle 30° \, \Omega$$

The loads are supplied with the balanced phase voltages:

$$V_{an} = 220 \langle 0° \, V$$
$$V_{bn} = 220 \langle -120° \, V$$
$$V_{cn} = 220 \langle 120° \, V$$

Calculate: (a) the phase currents, and (b) the line-to-line phasor voltages

SOLUTION

a. The phase currents are computed as:

$$I_{an} = \frac{220 \langle 0°}{44 \langle 30°} = 5 \langle -30° \, A$$

$$I_{bn} = \frac{220 \langle -120°}{44 \langle 30°} = 5 \langle -150° \, A$$

$$I_{cn} = \frac{220 \langle 120°}{44 \langle 30°} = 5 \langle 90° \, A$$

b. Applying the Equations (2.4), (2.5) and (2.6) the line-to-line voltages are obtained as:

$$V_{ab} = V_{an} - V_{bn} = 220 \langle 0° - 220 \langle -120° = 220\sqrt{3} \langle 30° \, V$$
$$V_{bc} = V_{bn} - V_{cn} = 220 \langle -120° - 220 \langle 120° = 220\sqrt{3} \langle -90° \, V$$
$$V_{ca} = V_{cn} - V_{cn} = 220 \langle 120° - 220 \langle 0° = 220\sqrt{3} \langle 150° \, V$$

2.3.2 POWER RELATIONSHIPS IN THREE-PHASE CIRCUITS

A three-phase Y-connected source is supplying a three-phase Y-connected balanced load, Figure 2.7, with three-phase sinusoidal phase voltages, as given here:

$$v_a(t) = \sqrt{2}V_P \sin(\omega t)$$
$$v_b(t) = \sqrt{2}V_P \sin(\omega t - 120°) \tag{2.16}$$
$$v_c(t) = \sqrt{2}V_P \sin(\omega t + 120°)$$

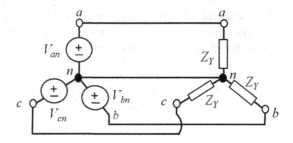

FIGURE 2.7   A Y-connected generator supplying a Y-connected load.

With current flowing through the load given by

$$i_a(t) = \sqrt{2}I_P \sin(\omega t - \phi)$$
$$i_b(t) = \sqrt{2}I_P \sin(\omega t - 120^o - \phi) \qquad (2.17)$$
$$i_c(t) = \sqrt{2}I_P \sin(\omega t + 120^o - \phi)$$

where ϕ is the phase angle between the voltage and current in each phase.
The instantaneous power supplied to one phase of the load, as in Figure 2.7, is:

$$pt) = v(t) \cdot i(t)$$

Therefore, the instantaneous power in each of the three phases of the load is:

$$p_a(t) = v_a(t) \cdot i_a(t) = 2VI \sin(\omega t)\sin(\omega t - \theta)$$
$$p_b(t) = v_b(t) \cdot i_b(t) = 2VI \sin(\omega t - 120^o)\sin(\omega t - 120^o - \theta) \qquad (2.18)$$
$$p_c(t) = v_c(t) \cdot i_c(t) = 2VI \sin(\omega t - 240^o)\sin(\omega t - 240^o - \theta)$$

The total instantaneous power flowing into the load is the expresses as:

$$p_{3\phi}(t) = v_a(t)i_a(t) + v_b(t)i_b(t) + v_c(t)i_c(t) \, \text{W} \qquad (2.19)$$

By substituting expressions for phase voltages and currents, Equations (2.16) and (2.17), respectively in Equation (2.19) and using additional trigonometric identity:

$$\cos(\alpha) + \cos(\alpha - 120^o) + \cos(\alpha - 240^o) = 0$$

Equation (2.19) then can be re-written as:

$$p_{3\phi}(t) = 3|V||I|\cos(\phi) = 3P \, \text{W} \qquad (2.20)$$

Here, $|V| = \sqrt{2}V_P$ and $|I| = \sqrt{2}I_P$ are the peak magnitude of the phase voltage and current. Equation (2.20) is a very important result. *In a balanced three-phase system, the sum of the three individually pulsating phase powers adds to a constant, non-pulsating total active power of magnitude three times the real (active) power in each phase.* However, one has to keep in mind that the Equation (2.20) is valid only for balanced conditions. The single-phase power equations apply to each phase of a Y- or Δ-connected three-phase load. The real, active and apparent powers supplied to a balanced three-phase load are:

$$P = 3V_\phi I_\phi \cos(\theta) = 3ZI_\phi^2 \cos(\theta)$$
$$Q = 3V_\phi I_\phi \sin(\theta) = 3ZI_\phi^2 \sin(\theta) \qquad (2.21)$$
$$S = 3V_\phi I_\phi = 3ZI_\phi^2$$

The angle θ is again the angle between the voltage and the current in any of the load phase, and the power factor of the load is the cosine of this angle. We can express the powers of Equation (2.21) in terms of line quantities, regardless the connection type (wye or delta), as:

$$P = \sqrt{3}V_{LL}I_L \cos(\theta)$$
$$Q = \sqrt{3}V_{LL}I_L \sin(\theta) \qquad (2.22)$$
$$S = \sqrt{3}V_{LL}I_L$$

We have to keep in mind that the angle θ in Equations (2.21) and (2.22) is the angle between the *phase voltage* and *the phase current*, not the angle between the line-to-line voltage and the line current.

Example 2.3   The terminal line-to-line voltage of three-phase generator equals 13.2 kV. It is symmetrically loaded and delivers an RMS current of 1.350 kA per phase at a phase angle of 24° lagging. Compute the power delivered by this generator.

SOLUTION

The RMS value of the phase voltage is

$$|V| = \frac{13.2}{\sqrt{3}} = 7.621 \text{kV/phase}$$

The per-phase active (real) and reactive power are given by:

$$P = 7.621 \cdot 1.350 \cdot \cos(24°) = 9.399 \text{MW/phase}$$
$$Q = 7.621 \cdot 1.350 \cdot \cos(24°) = 4.185 \text{MVAR/phase}$$

The instantaneous powers, in phase a, b and c, are pulsating and are given by:

$$p_a(t) = 9.399(1 - \cos(2\omega t)) - 4.185 \sin(2\omega t)$$
$$p_b(t) = 9.399(1 - \cos(2\omega t - 120°)) - 4.185 \sin(2\omega t - 120°)$$
$$p_a(t) = 9.399(1 - \cos(2\omega t - 240°)) - 4.185 \sin(2\omega t - 240°)$$

The total (constant) three-phase power is:

$$P_{3\phi} = 3 \times 9.399 = 28.197 \text{MW}$$

The fact that three-phase *active (real) power* is constant tempts us to believe that the *reactive power* in a three-phase is zero (as in a DC circuit). However, the reactive power is very much present in *each phase* as shown in Equations (2.57). The reactive power per phase is 4.185 MVAR.

Example 2.4   A three-phase load draws 120 kW at a power factor of 0.85 lagging from a 440 V bus. In parallel with this load, a three-phase capacitor bank that is rated 50 kVAR is inserted, find:

    a. The line current without the capacitor bank.
    b. The line current with the capacitor bank.
    c. The PF without the capacitor bank.
    d. The PF with the capacitor bank.

SOLUTION

    a. From the three-phase active power formula, the magnitude of the load current is:

$$I_{Load} = \frac{P}{\sqrt{3}V_L \times PF} = \frac{120 \times 10^3}{\sqrt{3}\,440(0.85)} = 185.25 \text{ A}$$

$$I_{Load} = 185.25\langle -\cos^{-1}(0.85) = 185.25\langle -31.8^\circ \text{ A}$$

    b. The line current of the capacitor bank (a pure reactive load) is:

$$I_{Cap} = \frac{50 \times 10^3}{\sqrt{3}\,440} = 65.6\langle 90^\circ \text{ A}$$

The line current is:

$$I_L = I_{Load} + I_{Cap} = 160.6\langle -11.5^\circ \text{ A}$$

    c. The PF without capacitor bank is PF = 0.85
    d. The PF with capacitor bank is

$$PF = \cos(11.5^\circ) = 0.98$$

2.4   POWER SYSTEM MAIN COMPONENTS

Interconnected power systems, divided in four main sections (generation, transmission, power distribution and end-users or consumers), are made of several components, working together to deliver power to the end-users or costumers, as and when needed. They include prime-movers (e.g. steam turbines, gas turbine, hydropower turbines), synchronous generators, transformers, transmission lines, substations, feeders, protection and monitoring devices, panel-boards, control centers, etc.

2.4.1   SYNCHRONOUS GENERATORS

The device converting mechanical energy to electrical energy is called an electric generator. Synchronous machines can produce high power reliably with high efficiency, and therefore, are widely used as generators in power systems. A generator serves two basic functions. The first one is to produce active power (MW), and the second function, frequently forgotten, is to produce reactive power (MVAR). The discussion on generators will be limited to the fundamentals related to these two functions. More details related to the dynamic performance of the synchronous generators can be found in the references at the end of this chapter or elsewhere in the literature. The mechanical structure of generators is out of the scope of this material. A simplified turbine-generator-exciter system is shown in Figure 2.8a. The turbine, or the prime-mover, controls the active power generation. For instance, by increasing the valve opening of a steam turbine, more active power can be generated and vice versa. The exciter, represented as an adjustable DC voltage source, controls the filed current that controls the internal generated voltage source, the so-called excitation voltage, E_f. In this way, the generator terminal voltage, $\bar{V}$, is controlled. The steady-state equivalent circuit of a synchronous generator can be drawn as an internal voltage source and its (direct-axis) synchronous reactance in series, with the armature resistance neglected (Figure 2.8b). The system is represented with an infinite bus, which holds a constant voltage. An infinite bus, by definition, is a large power system consisting of several (usually hundreds or more) large electric generators. The voltage and frequency of an infinite bus are constant (at least in inside the required and prescribed limits) and cannot be changed regardless of any action made by one or few generators or the connected or disconnected loads. An infinite bus can absorb or deliver any amount of active or reactive power without any changes in voltage or frequency. When a synchronous generator is connected to an infinite bus, as shown in Figure 2.9a, the frequency and voltage at infinite bus are constant and cannot be changed due to any change in the generator excitation current of speed of rotation. The generator terminal voltage, or system voltage, is usually chosen as the reference, therefore, a zero degree phase angle. Then, through a phasor representation, the generator internal generated voltage can be obtained as:

$$\bar{E}_f = \bar{I}\left(jX_d\right) + \bar{V} = E_f < \delta \qquad (2.23)$$

Here, the angle δ is called the generator power (torque) angle, and X_d is the synchronous reactance (Figure 2.8). Figure 2.9b is showing these quantities in the phasor notation, very useful in the generator power and characteristics calculations. The per phase analysis of the complex power injected into the power system by the synchronous generator can be calculated by:

$$\bar{S} = \bar{V} \cdot \bar{I}^* = \frac{VE_f}{X_d}\sin\delta + j\left[\frac{VE_f}{X_d}\cos\delta - \frac{V^2}{X_d}\right] = P + jQ \qquad (2.24)$$

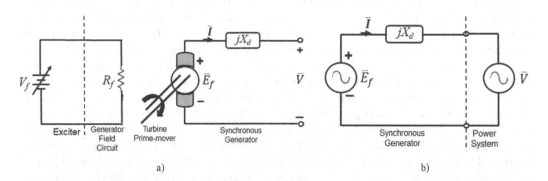

a)   b)

FIGURE 2.8 (a) Electrical representation of a simplified turbine-generator-exciter system, and (b) the per-phase steady-state equivalent circuit of a synchronous generator.

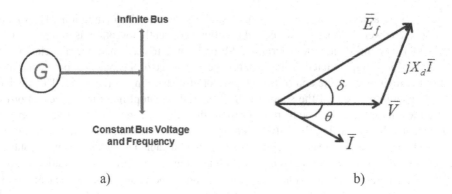

FIGURE 2.9   (a) Synchronous generator connected to an infinite bus; (b) phasor representation of the induced generator voltage.

Here, $\overline{I}^*$ is the complex conjugated armature current that can be expressed from Equation (2.24), after some algebraic manipulation, as:

$$\overline{I}^* = \frac{\overline{E}_f - \overline{V}}{\overline{X}_d} = \frac{E_f(\cos\delta + j\sin\delta) - V}{jX_d} \tag{2.25}$$

Substituting in Equation (2.24) yields the three-phase active (real) and reactive power delivered by a synchronous generator to an infinite bus, expressed by the following relationships:

$$P = 3\frac{E_f V}{X_d}\sin(\delta) \tag{2.26a}$$

And, respectively

$$Q = 3\frac{V}{X_d}(E_f\cos(\delta) - V) \tag{2.26b}$$

The maximum value of the active power, P_{max}, is referred as a steady-state stability limit, being calculated as the limit for a torque angle (δ) of 90°. However, at P_{max} the power angle is 90°, and the angle cannot be increased any further since the generator cannot maintain synchronism with the rest of the power system. It is worth mentioning that when the generator active power increases, the power angle also increases. Equation (2.26a) is showing that in order to increase the generator active power (P_{max} too), the excitation (filed) voltage, E_f, must be increased. However, the generator delivered power to the infinite bus is not increased unless the input generator mechanical power increases. Changing E_f is performed by controlling the excitation current (Figure 2.8a), which is controlled at the power plant level by the plant operator. Assuming negligible generator losses (a reasonable assumption for synchronous machines) in the input generator, mechanical power equals the electric active power delivered to the infinite bus. The mechanical power is controlled by the power plant operator and by increasing the input generator mechanical power, and the maximum delivered power can be increased. Therefore, by increasing mechanical power, the generator capacity to deliver more power also increases. Generator reactive power delivered to the infinite bus is also controlled by the magnitude of the excitation voltage, E_f, as shown in

Equation (2.26b). The reactive power can be positive, negative or zero depending on the generator excitation voltage level. If the excitation voltage is adjusted so $E_f \cos(\delta) > V$, the reactive power is positive, being the case of *the overexcited generator*, meaning that the generator is consuming reactive power and the current is lagging the infinite bus voltage. If the excitation voltage is adjusted so $E_f \cos(\delta) < V$, the reactive power is negative, the generator is supplying reactive power to the infinite bus and the current is leading the bus voltage, being the *underexcited generator* case. If the excitation is adjusted so $E_f \cos(\delta) = V$, there is no reactive power delivered or absorbed by the synchronous generator, and the current is in phase with the infinite bus voltage, the so-called *exact excited generator* case.

Power plants or generation stations are usually located in remote areas far from the load centers (cities, towns, industrial facilities, campuses, etc.) for reasons, such as proximity of the primary energy sources, transportation and storage facilities for fuels or possible pollutions from the power plant to be kept away from large urban and metropolitan areas. Generated electricity by power plant is transferred to load centers by a system of transmission lines. Figures 2.10a and 2.10b are showing the one-line diagram of a synchronous generator connected through a transmission line to an infinite bus, and the system equivalent circuit. The generator terminal voltage is V_g, the infinite bus voltage is V_O and the high voltage transmission line impedance, reactance and resistance are, Z_L, X_L and R_L. However, the transmission line resistance is much smaller than the line inductive reactance and is usually neglected in calculations. In this case, the power angle, δ, is defined as the angle between the infinite bus voltage, V_O, and the equivalent generator voltage, E_f. The equivalent voltage equation of this system is then expressed as:

$$E_f = \bar{V}_g + \bar{I}_a X_d = \bar{V}_O + \bar{I}_a X_d + \bar{I}_a X_L = \bar{V}_O + \bar{I}_a(X_d + X_L) \tag{2.27}$$

The relationships of equation can be interpreted by using the phasor diagrams of Figure 2.9, where the current is assumed to be lagging the infinite bus voltage. Introducing the total (series) impedance, neglecting the generator and line resistances, considering only the reactance of the generator and transmission line, $X = X_d + X_L$, Equation (2.27) is expressed in a short form as:

$$E_f = \bar{V}_O + \bar{I}_a X \tag{2.28}$$

The power equations for the system of generator, transmission line and infinite bus are the same as Equation (2.26a and b) with only difference being that the series reactance, X, is replacing the synchronous generator reactance, X_d. Hence, the real (active) and reactive powers are:

$$P_O = 3\frac{E_f V}{X}\sin(\delta) \tag{2.29a}$$

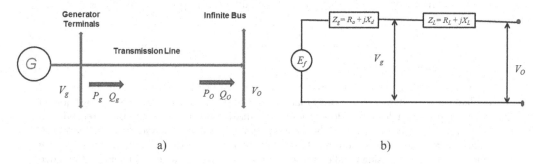

a)　　　　　　　　　　b)

FIGURE 2.10 (a) Generator connected to an infinite bus through a transmission line; and (b) system equivalent circuit.

And, respectively

$$Q_O = 3\frac{V}{X}(E_f \cos(\delta) - V) \tag{2.29b}$$

Notice that the reactive power at the generator terminals, as shown in Figure 2.7a, is:

$$Q_g = 3\frac{V}{X_d}(E_f \cos(\alpha) - V) \tag{2.30}$$

Here, α is the angle between the generator terminal voltage and the excitation voltage, E_f.

Example 2.5 A synchronous generator, having a 1.5 Ω reactance and a negligible armature resistance, is connected to an infinite bus through a transmission line, having a 1.2 Ω reactance and a negligible armature resistance. The generator excitation is adjusted so the line-to-line equivalent field voltage is 36 kV, and the generator delivered power to the infinite bus is 180 MW. If the infinite bus voltage is 30 kV, compute the terminal generator voltage and the reactive power at the infinite bus.

SOLUTION

In order to calculate the terminal generator voltage, the current must be first determined. From Equation (2.26a), the power angle is then calculated:

$$180 = \frac{36 \times 30}{1.5 + 1.2}\cos(\delta) \Rightarrow \delta = 26.8°$$

The current in transmission line is computed from Equation (2.25) as:

$$\overline{I}^* = \frac{\overline{E}_f - \overline{V}}{\overline{X}_d} = \frac{\dfrac{36}{\sqrt{3}} < 26.8° - \dfrac{30}{\sqrt{3}} < 0°}{2.7 < 90°} = 3.459 < -7.5° \text{ kA}$$

The generator terminal voltage is then calculated by using Equation (2.27):

$$\overline{V}_g = \overline{V}_O + \overline{I}_a X_L = \frac{30}{\sqrt{3}} < 0° + (3.459 < -7.7°)1 < 90° = 18.1 < 10.9° \text{ kV}$$

The line-to-line voltage is $\sqrt{3} \times 18.1 = 31.3$ kV and the reactive power at the infinite bus is:

$$Q_O = 3\frac{V}{X}(E_f \cos(\delta) - V) = \frac{\sqrt{3} \times 30}{2.7}\left(\frac{36}{\sqrt{3}}\cos(26.8°) - \frac{30}{\sqrt{3}}\right) = 21.369 \text{ MVAR}$$

The power system capacity is the maximum power that can be transmitted by the system. A power plant cannot generate more power than the transmission capacity, as expressed by Equation (2.29a), when power angle is 90°. However, the transmission line capacity can be increased if the excitation voltage, E_f, is increased or the total (system) reactance, X, is reduced. The two common methods to reduce the system reactance are: inserting a capacitor bank in series with the transmission line, and by using parallel transmission lines. When a capacitor, having a reactance, X_{cap}, is connected in series with the transmission line, total system reactance is $X = X_d + X_L - X_{cap}$. Parallel transmission lines are often used to increase the system transmission capacity. If the two transmission lines, having the reactances, X_{L1} and X_{L2}, are connected in parallel with a synchronous generator, the total reactance is:

$$X = X_d + \frac{X_{L1} \cdot X_{L2}}{X_{L1} + X_{L2}}$$

2.4.2 Power Transformers

A transformer consists of two or more windings that are magnetically coupled using a ferromagnetic core. For a two-winding transformer, the winding connected to the AC supply is referred to as the *primary* while the winding connected to the load is referred to as the *secondary*. A time-varying current passing through the primary coil produces a time-varying magnetic flux density within the core. According to Faraday Law, the time-changing flux passing through the secondary induces a voltage in the secondary terminals. The power distribution transformer reduces the primary voltage of the electric distribution system to the utilization voltage serving the customers. Power distribution transformers are static devices constructed with two or more windings used to transfer AC power from one circuit to another at the same frequency but with different values of voltage and current. Usually, power transformers are built as three-phase units and may have multiple winding patterns. Figure 2.11 is showing an ideal transformer and its symbol. We shall refer to the windings as HV and LV, the high- and low-voltage, respectively. Often additional windings (*tertiaries*, etc.) are added. A transformer is either core-type or shell-type construction. It can be seen from this figure that core-type construction the primary and the secondary windings are wound as a pair of concentric coils on each limb, whereas for a shell-type construction the primary and secondary windings form interleaved layers on a single limb. In all cases, the core will be of laminated construction in order to reduce iron losses to a minimum.

An ideal transformer is one with negligible winding resistances and reactances and no exciting losses, infinite magnetic permeability of the core and all magnetic flux remains into the transformer core. It consists of two conducting coils wound on a common core, made of high-grade iron. There is no electrical connection between the coils, with them being connected to each other through magnetic flux. The coil on input side is called the primary winding and that on the output side the secondary. Transformer action requires only the existence of time-varying mutual flux linking two windings, while the coupling between the windings can be made much more effective through the use of a core of ferromagnetic material because most of the flux will be confined within the core. When a time-varying AC voltage is applied to the primary winding, it causes an AC magnetic flux to appear in the transformer core, coupling the secondary winding. The resulting alternating magnetic flux magnitude depends on the voltage and the number of turns of primary winding. The alternating flux links the secondary winding and induces a voltage in

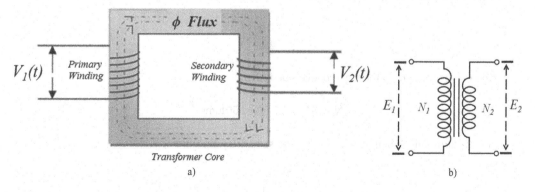

FIGURE 2.11   (a) Ideal transformer diagram, and (b) the ideal transformer symbol.

it with a value that depends on the number of turns of the secondary windings. The transformer operation is subject to Faraday's and Ampere's Laws. If the primary voltage is $v_1(t)$, the core flux $\phi(t)$ is established such that the counter-emf e(t) equals the impressed voltage (neglecting winding resistance), as:

$$v_1(t) = e_1(t) = N_1 \frac{d\phi(t)}{dt} \qquad (2.31)$$

Here, N_1 is the number of turns of the primary winding. The emf $e_2(t)$ is induced in the secondary by the alternating core magnetic flux $\phi(t)$, expressed as:

$$v_2(t) = e_2(t) = N_2 \frac{d\phi(t)}{dt} \qquad (2.32)$$

Taking the ratio of Equations (7.3) and (7.4), we obtain:

$$\frac{v_1}{v_2} = \frac{i_2}{i_1} = \frac{N_1}{N_2} = a \qquad (2.33)$$

Here, parameter a is the transformer turns ratio. If $a > 1$, the transformer is step-down, if $a < 1$, the transformer is step-up, while if $a = 1$, the transformer is the so-called impedance transformer, used to separate electric tow circuits. If a load is connected across the secondary terminals, as presented in Figure 2.7, it results in current i_2. This current will cause the change in the mmf in the amount $N_2 i_2$. Neglecting losses (ideal transformer assumption), the instantaneous power is conserved:

$$v_1 i_1 = v_2 i_2 \text{ or in pashor notation } V_P I_P = V_S I_S \qquad (2.34)$$

Example 2.6   A 220/20 V transformer has 50 turns on its low-voltage side. Calculate:

    a. The number of turns on its high side.
    b. The turns ratio a, when it is used as a step-down transformer.
    c. The turns ratio a, when it is used as a step-up transformer.

SOLUTION

Transformer, used as the step down unit, the turns ratio is

$$a_{SD} = \frac{220}{20} = 11$$

The number of turns in the high-voltage side is then:

$$N_P = aN_S = 11 \cdot 50 = 550 \text{ turns}$$

So, the turns ratio, when this transformer is used as step-up system, is:

$$a_{SU} = \frac{1}{a_{SD}} = \frac{1}{11} = 0.091$$

From Equations (2.33) and (2.34), it is clear that almost any desired voltage or current ratios or ratio of the transformation can be obtained by adjusting number of turns of the transformer windings. However, a magnetic material such as iron undergoes losses of energy due to the application of alternating voltage in the B-H loop. These losses are composed of two parts. The first one is called the *eddy-current loss*, and the second one is the *hysteresis loss*. Eddy-current loss is basically an I^2R loss due to induced current in magnetic materials of the core due to the alternating magnetic flux linking the windings. To reduce these losses, the magnetic core is usually made by a stack of thin iron-alloy laminations. For ideal transformers, the following assumptions are made: the resistances of the windings can be neglected, while the core reluctance (magnetic resistance) is negligible. All the magnetic flux is linked by all the turns of the coil, and there is no leakage of flux. The equations for the sinusoidal voltages in and ideal transformer are:

$$V_1 = 4.44fN_1\Phi_m \tag{2.35}$$

Here, Φ_m is the peak value of the magnetic flux. In a similar manner, the RMS value of the secondary emf will be given by:

$$V_2 = 4.44fN_2\Phi_m \tag{2.36}$$

Example 2.7 Suppose a coil having 100 turns is wound on a core with a uniform cross-sectional area of 0.25 m². A 5 A, 60 Hz current is flowing into this coil. If the maximum magnetic flux density is 0.75 T, find the mmf and the voltage induced into the coil.

SOLUTION

The calculation starts with: $mmf = NI = 100 \times 5 = 500\,At$
The induced voltage is computed by using Equation (2.35)

$$V = 4.44 \cdot 60 \cdot 100 \cdot 0.75 \cdot 0.25 \times 10^{-4} = 0.4995 \approx 0.5\ V$$

By dividing Equation (2.35) by (2.36), the transformer voltage relationship in phasor notation, the equivalent of Equation (2.33) is obtained. If we are considering now an arbitrary load (Z_2) connected to the secondary terminals of the ideal transformer as shown in Figure 2.12.
The input impedance seen looking into the primary winding is given by:

$$Z_1 = \frac{V_1}{I_1} = \frac{aV_2}{\frac{I_2}{a}} = a^2Z_2 \tag{2.37}$$

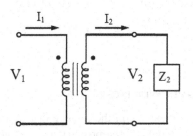

FIGURE 2.12   Load connected to an ideal transformer secondary.

The impedance seen by the primary voltage source of the ideal transformer is the secondary load impedance times the square of the transformer ratio (Figure 2.12). Using this property, the secondary impedance of the ideal transformer can be reflected to the primary. In a similar fashion, a load on the primary side of the ideal transformer can be reflected to the secondary.

$$Z_2 = \frac{V_2}{I_2} = \frac{\dfrac{V_1}{a}}{aI_1} = \frac{Z_1}{a^2} \tag{2.38}$$

Another important property of an ideal transformer that is derived from the above equations is the power conservation – *the primary and secondary apparent powers (volt-amperes) are equal in an ideal transformer*:

$$V_1 I_1 = V_2 I_2 \tag{2.39}$$

Example 2.8   Determine the primary and secondary currents for the ideal transformer, supplied by a 120 V, if the source impedance is Zs = (18–j4) Ω and the load impedance is Z_2 = (2+j1) Ω. The transformer ration is a = 4.

SOLUTION

The secondary voltage is:

$$V_2 = \frac{V_1}{a} = \frac{120}{4} = 30 \text{ V (RMS)}$$

The load impedance seen in the primary side is:

$$Z_2' = a^2 Z_2 = (4)^2(2 + j1) = 32 + j16 \ \Omega$$

The primary and secondary currents are computed as:

$$I_1 = \frac{V_1}{Z_S + Z_2'} = \frac{120\langle 0^\circ}{18 - j4 + 32 + j16} = 2.33\langle -13.5^\circ \text{ A (RMS)}$$
$$I_2 = aI_1 = 9.32\langle -13.5^\circ \text{ A (RMS)}$$

While the primary and the secondary voltages are:

$$V_1 = Z_2' I_1 = (32 + j16)2.33\langle -13.5^\circ = 83.50\langle 13.07^\circ \text{ V (RMS)}$$
$$V_2 = \frac{V_1}{a} = \frac{83.50\langle 13.07^\circ}{4} = 20.88\langle 13.07^\circ \text{ V (RMS)}$$

2.4.3   POLARITY OF TRANSFORMER WINDINGS

The operation of the transformer depends on the relative orientation of the primary and secondary coils. We mark one of the terminals on the primary and secondary coils with a dot to denote that currents entering these two terminals produce magnetic flux in the same direction within the

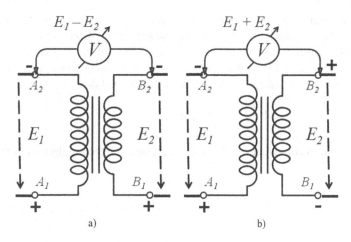

FIGURE 2.13   (a) Subtractive, and (b) additive polarity.

transformer core (as shown in diagrams of Figure 2.13). If either coil orientation is reversed, the dot positions are reversed and the current and voltage equations must include a minus sign. For power distribution transformers, the polarity is important only if the need arises to parallel transformers to gain additional capacity or to hook up three single-phase transformers to make a three-phase bank. The way the connections are made affects angular displacement, phase rotation, and the rotation direction of the connected motors. Polarity is also important when hooking up current transformers for relay protection and metering. Transformer polarity depends on which direction coils are wound around the core (clockwise or counterclockwise) and the leads. Transformers are sometimes marked at their terminals with polarity marks. Often, polarity marks are shown as white paint dots (for plus) or plus-minus marks on the transformer and on the nameplate.

More often, transformer polarity is shown simply by the American National Standards Institute (ANSI) designations of the winding leads as H1, H2 and X1, X2. By ANSI standards, if you face the low-voltage side of a single-phase transformer (the side marked X1, X2), the H1 connection will always be on your far left. If the terminal marked X1 is also on your left, it is subtractive polarity. If the X1 terminal is on your right, it is additive polarity. Additive polarity is common for small power distribution transformers. A transformer is said to have additive polarity if, when adjacent high- and low-voltage terminals are connected and a voltmeter placed across these terminals, the voltmeter reads the sum (additive) of the high- and low-voltage windings. Diagrams of Figure 2.13 are showing the high- and low-side voltage relationships for subtractive and additive polarity. For a given input, the primary voltage, output, secondary voltage of an ideal transformer is independent of the load attached to secondary. However, the output voltage of a real transformer depends on the load current. The voltage regulation (VR) is defined as the change in the secondary voltage as the load current changes from the no-load to the loaded condition.

$$VR(\%) = \frac{|V_S|_{NL} - |V_S|_{FL}}{|V_S|_{NL}} \times 100 \qquad (2.40)$$

Thus, the percentage voltage regulation may be written in terms of the reflected secondary voltages.

$$VR(\%) = \frac{|V_S'|_{NL} - |V_S'|_{FL}}{|V_S'|_{NL}} \times 100 \qquad (2.41)$$

The transformer equivalent circuit gives only the reflected secondary voltage. The actual loaded and no-load secondary voltages are equal to the loaded and no-load reflected secondary values divided by the transformer turns ratio. The prime quantities are the secondary voltages reflected into the primary. The reflected secondary voltage under no-load conditions is equal to the primary voltage, while the secondary voltage for the loaded condition is taken as the rated voltage:

$$\left|V_S'\right|_{NL} = V_P, \text{ and } \left|V_S'\right|_{FL} = \left|V_S'\right|_{rated}$$

Inserting the previous two equations into the percentage voltage regulation equation gives:

$$VR(\%) = \frac{\left|V_P\right| - \left|V_S'\right|_{rated}}{\left|V_S'\right|_{rated}} \times 100 \tag{2.42}$$

Example 2.9   Compute the voltage regulation of the transformer in example 2.8 for: (a) unit power factor, (b) 0.8 lagging power factor, and (c) 0.8 leading power factor.

SOLUTION

The impedance of the transformer is

$$Z_S = R_S + jX_S = 145.2 + j1210 \ \Omega$$

a. For unit power factor

$$V_1' = V_2 + I_2 Z_2 = 110000 + 9.09 \angle 0° \times (145.2 + j1210) = 111320 + j11100 \text{ V}$$

And

$$\left|V_1'\right| = \sqrt{1113206^2 + 11100^2} \approx 111870 \text{ V}$$

The voltage regulation, by using Equation (2.42) is:

$$VR = \frac{111870 - 110000}{110000} \times 100 = 1.7\%$$

b. For 0.8 lagging power factor

$$V_1' = V_2 + I_2 Z_2 = 110000 + 9.09 \angle -36.9° \times (145.2 + j1210) = 117805 + j8078 \text{ V}$$

And

$$\left|V_1'\right| = \sqrt{117805^2 + 8078^2} \approx 118070 \text{ V}$$

The voltage regulation, by using Equation (2.42) is:

$$VR = \frac{118070 - 110000}{110000} \times 100 = 7.3\%$$

c. For 0.8 leading power factor

$$V_1^1 = V_2 + I_2 Z_2 = 110000 + 9.09\langle 36.9^\circ \times (145.2 + j1210) = 104355 + j9662 \text{ V}$$

And

$$|V_1^1| = \sqrt{104355^2 + 9662^2} \simeq 104800 \text{ V}$$

The voltage regulation, by using Equation (2.42) is:

$$VR = \frac{104800 - 110000}{110000} \times 100 = -4.7\%$$

2.4.4 TRANSFORMER RATINGS, CATEGORIES, TYPES AND TAP CHANGERS

Transformer ratings are given on their nameplates, indicating the normal operating conditions. The nameplate includes parameters such as: primary-to-secondary voltage ratio, design operation frequency and apparent rated output power, etc. Transformers ratings are related to the primary and secondary windings, referring to the apparent power (kVA) and primary and secondary voltages. A rating of 10 kVA, 1100/110 V means that the primary is rated for 1100 V while the secondary is rated for 110 V ($a = 10$). Power generated at a generating station (usually at a voltage in the range of 11 kV to 25 kV) is stepped-up by a transformer to a higher voltage (220, 345, 400 or 765 kV) for transmission purpose. Transformers are one of the most important power system components of the power system. Generator transformers are designed with higher losses since the cost of supplying losses is the cheapest at a generating station. Generator transformers are usually provided with off-circuit tap changer with a small variation in voltage (e.g. ±5%) because the voltage can always be controlled by generator field current. Unit auxiliary transformers are step-down transformers with primary connected to generator output directly. Its secondary voltage is of the order of 6.9 kV for fitting to the power requirements of the auxiliary generating station equipment. Station transformers are required to supply auxiliary equipment during setting-up of the generating station and subsequently during each start-up operation. The rating of these transformers is small, and their primary is connected to a high-voltage transmission lines, resulting in smaller conductor size for HV winding, necessitating special measures for increasing the short-circuit strength. Interconnecting transformer are usually autotransformers used to interconnect two grids/power systems operating at two different system voltages. They are normally located in the transmission system between the generator transformers and receiving end transformers, reducing the transmission voltage (400 or 345 kV) to the sub-transmission level (220 or 138 kV). In autotransformers, there is no electrical isolation between primary and secondary windings, some volt-amperes are conductively transformed and remaining is inductively transformed. Autotransformer design becomes more economical as the ratio of secondary voltage to primary voltage approaches unity. These are characterized by a wide tapping range and an additional tertiary winding which may be loaded or unloaded. Unloaded tertiary acts as a stabilizing winding by providing a path for the third harmonic currents. Synchronous condensers or shunt reactors are connected to the tertiary winding, if required, for reactive power compensation. In the case of an unloaded tertiary, adequate conductor area and proper supporting arrangement are provided for withstanding short circuit forces under asymmetrical fault conditions.

Example 2.10   Determine the turns ratio and the rated currents of a transformer from its nameplate data, 480 V/120 V, 48 kVA and 60 Hz.

SOLUTION

Assuming ideal transformer, the transformer ration is:

$$a = \frac{480}{120} = 4$$

The primary and the secondary currents are:

$$I_P = \frac{|S|}{V_P} = \frac{48000}{480} = 100 \text{ A}$$

$$I_S = \frac{|S|}{V_S} = \frac{48000}{120} = 400 \text{ A}$$

Receiving station transformers are basically step-down transformers reducing transmission/ sub-transmission voltage to primary feeder level (e.g. 33 kV). Some of these may be directly supplying an industrial plant. Loads on these transformers vary over wider limits, and their losses are expensive. The farther is the location of transformers from the generating station, the higher the cost of supplying the losses. Automatic tap changing on load is usually necessary, and tapping range is higher to account for wide variation in the voltage. A lower noise level is desirable if they are close to residential areas. Distribution transformers are used to adjust the primary feeder voltage to the actual utilization voltage (~415 or 460 V) for domestic or industrial use. A great variety of transformers fall into this category due to many different arrangements and connections. Load on these transformers varies widely, and they are often overloaded. A lower value of no-load loss is desirable to improve all-day efficiency. Hence, the no-load loss is usually capitalized with a high rate at the tendering stage. Since very little supervision is possible, users expect the least maintenance on these transformers. The cost of supplying losses and reactive power is highest for these transformers. Classification of transformers as above is based on their location and broad function in the power system. Power distribution transformers for utility applications, single-phase or three-phase service have fixed voltage ratings, having no way to adjust the transformer voltage ratio to allow for applications in which the system voltage is slightly off the nominal values. When a transformer is required to give a constant load voltage despite changes in load current or supply voltage, turns ratio of the transformer must be altered. In such situations where the system voltage differs from the nominal, voltage taps are used to accomplish such tasks, via a tap-setting mechanism. The taps are usually on the high-voltage windings to adjust for the variations in the supply voltage. This is the function of a tap changer and the two basic types are: a) *off-load*, and b) *on-load*. For tap operation understanding, let's consider how a standard tap changer works by considering a common example.

Example 2.11   A 13800V/4160 V transformer has five taps on the primary winding giving –5%, –2-1/2%, nominal, +2-1/2% and +5% turns. If, on-load, the secondary voltage reduces to 4050V then, which tap, should be used to maintain 4160V on-load (assuming the supply voltage remains constant)?

SOLUTION

To keep the secondary voltage at (or as close as possible to) 4160 V, either primary supply voltage or the HV winding tap position must be altered. Examining the transformer relationship, indicates that in order to keep the equation in balance with primary voltage and secondary winding turns fixed, either V_2 or N_1 must be adjusted. Since the objective is to raise V_2 back to nominal, then N_1 must be reduced. To raise V_2 from 4050 V to 4160V requires an increase in secondary volts of:

$$Tap = \frac{4160}{4050} = 1.027 \text{ or } 102.7\%$$

N_1 must be reduced to $1/1.027 = 0\ 974$. Therefore, N_1 must be reduced by $(1 - 0.974) = 0.026$ or 2.6%.

Reducing $N1$ by 2.6% is accomplished by the increase in secondary voltage output. The nearest tap to select is $-2\text{-}1/2\%$, as specified in the example statement.

Most transformers associated with medium-level voltage distribution for stations have off-circuit tap *changers. With this tap changer type, the transformer has to be switched out of circuit before any tap changing.* The tap changer contacts are not designed to break any current, even the no-load current. If an attempt is made to change the tap positions while on-line, severe arcing results which may destroy the tap changer and the transformer. On-load tap changers permit tap changing and voltage regulation with the transformer on-load. Tap changing is usually done on the HV winding for two reasons. First, because the currents are lower, the tap changer contacts, leads, etc., can be smaller. Second, as the HV winding is wound outside the LV winding, easier to get the tapping connections out to the tap changer. Figure 2.14 shows the connections for an on-load tap changer that operates on the HV winding of the transformer.

The tap changer has four essential features. Selector switches that select the physical tap position on the transformer winding and, because of their construction, cannot and must not make or break the load current. The load current must never be interrupted during a tap change. Therefore, during each tap change, there is an interval where two voltage taps are spanned. Reactors (inductors) are used in the circuit to increase the impedance of the selector circuit and limit the current circulating due to this voltage difference. Under normal load conditions, equal load current flows in both halves of the reactor windings and the fluxes balance out giving no resultant flux in the core. With no flux, there is no inductance and, therefore, no voltage drop due to inductance. There will be, however, a very small voltage drop due to resistance. During the tap change, the selector switches are selected to different taps and circulating current flows in the reactor circuit, creating a magnetic flux and the resulting inductive reactance limits the circulating current. The vacuum switch device performs the duty of a circuit breaker that makes and breaks current during the tap changing sequence. The bypass switch operates during the tap changing sequence it does not make or break load current, though it only makes "before break" each connection.

Single-phase transformers can be connected in several configurations. Two single-phase transformers can be connected in four different combinations provided that their polarities are

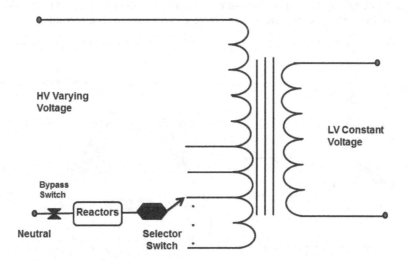

FIGURE 2.14 High-voltage tap changer.

observed. When transformer windings are connected in parallel, the transformer having the same voltage and polarity are paralleled. When connected in series, windings of opposite polarity are joined in one junction. Coils of unequal voltage may be series-connected with polarities either adding or opposing. In many sections of the power system, three-winding transformers are used, with the three windings housed on the same core to achieve economic savings. Three-phase (3-phase supplies) are used for electrical power generation, transmission and distribution, as well as for all industrial uses. Three-phase supplies have many electrical advantages over single-phase power and when considering three-phase transformers, we have to deal with three alternating voltages and currents differing in phase-time by 120 degrees. A transformer cannot act as a phase changing device and change single-phase into three-phase or three-phase into single-phase. To make the transformer connections compatible with three-phase supplies, we need to connect them together in a particular way to form a three-phase transformer configuration. A three-phase transformer can be constructed either by connecting together three single-phase transformers, thereby forming a so-called three phase transformer bank, or by using one pre-assembled and balanced three-phase transformer which consists of three pairs of single-phase windings mounted onto one single laminated core. The advantages of building a single three-phase transformer is that for the same kVA rating it will be smaller, cheaper and lighter than three individual single-phase transformers connected together because the copper and iron core are used more effectively. The methods of connecting the primary and secondary windings are the same, whether using just one three-phase transformer or three separate single-phase transformers. The primary and secondary windings of a transformer can be connected in different configurations, as shown, to meet practically any requirement. In the case of three-phase transformer windings, three forms of connection are possible: "star" (wye), "delta" (mesh) and "interconnected-star" (zig-zag). The combinations of the three windings may be with the primary delta-connected and the secondary star-connected, or star-delta, star-star or delta-delta, depending on the transformers use. When transformers are used to provide three or more phases, they are generally referred to as a poly-phase transformer. But what do we mean by "star" (also known as Wye) and "delta" (also known as Mesh) when dealing with three-phase transformer connections? A three-phase transformer has three sets of primary and secondary windings. Depending upon how these sets of windings are interconnected, it determines whether the connection is a star or delta configuration. The three available voltages, which themselves are displaced from each other by 120 electrical degrees, not only decides on the type of the electrical connections used on both the primary and secondary sides but also determine the flow of the transformers' currents. With three single-phase transformers connected together, the magnetic flux in the three transformers differs in phase by 120 time-degrees. With a single three-phase transformer, there are three magnetic fluxes in the core differing in time-phase by 120 degrees.

The standard method for marking three-phase transformer windings is to label the three primary windings with capital (upper case) letters A, B and C, used to represent the three individual phases of RED, YELLOW and BLUE (see Figure 2.15 for details). The secondary windings are

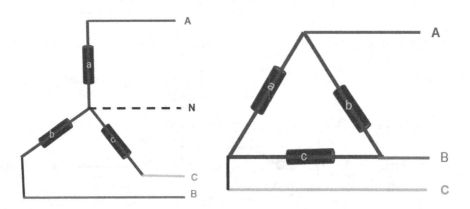

FIGURE 2.15   Star (Y) and delta (Δ) transformer connections.

labeled with small (lower case) letters a, b and c. Each winding has two ends normally labeled 1 and 2, so that, e.g. the second winding of the primary has ends which will be labeled B_1 and B_2, while the third winding of the secondary will be labeled c1 and c2 as shown. There are four ways in which three single-phase transformers may be connected together. The configurations are delta-delta, star-star, star-delta and delta-star. Transformers for high voltage operation with the star connections has the advantage of reducing the voltage on an individual transformer, reducing the number of turns required and an increase in the size of the conductors, making the coil windings easier and cheaper to insulate than delta transformers. The delta-delta connection nevertheless has one big advantage over the star-delta configuration, in that if one transformer of a group of three should become faulty or disabled, the two remaining ones will continue to deliver three-phase power with a capacity equal to approximately two-thirds of the original output from the transformer unit. One disadvantage of delta-connected three-phase transformers is that each transformer must be wound for the full-line voltage (in our example above 100V) and for 57.7% line current. The greater number of turns in the windings and the required insulation between turns necessitate larger and more expensive coils than the star connection. Another disadvantage with delta-connected three-phase transformers is that there is no "neutral" or common connection. In the star-star arrangement (Y-Y, or wye-wye), each transformer has one terminal connected to a common junction, or neutral point with the three remaining ends of the primary windings connected to the three-phase mains supply. The number of turns in a transformer winding for star connection is 57.7%, of that required for delta connection. The star connection requires the use of three transformers, and if any one transformer becomes fault or disabled, the whole group might become disabled. Nevertheless, the star-connected three-phase transformer is especially convenient and economical in electrical power distributing systems, in that a fourth wire may be connected as a neutral point (n) of the three star-connected secondaries as shown.

2.4.5 Transformer Efficiency

The efficiency (η) of a transformer is defined as the ratio of the output power (P_{Out}) to the input power (P_{In}). The output power is equal to the input power minus the transformer losses. The transformer losses have two components: core loss (P_{Core}) and so-called copper loss (P_{Cu}) associated with the winding resistances. The transformer efficiency in percent is given by:

$$\eta = \frac{P_{Out}}{P_{In}} \times 100 = \frac{P_{Out}}{P_{Out} + Losses} \times 100 = \left(1 - \frac{Losses}{P_{In}}\right) \times 100 \tag{2.43}$$

Or

$$\eta = \frac{P_{Out}}{P_{Out} + P_{Core} + P_{Cu}} \times 100 \tag{2.44}$$

Assuming a relatively constant voltage source on the primary of the transformer, the core loss can be assumed to be constant and equal to power dissipated in the core loss resistance of the equivalent circuit for the no-load test. The copper loss in a transformer may be written in terms of both the primary and secondary currents, or in terms of only one of these currents based on the relationship (2.33). The copper losses are a function of the load current, while the core losses depend on the peak core magnetic flux density, which in turn depends on the transformer applied voltage. The core losses are in fact constant because the supply voltage is constant. From the equivalent circuit, one can show that the transformer efficiency depends on the load current (I_2) and the load power factor (θ_2), expressed as:

$$\eta = \frac{V_2 I_2 \cos(\theta_2)}{V_2 I_2 \cos(\theta_2) + P_c + R_{2eq} I_2^2} \times 100 \tag{2.45}$$

Example 2.12   Compute the efficiency of a transformer that has core losses of 1 kW, load current 17.7 A and load phase angle 30°. The output voltage is 900 V, and the equivalent resistance from the secondary side is 1.2 Ω.

SOLUTION

The copper losses are:

$$P_{cu} = R_{2eq}I_2^2 = 1.2 \cdot (17.7)^2 = 375.95 \text{ or } 376 \text{ W}$$

The output power is:

$$P_{out} = V_2 I_2 \cos(30°) = 900 \cdot 17.7 \cdot 0.866 = 13795.4 \text{ W}$$

Using Equation (7.25) the efficiency is:

$$\eta = \frac{V_2 I_2 \cos(\theta_2)}{V_2 I_2 \cos(\theta_2) + P_c + R_{2eq}I_2^2} = \frac{13795.4}{13795.4 + 1000 + 376} = 0.909 \text{ or } 90.9\%$$

By taking the derivative of the efficiency vs. load current, the transformer maximum efficiency can be determined. The maximum efficiency condition is that the copper losses are equal to the core losses. Power distribution transformers usually operate near maximum capacity over 24-hour interval and they are taken out when are not required. To account for their efficiency performance, a merit figure is used, the so-call "all-day" or "energy" power distribution transformer efficiency, expressed as:

$$\eta_{All-day} = \frac{24\text{-hour energy output}}{24\text{-hour energy intput}} \qquad (2.46)$$

2.4.6   TRANSMISSION LINES

Bulk electric power is transmitted through three-phase transmission systems. Each phase conductor has resistance, inductance and capacitance. In transmission line models, the line parameters, resistance, inductance and capacitance are expressed per mile or per kilometer, as lumped models that are used in the analysis of power grids. Any transmission line is designed to carry current and at a given designed voltage, based on the amount of power (usually specified in MVA) that lines must carry. The line current rating is based on the conductor size and by its thermal rating. Overhead transmission lines are suspended from insulators which are supported by towers or poles. The span between two towers (poles) is dependent upon the allowable line sag, which for the steel towers with very-high-voltage transmission lines; the span is usually between 350 to 460 m (1100 to 1500 ft.). When specifying towers and transmission lines, ice and wind loadings are considered into calculations, as well as the extra forces due to a break in the line on one side of the tower. For LV and power distribution lines, wooden, reinforced concrete poles and glass fiber tubes are commonly used with conductors supported in horizontal formations. The conducting material for overhead lines is usually aluminum (a lightweight material), reinforced with stranded steel to increase the mechanical strength. Undergrounded cables are usually made of copper, better conductor than aluminum but more expensive. The parameters of interest in transmission line analysis are inductance, capacitance, resistance and leakage resistance. The representations of the transmission lines and cables depend mainly on their length and the required accuracy. There are three broad classification of length: short, medium and long transmission lines. The actual transmission lines or cables are a distributed-constant electric circuit, i.e. their resistance, inductance, capacitance and leakage resistance is evenly distributed along the line.

Conductors are the physical medium, part of a transmission line to carry electrical energy form one place to other, being an important component of overhead and underground electrical transmission and distribution systems. The choice of conductor depends on the cost and efficiency. An ideal conductor has following features: good electric conductivity; high tensile strength; lower specific gravity, i.e. weight per unit volume and cost without sacrificing other factors. In the early days of the power systems, the conductors used on transmission lines were usually copper, but aluminum conductors have completely replaced copper because of the much lower cost and lighter weight compared with a copper conductor of the same resistance. The fact that aluminum conductor has a larger diameter than a copper conductor of the same resistance is also an advantage. With a larger diameter, the lines of electric flux originating on the conductor will be farther apart at the conductor surface for the same voltage. This means a lower voltage gradient at the conductor surface and less tendency to ionize the air around the conductor. Ionization produces the undesirable effect called corona effect. Aluminum Conductor Steel Reinforced (ACSR), the most used type is concentrically stranded conductor with one or more layers of hard drawn 1350-H19 aluminum wire on galvanized steel wire core. The core can be single wire or stranded depending on the size. ACSR conductors are recognized for their record of economy, cost, dependability and favorable strength per weight ratio. ACSR conductors combine the light weight and good conductivity of aluminum with the high tensile strength and ruggedness of steel. In power transmission line design, this can provide higher tensions, less sag and longer span lengths than obtainable with most other types of overhead conductors. Bundle conductor is a conductor which consists of several conductor cables which connected. Bundle conductors also will help to increase the current carried in the transmission line. The main disadvantage of transmission line is its high wind load compare to other types of conductors. The combination of more than one conductor per phase in parallel suitably spaced from each other used in overhead transmission line is defined as conductor bundle.

The electric parameters of transmission lines (i.e. its resistance, inductance and capacitance) can be determined from the specifications for the line conductors, and from the geometric arrangements of the conductors. Every transmission line has three basic electrical parameters: the conductor resistance, inductance and capacitance. As the transmission line is a set of conductors being run from one place to another, it is supported by transmission towers and the parameters are distributed uniformly along the transmission line. Whatever may be the category of transmission line, the main aim is to transmit power from one end to another. Like other electrical system component, the transmission network also will have some power losses and voltage drops during transmitting power from sending end to receiving end. The power losses are proportional with the line resistance, determined by the transmission line length and type. Utilities strive to maintain constant voltage at receiving end of the transmission lines. An important characteristic of a transmission line is the *thermal limit*. Thermal limits on conductors, transmission lines and equipment depend on the conductor insulation materials. The RI^2 losses are converted into heat, which increases the insulation material temperature, and if it passes some threshold, may deteriorate the transmission lines or equipment. The power losses change with the load changes, meaning that the rated load is basically set be the temperature limits. The performance of transmission line can also be determined by its efficiency and voltage regulation. The transmission lines are categorized as three types: (1) short transmission line, if the line length is up to 80 km or 50 miles; (2) medium transmission line, if the line length is between 80 km and 160 km, or from 50 to about 100 miles; and (3) long transmission line, if the line length is more than 160 km or 100 miles. Frequency used in power transmission is either 50 Hz or 60 Hz, depending on the country. The voltage or current wavelength is then determined by $c = f \times \lambda$, leading to wavelengths of about 6000 km. For this reason, the transmission line, with length less than 160 km, the parameters are assumed to be lumped and not distributed. Such lines are known as electrically short transmission line. This electrically short transmission lines are again categorized as short transmission line (length up to 80 km) and medium transmission line (length between 80 and 160 km). The capacitive parameter of short transmission line is ignored whereas in case of medium length line, the capacitance is assumed to be lumped at the middle of the line or

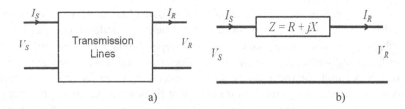

FIGURE 2.16   (a) Two-port network model; (b) short transmission line diagram.

half of the capacitance may be considered to be lumped at each ends of the transmission line. Lines with length more than 160 km or 100 miles, the parameters are considered to be distributed over the line. A major section of power system engineering deals in the transmission of electrical power from one particular place (e.g. generating station) to another like substations or power distribution units with maximum efficiency. It is of substantial importance for power system engineers to be thorough with its mathematical modeling. Thus, the entire transmission system can be simplified to a two-port network for the sake of easier calculations (Figure 2.16a). For short length, up to 80 km or up to 50 miles, the shunt capacitance of this type of transmission line is neglected and other parameters like resistance and inductance of these short lines are lumped, hence the equivalent circuit is represented as given in Figure 2.16b. In a later chapter, a short transmission line is used for voltage drop calculations. The relationships between the voltages and currents of the sending end (source) and receiving ends (load) are given below. Notice that a two-port model of a network simplifies the network solving technique. Mathematically, a two-port network can be solved by 2 by 2 matrixes.

$$V_S \cong V_R + I_R(R\cos\phi + jX\sin\phi) \tag{2.47a}$$

And

$$I_S \cong I_R \tag{2.47b}$$

Here, ϕ is the phase angle, assumed the same for generating and receiving ends.

2.4.7   AC ELECTRIC MOTORS

Electric motors can be found in almost every production process today. An electrical motor is an electromechanical device which converts electrical energy into a mechanical energy. Electric motors are essentially inverse generators: a current through coils of wire causes some mechanical device to rotate. The core principle underlying motors is electromagnetic induction. By Ampere's law, the current induces a magnetic field, which can interact with another magnetic field to produce a force, and that force can cause mechanical motion. Electrical motors are exploiting the force which is exerted on a current-carrying conductor placed in a magnetic field, depending on the current in the wire, and the strength of the magnetic field and the angle between them. The force on a wire of length l, carrying a current I and exposed to a uniform magnetic flux density B throughout its length is given by:

$$F = B\vec{I} \times \vec{l} \tag{2.48}$$

Where F is in newton (N), B is in tesla (T), I in ampere (A) and l in meter (m). In electrical motors, we intend to use the high magnetic flux density to develop force on current-carrying conductors.

Electric motors are estimated to now consume over 25% of US electricity use (though some estimates are even higher, to up to 50%, and over 20% of US total primary energy). While large electric motors are very efficient at converting electrical energy to kinetic energy (efficiency higher than 90%), those efficiencies are only achieved when motors are well-matched to their loads. Small electric motors are also inherently less efficient (more like 50% or so). Motor design and, even more importantly, motor choice and use practices are an important area of potential energy conservation. Before we can examine the function of a drive, we must understand the basic operation of the motor. It is used to convert the electrical energy, supplied by the controller, to mechanical energy to move the load. There are really two types of motors: AC and DC. The basic principles are similar for both. Magnetism is the basis for all electric motor operation. It produces the force required to run the motor. There are two types of magnets: the permanent magnet and the electromagnet. Electromagnets have the advantage over permanent magnet in that the magnetic field can be made stronger. Also, the polarity of the electromagnet can be easily reversed. When a current passes through a conductor, lines of magnetic force (flux) are generated around the conductor. The direction of the flux is dependent on the direction of the current flow. If you are thinking in terms of conventional current flow (positive to negative), then using your right hand point your thumb in the direction of the current flow and your fingers will wrap around the conductor in the same direction of the flux lines. There are basically two types of AC motors: *synchronous* and *induction*.

2.4.7.1 Synchronous Motors

There is no fundamental difference between a synchronous motor and a synchronous generator. In a motor, the magnetic axis of the rotating magnetic field is ahead of the magnetic axis of the rotor, resulting in a positive torque that depends on the displacement between the two axes. In a synchronous generator, the displacement is reversed: the magnetic axis of the rotor is ahead of the magnetic axis of the rotating field, so the torque is negative. Most of the AC generators in electric power systems are synchronous machines. High-speed turbine generators normally have two poles. The rotor is made from a cylindrical steel forging, with the field winding embedded in slots machined in the steel. Apart from the slots for conductors, the active surfaces of the stator and rotor are cylindrical, so these are uniform air-gap or non-salient machines. Low-speed hydro generators have many poles. These are salient-pole machines, where the poles radiate like spokes from a central hub. The circuits in Figure 2.17 represent one phase of a three-phase synchronous machine. The voltage V is the phase voltage at the machine terminals, and the current I is the corresponding phase current. Other elements in the circuit have the following significance: The voltage E is termed the excitation voltage. It represents the voltage induced in one phase by the rotation of the magnetized rotor, so it corresponds to the rotor magnetic field. The reactance X_S is termed the synchronous reactance. It represents the magnetic field of the stator current in the following way: the voltage jX_SI is the voltage induced in one phase by the stator current. This voltage corresponds to the stator magnetic field. The resistance R_a is the resistance of one phase of the stator, or armature, winding. The resistance R_a is usually small in comparison with the reactance X_S, usually being neglected in most calculations

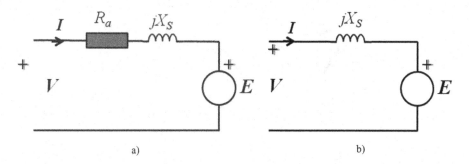

FIGURE 2.17 (a) Synchronous machine equivalent circuits; (b) approximate equivalent circuit.

from the equivalent circuit (diagrams of Figure 2.17). The voltage V represents the voltage induced in one phase by the total magnetic field, and is expressed as:

$$V = E + \left(R_a + jX_s\right)I \tag{2.49}$$

The approximate equivalent circuit is shown in Figure 2.17b and is described by the simplified equation:

$$V = E + jX_s I$$

For operation as a motor, V leads E by an angle δ (the torque or power angle), as shown in Figure 2.18. The phase angle is now less than 90°, indicating a flow of electrical power into the machine. The motor developed torque is thus given by:

$$T_d = \frac{3VE}{\omega_s X_s} \sin(\delta) \tag{2.50}$$

This has a maximum value when the torque angle δ = 90°, given by:

$$T_{dMax} = \frac{3VE}{\omega_s X_s} \tag{2.51}$$

Then the three-phase synchronous motor power is given by:

$$P_d = \frac{3VE}{X_s} \sin(\delta) \tag{2.52}$$

The magnitude of internal generated voltage induced in a given stator is:

$$E = K\Phi\omega \tag{2.53}$$

Since the magnetic flux, Φ, in the machine depends on the field current through it, the internal generated voltage is a function of the rotor field current. If the mechanical load on a synchronous motor exceeds T_{dMax}, the rotor is pulled out of synchronism with the rotating field, and it stalls. T_{dMax} is therefore known as the pullout torque. The synchronous motor torque characteristic has an important practical consequence. If the rotor loses synchronism with the rotating field, the load angle will change continuously. The torque will be alternately positive and negative, with a mean value of zero. Synchronous motors are therefore not inherently self-starting. Induction machines,

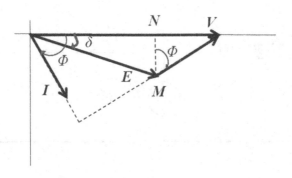

FIGURE 2.18 Synchronous motor phasor diagram.

considered in the next chapter subsection do not have such limitation, and the induction principle is generally used for starting synchronous motors. If the mechanical load is removed from an over-excited synchronous motor, then $\delta = 0$, and the phase angle is now 90°, so the machine behaves as a three-phase capacitor. In this condition, it is known as a synchronous compensator. The magnitude of the current, and hence the effective value of the capacitance, depends on the difference between E and V. The ability of a synchronous motor to operate at a leading power factor is extremely useful. It will be shown in the next chapter subsection that induction motors always operate at a lagging power factor. Many industrial processes use large numbers of induction motors, with the result that the total load current is lagging. It is possible to compensate for this by installing an over-excited synchronous motor. This may be used to drive a large load such as an air compressor, or it may be used without a load as a synchronous compensator purely for power factor correction.

Example 2.13 **A synchronous generator stator reactance is 190 Ω and the internal voltage (open circuit) generated is 35 kV. The machine is connected to a three-phase bus whose line-to-line voltage is 35 kV. Find the maximum possible output power of this synchronous generator.**

SOLUTION

The line-to-neutral input voltage is:

$$V_{in} = \frac{V_{LL}}{\sqrt{3}} = \frac{35\ kV}{\sqrt{3}} = 20.2 kV$$

The maximum power is when the torque angle is 90°, so from Equation (2.52):

$$P_{dMax} = \frac{3VE}{X_S} = \frac{3 \times 20.2 \times 20.2}{190} = 6.3\,MW$$

Efficiency of a synchronous motor is computed by using a well-known relationship. It is defined in the normal way as the ratio of useful mechanical output power P_{OUT} to the total electrical input power P_{IN}:

$$\eta = \frac{P_{OUT}}{P_{IN}} = 1 - \frac{Losses}{P_{IN}} \tag{2.54}$$

Example 2.14 **A 1492 kW, unity power factor, three-phase, star-connected, 2300 V, 50 Hz, synchronous motor has a synchronous reactance of 1.95 ohm/phase. Compute the maximum torque in Nm which this motor can deliver if it is supplied from a constant frequency source and if the field excitation is constant at the value which would result in unity power factor at rated load. Assume that the motor is of cylindrical rotor type. Neglect all losses.**

SOLUTION

Rated three-phase apparent power, at PF = 1 is $S_{3-\phi} = 1492$ kVA, while the rated per-phase apparent power is $S = 1492/3 = 497.333$ kVA. The rate voltage per phase is:

$$V_{ph} = \frac{V_{LL}}{\sqrt{3}} = \frac{2300}{\sqrt{3}} = 1327.906\,V$$

The rated per-phase current is:

$$I_{ph} = \frac{S}{V_{ph}} = \frac{497,333}{1327.906} = 374.52\,\text{A}$$

The induced per-phase voltage for unit power factor is:

$$E = \sqrt{V_{ph}^2 + (X_S I_{ph})^2} = 1515.489\,\text{V}$$

The maximum power is computed using Equation (2.24) for a torque angle of 90°:

$$P_{Max} = \frac{EV_{ph}}{X_S} = \frac{1515.489 \times 1327.906}{1.95} = 1032.014\,\text{kW/phase}$$

The maximum per-phase torque computed is then:

$$\tau_{Max} = \frac{P_{Max}}{\omega_m} = \frac{1032,014}{2\pi50} = 3285\,\text{N}\cdot\text{m/phase}$$

The three-phase maximum torques of this synchronous motor is 9855 Nm.

In general, larger synchronous machines have higher efficiencies because some losses do not increase with machine size. Losses are due to: rotor resistance; iron parts moving in a magnetic field causing currents to be generated in the rotor body; resistance of connections to the rotor (slip rings), stator; resistance, magnetic losses (e.g. hysteresis and eddy current losses), mechanical losses (windage, friction at bearings, friction at slip rings) and stray load losses due to non-uniform current distribution. Synchronous motors are usually used in large sizes because in small sizes they are costlier as compared with induction machines. The principal advantages of using synchronous machine are as follows:

1. Power factor of synchronous machine can be controlled very easily by controlling the field current.
2. It has very high operating efficiency and constant speed.
3. For operating speeds less than about 500 rpm and for high-power requirements (above 600 kW), synchronous motor is cheaper than induction motor.

In view of these advantages, synchronous motors are preferred for driving the loads requiring high power at low speed, e.g. reciprocating pumps and compressor, crushers, rolling mills, pulp grinders etc. Synchronous motors are used for constant speed, steady loads. With high power factor operations, these motors are sometimes exclusively used for power factor improvement. These motors find application in driving low-speed compressors, slow speed fans, pumps, ball mills, metal rolling mills and process industries.

2.4.7.2 Poly-Phase Induction Motors

An **induction** or **asynchronous motor** is an AC electric motor in which the electric current in the rotor needed to produce torque is obtained by electromagnetic induction from the magnetic field of the stator windings. An induction motor can therefore be made without electrical connections to the rotor as are found in universal, DC and synchronous motors. An induction motor's rotor can be either wound type or squirrel-cage type. A three-phase induction motor consists of the fixed stator or frame, a three-phase winding supplied from the three-phase mains and a turning rotor. The currents in the rotor are induced via the air gap from the stator side. Stator and rotor are made of highly magnetizable core sheet providing low eddy current and hysteresis losses. The *stator winding* consists of three individual windings which overlap one another and are offset by an electrical angle of 120°. When it is connected to the power supply, the incoming current will first magnetize the stator. This magnetizing current generates a rotary field which turns with synchronous speed N_S. The *rotor* in induction machines with squirrel-cage rotors consists of a slotted cylindrical rotor core sheet package with aluminum bars which are joined at the front by rings to form a closed cage. The stopped induction motor acts like a transformer shorted on the secondary side. The stator winding thus corresponds to the primary winding, the rotor winding (cage winding) to the secondary winding. Because it is shorted, its internal rotor current is dependent on the induced voltage and its resistance. The interaction between the magnetic flux and the conductor currents generates a torque that corresponds to the rotary field. The cage bars are arranged in an offset pattern to the axis of rotation in order to prevent torque fluctuations. The induction motor operates much in the same way that the synchronous motor does. It uses the same magnetic principles to couple the stator and the rotor. However, one major difference is the synchronous motor uses a permanent magnet rotor and the induction motor uses iron bars arranged to resemble a squirrel cage. As the stator magnetic field rotates in the motor, the lines of flux produced will cut the iron bars and induce a voltage in the rotor. This induced voltage will cause a current to flow in the rotor and will generate a magnetic field. This magnetic field will interact with the stator magnetic field and will produce torque to rotate the motor shaft; which is connected to the rotor. The torque available at the motor shaft is determined by the magnetic flux acting on the rotor (developed force) and the distance from the center of rotation. Another factor determining torque and another difference between the induction motor and the synchronous motor is slip. Slip is the difference between the stator magnetic field speed and the rotor speed. In order for a voltage to be induced into a conductor, there must be a relative motion between the conductor and the magnetic lines of flux. If the induction motor ran at synchronous speed no torque is produced. As we apply a load, the rotor begins to slow down which creates slip. At about 10% slip, we are getting maximum torque and power transfer from the motor, the best place on the curve to operate the motor. Vector control (slip control) can be used to keep the motor operating at this optimum point on the torque-speed curve. The RPM (rotation per minute) speed and the torque of an induction motor are given by the following equations:

$$N_S = \frac{120 f}{P} \tag{2.54}$$

And

$$T = K_{IM} I_{RMS} \tag{2.55}$$

Here, T is the motor torque, K_{IM} is the torque constant, I_{RMS} is the RMS motor current, N_S is the motor synchronous (rpm), f is the frequency of stator current (supply frequency) and P is the motor number of poles. The rotor magnetic field produced by the induced voltage is alternating in nature. The rotor starts running in the same direction as that of the stator flux and tries to catch up with the rotating flux. An essential feature of induction motors is the speed difference between the rotor and the rotating magnetic field, which is known as slip. There must be slip for currents to be induced in the rotor conductors, and the current magnitude increases with the slip. It follows that the developed torque varies with the slip, and therefore with the rotor speed. The rotor runs slower than the stator field, at the so-called the base or mechanical speed (N_m). The slip, s the difference between N_S (or angular synchronous velocity, ω_{syn}) and N_m (or actual angular velocity of the motor, ω_m) is expressed as:

$$s = \frac{N_S - N_m}{N_S} = \frac{\omega_{syn} - \omega_m}{\omega_{syn}} \tag{2.56}$$

The slip varies with the load. An increase in load causes the rotor to slow down or increase slip. A decrease in load causes the rotor to speed up or decrease slip. The frequency of the rotor currents is proportional to the slip speed, therefore to s, given by:

$$f_r = sf_s \tag{2.57}$$

Here, f_r is the frequency of rotor currents and f_s is the stator supply frequency.

Example 2.15   An induction motor has four poles, the supply frequency is 60 Hz and the actual speed of rotation is 1775 RPM. Determine the synchronous speed, the slip and the rotor frequency.

SOLUTION

The synchronous speed is

$$N_S = \frac{120f}{P} = \frac{120 \times 60}{4} = 1800 \, \text{RPM}$$

Then the slip is:

$$s = \frac{1800 - 1775}{1800} = 0.014 \, \text{or} \, 1.4\%$$

By using Equation (2.57) the rotor frequency is:

$$f_r = 0.014 \times 60 = 0.833 \, \text{Hz}$$

The rotating magnetic field exerts a torque T_d on the rotor, and does work at the rate of $\omega_s T_d$, being the input power to the rotor. The rotor revolves at an angular speed ω_r and therefore does work at the rate of $\omega_r T_d$, which is the rotor output power. Their difference represents power lost in the rotor resistance. As in any electrical motors, there are mechanical losses in the motor, so the shaft torque T is less than T_d. We have the following set of rotor power relationships.

$$P_{em} = \omega_{syn} T_d \tag{2.58}$$

Then the rotor power output is:

$$P_{rot} = \omega_r T_d = (1 - s)\omega_{syn} T_d \tag{2.59}$$

So, the rotor copper loss is then given by the difference between (2.56) and (2.57):

$$P_{Loss} = (\omega_{syn} - \omega_r)T_d = s\omega_{syn} T_d \tag{2.60}$$

Thus, a fraction $(1 - s)$ of the rotor electromagnetic input power is converted into mechanical power, and a fraction s is lost has heat in the rotor conductors. The quantity $(1 - s)$ is termed the rotor efficiency. Since there are other losses in the motor, the overall efficiency must be less than the rotor efficiency. For high efficiency, the fractional slip s should be as small as possible. In large motors, with power ratings of 100 kW or more, the value of s at full load is about 2%. For small motors, with power ratings below about 10 kW, the corresponding value is about 5%. When the rotor is stationary, the induction motor behaves as a three-phase transformer with a short-circuited secondary. When the rotor moves, the voltage induced in the rotor depends on the relative motion, which can be represented by a simple change to the equivalent circuit: the secondary

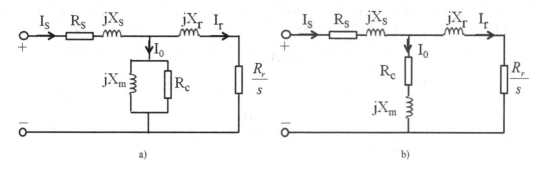

FIGURE 2.19 (a) Induction motor equivalent circuit; and (b) modified equivalent circuit.

resistance is not constant but depends on the fractional slip s. The circuit for one-phase takes the form shown in diagram of Figure 2.19a. The parameters in Figure 2.19 have the same significance as in a transformer: R_S is the stator winding resistance, x_S is the stator leakage reactance, representing stator flux that fails to link with the rotor, R_r is the rotor resistance referred to the stator, x_r is the rotor leakage reactance referred to the stator, R_C represents core loss mainly in the stator and X_m is the magnetizing reactance. With a transformer, it is possible to simplify the equivalent circuit by moving the shunt elements to the input terminals. However, this is a poor approximation with an induction motor, because the magnetizing reactance X_m is much smaller in comparison with the leakage reactances x_S and x_r. The reason for this is the presence of an air-gap between the stator and the rotor, which increases the reluctance of the magnetic circuit. The developed torque is obtained by equating the power absorbed in the resistance Rr/s to the rotor input power from Equation 2.58, giving the result:

$$T_d = \frac{3}{\omega_{syn}} \frac{R_r}{s} I_r^2 \text{ N·m} \tag{2.61}$$

It is necessary to solve the equations of the equivalent circuit (Figure 2.19a) for the currents, I_S and I_r, and hence determine the torque from Equation 2.34. This process is simplified by transforming the equivalent circuit to the form shown in Figure 2.19b. Here, the parallel combination of X_m and R_C has been replaced by the series combination of x_m and r_C. The series elements in Figure 2.20b are related to the parallel elements in Figure 2.16a by the following equations:

$$r_C = \frac{X_m^2}{R_C^2 + X_m^2} R_C$$
$$x_m = \frac{R_C^2}{R_C^2 + X_m^2} X_m \tag{2.62}$$

The value of rc depends on X_m, and therefore on the frequency. When the speed of an induction motor is controlled by varying the frequency, the resistance R_C is approximately constant. Under these conditions, r_C is proportional to the square of the frequency, so the modified equivalent circuit is less useful.

Example 2.16   A four-pole 3.6 kW, wye-connected induction motor operates from a 50 Hz supply with a line voltage of 400 V. The equivalent-circuit parameters per phase are as follows:

$$R_S = 2.27\Omega, \quad R_r = 2.28\Omega, \quad x_S = x_r = 2.83\Omega$$
$$X_m = 74.8\Omega, \quad r_C = 3.95\Omega$$

If the full-load slip is 5%, determine: (a) the no-load current, (b) the full-load stator current, (c) the full-load rotor current, (d) the full-load speed in rev/min and (e) the full-load developed torque.

SOLUTION

The per-phase voltage is:

$$V_p = \frac{400}{\sqrt{3}} = 231V$$

The motor impedances are:

$$Z_S = R_S + jx_S = 2.27 + j2.83\Omega$$

$$Z_r = \frac{R_r}{s} + jx_r = \frac{2.28}{0.05} + j2.83\Omega = 45.6 + j2.83\Omega$$

$$Z_m = r_C + jx_m = 3.95 + j74.8\Omega$$

$$Z_P = \frac{Z_m Z_r}{Z_m + Z_r} = 31.1 + j20.3\Omega$$

a. The no-load current is:

$$I_0 = \frac{V_p}{|Z_S + Z_m|} = \frac{231}{|6.22 + j77.6|} = \frac{231}{77.9} = 2.97 \text{ A}$$

b. The full-load current and its magnitude are:

$$I_S = \frac{V_p}{Z_S + Z_P} = \frac{231}{33.4 + j23.1} = 4.68 - j3.23\text{A}$$

$$|I_S| = |4.68 - j3.23| = 5.69\text{A}$$

c. The full-load rotor current and its magnitude are:

$$I_r = \frac{Z_P I_S}{Z_r} = \frac{211 - j5.89}{45.6 + j2.83}$$

$$|I_r| = \frac{|211 - j5.89|}{\left|\dfrac{211 - j5.89}{45.6 + j2.83}\right|} = \frac{211.1}{45.7} = 4.62\text{A}$$

d. The synchronous speed and the supply angular speed are:

$$N_s = \frac{120f}{P} = \frac{120 \times 50}{4} = 1500\text{RPM}$$

$$N_m = N_S(1 - s) = 1500(1 - 0.05) = 1425\text{RPM}$$

$$\omega = 2\pi f = 2 \times 50 \times \pi \approx 314\text{rad/s}$$

e. The developed toque is then:

$$T_d = \frac{3}{\omega_s} \frac{R_r}{s} I_r^2 = \frac{3 \times 2.28 \times (4.62)^2}{314 \times 0.05} = 37.2\text{N} \cdot \text{m}$$

Notice: The maximum torque is known as *the breakdown torque*. If a mechanical load torque greater than this is applied to the motor, it will stall. The torque is zero at the synchronous speed of 1500 RPM. When the rotor is stationary, the fractional slip is s = 1. When the rotor is running at the synchronous speed, the fractional slip is s = 0. The rotor frequency is:

$$f_r = sf = 0.05 \times 50 = 2.5\text{Hz}$$

Most induction motors are started by connecting them straight to the AC mains supply. This is known as direct on line starting; however, it may result in a large starting current. Direct on line starting may be unacceptable, either because the supply system cannot support such a large current or because the transient torque could damage the mechanical system. In-depth presentation of the induction motor starting methods is discussed later in this book. The efficiency of an induction motor, as well as of any power system equipment is of great importance to the user. It is defined in the normal way as the ratio of useful mechanical output power P_{OUT} to the total electrical input power P_{IN}:

$$\eta = \frac{P_{OUT}}{P_{IN}} = 1 - \frac{Losses}{P_{IN}} \tag{2.63}$$

The induction motor losses are considered to have five components as follows: stator I^2R loss (stator copper loss), P_{SCL}; rotor I^2R loss (rotor copper loss), P_{RCL}; core losses; friction and windage loss (rotational loss), P_{w+f}; and stray load loss, P_{Stray}.

The total loss is the sum of items 1–5. Core loss is the eddy current and hysteresis loss in the magnetic core of the machine, mostly in the stator, which is represented by the resistance r_C. Friction and windage loss is the total mechanical power loss within the motor, from bearing friction and aerodynamic drag on the rotor. Stray load loss is an additional loss under load, which is not included in the other four categories. It may be attributed to departures from a purely sinusoidal winding distribution and to effects of the stator and rotor slot openings on the magnetic field distribution in the machine. The full set of power relationships for poly-phase induction motors are:

$$P_{IN} = \sqrt{3}V_{LL}I_L \cos\theta = 3V_{ph}I_{ph} \cos\theta$$
$$P_{SCL} = 3R_S I_S^2$$
$$P_{RCL} = 3R_r I_r^2$$
$$P_{AG} = P_{IN} - (P_{SCL} + \text{CoreLosses})$$
$$P_{Conv} = P_{AG} - P_{RCL}$$
$$P_{OUT} = P_{Conv} - (P_{w+f} + P_{Stray})$$

The parameters of the induction motor equivalent circuit in Figure 2.14b are determined from three tests: (a) a DC measurement of the stator phase resistance, (b) no-load test for efficiency determination and (c) a locked-rotor (or blocked-rotor) test, where the rotor is prevented from revolving. These tests resemble the open-circuit and short-circuit tests for determining the equivalent circuit parameters of the transformer.

2.5   PER-UNIT SYSTEM, SINGLE-LINE DIAGRAMS AND LOAD MODELS

The per-unit (p.u.) value representation of electrical variables in power system and electric machine computation is a common and useful practice. In the power systems analysis field of electrical engineering, a per-unit system is the expression of system quantities as fractions of a defined base unit quantity. An interconnected power system typically consists of many different voltage levels given a system containing several transformers and/or rotating machines. The *per-unit system* simplifies the analysis of complex power systems by choosing a common set of base parameters in terms of which all systems quantities are defined. The different voltage levels disappear and the overall system reduces to a set of impedances. Calculations are simplified because quantities expressed as per-unit do not change when they are referred from one side of a transformer to the other. This is an advantage in power system analysis where large numbers of transformers may be encountered. Moreover, similar types of apparatus will have the impedances lying within a narrow numerical range when expressed as a per-unit fraction of the equipment rating, even if the unit size varies widely. Conversion of per-unit quantities to volts, ohms or amperes requires knowledge of the base that the per-unit quantities were referenced to. The idea of per-unit system is to absorb large difference in absolute values into base relationships. Representations of system elements with per unit values become more uniform. The per-unit numerical value of any quantity is the ratio of its value to the chosen base quantity of the same dimension. A per-unit value is a normalized quantity with respect to the chosen base value. There are several reasons for using a per-unit system:

- Similar apparatus (generators, transformers, transmission lines) have similar per-unit impedances and losses expressed on their own rating, regardless of their absolute size, so per-unit data can be checked rapidly for gross errors. A per-unit value out of normal range is worth looking into for potential errors.
- The per-unit values for various components lie within a narrow range regardless of their ratings.
- Manufacturers usually specify the apparatus impedance in per-unit values.
- Use of the constant $\sqrt{3}$ is reduced in three-phase calculations.
- Per-unit quantities are the same on either side of a transformer, independent of voltage level
- By normalizing quantities to a common base, both hand and automatic calculations are simplified.
- It improves numerical stability of automatic calculation methods.
- Per-unit data representation yields important information about relative magnitudes.
- Ideal for computer simulations and modeling.

The definition of the per-unit value of a quantity is:

$$p.u.\,\text{value} = \frac{\text{Actual value}}{\text{Base (reference) value of the same dimension}} \tag{2.64}$$

The complete characterization of a per-unit system requires that all four base values be defined. Given the four base values, the per-unit quantities are defined as:

$$V_{p.u.} = \frac{V}{V_{base}}; \; I_{p.u.} = \frac{I}{I_{base}}; \; S_{p.u.} = \frac{S}{S_{base}}; \; Z_{p.u.} = \frac{Z}{Z_{base}} \tag{2.65}$$

The per-unit system was developed to make manual analysis of power systems easier. Although power system analysis is now done by computers, results are often expressed as per-unit values on a convenient system-wide base. The base value is always a real number and the per-unit value is dimensionless. Five quantities are involved in this calculation: the current, the voltage, the complex power, the impedance and the phase angle. Phase angles are dimensionless, the other four quantities are completely described by the knowledge of only two of them. Usually, the nominal line or equipment voltage is known as well as the apparent (complex) power, so these tow quantities are often selected for base value calculation. For example, considering a single-phase system, the expression of base current is:

$$I_{base} = \frac{S_{base(1-\phi)}}{V_{base(LN)}} \tag{2.65}$$

The expression of the base impedance is:

$$Z_{base} = \frac{V^2_{base(LN)}}{S_{base(1-\phi)}} = \frac{V_{base(LN)}}{I_{base}}$$

(2.66)

The magnitude of the base current in a three-phase system can be calculated as:

$$I_{base} = \frac{S_{base(3-\phi)}}{\sqrt{3}V_{base(LL)}}$$

(2.67)

The base impedance can be calculated as:

$$Z_{base} = \frac{V^2_{base(LL)}}{S_{base(3-\phi)}} = \frac{V_{base(LL)}}{\sqrt{3}I_{base}}$$

(2.68)

Per unit quantities obey the circuit laws, thus:

$$S_{p.u.} = V_{p.u.}I^*_{p.u.}$$
$$V_{p.u.} = Z_{p.u.}I_{p.u.}$$

(2.69)

Far a three-phase system the phase impedance in per-unit is given by:

$$Z_{p.u.} = \frac{V^2_{p.u.}}{S^*_{L(p.u.)}}$$

(2.70)

Example 2.17   Assuming that line voltage of 735 kV for 120 MVA transmission line with the impedance:

$$Z = 4.50 + j75.30\Omega$$

Calculate the per-unit transmission line resistance, reactance and impedance.

SOLUTION

$$Z = \sqrt{4.5^2 + 75.3^2} \left\langle \tan^{-1}\left(\frac{75.3}{4.5}\right) = 75.4\langle 86.6° \, \Omega \right.$$

The base impedance is:

$$Z_{base} = \frac{V^2_{base}}{S_{base}} = \frac{(735\times10^3)^2}{120\times10^6} = 4502\Omega$$

The per-unit transmission line resistance, reactance and impedance are computed as:

$$R_{p.u.} = \frac{4.5}{4502} = 9.996 \times 10^{-4} \text{ p.u.}$$

$$X_{p.u.} = \frac{75.3}{4502} = 0.01673 \text{p.u.}$$

$$Z_{p.u.} = \frac{75.4}{4502} = 0.01675 \text{p.u.}$$

Usually, if none are specified, the p.u. values given are on nameplate ratings as base. There are situations when the base for the system is different from the base for each particular generator or transformer, hence it is important to be able to express the p.u. value in terms of different bases. The rule for the impedances is:

$$Z_{p.u.(new)} = Z_{p.u.(old)} \frac{S_{base(new)}}{S_{base(old)}} \cdot \frac{V_{base(new)}^2}{V_{base(old)}^2} \tag{2.71}$$

Example 2.18   Convert the impedance value of the Example 2.17 to the new base 240 MVA and 345 kV.

SOLUTION

We have

$$Z_{p.u.(old)} = 9.996 \times 10^{-4} + j0.01673$$

for a 120 MVA, and 735 kV base. With a new base 240 MVA and 345 kV, by using the impedance conversion relationship (3.30):

$$Z_{p.u.(new)} = Z_{p.u.(old)} \left(\frac{240}{120} \right) \cdot \left(\frac{735}{345} \right)^2 = 9.0775 \cdot Z_{(p.u.)(old)}$$

And

$$Z_{p.u.(new)} = 0.0091 + j0.1519$$

The three-phase circuits, loads and grid elements can be represented as shown in Figure 2.21. However, the source and the load are not always connected in the same manner. For example, the load can be Δ-connected and the source Y-connected, or vice versa. In either case, attention must be given to the calculation of the line and phase quantities. The phase and line voltage and current relationships established in previous subsections apply straight here. A more convenient approach is to employ single-line diagram to represent a three-phase system with its three-phase loads. Figure 2.20a is showing a Y-connected generators supplying power to a three-phase Δ-connected loads, while Figure 2.20b, its single-line diagram. The single-phase power equations are applied to each phase of Y- or Δ-connected three-phase loads. The real, active and apparent powers supplied to a balanced three-phase load are:

$$S = 3V_\phi I_\phi^* = P + jQ$$
$$P = S \cos(\theta)$$
$$Q = S \sin(\theta)$$

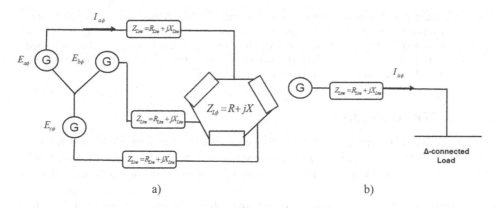

a)   b)

FIGURE 2.20   (a) A Y-connected generator supplying a Δ-connected load; (b) single-line diagram.

In a single-line (one-line) diagram, the voltages are given as line-to-line voltage and power consumption is specified for all three phases. A three-phase system consists of three single-phase systems, distribute three times more power that the single-phase counterpart. The apparent, active and reactive powers are expressed by the relationships, previously discussed:

$$S_{3\phi} = 3S_\phi = 3V_\phi I_\phi^* = \sqrt{3}V_{L-L}I_{L-L}^* = P_{3\phi} + jQ_{3\phi} \qquad (2.72)$$

For three-phase Y- or Δ-connected systems, the active and reactive powers are:

$$P_{3\phi} = \sqrt{3}V_{L-L}I_{L-L}\cos\theta = S_{3\phi}\cos\theta$$
$$Q_{3\phi} = \sqrt{3}V_{L-L}I_{L-L}\sin\theta = S_{3\phi}\sin\theta \qquad (2.73)$$
$$|S_{3\phi}| = \sqrt{P_{3\phi}^2 + Q_{3\phi}^2}$$

Here, θ is the phase angle, and the power factor PF is equal to cosθ. For lagging power factor Q is positive, the load is of inductive type and is absorbing power from the grid, while for leading power factor Q is negative, the load is capacitive type and is providing reactive power to the grid. Most of the loads connected to a power system are of inductive type, and large part of the loads is induction motors. In single-line diagrams, the inductive loads are represented as in Figure 2.21a, and the capacitive loads as in Figure 2.21b, showing the flows of the active and reactive powers. In phasors diagrams the voltage is take as reference, so $V_L = |V_L| < \pm 0$ with positive sign for inductive loads and the negative sign for capacitive loads. In recent years, more variable-speed

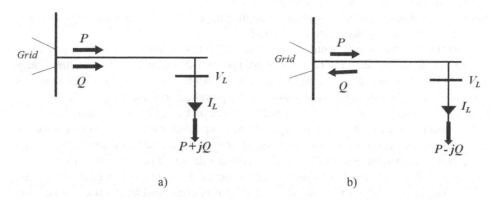

a)   b)

FIGURE 2.21   Single-line diagrams: (a) inductive load and (b) capacitive loads.

drive systems, controlled by electronic power converters are controlling various types of electric motors. In addition, more and more power electronic loads are connected to power systems. These types of loads and equipment are nonlinear types of loads, acting as inductive and capacitive loads during their transient and steady-state operations. The power factor correction, voltage control and stability are active research areas for smart grids.

2.6   CHAPTER SUMMARY

The main four sections of any power system are generation, transmission, loads and end-users. The voltages and currents are varying along the power systems in order to minimize the losses and costs, transfer power at maximum capacity and to accommodate the section constrains and requirements. Most of today's power system, from generation, transmission and large part of the power distribution, is in the form of three-phase systems. In the earlier days of the power generation, Tesla not only led the battle of whether the power system used DC or AC but also proved that the three-phase electric power was the most efficient way to generate, transfer and use electricity. Main components of any power systems include: prime-movers, synchronous generators, transmission lines and a large variety of power transformers, induction motors, capacitor banks, equipment and devices for monitoring, control, metering, communication and protection. The essential feature of a three-phase system is: although all currents are sinusoidal alternating waveforms, if the system is balanced, then the total instantaneous power of the system is constant. A poly-phase electric generator or motor converts power form one form to another without fluctuations or pulsations, with constant energy stored in the electromagnetic field as the single-phase counter-parts. The *per-unit system* simplifies the analysis of complex power systems by choosing a common set of base parameters in terms of which all systems quantities are defined. The different voltage levels disappear and the overall system reduces to a set of impedances. Standard frequency used in North America is 60 Hz, while Europe and the rest of the world are using 50 Hz frequency. Three-phase systems are often analyzed by using single-line diagrams for the sake of simplicity. The loads presented in power systems are most often inductive type, absorbing reactive power from the system. Over the last few years, there is strong penetration into the grid of variable-speed drive, controlled by power converters and power electronic loads, acting as nonlinear, inductive and capacitive loads during their operation, affecting the power system operation and power quality, being hot research areas of the smart grid.

2.7   QUESTIONS AND PROBLEMS

1. What types of connections are possible for three-phase sources and loads?
2. A Y-connected load has a 460 V voltage applied to it. What is its phase voltage?
3. If the load in previous question is Δ-connected, what will be its phase voltage?
4. What are the advantages of the three-phase systems?
5. List the advantages and disadvantages of high-voltage and low-voltage transmission lines.
6. What is function of power plant generators?
7. A 60 Hz, four-pole synchronous generator has a 2.5 Ω synchronous reactance and a negligible armature resistance. The stator windings are Y-connected, the line-to-line voltage is 23 kV and the line generator current is 1.2 kA at 0.9 lagging power factor. Compute the equivalent filed voltage, E_f and real and reactive power delivered to the grid.
8. A synchronous generator, having a 1.8 Ω reactance and a negligible armature resistance, is connected to an infinite bus through a transmission line having a 1.2 Ω reactance and a negligible armature resistance. The generator excitation is adjusted so the line-to-line equivalent field voltage is 30 kV, and the generator delivered power to the infinite bus is 120 MW. If the infinite bus voltage 25 kV, compute the terminal generator voltage, the reactive power at the infinite bus, the reactive power consumed by the transmission line and the reactive power at the generator terminals.

9. For the previous problem, compute the system capacity (the maximum system power). If a capacitor is connected in series with the transmission line to increase the transmission capacity by 1/3, compute its reactance.

10. A transformer is rated at 500 kVA, 60 Hz and 2400/240 V. There are 200 turns on the 2400 V winding. When the transformer supplies rated load, find (a) the ampere-turns of each winding, and (b) the current in each winding.

11. A transformer is made up of a 1200-turn primary coil and an open-circuited 80-turn secondary coil wound around a closed core of cross-sectional area 45 cm². The core material can be considered to saturate when the RMS applied flux density reaches 1.50 T. What maximum 60 Hz RMS primary voltage is possible without reaching this saturation level? What is the corresponding secondary voltage? How are these values modified if the source frequency is lowered to 50 Hz?

12. A single-phase transformer has 1200 turns on primary and 400 turns on secondary. The primary winding is connected to 240 V supply and the secondary winding is connected to a 6.40 kVA load. If the transformer is considered ideal, determine: (a) the load voltage, (b) the load impedance and (c) the load impedance referred to the primary side.

13. A single-phase core-type transformer is designed to have primary voltage 33 kV and a secondary voltage 6.6 kV. If the maximum flux density permissible is 1.24 Wb/m²and the number of primary turns is 1350, calculate the number of secondary turns and the core cross-sectional area when operating at 50 Hz frequency.

14. A transformer is to be used to transform the impedance of a 75 Ω resistor to an impedance of 225 Ω. Calculate the required turns ratio, assuming the transformer to be an ideal one.

15. Prove that the maximum efficiency of a transformer is when the core-losses are equal to RI^2 copper losses (due to the winding resistance) if the secondary voltage and the power factor are assumed constant.

16. A single-phase step-up transformer, rated at 1800 kVA, 60 Hz and 13.5/135 kV. Transformer equivalent resistance and reactance are 1.560 Ω and 15.6 Ω, respectively. Compute the percent resistance and reactance, as seen from primary and secondary transformer sides.

17. The high-voltage side of a step-down transformer has 800 turns, and the low-voltage side has 100 turns. A voltage of 240 V is applied to the high side, and the load impedance is 3Ω (low side). Find:
   a. The secondary voltage and current.
   b. The primary current.
   c. The primary input impedance from the ratio of primary voltage and current.
   d. The primary input impedance.

18. Compute the voltage regulation of the transformer in problem 7 for: (a) unit power factor, (b) 0.85 lagging power factor and (c) 0.85 leading power factor.

19. A single-phase transformer with a nominal voltage ratio of 4160 V to 600 V has an off-load tap changer in the high voltage winding. The tap changer provides taps of 0, ±2½ and ±5%. If the low voltage is found to be 618 V, what tap would be selected to bring the voltage as close to 600 volts as possible?

20. Determine the rotor speed (RPM) of the following three-phase synchronous machines: (a) $P = 4$ and $f = 60$ Hz, (b) $P = 12$ and $f = 50$ Hz and (c) $P = 4$ and $f = 400$ Hz.

21. If the torque angle of Example 7.5 is limited to 45 degrees, find the generator power output.

22. What is the speed (RPM) of a 30-pole, 60 Hz 440 V synchronous motor? Is this motor classed as a high- or low-speed motor.

23. A synchronous motor with an input of 480 kW is added to a system that has an existing load of 720 kW at 0.82 lagging power factor. What are the new system active power, apparent power and power factor if the new motor is operated at: (a) 0.85 lagging power factor, (b) unit power factor and (c) 0.85 leading power factor?

24. A three-phase induction motor 50 Hz has a synchronous speed of 1500 RPM and runs at 1450 RPM at full load. (1) How many poles do the motor have? (2) What is the slip when it runs at full load? (3) What is the rotor frequency?

25. A 208 V, 10 HP, four-pole, 60 Hz, Y-connected induction motor has a full-load slip of 5%. Find: (a) the motor synchronous speed, (b) the actual motor speed at rated load, (c) the rotor frequency and (d) the shaft torque at rated load.

26. What is the rotor frequency of an eight-pole 60 Hz squirrel cage motor operating at 850 RPM?

27. A 50 HP, 230 V three-phase induction motor require a full-load current of 130 A per terminal at a power factor of 0.88. What is its full-load efficiency?

28. A 480 V, 60 Hz, three-phase induction motor is drawing 60 A at 0.85 PF lagging. The stator copper losses are 2 kW, the rotor copper losses are 0.7 kW, the windage and friction losses are 0.6 kW, the core losses are 1.8 kW, while the stray losses are 0.2 kW. Find the air gap power, the converted power, the output power and the motor efficiency.

29. A two-pole, 50 Hz induction motor supplies 15kW to a load at a speed of 2950 RPM.
    a. What is the motor's slip?
    b. What is the induced torque in the motor in Nm under these conditions?
    c. What will be the operating speed of the motor if its torque is doubled?
    d. How much power will be supplied by the motor when the torque is doubled?

30. What is meant by the term "balanced" in a three-phase system?

31. What is a phase sequence? What is its importance?

32. In a balanced Δ-connected circuit with all resistive loads, what is the led or lag of the line currents with respect to line voltages?

33. What are the relationships between the phase and the line voltages and current for a Y-connection?

34. What are the relationships between the phase and line voltages and current for a Δ-connection?

35. Write the relationships for active, reactive and apparent powers in three-phase circuits in terms of both line and phase voltages and currents.

36. A Y-connected balanced three-phase source is supplying power to a balanced three-phase load. The source phase voltage and current are given by:

$$v(t) = 420\sin(377t + 30°)\,\text{V}$$
$$i(t) = 90\sin(377t + 15°)\,\text{V}$$

Calculate: (a) the RMS and line-to-line voltages and currents; (b) supply frequency; (c) power factor at the source side; (d) three-phase active, reactive and apparent powers supplied to the load; and (e) the load impedance if the load is balanced and Y-connected.

37. A 120 HP, three-phase, 480 V induction motor operates at 0.88 PF lagging. Find: (a) active, reactive and apparent powers consumed per phase and (b) suppose that the motor is supplied from a 460 V source through a feeder whose impedance is 0.3+j0.5 Ω per phase. In this case, calculate the motor side voltage, the source power factor and the transmission efficiency.

38. Repeat problem 7 if the motor's efficiency is 85%.

39. A three-phase wye-connected generator has the phase voltage 230 V. Calculate the phase voltages with angles and the line voltages.

40. The magnitude of each phase voltage of an unbalanced load is 220 V RMS. The load impedances are:

$$Z_A = 6 + j8\,\Omega$$
$$Z_B = 4 + j6\,\Omega$$
$$Z_A = 3 + j4\,\Omega$$

Calculate: (a) line currents and (b) the neutral current.

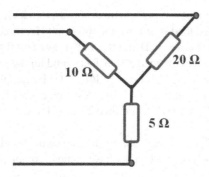

FIGURE P2.1

41. The per-phase reactance, 10 kVA, 120 V, Y-connected synchronous generator is 12 Ω. Determine the per-unit reactance, considering the base values are 10 kVA and 120V.

42. The per-phase load impedance of a three-phase delta-connected load is 4+j6 Ω. If a 480 V, three-phase supply is connected to this load, find the magnitude of: (a) phase current and (b) line current

43. Calculate the RMS value, supply frequency and the phase shit in degrees for the AC voltage given by:

$$v(t) = 180\sin(300t + 0.866)\,\text{V}$$

44. For the balanced circuit in Figure P2.1, the magnitude of voltage is 270 V. Calculate the power delivered to each of the resistors of the unbalanced load.

45. A single-phase load absorbs a 200 kVA power from a 35 kV busbar. Determine the base current and base impedance by considering those quantities as base values.

46. A three-phase, 460 V, wye-connected source is supplying power to a wye-connected balanced load. Calculate the load impedance, if the load current I_a is 10 A and in phase with the line-to-line voltage, V_{bc}.

47. The base quantities of a system are 450 kVA and 35 kV. The per-unit impedance of this system is 0.035. Determine the actual impedance value.

48. The voltage and current measured for a wye-connected load are:

$$\bar{V}_{AB} = 230\langle 45°\,\text{V}$$
$$\bar{I}_C = 6\langle 130°\,\text{A}$$

   a. Calculate the power factor angle
   b. Calculate the real power consumed by the load

49. A balanced three-phase load is connected to a 480 V feeder. The line current I_A, in phase with the line-to-line voltage V_{BC} is equal to 12.5 A. Compute the load impedance if: (a) the load is wye-connected and (b) the load is delta-connected.

50. Three loads are connected in parallel across a 12.47 kV power supply. One is resistive 63 kW load, the second one is an induction motor of 72 kW and 63 kVAR, and the last one is a capacitive load drawing 180 kW at 0.85 PF. Find the total apparent power, power factor and power supply current.

51. A single-phase transformer is rated 220/4400 V and 5.0 kVA. The reactance of the transformer is 0.1 Ω measured from the low voltage side. Determine the reactance of the transformer on its own base.

52. Repeat Problem 19 using a new base: apparent power of 10 kVA and a voltage of 440 V.
53. A balanced star-connected load is fed from a 460 V, 60 Hz, three-phase supply. The resistance in each phase of the load is 30 Ω and the load draws a total power of 15 kW. Calculate (a) the line current drawn, (b) the load power factor and (c) the load inductance.
54. A three-phase load is connected in star to a 460 V, 60 Hz supply. Each phase of the load consists of a coil having inductance 0.2 H and resistance 45 Ω. Calculate the line current.
55. If the load specified in Problem 28 is connected in delta, determine the values for phase and line currents.
56. A balanced Y-connected load have a phase impedance of 8+j6 Ω is connected to a three-phase supply of 460 V. Determine: (a) the phase voltage; (b) the phase and line currents; (c) the power factor at the load; and (d) the power consumed by the load.

3 Smart Grid Basics, Benefits and Challenges

3.1 INTRODUCTION, SMART GRID CONCEPTS AND EVOLUTION

A *smart grid* (SG), known also as the smart electrical grid, intelligent grid, inter- or intra-grid, is a 20th century power grid enhancement. The traditional power grids are designed to carry power from a few central and large power stations to a large number of end-users or customers. In contrast, the SG uses two-way flows of electricity and information to create an automated and distributed advanced energy delivery network. The smart grid is recognized as a key strategic and critical infrastructure needed for consumers, utilities, service providers, operators, communities and every country as a whole. Smart grid technologies offer a long list of benefits, including a more efficiently operated electricity system, sustainability and reduced operational costs. The SG term refers to a modernization of the electricity delivery systems, the electric grids in order to provide better and diversified services, fully monitor, better protect, automatically optimizing and controlling the operation of its interconnected electric grid components and of the grid as a whole. Smart grids affect all power system sections and components, from the central and distributed generators through the high-voltage transmission networks and power distribution systems, to industrial and commercial users and building automation systems, to energy storage installations and to end-use consumers and their heat, how water and air conditioning controllers, electric vehicles, appliances and other household devices and equipment. The smart grid is a framework for solutions, both revolutionary and evolutionary in its nature, because it can significantly change and improve the way that electrical system operates today, while providing for ongoing enhancements in the future. Traditionally, the operational objectives of electric utilities have been to provide reliable energy at minimum cost, stated as: (1) to supply quality power at constant voltage and frequency; (2) to minimize adverse effects on people and the environment; (3) to maintain adequate system security and reliability and (4) minimize energy losses. Recent technological advances in small generators, power electronics and energy storage devices have provided new opportunities at the distribution level. These advancements resulted from economic incentives to decentralize, expand and improve the existing power systems. The current electric grid is operated by complex software programs and automation routines and protected by microprocessor-based relays, computer-based and automatic devices and equipment. Though true in some parts of the world, the evolution of electric infrastructure has been slow, hampered by economics, demographics, regulation and other factors. However, what is similar across the globe are the challenges of operating electric systems to deliver power reliably and cost effectively to consumers. Among others, the SG vision is to give much greater visibility to lower voltage networks and power distribution and to enable the participation of customers in the operation of the power system, particularly through smart meters, communication, information technology and smart homes. The smart grid supports the improved energy efficiency, resilience and service restoration in the case of extreme events, while allowing larger uses of the renewable energy sources, distributed generation and energy storage units.

Smart grid technologies are playing a central role in social, technical and economic developments at all scales. Furthermore, energy production is associated with environmental pollution and SG technologies are alleviating such problems by integrating more renewable energy sources and energy storage into the power grid. The major drivers for power system evolution are the needs to meet rising demands for electricity, more secure and reliable energy supply, while reducing operation costs and pollutant emissions to avoid irreversible changes to the environment. All of these

DOI: 10.1201/9780429174803-3

must be achieved without compromising the supply reliability and the cost of electricity on which the world's economies are increasingly dependent. In fact, the smart grid is a system or a structure of systems, meaning a complex ecosystem of heterogeneous (possibly) cooperating entities that interact in order to provide the envisioned functionality. It is a complex infrastructure depicting system of system characteristics, such as interdisciplinary nature, operational and managerial independence of its elements, geographical distribution, high heterogeneity of the networked systems and the emergent behavior and evolutionary development. This new infrastructure is heavily relying on modern information and communication technologies (ICTs) to achieve its expected functionalities and services. Utility companies are turning to IT solutions to monitor and control the electrical grid in real time. These solutions can prolong the useful life of the existing grid, delaying major investments needed to upgrade and replace current infrastructure. Until now, monitoring has focused only on high-voltage transmission grids. Increasing overall grid reliability and utilization, however, will also require enhanced monitoring of medium- and low-voltage distribution grids. From the business side, distributed business processes are needed to empower the new interactions. The traditional static customer processes will increasingly be superseded by a very dynamic, decentralized and market-oriented process where a growing number of providers and consumers interact. The energy needs and contributions of the *end-user, consumer, customer* or other related terms are figuring prominently in *smart energy programs*. In the United States, EU and many other countries' policy of deploying smart grids promises new possibilities for self-managing energy consumption, improved energy efficiency among final consumers and transition to more consumer-centric energy systems.

While minor upgrades have been made to meet increasing demand, the power grid is still operating on a large extent, the way it did almost 100 years ago. The energy flows over the grid from central power plants to consumers and reliability is ensured by maintaining excess capacity. The result is an inefficient and environmentally wasteful system that is a major emitter of greenhouse gases, consumer of fossil fuels and not well suited to distributed, renewable solar and wind energy sources. In addition, the grid may have insufficient capacity to meet future energy demands. A new, more intelligent electric system, the *smart grid* is required, combining information technology (IT) with distributed generation to significantly improve how electricity is produced, transferred, delivered and consumed. Smart grid has been advocated in both developing and developed countries over last two decades to deal with the bottlenecks and issues of feeding large requirement in energy consumption as the growing of industries, transportation and residential sectors. As a new concept for power delivery system, smart grid involves plenty of advanced technologies, outstanding methodologies, novel algorithms and creative architectures in service, business and operation to solve problems like carbon emission deduction, resources allocation optimizations, grid security and reliability enhancement and deliver power energy in a more efficient, reliable and optimal way. Expensive power outages can be avoided if proper action is taken immediately to isolate the cause of the outage. Utility companies are installing sensors to monitor and control the electrical grid in near-real time (seconds to milliseconds) to detect faults in time to respond. These monitoring and control systems are being extended from the point of transmission down to the distribution grid. Grid performance information is integrated into utility companies' supervisory control and data acquisition (SCADA) systems to provide automatic, near-real-time electronic control of the grid. Smart grid deployment covers a broad array of electricity system capabilities and services enabled through pervasive communication and information technology, with the objective of improving reliability, operating efficiency, resiliency to threats and our impact on the environment. Smart grids are providing the utilities with near-real-time information to manage and operate the entire electrical grid as an integrated energy system, actively sensing, monitoring and responding to any changes in the power demand, energy supply, transmission and distribution, costs and emissions, from rooftop PV solar panels on homes to remote, unmanned wind farms or to energy-intensive facilities. Such complex electricity infrastructure is expected to be pervasive, ubiquitous and service-oriented. Service-oriented architectures and/or platforms, methods and tools focusing on a network-centered approach are needed to be developed to support the networked energy and information enterprise. Understanding and managing such critical infrastructure complexity, as in the energy sectors is crucial, implying systemic risks, in-depth analysis, resilient distributed information and process control frameworks. Many of the electric grid assets used to generate and transmit electricity are vulnerable to terrorist attacks and natural disasters. Substations, transformers and power lines are being connected to data networks, allowing utility companies to monitor their security using live video, tamper sensors and active monitoring. The bidirectional information exchange set the basis for cooperation among the different grid entities, as they are able to access and correlate information and data that up to now was either only available in a limited fashion (and thus unusable in large scale approaches) or extremely costly to fully integrate. Advanced electric grid business services are envisioned that will take advantage of the near-real-time information flows among all electric grid participants. These real-world energy services are going way beyond the present ones, enabling the stakeholders not only to become more energy aware, but also to optimally manage the electricity uses. Smart grid prosumers can take into account real-time information and engage in buying and selling energy, as well as having their environment being automatically adjusted to their behavior and their energy saving goals. The smart grids are dealing with multiple stakeholders, capture monitoring information and provide control capabilities for a large scale complex heterogeneous infrastructure. Moreover, the smart grid concepts have been discussed, expanded, developed by important organizations, research institutes and government departments around the world. However, there is no smart grid agreed definition; even different countries has different concepts on the smart grids, the future electric grids.

Energy efficiency is arguably the fastest, most sustainable and cheapest way of reducing greenhouse gas emissions. And there are other advantages. Not only are energy-efficient technologies already available (and have been for some time), investment payback times are short and they enable energy savings to be made without compromising economic development. But the applications of energy-efficiency measures are not limited to end users of electricity. They can be implemented at every stage of the power system, including generation and transportation through the world's transmission and distribution networks. While the environmental benefit of reducing our dependence on fossil fuels is clear, the large-scale integration of wind farms and solar plants into the grid will have a severe impact on the stability of electricity supplies, unless we begin to make changes. The greatest challenge stems from the erratic nature of renewable energy. With the exception of hydropower, the availability of renewable resources can quite literally change with the wind. Power generation in wind farms is characterized by periods of high productivity followed by lulls in calmer weather and the performance of solar plants wanes during cloudy weather and at night. A further challenge is the location of renewable energy sources. Large-scale sources are often far from the centers of demand (offshore or out in the desert), and small-scale producers are often in light-industrial or residential areas where the local distribution grid is not set up to receive as well as deliver electricity. The mix of renewable, thermal and nuclear power plants will introduce new variation in the quality of power in the grid. Weather patterns affect the availability of wind and solar power, and the emergence of distributed power generation (rooftop solar panels, for example) will complicate matters further, requiring local networks to receive as well as deliver power. The existing power supply infrastructure is unable to manage such complexity and needs to change. It needs to be equipped with advanced communications and information technologies to monitor, analyze and organize the supply and demand of electricity. The development of more intelligent power systems will directly support these two objectives. In a smart grid, advanced technologies improve energy efficiency by managing demand so that it matches the availability of electricity, and they feed renewable energy into the network without letting changes in weather patterns affect the stability or reliability of the supply.

Over the past three decades, almost every industry changed radically, due to IT, computing, modern communication and Internet. From package delivery to consumer banking, retail sales, online education, airline travel, all have seen substantial changes in their infrastructure, the way they deliver services, the way they interact and engage consumers and the way they conduct business. The changes in these industries rely not just on more information, but on better information, more accurate, more timely and two-way communication that allows the "system operator" at the enterprise level to plan, design and operate faster, smarter and more efficiently. The end result is improved services, more convenience, lower costs and happier customers. The same principles apply to the electric power sector, where with better information comes the opportunity for rapid innovation in utility operations, while the utilities can plan future requirements much more accurately, can design the system more efficiently and with closer tolerances and can operate it in entirely new and more efficient ways. It also enables the options of empowering consumers by providing information on when and how they use energy in more effective and efficient ways, giving them the ability to align their use with optimal system performance and diverse personal preferences. Several studies over the last two decades have described the needs to transform the electricity power industry, emphasizing on the lack of investment in basic infrastructure, including the transmission and distribution network, the need for innovation, the aging equipment and workforce, the cost of power interruptions, the need for emission reductions and more. In the coming years, aging grid components will need to be replaced, and new generation and transmission infrastructure will be built to meet increasing demand and clean energy goals. Yet, shifting priorities, evolving technologies and the experience of other industries are suggesting that we won't replace old systems with old systems. As described in this chapter and in other book chapters, the 21st-century grid, the smart grid, will include a complex network of technologies and systems, hardware and software, communications and controls that taken together will provide both producers and consumers a high level of visibility and control. Being inherently flexible and adaptable to unanticipated future changes, it will include technologies that we already know and technologies yet to be created and designed.

The smart grid is a summary of concepts, technologies, approaches and operating practices intended to bring the electric grid into the 21st century. Defining the smart grid is difficult for several reasons. First, there is not a single template that defines exactly what the smart grid looks like or how it operates in any given service area, region or country. Without a consensus template, the community and professionals of power, energy and communication industries are tending to construct their own mental vision of how the concepts, technologies, systems and customers will interact. There is also the reality that customer mix, geography, weather and other factors will almost certainly make the smart grid in each service area a little unique. The second factor contributing to the uncertain definition is the fact that a fully implemented and in operation smart grid does not yet exist. Several engineering systems and technologies expected to become a part of smart grid either haven't been fully developed yet or are in prototype, design or early stages of testing and implementation. A smart grid is often defined as an electricity network that can intelligently integrate the actions of all users connected to it, generation units, consumers and those that do both, generation and use, in order to efficiently deliver sustainable, economic and secure electricity supplies. The smart grid, as described by Institute of Electrical and Electronics Engineers (IEEE) is a next-generation electrical power system that is typified by the increased use of communications and information technology in the generation, delivery and consumption of electrical energy. Comparing to the conventional power system, smart grid is the next generation of power delivery system, which includes plenty of the creative and innovative features and several new energy, IT, computing, communication and monitoring technologies. Regardless of which is the most accurate definition, the smart grid concept includes following information at least: smart grid combines digital technologies throughout the whole power systems from generation to end-users; it improves reliability, security and efficiency of the power delivery systems; and finally it contains both bulk generations and distributed generations, non-renewable energy conversion and renewable energy conversion. Therefore, it clear that it is quite difficult to define something that does not exist in full development and/or operation or it is in very diffuse and inceptive format. While these factors attach a degree of uncertainty to smart grid, there are several foundational concepts, technologies and regulatory practices that define smart grid expectations. Moreover, all the smart grid elements and components are including important engineering, economic and policy issues and need to be reflected into its definitions. On the other hand, the smart grid is often defined as the implementation of various enabling power system automation, communication, protection, management and control technologies and practices allowing a real-time interoperability between end-users and energy providers, in order to enhance efficiency in utilization of decision-making based on resource availability and economics. From a regulatory perspective, a clear definition of smart grid is important for two reasons: it helps if consumers, utilities, vendors and regulators start from a common smart grid understanding, while the way that the smart grid is defined establishes the framework to guide expectations, resource allocation decisions and implementation priorities. Smart grids that incorporate demand management, distributed electricity generation and grid management allows for a wide array of more efficient, "greener" systems to generate and consume electricity. In fact, the potential environmental and economic benefits of a smart grid are significant. A summarized smart grid definition can be expressed as: smart grid is defined as any combination of enabling technologies, hardware, software or practices that collectively make the delivery infrastructure or the grid more reliable, more versatile, more secure, more accommodating, more resilient and ultimately more useful to consumers. Moreover, a useful definition of the smart grid must encompass its ultimate applications, uses and benefits to society at large. In this sense, the smart grid must include a number of key features and characteristics including, but not limited to, the following ones:

1. It must facilitate the integration of diverse supply-side energy resources including increasing levels of intermittent and non-dispatchable renewable energy resources;
2. It must facilitate and support the integration of distributed and on-site electricity generation on the customer side of the meter;
3. It must allow and promote more active engagement of demand-side resources and participation of customer load in the operations of the grid and electricity market operators;
4. It must allow and facilitate the prices-to-devices revolution that consists of allowing wide spread permeation of dynamic pricing to beyond-the meter applications, enabling intelligent devices to adjust usage based on variable prices and other signals and/or incentives;
5. It must ultimately turn the grid from a historically one-way conduit that delivers electrons from large central stations to load centers, to a two-way intelligent conduit, allowing power flows in different directions, at different times, from different sources to different sinks;
6. It must allow for broader participation of energy storage devices on customers premises or centralized devices to store increasing levels of energy when it is plentiful and inexpensive, to be utilized during times when the reverse is true and accomplish this intelligently and efficiently;
7. It must allow distributed generation as well as distributed energy storage to actively participate in balancing generation and load;
8. It must encourage more efficient utilization of the supply-side and delivery "network" through efficient and cost-effective implementation of dynamic pricing and similar concepts;
9. It must allow storage devices on customer side of the meter, including electric batteries and similar devices, to feed the stored energy back into the grid when it makes economic sense to do so;
10. It must facilitate any and all concepts and theories encouraging greater participation by customers and loads in balancing supply and demand in real time through concepts broadly defined as demand response;
11. It must make the electric grid, the complex generation, transmission and distribution networks more robust, reliable and more secure to interruptions and less prone to accidents or attacks of any kind; and
12. Of course, it must accomplish all these while reducing the costs of the network operation and maintenance, with commensurate savings to ultimate consumers.

If this sounds like an all-encompassing smart grid definition, it is intentional and by purpose of drawing reader attention. If it sounds like a tall order, that is also intentional as a glance at the book is suggesting. Moreover, everything from the improved building of energy efficiency to effective implementation of transportation electrification or to higher penetration of renewable energy sources are enhanced through the effective smart grid implementation. Smart grids allow renewable energy sources to be associated safely to the power systems to incorporate the power supply with distributed generation and energy storage units. According to most of the current view and understanding, the *smart grid* generally and usually refer to a class of technologies that are used to bring utility of electricity delivery systems into the 21st century using computer-based remote control and automation. These systems are made possible by two-way digital communications technologies and computer processing that has been used for decades in other industries. They are used on electricity networks, from the power plants and wind farms all the way to the end-users of electricity, offering several benefits to utilities and consumers, mostly seen in huge improvements in energy efficiency and reliability on the electricity grid and in electricity uses. In fact, the smart grid is understood as a power transmission grid that integrates *sophisticated and advanced sensing, monitoring, metering, communication, IT and computing with cutting-edge power engineering*, essentially superimposing the Internet on the power grid. The smart grid operation covers an extensive variety of breaking points, organizations of energy systems empowered through communication, computing, monitoring and information innovations, with the purpose of enhancing power quality, supply security, gird resilience and efficiency, operational profitability, grid flexibility, while reducing our impacts on Earth's environment. In summary, a smart grid is an electrical system that can make intelligent the incorporation of the activities of all stallholders associated with it – providers, customers and individuals who do both – so that the ultimate goals of vital support, protection, safety and economy.

Fundamentally, smart grid is a combination of information and communication applications that link together generation, transmission, distribution and customer end-use technologies. The need to integrate all of the systems that generate and supply electrical energy with customer usage is one of the very certain design principles of smart grid. System integration is achieved by using information and communication systems. Smart grid is not necessarily a specific combination of parts as much as it is a process for using information and communications to integrate all the components that make up each electric system. As a result of these technological changes and advances, the smart grid differs in several ways from the conventional power grids. The smart grid wastes less energy because it quickly alleviates problems like congestion and disturbances, preventing electricity losses. Smart grid is able to connect new generators to the transmission system, allowing greater renewable energy incorporation. In fact, the smart grid even has the potential to accommodate homeowners who want to sell energy that they produce from renewable energy sources to utilities. The smart grid has to deal with multiple stakeholders, capture monitoring information and provide control capabilities for a large-scale complex heterogeneous infrastructure. This is going to be challenging not only from a technical perspective but also from a political one. Today, we see a race among companies to place themselves as leaders in this emerging infrastructure and provide solutions. However, as many of them gear towards quickly providing patches so that their existing product offers may be characterized as *smart-grid-ready*, not enough focus is given to openness, interoperability and cooperation. It is also important to understand that to fully achieve the benefits of the smart grid requires a strong partnership between utilities, government, regulators and the public and new energy use behaviors. The smart grid effort still has a long way to go and will continue to evolve, driven by the economies and regulations across the globe. The chapter aims to provide a basic discussion of the smart grid concepts, evolution, issues and components, and then in some detail, to describe the technologies that are required for its realization. Even though the SG concepts are not yet fully defined and there is no fully acted single definition, the chapter will be helpful in describing the key enabling technologies and thus allowing the reader to play a part in the debate over the future of the smart grids.

In summary, a smart grid is closer to what it is understood as – a perfect or ideal power system. Some of the major attributes and capabilities of an ideal (perfect) power system are similar to those of the smart grid as discussed previously. For example, in order to achieve a quasi-ideal power system, the power system must meet the following goals: it must be *smart, self-sensing, secure, self-correcting* and *self-healing*. It must sustain the failure of individual components without interrupting the service, while focusing on regional specific area needs, being able to meet consumer needs at a reasonable cost with minimum resource utilization and minimal environmental impacts, enhancing life quality and improving economic productivity. The development of the better power system is based on integrating devices (smart loads, local generation and storage devices), then buildings (building management systems and micro CHP, smart appliances and equipment), followed by construction of an integrated power distribution system (shared resources and energy storage) and finally to set up a fully integrated power system (energy optimization, market systems and integrated operation). A smart grid encourages home and building owners to invest in high-efficiency, low-emission micro-generation devices to meet their own needs and to sell excess energy back to utility companies to offset peak demands on the electrical grid. This reduces the need for new, large-scale power plants. Virtual power plants can also be created that include both distributed generation and energy efficiency measures. In addition, a smart grid accelerates plug-in hybrid electric vehicle (PHEV) uses to act as electricity storage devices, as well as provide incremental energy generation to offset peak demand on the grid. Intelligence within the smart grid will be required to maintain reliability and stability once tens of thousands of micro-generation devices and PHEVs are brought online. To realize these power system attributes and functionalities, an integrated energy and communication systems architecture should be first developed, implemented and structured. Such electric grid architecture enables the automated monitoring and control of the power delivery system, increasing the capacity of the power delivery system and significantly enhancing the performance and connectivity of the end-users (consumers).

3.2 SMART GRID DRIVERS, ISSUES, CHALLENGES AND BENEFITS

For any power system, main objectives are to provide continuous service and reliable electricity at affordable costs. In technical terms this can be stated as: (1) supply quality power at constant voltage and frequency; (2) generate power within pollution limits set by environmental regulations; (3) provide adequate system security and reliability and (4) meet the load or customer demands at the lowest possible costs. The term *continuous service* can be translated to *secure and reliable service*. Secure service means that upon occurrence of a contingency the system is able to recover to its previous state, and reliability is the ability of the system to supply power as the load changes. Notice that in a mixed, non-standardized and highly complex infrastructure such as the smart grid, business applications have a very hard way to dynamically discover, integrate and interact with the sensing or actuating devices (e.g. smart home appliances, smart meters, smart sensing or electric cars) and services. By abstracting from conventional hardware and communication-driven interaction and focusing on the information available via services, smart grids can move towards service-driven interactions, in which services can be dynamically discovered, combined and integrated in mash-up applications. By accessing the isolated information and making the relevant correlations, business services could evolve, and acquire not only a detailed view of the interworking of their processes but also take real-time feedback from the real physical-domain services and flexibly interact with them. A service-based smart grid, where all of its functionalities are offered as a service, seems to be a promising way to go. The interrelationship between those various objectives has been defined as shown in Figure 3.1. The direction of the arrows indicates the priority in which the objectives are implemented. The dotted line indicates that if the operation of the system violates the environmental constraints, then the operation of the system may have to be altered or curtailed. However, the realization of the full benefits of the smart grid goes beyond near-term steps like installing meters and improving delivery efficiency. It also depends on a series of long-term moves such as

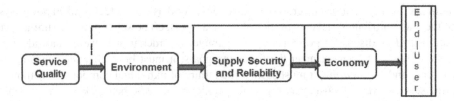

FIGURE 3.1 Interrelated operation objectives of a power system.

increasing the ability to transmit electricity across states and regions in order to tap renewable energy sources. Technologies are coming into the marketplace at different rates too. Some smart grid elements, like smart meters, are moving fast, while other essential components, like grid-level energy storage units, are advancing more slowly. Moving forward requires change across many elements of the grid. Given increasing demands on power generation, the smart grid offers the only gateway to increased use of renewable energy and decreased reliance on fossil fuels. Hydroelectric, wind and solar power offer tremendous potential, but unlike natural gas or coal-fired plants, these energy sources are often located far from transmission infrastructure and are non-dispatchable. Tapping into renewable energy requires more advanced and flexible transmission and distribution. The implementation of the smart grid is a multi-faceted challenge and enterprise, involving aspects such as: investment, regulation, business models, consumer education, cybersecurity and even space weather are leading factors.

Reducing congestion and conserving energy may provide additional benefits to consumers. The smart grid also alleviates congestion and by allowing customers, end-users and the utility to better manage electricity demands and services. The energy industry is at the center of this transformation and optimization of the integration of existing and future energy supplies, including the connection of renewables; leveraging demand side resources such as dynamic pricing and direct load control of customer appliances; and offering customers access to enhanced energy information and tools to lower energy use to help keep supply and demand in balance. While the present power grid is a marvel and one of the most amazing achievements, not only in the engineering design but also in the human history, it has yet to be transformed into a modern power and energy grid and infrastructure. A modern smart grid would include sustainable concepts that leverage proven, cleaner, cost-effective technologies available today or under development. Via the smart grid, consumers can monitor their electricity consumption using smart meter and advanced metering infrastructure. Smart meters are similar to meters currently used by electric companies to track consumers' electricity use and are installed on the outside of a home or apartment complex. Smart meters are more advanced than traditional meters, however, because they can track energy use daily, hourly, monthly and even instantaneously, and send that data to power companies. When demand is particularly high, smart meters allow utility companies to communicate with consumers, likely via e-mail or another form of electronic notification, so that customers can reduce their electricity use and help decrease overall congestion. To benefit their customers, some utilities even offer consumers online access to their power consumption, allowing them to view their electricity use instantaneously. More likely, the structure of electricity rates will change, and the fixed electricity prices that most consumers currently pay will be replaced by rates that vary depending on consumer demand. Smart Grid has the ability to decrease fossil fuel consumption by replacing electricity generated from oil with electricity generated from alternative energy sources. Current transmission and distribution grids were not designed with the smart grid requirements and expectations, being designed for the cost-effective, rapid electrification of developing economies. The requirements of smart grid are quite different, and, therefore, the reengineering of the current grid is imminent. This engineering endeavor takes many forms including existing grid extensions and enhancements, inspection and maintenance activities, preparation for distributed generation and energy storage and the development and deployment of an extensive two-way communications system. Since the invention of electric power technology and the establishment of centralized generation facilities, the greatest changes in the utility industry have been driven not by innovation but by system failures and regulatory/government reactions to those failures.

Smart grid technologies are generating tremendous amounts of real-time and operational data with the increase in sensors and the need for more information on the operation of the system, while decentralized information technology is changing the rules in this future electric utility industry. The smart grid and similarly denominated programs have been proposed as an effort to integrate the three critical developments in the future power grid: expansion of the power grid infrastructure to accommodate renewable energy resources and microgrids; penetration of information technology and computing to implement the full digital control and monitoring into the power generation, transmission and distribution systems; and development of new applications. Further, the smart grid programs respond to the political, public and scientific community requests to deploy higher percent of low pollutant and CO_2-emitting renewable energy resources and distributed generation systems. The smart grid and similarly programs have been proposed as an effort to integrate the three critical developments in the future grid: grid infrastructure expansion to accommodate renewable resources and microgrids; penetration of information and communication technologies to implement fully digital control and monitoring in generation, transmission and distribution subsystems; and new applications. Among others, the main smart grid tenets and characteristics have the following goals: enabling active participation by consumers in grid operation; accommodating all power generation and energy storage options; enabling new products, services and markets; providing power quality (PQ) and services for 21st-century needs; optimizing assets and operating efficiently; self-healing capabilities and functions from power disturbance events; operating resiliently against physical and cyber-attacks; and faster service restoration.

The current electric system is based on a one-way flow of energy and information from the sources to the end-users; the smart grid provides multiple paths for the flow of electricity, particularly data and information about that flow, throughout the power systems. Three dominant factors are impacting the electric systems: the government policies, efficiency and power quality needs and requirements of the consumers and the introduction of new communication, computing, IT and intelligent hardware technologies. In addition, environmental concerns have created governmental policies, including at the federal and state levels, which are driving the entire energy industry to efficiency, conservation and renewable energy sources of electricity. These factors are the main drivers that are expanding the use of all sorts of new renewable energy and storage technologies, new energy efficiency and conservation technologies. Consumers are becoming more proactive and are being empowered to engage in the energy usage decisions affecting their daily lives. At the same time, they are expanding their energy needs, requirements and expectations. The electric energy systems need to address all these needs and concerns by using advanced technologies to create a smarter, more efficient and sustainable grid. For example, the active participation of consumers and end-users in electricity markets brings tangible benefits to both the power grid and the environment. The smart grids give end-users information, control, choices and options that allow them to engage into the new electricity markets. The ability to reduce or shift peak demands is allowing the utilities to minimize capital expenditures and operating expenses while also providing substantial environmental benefits by reducing losses and minimizing the operation of inefficient and high-cost peak power plants. In addition, emerging products like the plug-in hybrid electric vehicles (PHEVs) and electric vehicles (EVs) result in substantially improved load factors while providing significant environmental benefits. Smart grids accommodate all generation and energy storage options and seamlessly integrate all types and sizes of electrical generation and energy storage systems using simplified interconnection processes, approaches and universal interoperability standards to support plug-and-play level of convenience. There are several objectives for developing and deploying smart grid technologies. Smart grid's key objectives and goals are including among others:

1. Improved efficiency and economy in energy conversion, transmission, distribution, storage and uses.
2. Security and safety enhancements and improvements into the power system operation by increasing the power grid observability and controllability.
3. Improved energy supply reliability and availability to the consumers.
4. Enabling and promoting renewable energy integration, creating sustainable energies.
5. Enabling and facilitating demand side participation to increase asset uses and return on investment.
6. Maintaining and improving power quality to increase the uses of digitally based and electronic loads.

3.2.1 SMART GRID MAJOR DRIVERS

Existing power grids were designed essentially as a unidirectional power flow system to transmit vast amounts of power generated primarily at a limited number of large central power stations to the load centers. The original thinking and concept behind the power networks, from generation to transmission and distribution, was based on the outdated premises that customer load and energy demands are a given, which requires only the generation to be adjusted to meet them. The balancing of supply and demand in real time was routinely accomplished by adjustments on the energy supply side. Until recently, customer demands were not subject to control or management, with virtually no means or incentives for the end-users to play any active role in power system operation and management. In this regard, the *smart grids* refer to a class of technologies that are used to bring the electricity delivery systems into the 21st century, using *computer-based remote control automation and consumer participation*. These systems are made possible by *two-way power flows and digital communications* technologies and computer processing that has been used previously and on large extent in other industries. They are now used in electricity networks, from the power plants and wind farms all the way to the consumers of electricity in homes and businesses. They offer many benefits to utilities and consumers, mostly seen as significant improvements of the energy *efficiency, reliability and uses*. A description of conventional electricity delivery system is represented in Figure 3.2. Usually, a power system is broken into mostly isolated components: *generation, transmission, substation, distribution and the consumers*. Key characteristics of the conventional power system that are most strongly impacted by the smart grid implementation are the following attributes: centralized generation, unidirectional energy flows, passive participation by the customers (consumer knowledge of electrical energy usage is limited to a monthly bill received after the fact), real-time monitoring and control is mainly limited to generation and transmission and only at some utilities does it fully extend to the power distribution system, the system is not flexible so that it is difficult to either inject electricity from alternative sources at any point along the grid or to efficiently and sustainably manage new services desired by the users of electricity. These conventional attributes have adequately served the needs of electric utilities and their customers in the past. However, the new needs of more energy knowledgeable, computer and IT savvy and environmentally conscious consumers, combined with regulatory changes, promoting sustainability and energy independence, availability of intelligent (smart) technologies and equipment and greater demands for enough energy to drive the global economy, require an electric energy system of the future that is fundamentally different in all areas listed before. A general schematic of the smart electric energy system is presented in Figure 3.2. The key requirements of the smart system are addressing the following transformational functionalities: allowing the integration of renewable energy resources, active customer participation to enable far better energy conservation and uses, cyber-secure communications systems to address system safety, better utilization of existing assets to address long term sustainability, optimizing the energy flows to reduce losses and lower the cost of energy, integration of electric vehicles to reduce dependence on hydrocarbon fuels, the

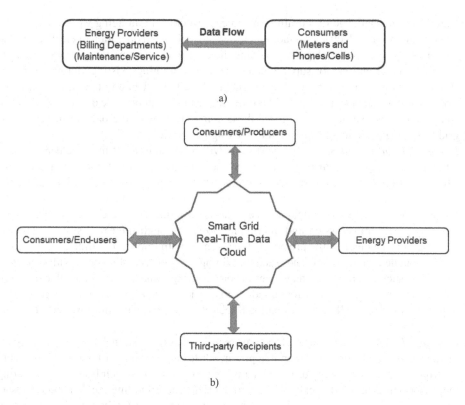

FIGURE 3.2   (a) Existing power grid data flows and (b) smart grid data flow.

management of distributed generation and energy storage to eliminate or defer system expansion to reduce the overall cost of energy and the integration of communication and control across the energy system to promote interoperability and open systems and to increase safety and operational flexibility.

The smart grid objective is essentially to enable a power system future that is prosperous, efficient, resilient and sustainable. All stakeholders and interested parties must be aligned around a common vision to fully modernizing today's electric grid. Throughout the 20th century, the electric power delivery infrastructure has served many countries well to provide adequate, affordable energy to homes, businesses, schools, colleges, agencies, commercial facilities, transportation and factories. Once, a technologically state-of-the-art system – the electricity grid – brought a prosperity level unmatched by any other technology in the world and human history. But a 21st-century economy cannot be built on a 20th-century electric grid technology and structure. There is an urgent need for major improvements in the world's power delivery system and in the technology areas needed to make these improvements possible. A number of converging factors will drive the energy industry to modernize the electric grid. These factors can be combined in five major groups, as follows:

1. *Policy, Governmental and Legislative Driver:* these are referring to the electric market rules that create comparability and monetize benefits, electricity pricing and access to enable smart grid options, state and federal regulations to allow smart grid deferral of capital and operating costs and compatible federal and state policies to enable full integration of smart grid benefits and advantages.
2. *Economic Competitiveness:* smart grids are creating new businesses and new business models for energy and power industries, while adding "green" jobs, technology regionalization,

alleviating the challenge of a drain of technical resources in an aging workforce. On the other hand, the new electric grid facilitates the connection and operation of electric generators of all sizes and technologies, providing the consumers with better and more accurate information and options for supply choice, while maintaining existing and new services efficiently. Moreover, electric grid infrastructure here in the United States and around the world is aging, and at some point will require the replacement of some of the power system components, and we may well replace those components with the ones from new smart grid technologies, reducing costs and improving the services.

3. *Energy Reliability and Security:* smart grids will bring significant improvements of grid reliability and supply security through decreased outage duration, frequency and effects, while reducing the labor costs, such as manual meter reading and field maintenance, etc. The advanced smart monitoring sensing and IT infrastructure in the smart grids reduce the on-labor costs, e.g. the use of field service vehicles, insurance, damage, etc. Smart grids are reducing power transmission and distribution losses through improved system planning and asset management, while protecting the revenues with improved billing accuracy, prevention and detection of theft and fraud, providing new sources of revenue with consumer programs, such as energy management or load shifting. Smart grids also defer the capital expenditures as a result of increased electric grid efficiencies and reduced generation requirements. They will fulfill our national security objectives; and improve the wholesale market efficiency.

4. *Customer Empowerment:* respond to consumer demand for sustainable energy resources, increasing demands for uninterruptible power, while empowering customers to have more control over their own energy usage with minimal compromise in their lifestyle, facilitating the performance-based rate behavior. Smart grids are incorporating the demand response, demand-side energy resources, high-efficient equipment, while facilitating the integration of smart appliances and intelligent devices, providing the consumers with timely information, communication, monitoring and smart control options.

5. *Environmental Sustainability:* in response to governmental mandates, the power systems must be less harmful to the environment, while at the same time must support the addition of renewable energy sources and distributed generation units to the grid. By deploying and integrating advanced energy storage and peak shaving technologies, including thermal storage air conditioning, plug-in and hybrid vehicles, smart grids will reduce overall fossil fuel uses, making the power systems more efficient and economic viable will also have less harmful effects on the environment.

Gradual application of emerging technologies for advanced power grid management, operation, control and protection represents an effective SG transition, which from the perspective of different authors and researchers may assume different characteristics, but in general shares the following functional properties:

- Ability to resiliently recover to the extent possible from the effects of damaging or disruptive disturbances (self-healing);
- Providing opportunities for consumer participation in energy management and demand response (often via advanced metering options which may provide additional support and information, both for the utility and the customer, during normal operation);
- Ability to respond to, cope with and resiliently enhance itself against physical and cyber-attacks;
- Providing power quality for modern equipment anticipated to be needed in the future;
- Accommodating all generation and storage options;
- Enabling new products, services and markets; and
- Optimizing assets and operating efficiently.

The transition to smart grid may in some cases be spontaneously driven by obvious benefits and cost effectiveness, while in others may be supported by regulatory and policy actions.

3.2.2 SMART GRID MAJOR BENEFITS AND CHALLENGES

The major drivers for the evolution of the power system is the need to meet rising demand for electricity while reducing carbon and other pollutant emissions to avoid irreversible changes to the earth's environment, while improving the supply security, electric services, overall grid efficiency, grid stability, affordability and power quality. However, regardless of regulatory structure, investments in smart grids must make economic sense. There is considerable evidence that the benefits of smart grid technologies consistently outweigh the costs. Any decisions regarding smart grid technology adoption should involve a comprehensive look at the many costs and benefits. Most smart grid projects, especially those that are enabling renewable energy uses and integration, provide socio-economic benefits that accrue not solely to the utility system, but also to customers and the local or global community. These broader benefits include economic gains from greater reliability, improved public health due to emissions reductions and long-term environmental and economic gains from low-carbon electricity. All these must be achieved without compromising the reliability of electricity supplies on which the world's economies are increasingly dependent. Energy efficiency is arguably the fastest, most sustainable and cheapest way of reducing greenhouse gas emissions. And there are other advantages. Not only are energy-efficient technologies already available (and have been for some time), investment payback times are short and they enable energy savings to be made without compromising economic development. Smart Grid is a concept and a range of functionalities: it is designed to be inherently flexible, accommodating a variety of energy production sources and adapting to and incorporating new technologies as they are developed. It allows for charging variable rates for energy, based upon supply and demand at the time. In theory, this will incentivize consumers to shift their heavy uses of electricity (such as for heavy-duty appliances or processes that are less time sensitive) to times of the day when demand is low. As an example of these range of functionalities, in 2008, US Department of Energy (DOE) defined functions of a smart grid as:

- "Self-healing" from power disturbance events;
- Enabling active participation by of consumers;
- Operating resiliently against physical and cyber-attacks;
- Providing power quality for 21st-century needs;
- Making large-scale energy storage a realty;
- Accommodating all generation and energy storage options;
- Allowing seamless integration of renewable energy sources;
- Enabling new products, services and markets; and
- Optimizing assets and operating efficiently.

The ultimate goals for the future end-to-end electric power system (from fuel sources, to power generation, transmission, distribution and end-users) are: allow secure and real-time two-way power and information flows, enabling integration of intermittent renewable energy sources and help decarbonize power system, enabling effective demand management, customer choice, secure and efficient operation of the grid and enabling the secure collection and communication of detailed data regarding energy usage to help reduce demand and increase efficiency. Several new energy services are to exist in the smart grid infrastructure. These are to act as auxiliary services that can be integrated into applications and traditional systems in order to enhance their functionality, e.g. timely energy monitoring, control and management. Fine-grained capabilities are expected to provide management functionalities that go beyond the conventional approaches, energy brokering, real-time analytics and value added services, community energy management services and energy application

Stores. Smart grids rely on several advanced technological applications to deliver benefits to customers, grid operators and stakeholders. Here, the applications are referring to any programs, algorithms, computing applications, calculations, data analysis and other programs and software that are processing historical, real-time or forecast data to provide outputs that are used to change the operating state of the power system with the objectives to improve security, efficiency, reliability and economy. The integration of smart sensing, communications and control technologies with field devices in power distribution is improving reliability and efficiency. Smart grid applications enable utilities to automatically locate and isolate faults, reducing outages, dynamically optimize voltage and reactive power levels for efficient power uses and to monitor asset health to guide the maintenance, while deployment of advanced and smart sensors and high-speed communication on transmission systems is advancing the ability to monitor and control operations at high-voltage substations and across the transmission subsystems. According to the National Institute of Standards and Technology (NIST), the major smart grid major benefits include the followings:

- Improving power reliability and quality, by better, smart and improved monitoring using sensor networks and communications; and better and faster balancing of supply and demand.
- Minimizing the need to construct back-up (peak load) power plants, by better demand side management and the use of advanced metering and monitoring infrastructures.
- Enhancing the capacity and efficiency of existing electric grid, by better monitoring using sensor networks and communications, and consequently, better control and resource management in real-time.
- Improving resilience to disruption and being self-healing, by better monitoring using sensor networks and communications, and by using distributed grid management and control.
- Expanding deployment of renewable and distributed energy sources, by better monitoring using sensor networks and communications, and consequently, better control and resource management in real-time, better demand side management and better renewable energy forecasting models, and by providing the infrastructure and/or incentive.
- Automating maintenance and operation, through better monitoring using sensor networks and communications, and through distributed grid management and control.
- Reducing greenhouse gas emissions, by supporting and/or encouraging the use of electric vehicles, by extended uses of renewable power generation with low carbon footprint and by reducing oil consumption, by supporting/encouraging the use of electric vehicles, and renewable power generation with low carbon footprint, and better demand side management.
- Enabling transition to plug-in electric vehicles, which can also provide new storage opportunities.
- Increasing consumer choice, by the use of the use of advanced metering infrastructures, home automation, energy smart appliances and better demand side management.

The key challenges for the smart grids include, among others, generation and transmission capacity, increased resilience, shorter service restoration in the case of extreme, cost and efficient assets uses, enhanced power quality and offered services. Strengthening the grid ensures that there is sufficient transmission capacity to interconnect energy sources, especially renewable energy resources. Smart grid must move some of the generation offshore, by developing the most efficient connections for offshore wind farms and for other marine energy technologies. Developing decentralized architectures, enabling smaller scale electricity supply systems to operate harmoniously with the total system. Adequate communication infrastructure, delivering in real-time information and data, allows potential huge number of parties to operate and trade in single market. Smart grid must feature active demand side, enabling all consumers, with or without their own generation, to play an active role in the operation of the system. Smart grid must integrate all kind of intermittent generation systems, by finding the best ways of integrating intermittent generation including residential micro-generation. Smart grid must enable enhanced intelligence of generation, demand and most notably in the grid. Smart grid must have capabilities for integrating electric vehicles, whereas SGs must accommodate the needs of all consumers, electric vehicles are particularly emphasized due to their mobile and highly dispersed character and possible massive deployment in the next years, what would yield a major challenge for the future electricity networks. On the other hand, in order to manage the ever-increasing demands for energy trading and security of supply, the existing transmission and distribution networks require improved integration and coordination. To control electric power flows across state borders, advanced applications and tools, that are already available today, should be deployed to manage the complex interaction of operational security and trading and to provide active prevention and remedy of disturbances.

3.2.3 SMART GRID MAJOR COMPONENTS, OPERATION, MANAGEMENT AND PRACTICE

Monitoring/sensing, communication and control are the three fundamental building blocks that will convert a power distribution system into a smart grid. The smart grid could thus be thought of as a combination of the electric power infrastructure and the communications infrastructure. Many smart grid applications could be developed simply using the existing infrastructures for both. However, the power grid complexity and interconnectivity is increasing with distributed integration of renewable energy sources and energy storage systems of all kinds, sizes and types. This growing complexity requires different approaches to traditional modeling, control and optimization in power systems. These new approaches need either to be augmented with existing ones, or completely replaced in some cases, providing capabilities for rapid adaptation, dynamic foresight, sense-making of situations, fault-tolerance and robustness to disturbances and randomness. Monitoring and sensing will have the ability to detect malfunctions or deviations from normal operational ranges that would warrant actions. Further, since in a smart grid, a point of electricity consumption can also become a point of generation, the sensing process will be closely linked with the metering process. Communications will allow inputs from sensors to be conveyed to the control elements in the smart grid which will generate control messages for transmission to various points in the smart grid resulting in appropriate actions. The communication infrastructure has to be robust enough to accept inputs from a user and make it an integral part of the process. By the same token, the user must be capable of getting the appropriate level of information from the smart grid. Smart grid, from an architectural perspective, is comprised of three high-level layers, i.e. the physical power layer, the data transfer and control layer and the applications layer. Each of these high-level layers breaks down further into sub-layers. The four major systems within SG from a technical perspective are

a. *Smart infrastructure system*: the smart infrastructure system is the energy, information and communication infrastructure underlying of the SG that supports: (1) advanced generation, delivery and consumption; (2) advanced metering, monitoring and management; and (3) advanced communication. An integrated, standards-based, SG two-way communications infrastructure will provide an open architecture for real-time information, data communication and control to every endpoint on the electric grid.

b. *Advanced metering infrastructure (AMI)*: the AMI system comprises smart meters, communication networks and information management systems, is enhancing the operational efficiency of utilities and providing electricity customers with information to more effectively manage their energy use. Smart sensing and metering technologies are providing faster and more accurate response for consumer options such as remote monitoring, time-of-use pricing, and demand-side management. AMI introduces the two-way communication control to allow customers and public services to get the real-time cost and energy consumption. AMI specifies the energy losses and the location of the electrical theft. AMI provides customers with the data needed to decide on smart options, the ability to decide on those options and various options for the benefit of the customer. Moreover, AMI provides a fundamental link fixed between the networks, consumers with their loads and generation as well as data storage space resources through the merging of different technologies such as intelligent evaluation, starting zone systems, coordinated exchanges, data management applications and institutionalized programming interfaces

c. *Smart management system*: the smart management system is the subsystem in SG that provides advanced management and control services. This is consisting of software system architecture with improved interfaces, decision support, analytics and advanced visualization that enhances human decision-making, effectively transforming grid operators and managers into knowledge workers. In the smart grid framework, the consumer-side coordination is updated to monitor and optimize the electrical energy consumption at the user level, such as industrial, tertiary and residential levels. Demand Side Management (DSM) and energy efficiency options developed for effective means of modifying the customer demand to cut operating expenses from expensive generators and suspend capacity addition. DSM options provide reduced emissions in fuel production, lower costs and contribute to reliability of generation. These options have an overall impact on the utility load curve. Electric power companies are required to maintain constant frequency levels and the instantaneous balance between demand and electricity supply by adjusting output through the use of thermo-electric and pumped hydro storage generation. In order to resolve this issue, optimal demand-supply control technologies are required to develop methods to control not only conventional electrical generators but also distributed generation and energy storage units. In summary, all DMS functions are through the processing of real-time information, the advanced sensor system, smart metering and monitoring for reducing of interruption and repair times, periods, keeping the voltage level into the required range, detect the faulty positions/locations, improve the overall energy management, reset the power supplies automatically, optimize the voltage and reactive power flows and provide an advanced control of the distributed generation units.

d. *Smart protection and control systems*: the smart protection system is the subsystem in SG that provides advanced grid reliability analysis, failure protection and security and privacy protection services. An advanced control methods that monitor critical components, enabling rapid diagnosis and precise responses appropriate to any event in a *self-healing* manner, significantly improving grid resilience and service restoration in any events. Advanced control systems monitor and control essential elements and components of the SG, while computer-based algorithms are allowing efficient data collection and analysis, provide solutions to the operators and are also able to act autonomously. Faults can be detected much faster than in conventional power grids, while outage and interruption times can be significantly reduced. To fulfill such purposes and objectives, the smart grid communication and information infrastructure has to integrate enabling networking and computing technologies.

In many ways, the smart grid is a system of systems i.e. it is a complex ecosystem of heterogeneous (possibly) cooperating entities that interact in order to provide the envisioned functionality. It is a complex infrastructure depicting system of system characteristics, e.g. interdisciplinary nature, with its elements of operational and managerial independence, geographical distribution, high heterogeneity of the networked systems as well as emergent behavior and evolutionary development. The new infrastructure will heavily rely on modern information and communication technologies to achieve its expected functionality. The smart grid evolves the architecture of legacy grid can be characterized as providing one-way flow of centrally generated power to end-users (consumers) into a more distributed, dynamic system characterized by two-way flow of power and information. Information and communication technologies play crucial and critical roles in the

SG development, management, control and operation. The SGs involve networking vest numbers of sensors in transmission and distribution facilities, smart meters, back-office systems as well as home devices which will interact with the grid. Large amount of data traffic is generated by smart meters, sensors, monitoring units and synchrophasors. While networking technologies have been greatly enhanced, the SG faces challenges in terms of reliability and security in both wired and wireless communication environments. The integration of sensing, communications and control technologies with field devices in distribution is improving reliability and efficiency, enabling utilities to automatically locate and isolate faults, reducing the outages, optimizing voltage and reactive power levels for more efficient power use and continuously monitoring the asset health for optimum maintenance. Interoperability between smart grid components is paramount. A SG framework can define the components at three levels: the electricity infrastructure level, the smart infrastructure level and the smart grid solution level. At each of these levels, different applications exist that need to interoperate among themselves (horizontally) and with the levels above or below (vertically). In smart grid applications, the information flow between the different parts is controlled by an energy management system (EMS) which communicates with individual smart meters located at residential, commercial and industrial customer sites. The main components of the SG are: new and advanced grid components, smart devices, loads and metering, advanced monitoring infrastructure, smart sensors, integrated communication technologies, programs and software packages for decision support, management and human interfaces, advanced control and protection systems.

The current and future smart grid architecture advocates a synergy of computing and physical resources and envisions a trustworthy middleware providing services to grid applications through message passing and transactions. The architecture also accounts for a power system infrastructure operating on multiple spatial and temporal scales. This infrastructure must support growing penetration of distributed energy resources, energy storage units, communication, monitoring, data and information subsystems, smart equipment, devices and appliances, etc. There will also be thousands of sensors and actuators that are connected to the grid and to its supporting monitoring, control information and data networks. Data communication systems are essential in any modern power system and their importance are only increasing, as the smart grid is developed and implemented. The SG operation is relying heavily on two-way communication for the exchange of information and data. Real-time information and data must flow all the way to and from the large central generators, transmission networks, substations, end-user (customer) loads and the distributed generators and energy storage units. With tens of millions of customers as part of the smart grid, the information and communication infrastructure will use different communication technologies and network architectures that may become vulnerable to theft of data or malicious cyber-attacks. Ensuring the smart grid information and data security is a much more complex task than in conventional power systems or in other industries because the power systems are so extensive and integrated with other networks. Potentially sensitive personal data is transmitted, and, in order to control costs, public ICT infrastructure such as the Internet will be used. Obtaining personal information and data about customers' loads could be of interest to unauthorized parties and could infringe the privacy of customers. The ability to gain access to electricity use data and account numbers of customers opens up numerous avenues for fraud. Breaching the security of power system operating information by an unauthorized person has also obvious dangers for the power system (smart grid) operation. In such regards, the smart grid cyber-security is somewhat different than cyber-security for other sectors because of the need to protect a wide array of devices connected to and enabling the grid. Encryption, limited physical access and "white hat" techniques (barring access to all but safe users) are common to military or financial institutions cyber-security, for example. Part of the smart grid cyber-security challenge will be to implement standards at all levels and in all domains, from the consumer to the Supervisory Control and Data Acquisition (SCADA). Over the last decade, NIST has approached the process by examining the grid as a whole and assessing risk across its different domains. Key areas identified include electric transportation, electric storage, wide area situational awareness, demand response, advanced metering infrastructure and distribution grid management. Moreover, a significant number of documents outline smart grid cyber-security standards.

3.3 STANDARDS FOR SMART GRIDS

The power grid is too large an entity and there are too many entities involved for it to be constructed, operated and structured without standards to guide how the parts fit and operate together. Power electronics, communication, protection, operation, grid integration of distributed generation and management are requiring standards and guidelines in order to ensure that components and subsystems are interoperable and the architecture takes shape in the desired and proper manner. Full standardization of DG and SG devices and protocols is necessary to guarantee proper operations with high efficiency, reliability and security. Indeed, efficient and reliable operations can be achieved only by adhering to standards. Because of the complexity and the wide spectrum, several distinct standards have been proposed and often revised to fit with the new DG developments, while others are under study to cover new aspects and issues. Important standards, regarding the characteristics of DG and SGs are IEEE 1547, IEC 61850-7-420, IEC 61400-25, IEEE 1379, IEEE 1344, IEEE C37.118 and IEEE 519. IEEE 1547 is the standard for interconnection of distributed energy resources (DERs) and was published in 2003. It gives a set of criteria and requirements for the interconnection of DG resources into the power grid. Currently, there are six complementary standards designed to expand upon or clarify the initial standard, two of which are published, while the other four are still drafts. IEC 61850-7-420 represents the standard related to communication and control interfaces for all DER devices. This standard defines the information models to be used in data exchange among DERs, which comprise DG devices and storage devices, including fuel cells, microturbines, PVs and combined heat and power. Where possible, it utilizes existing IEC 61850-7-4 logical nodes and defines DER-specific logical nodes where needed. Such standards allow significant simplifications in implementation, reduction of installation costs, sophisticated market-driven operations and simplification of maintenance, thus improving the overall reliability and efficiency of power system operations. IEEE P2030 is the first IEEE standard providing a roadmap and guidelines for better understanding and defining smart grid interoperability. Hence, IEEE P2030 has defined the SG interoperability reference model (SGIRM) to expand the current knowledge base, characteristics and principles for SG interoperability and power grid architectural designs and operations to provide more reliable and flexible power system. In general, IEEE P2030 focuses on three integrated architectural perspectives, e.g. power systems, communication technology and information technology. There exist two additional complementary standards based on the IEEE P2030 standard. IEEE P2030.1 provides guidelines on the knowledge base addressing terminology, methods, equipment and planning requirements for transportation electric-sourced, road-based personal and mass transportation applications, which can be used by utilities, manufacturers, transportation providers, infrastructure developers and end-users of electric-sourced vehicles. IEEE Standard P2030.2 focuses on discrete and hybrid energy storage systems integrated with the electric power infrastructure. Notice that the *interoperability* refers the capability of two or more networks, systems, devices, applications or components to externally exchange and readily use information securely and effectively.

IEC 61400-25 defines the communication needed for monitoring and control of wind power plants. It is a subset of IEC 61400, a set of standards for the design of wind turbines. This standard allows information control and monitoring from different wind turbine vendors in a homogeneous manner. The information is hierarchically structured and covers, e.g. common information regarding the rotor, generator, converter, grid connection and the like. It covers all components required for the wind power plant operation, including the meteorological and electrical subsystems and wind power plant management system. IEEE 1379 is the recommended practice for data communications between remote terminal units and intelligent electronic devices in a substation. This standard, published in 2000, provides a set of guidelines for communications and interoperations of remote terminal units (RTUs) and intelligent electronic devices (IEDs) in a substation. It does cover two widely used protocols for Supervisory Control and Data Acquisition (SCADA) systems: IEC 60870-5 and DNP3. IEC standard 60870-5 deals with tele-control, tele-protection and the associated telecommunications for electric power systems. It provides a communication profile for sending tele-control messages between two power substations and defines operating conditions, electrical interfaces, performance requirements and data transmission protocols. DNP3 (Distributed Network Protocol) is a set of communication protocols used between components in process automation systems. Recently, IEEE adopted DNP3 as IEEE standard 1815-2010 in 2010 and then modified it in the present standard 1815-2012. In addition, using 3G cellular communication system as the backhaul network has also been recommended by IEEE P2030. Compared with WMN, the biggest strength of cellular communication systems is the pervasiveness of this mature technology. Three wireless communication technologies based on IEEE 802.15.4 protocol stack are recommended to be used in SG.

Standard IEEE C37.118 is the current standard for measurement systems of synchronized phasors. Synchronization is fundamental for ensuring proper operations and for eliminating potential faults and failures. A phasor measurement unit (PMU), which can be a stand-alone physical unit or a functional unit within another physical unit that estimates an equivalent synchro-phasor for an AC waveform, is introduced. The total vector error compares both the magnitude and the phase of the PMU estimate with the theoretical phasor equivalent signal for the same instant of time. It provides an accurate method for evaluating the PMU measurement and establishes compliance requirements under steady-state conditions.

The latter defines the levels for phasor frequency, magnitude and angle measurements, harmonic distortion and out-of-band interference. It is worth noticing that IEEE C37.118 does not establish compliance requirements under dynamic conditions or other tests during which the amplitude or frequency of the signals varies, even if several dynamic tests are described in this standard. IEEE 519 establishes the limits on harmonics amplitudes for currents and voltages at the PCC or at the point of metering in an EPS. The limits assure that the electric utility can deliver relatively clean power to all customers and protect its electrical equipment from overheating, avoid loss of life from excessive currents harmonics and prevent excessive voltage stress because of excessive voltage harmonics. IEEE P2030 deals with interoperability of energy technology and information technology operation with the EPS and customer-side applications. It is responsible for bidirectional data transfer for electricity generation and reliable power delivery. IEC 62351 is a standard for cyber security and protection of communication protocols from hacker attacks. Other standards have been defined for communications in WANs, FANs and HANs, e.g. G3-PLC, HomePlug, PRIME, U-SNAP, IEEE P1901, Z-Wave, IEC 61970 and IEC 61969 and IEC 60870–6. Finally, vehicle-to-grid (V2G) operations are regulated by SAE J2293, which provides the requirements for EVs and electric vehicle supply equipment and SAE J2836 with regard to the communication between plug-in electric vehicles (PEV) and power grid, while SAE J2847 is specific for communications between PEVs and grid components. In conclusion, numerous international standards regulate almost all aspects of DG systems and SGs, thus requiring equipment producers and users (service providers, generator builders and final users) to fulfill standard requirements to guarantee smooth and correct operation of the whole SG. M441 is the mandate of standards and technical documents to the organizations European Committee for Standardization (CEN), European Committee for Electrotechnical Standardization (CENELEC) and European Telecommunications Standards Institute (ETSI) for any type of smart meter's functionalities and communication issued by the European Union. The main objective is providing reliability, interoperability and security for smart metering deployments.

3.4 DIFFERENCES BETWEEN CONVENTIONAL ELECTRIC GRID AND SMART GRID

The three dominant factors that are influencing the future electric systems in the United States and abroad are: government policies, higher demands and needs for improved efficiency and power quality needs and the introduction and the application of new intelligent computer, communication and hardware technologies. In addition, environmental concerns have created governmental policies around the world, including at the national, federal and state levels, which are driving the entire energy system to efficiency, conservation and the use of renewable energy sources and distributed generation the electricity production. These factors are the main drivers that are expanding the use of all sorts of new smart and adaptive control, power electronics, intelligent energy management, the use of the renewable energy and electricity storage technologies, on the one hand, and new energy efficiency, conservation techniques and methods, on the other hand. Consumers are becoming more proactive and are being empowered to engage in the energy consumption decisions affecting their day-to-day lives. At the same time, they are expanding their energy needs. For example, consumer participation ultimately includes extensive use of electric vehicles (both cars and trucks), remote control of in-home appliances to promote energy conservation and efficient uses, ownership of distributed generation from evermore renewable energy sources and management of electricity storage to locally match supply to demand. The intelligent electric energy system – the smart grid – needs to address all these needs and concerns by using advanced technologies to create a smarter, more efficient and sustainable electric grid. It is worth to be noted that the smart grid, as characterized above and in previous chapter sections, does not replace the existing electric system (power grid) but rather builds on the available power, communication and IT infrastructure to increase the utilization of existing assets and to empower the implementation of the new functionality. For example, centralized energy sources (power plants) of generation will still play a major role in the smart grid, and large-scale wind and solar generation, wherever cost justified, will become major parts of the electricity generation mix. Availability of a two-way, cyber-secure, end-to-end communications system will provide consumers with the knowledge of their energy usage necessary to allow them to locally and/or remotely control their smart appliances, monitoring, air conditioning, heat and temperature settings. In summary, there are three primary differences in the function of the existing (conventional) electrical grid and the smart grid. They are energy flow, data flow and the fuel mix. Smart grid technologies allow energy to flow in a loop, allowing exchanges between stakeholders.

Consumers have the ability to join energy providers in an enhanced role as energy consumer and/or producer. Figure 3.2 illustrates the conventional and smart grid data flows. Data flows in one direction from the consumer to the provider, with no actionable real-time feedback. Energy providers manually or remotely read aggregate usage from consumer meters, and consumers receive an after-the-fact billing invoice. Consumers participate passively in current data flow because they do not have the ability to identify specific energy usage patterns in real-time. Data resides in a cloud in real-time, moves in multiple directions and is accessible by consumers as well as energy companies as actionable information. Measurements taken using smart grid communications technologies are precisely time-synchronized, taken many times per second and displayed digitally, thereby facilitating consumer action by allowing consumers to actively respond by adjusting consumption. Smart grid technologies automatically read meters, record power consumption habits and support remote control. Consumers are provided with options to choose low-cost electricity, level electricity consumption amounts between peak and low hours, services and reserve electricity for emergency use. Real-time power costs, energy consumption habits and energy use suggestions are all accessible through smart grid technologies. The installation of a smart device, a net meter, allows independent generation sources to be connected to the grid, resulting in reduced dependence on external sources. Consequently, power producers of all sizes, including commercial consumers, have the opportunity to produce electricity with various energy sources and methods of their choice. Smart grid technologies facilitate exchanging and selling self-generated power to the smart grid by automatically balancing the electrical power load on the grid.

A general description of the conventional electric delivery system (power system) consists of mostly and relatively isolated components: generation, transmission, substation, distribution and the consumer (end-users). Key characteristics of such conventional power system that will be most strongly impacted by the changes required to implement the smart grid are the following attributes or functionalities: centralized sources of power generation, unidirectional flow of energy from the source (generation end), passive participation by the customers (consumer knowledge of electrical energy usage is limited to a monthly bill received, after the fact), real-time monitoring and control is mainly limited to generation and transmission and only at some utilities does it extend to the power distribution system, and power system is not flexible, being difficult to either inject electricity from alternative energy sources at any point along the grid or to efficiently and sustainably manage new services desired by the electricity users of electricity. A general schematic of the smart grid is shown in the diagram of Figure 3.3. The key requirements of this electric system are addressing the following transformational functionalities: allow for the integration of renewable energy resources to address supply security, diversification and environmental issues, allow for active customer participation to enable far better energy conservation, allow for cyber-secure communication systems to address system safety, allow for better uses of the existing assets to address long term sustainability, allow for optimized energy flow to reduce losses and lower the cost of energy, allow for the integration of electric and hybrid vehicles to reduce dependence on hydrocarbon fuels, allow for the management of distributed generation and energy storage to eliminate or defer system expansion to reduce the overall energy costs, finally allowing integration of communication and intelligent control across the entire energy system to promote interoperability and open systems and to increase power quality, safety, resilience, faster service restauration and operational flexibility.

As the next-generation intelligent electricity delivery system – the smart grid – optimizes the energy efficiency by grafting information technologies onto the existing network and exchanging real-time information between electric suppliers and customers. Besides the benefits to every group of stakeholders from smart grid, there are many driving forces of the smart grid implementation. Firstly, the conventional grid is aging, old-designed and with poor reliability, one example is the blackout occurring in many countries. Table 3.1 summarizes the differences between the smart grid and conventional power grid, as found in the literature. Smart grid is applying two-way communication technologies to enable customers to participate the grid actions and management. For instance, photovoltaic solar panels, installed in the customers' houses or facilities, could generate electricity in daytime and selling the excess of electrical energy to the electric grid or local utilities, while during the nighttime, when solar panels cannot generate energy, the utility is supplying the electricity to the load at home as usual. Besides, new technologies such as distributed generation, electric and hybrid vehicles charging and discharging, Flexible Alternating Current Transmission Systems (FACTS) technologies and so on are applied to the grid to enhance energy efficiency and reduce emissions. Monitoring and control of the electric system components will provide the utility with the real-time status of the power system. This real-time data use, combined with integrated system modeling and powerful new diagnostic tools and methods, providing detection of incipient failures

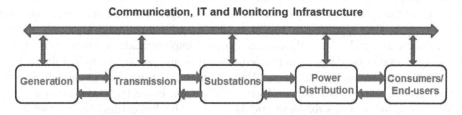

FIGURE 3.3   The smart grid structure, energy, data and information flows.

TABLE 3.1

Comparison between Conventional Electric Grid and Smart Grid

Conventional Electric Grid	Smart Grid	Functionalities and Aspects
Passive consumers	Consumers actively participating in grid operation	Grid-consumer interactions
Challenges of renewable energy grid integration	Integration with renewable resources enhancement	Renewable energy Integration
Limited market choices for customers	With digital market trading, more choice for customer	Options for customers
No or limited choices on power quality, no price plan options	Different power quality and price levels for different consumers	Power quality and price options
Ageing power assets, no efficient operation	Assets operating optimization, less power losses	System operation
Limited protection and service restoration	Have capability of self-healing, less damage affected by fault	Protection
Susceptible to physical and cyber Attack	More reliable for national security and human safety	Reliability and Security

in order to drive preventive maintenance and dynamic work management systems. Automatic reconfiguration of the system, powered by sophisticated, adaptive, autonomous optimization controllers maintains the energy flow without interruption when equipment failures do happen. DG, energy storage resources and remotely controlled equipment will also play an important role in the management of the smart grid energy system not only to address contingency needs but also to optimize power flow, eliminate load pockets and minimize system losses. Fully implemented smart grids have the following characteristics, not available in conventional electric systems: secure two-way communications covering the entire system and main components; sensed and variances detected, e.g. cables, joints, terminations, transformers, consumer usage, power quality, etc., monitored in real time. Such characteristics are providing massive amounts of incoming data and information that must be converted into grid state situational awareness. However, new problems and challenges are appearing or getting worse with some new applications deployed in smart grids. The availability of new and intelligent technologies such as abundant and aware SCADA sensors, secure two-way communications, integrated data management and intelligent, autonomous controllers has opened up opportunities that did not exist even a decade ago, shaping and structuring the electric grids in significantly different way as the conventional one.

The future smart grid is information driven and relies on services to empower the interactions among its stakeholders at multiple layers. The smart grid is a system of systems, i.e. it is a complex ecosystem of heterogeneous (possibly) cooperating entities that interact in order to provide the envisioned functionality. It is a complex energy and information infrastructure depicting system of system characteristics and features, such as interdisciplinary and diverse nature, operational and managerial independence of its elements and subsystems, geographical distribution, high level of heterogeneity of the networked systems, as well as the emergent behavior and evolutionary development. As any modern and advanced technological brand and revolutionary new concept and with plenty of creative and state-of-the art technological implementations and novelties, the smart grid is facing several of challenges and problems. Security and privacy are two of the major challenges of smart grid technology adoption. With communication network integration into power grid, smart grid also brings the issues, which never happen, in the traditional electric networks. The primary cyber-security risk for smart grid data is posed by data existing on, and moving in and out of, a cloud, as illustrated in Figure 3.2b. The cloud is a central concept in information system virtualization. As formerly isolated systems are interfaced with smart grid information communication systems, there is much to be considered. From a national security perspective, the surest way to bring a country to its knees is to compromise its electrical supply. For this reason, US federal and state governments have provided billions of dollars of stimulus funds for the development of smart grid technologies and for ensuring the data and information security and privacy of the smart grid systems. For such reasons, the cyber-security issues need to be taken care of greatly in order to prevent power grid from the operation modification disruption or wrong message inserting. Targets for the power grid self-healing technologies deployment also need to be carefully considered against natural disasters and physical attacks. Safety and security issues are being more stringent and critical in the smart grid as they are in the conventional power grid due to the data and information flows and collection into the former one. Communication network integration brings reliability problem to the power system networks. There is no doubt that the communication system could deliver message efficiently which can make power system operators respond faster when facing some critical situation. However, the wrong messages produced by hackers sent to the power network may be accompanied by serious consequences, and ultimately result in power blackouts. Notice that the distinction between physical security and cyber-security may not have been recognized during the design process, potentially resulting in systems that are woefully inadequate to assure that smart grid data is not vulnerable to cyber-attack. There is also potential liability in existing systems being, patched in an effort to keep up with smart grid technologies, without complete knowledge of smart grid technology data implications, a situation that is difficult to remedy because development of smart grid technologies is proceeding at a blindingly fast pace.

As we move from the conventional data storage model to a model that includes data collection, transport and analysis, the potential for cyber-attacks increases exponentially. In all industries, the information systems are virtualized in order to cut costs on hardware and software design and maintenance, but this is the case to coin an old phrase, where an ounce of prevention is most certainly worth a pound of cure. System virtualization opens up a plethora of cyber-security risks. Moreover, as smart grid data becomes more mobile through smart grid technologies, meaning the more the data and information move in and out of a cloud, consumer entities, via smart meters, electric provider entities and lodging energy management information systems, the more the potential for cyber breaches increase dramatically because of proliferation of the data and information movement and processing. Coupling this movement with the vulnerability inherent in the expected massive data volume serge (commonly referred to as the tsunami in smart grid information systems circles) accompanying the smart grid technology implementation and the potential liability inherent in handling smart grid data are staggeringly apparent. From a privacy perspective, there is increasing evidence that consumers, and the regulatory entities that are protecting them, will be in some way reluctant to participate in the smart grid boon if the data privacy and security are not guaranteed. So, the data privacy goes beyond the traditional IT security, privacy and touches on issues such as corporate culture, data collection policies and data quality initiatives. Organizations, ignoring these additional dimensions, increase their risk of having data breaches resulting in financial penalties and lost consumer trust. The current smart grid plan is a centralized data system. The more centralized the system is, the harder it will be to secure, and the more vulnerable it will be to threats. Opposition to centralized smart grid implementation is being voiced by a rising wave of smart grid technology proponents who support widespread adoption of local, or decentralized, smart microgrids, but the tide has not yet turned.

Another major SG difference consists of the fact that the generation, transmission and distribution is controlled by a new generation of cyber-enabled and cyber-secure energy management systems (EMSs) with a high-fidelity supervisory control and data acquisition (SCADA) frontend. Notice that the existing utility grids may or may not include SCADA sensors, computing and communications to monitor grid performance. Utility systems may depend instead on separate reporting systems, periodic studies and on the standalone outage's management applications. Information to the customer is generally limited to a periodic bill for services consumed in a prior time interval or billing cycle. Sensors, remote monitoring, automated switches, reclosers, upgraded capacitor banks and other equipment may be integrated into the grid to provide end-to-end monitoring and control of the transmission and distribution network. Equivalent additions on the customer side of the meter would include automated control systems and smart appliances with embedded price and event-sensing and energy management capability. Sensors are providing the information to better understand grid operation, while control devices provide options to better manage system operation. The information and data networks of the future are merging the capabilities of traditional EMSs and SCADA with the next generation of substation automation solutions. It will enable multiscale networked sensing and processing, allow timely information and data exchanges across the electric grid and facilitate the closing of a large number of control loops in real time. This will ensure the responsiveness of the command and control infrastructure in achieving overall system reliability and performance objectives. The key to the success of the smart grid infrastructure is an underlying system of data acquisition, data validation and data processing that will provide accurate and reliable data as well as extracted information to all the implied applications. The future smart management system is a SG subsystem that provides advanced management and control services, making smart grid a more efficient and better energy delivery infrastructure. The smart management system is the subsystem in SG that provides advanced management and control services and functionalities. The key reason why SG revolutionizes the grid is the explosion of functionality based on its smart infrastructure. With the development of new management's applications and services that can advantage the technology and capability upgrades enabled by this advanced infrastructure, the grid keeps becoming *smarter*. The smart management system takes advantage of the smart infrastructure to pursue various advanced management objectives. Thus far, most of such objectives are related to energy efficiency improvement, supply and demand balance, emission control, operation cost reduction and utility maximization.

3.5 CHAPTER SUMMARY

A smart grid, also called smart electrical/power grid, intelligent grid, intelligrid, future electric grid, intergrid or intragrid, is an enhancement of the 20th-century power grid. The traditional power grids are generally used to carry power from a few central generators to a large number of users or customers. In contrast, the SG uses two-way flows of electricity and information to create an automated and distributed advanced energy delivery network. The smart grid has to deal with multiple stakeholders, capture monitoring information and provide control capabilities for a large-scale, complex and heterogeneous infrastructure. Smart grid technologies support both direct and indirect energy efficiency efforts, as they can judiciously determine adequate supply margins in real-time, improve the operational efficiency of the distribution network and allow for a host energy efficient applications (such as demand response), etc. The primary driver of successful smart grid technology adoption on the national and global stage are the recognition and requirements of the necessity of energy efficiency, less environmental impacts, better power quality and services implemented by governments and the industry that possess the power to override individual inclinations and fund technological innovations in accordance with central policies. Smart grids are encompassing the entire power systems, generation, transmission, power distribution and utilization cycles. It consists of advanced actuators, sensors, including intelligent sensors, smart control, communication, monitoring, metering, IT and computing infrastructures. It is regarded as the next-generation power grid, uses two-way flows of electricity and information to create a widely distributed automated energy delivery network. By utilizing modern information technologies, the SG is capable of delivering power in more efficient ways and responding to wide ranging conditions and events. Broadly stated, the SG could respond to events that occur anywhere in the grid, such as power generation, transmission, distribution and consumption and adopt the corresponding strategies. More specifically, the SG can be regarded as an electric system that uses information, two-way, cyber-secure communication technologies and computational intelligence in an integrated fashion across electricity generation, transmission, substations, distribution and consumption to achieve a system that is clean, safe, secure, reliable, resilient, efficient and sustainable. Data security, safety and privacy are major issues in the smart grid due to the new structure of the data and information flows. This description covers the entire spectrum of the energy system from the generation to the end points of consumption of the electricity. In summary, there is no doubt that the emergence of SG will lead to a more environmentally sound future, better power supply services, and eventually revolutionize our daily lives. However, we still have a long way to go before this vision comes true and fully practical way.

3.6 QUESTIONS

1. What are the smart grid main benefits?
2. List the major smart grid drivers. Briefly discuss them.
3. Define the smart grid concept in your own words.
4. List the main smart grid features.
5. List of few challenges of the smart grid implementation and development.
6. List the main attributes of an ideal power system.
7. Explain in your own words what there no accepted smart gird definition.
8. Briefly discuss the major changes to the future power systems.
9. List the major benefits of the smart grid development.
10. Briefly describe the three major smart grid components form technical perspective.
11. Briefly describe the major smart grid components.
12. List the consumer's smart meter functionalities.
13. What are the utility smart meter functions?
14. List and briefly describe the major smart grid drivers.
15. List the differences between the conventional electric grid and smart grid
16. Briefly discuss these differences.
17. Briefly discuss the new smart grid functionalities and attributes.
18. In your own words what are the importance of standards for power systems.
19. Briefly describe the differences between the existing electric grid and smart grid energy flows.
20. Briefly describe the differences between the existing electric gird and the smart grid data flows.
21. What are the major security issues of the smart grid?

4 Renewable Energy, Distributed Generation in Smart Grids

4.1 INTRODUCTION

Economic, technological and environmental incentives and issues are changing the face of the electricity generation and transmission. Centralized energy generation structures, needing very complex and bulk power networks to produce and to transfer energy to the consumers, have high investment, operation and maintenance costs, while the overall system efficiency is low due to the large losses in these networks and systems. At the same time, the existing electricity and energy infrastructures are becoming older, and in addition, there are new energy supply security and environmental issues with the construction of new network components (e.g. power plants or high-voltage transmission lines). Other critical issues of the power distribution are maintaining the required power quality and supply stability, while the customers have increased power quality and supply stability demands, due to the extended use of sensitive or critical loads, which may also significantly affect the power quality. For such reasons the centralized power generation is giving way to smaller and distributed energy supplies. Distributed generation (DG) is loosely defined as a small-scale electricity generation, often located at the consumption points. Usually, the connection is to the distribution networks or on the meter customer side. For most DGs the customer uses all of the energy output, and any surplus is delivered to the main grid or stored. If the customers require additional power than the DG generation, the power is taken from the grid. Distributed generation has become more present into the power systems for reasons such as: an alternative to the construction of large power plants, constraints on the construction of new transmission lines, higher power quality and supply stability demands. It is also economically attractive, as the cost of small-scale generation keeps decreasing with technological advances, changing the economic and regulatory environment and the electricity market liberalization. The new electricity distribution networks are consisting of large DG and RES penetration, energy storage units, electric vehicles and customers with smart meters and controllable loads. DERs are encompassing a wide range of technologies, such as: wind turbines, gas turbines, micro-turbines, PV systems, fuel cells or energy storage units. These technologies have lower emissions and higher potential to lower the overall costs. Their applications include substation power support, deferrals of transmission and distribution upgrades, highly efficient combined heat and power generation. They are also characterized by higher power quality and the possibility of smarter power distribution. In present environment it is quite unlikely that the traditional power grids can expand rapidly and perform well enough to meet the future economic needs and the expanded electricity uses. Distributed energy resources (DERs) are vital to meet the growing needs of the electricity uses, and are pushing the limits of affordable power quality and grid expansion. While grid dependency has intensified, smaller generation using a broad spectrum of the generation mix has emerged as competitive alternatives to large central generation stations. Distributed generation (DG) is not a new phenomenon and application. Prior to the advent of alternating current and large-scale steam turbines, during the power industry beginnings, most of, if not all, the energy requirements – heating, cooling, lighting and motive power – were supplied at or near their point of use. DG usually refers to small-scale systems that generate electricity and often heat close to the point where the energy is actually used. Distributed energy resources, dispersed or distributed generation systems are becoming more important to the electricity generation mix. DG technologies can run on renewable energy resources, fossil fuels or waste heat. Equipment ranges in size from less than a kilowatt to tens of megawatts, in order to meet all or part of a customer power needs.

DOI: 10.1201/9780429174803-4

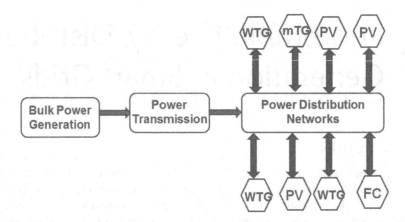

FIGURE 4.1   Distributed energy resources on the power distribution network.

Historically, the type of technologies employed has varied, but are generally limited to small power units. Recently, renewable resources such as solar photovoltaic (PV) systems, small power-hydro units and wind energy conversion systems are included into DG mix. However, the small-scale, fossil-fired generation is still used, as primarily providing reliable, back-up power in the event of grid interruptions.

The pressure of improving the overall power system efficiency, power quality, energy supply security, stability and environmental impacts has forced the energy industries to answer to these issues. There are also significant increases in the electricity and energy demands, while the DG and RES seems one way to cope with these demands and grid issues. DER and RES integration into the existing energy networks can result in many benefits, in addition to the reduced grid losses and environmental impacts, such as relieved transmission and distribution (T&D) congestion, peak demand shavings, voltage support, reduced price fluctuations and the deferred investments to upgrade existing systems. DERs have and are expected to have even more impacts in the future on the energy market. Modern power systems generate and supply electricity through a complex process and system. Usually, a power system consists of the electricity generation in large power plants, usually located close to the primary energy sources (e.g. coal mines, water reservoir), far away from the large consumer centers, delivering the electricity by a large passive but complex distribution infrastructure, involving high-voltage (HV), medium-voltage (MV) and low-voltage (LV) electric networks. Power distribution operates mostly radially; power is flowing from HV levels down to customers, along the distribution feeders. Nowadays, with the technological advancements, environmental policies and the expansion of the electricity markets are promoting significant changes into the electricity industry. New technologies allow generation in smaller size units or in DG units, located in the MV and LV grid sections, closer to the end-users (consumers), as shown in Figure 4.1. Moreover, the increasing RES use in order to reduce the environmental impacts and diversify the supply leads to a new electricity supply scheme. In this new paradigm, the power production is not exclusive to the generation end but is shifted to the MV and LV networks, with part of the energy supplied by the centralized generation and part is produced by the DG units, closer to the customers. Large-scale DG integration is a main trend into modern power systems. These generators are of considerable smaller size compared to the traditional generation units.

4.2   RENEWABLE ENERGY SOURCES

Sun, our only primary source of energy, emits continuously energy as electromagnetic radiation at an extremely large and relatively constant rate. The rate at which this energy is the one emitted by a blackbody at temperature of about 6000°K (10340°F). If we are able to harvest the energy coming from just 10 hectares (25 acres) of the Sun's surface, it will be enough to supply the world's current energy demands. However, this cannot be done due to three reasons. First, the Earth is displaced from the Sun, the solar energy decreases with inverse of the distance squared, only a small fraction of the energy leaving the Sun's surface reaches an equal area on the Earth. Second, Earth rotates about its polar axis, so any collection device located on the Earth receives the Sun's radiant energy for only about one-half of each day. The third and least predictable factor are the atmosphere conditions, accounting for about 30% reductions in the Sun's energy reaching the ground. However, the weather conditions can stop all but a minimal amount of solar radiation from reaching the Earth's surface for many days in a row. The *rate* at which solar energy reaches a unit area at the Earth surface is called the *solar irradiance* or *insolation* (W/m^2). Solar irradiance is in fact a power density of the solar radiation and is varying over time. The maximum solar irradiance value is used in the solar energy system design to determine the peak rate of energy input into the system. Solar thermal energy systems and photovoltaics are increasingly being used for electricity generation, both for residential (low and medium power settings) and power system applications.

The use of wind energy has its roots in antiquity, and for long time it was the major source of power for pumping water, grinding grain or long-distance transportation (sailing ships). The farm's windmills were instrumental in the Great Plains settlement during 19th century. Among the wind energy advantages are: renewable, ubiquitous and does not require water for the generation of electricity. The disadvantages are: variable and low power density, which means high initial investment costs. Wind turbines convert the wind kinetic energy into mechanical and eventually into electrical energy that can be used for a variety of tasks. Regardless of the task, wind offers an inexpensive, clean and reliable form of mechanical power. It represents an important energy source of new power generation trends and an important player in the energy market. As a leading renewable energy technology, wind power's technical maturity and speed of deployment is acknowledged, along with the fact that there is no practical upper limit to the percentage of wind that can be integrated into the electricity system. It is estimated that the total solar power received by the Earth is approximately 1.8×10^{11} MW. Of this solar input, only 2% (i.e., 3.6×10^9 MW) is converted into wind energy and about 35% of wind energy is dissipated within 1000 m of the Earth's surface. Therefore, the available wind power that can be converted into other forms of energy is approximately 1.26×10^9 MW, value that represents 20 times the rate of the present global energy consumption, meaning that wind energy in principle could meet entire energy needs of the world. Worldwide development of wind energy expanded rapidly starting in the early 1990s. The average annual growth rate from 1994 to 2015 of the world's installed capacity of wind power has been over 35%, making the wind industry one of the fastest growing in the field. Unlike the last surge in wind power development during 1970s which was mainly due to the oil embargo of the OPEC countries, the current wind energy development is driven by many forces that make it favorable. These include its tremendous environmental, social and economic benefits as well as its technological maturity, the deregulation of electricity markets, public support and government incentives. Even among other applications of renewable energy technologies, power generation through wind has an edge because of its technological maturity, good infrastructure and relative generated energy cost competitiveness.

Both wind and solar energy are highly intermittent electricity generation sources. Time intervals within which fluctuations occur span multiple temporal scales, from seconds to years. These fluctuations can be subdivided into periodic fluctuations (diurnal or annual fluctuations) and non-periodic fluctuations related to the weather change. Wind and solar energy are complementary to each other in time sequence and regions. In the summer, sunlight is intensive and the sunshine duration is long, but there is less wind. In the winter, when less sunlight is available, wind becomes strong. During a day, the sunshine is strong while wind is weak. After sunset, the wind is strengthened due to large temperature changes near the ground. It has been reported that the effects of complementarity are more dramatic in certain periods and locations in many areas of the globe. Because the major operating time for wind and solar systems occurs at different periods of time, wind-solar hybrid power systems can ensure the reliability of electricity supply. The applications of wind-solar hybrid systems range extensively from residential houses to municipal and industrial facilities, either grid-connected or standalone configurations.

Geothermal energy is an inexhaustible source of thermal and electrical energy on a human time scale. Its utilization is friendly to the environment and supplies base-load energy. It powers the movement of the continents across the planet surface, it melts rock that erupts as volcanoes and it supplies the energy that supports life in the ocean depths. It has been present for 4,500 million years and will be present for other few billions of years. It flows through the Earth constantly and continuously, having the potential to provide power to every nation in the world. In the United States alone, it has been noted that the amount of geothermal energy available for electricity generation exceeds by several times the total electrical power consumption of the country. All of this is possible and with minimal environmental consequence. Our planet, from its center to its surface, is a massive thermal energy storehouse. Earth is formed from a core of molten metal which is slowly transferring heat to the outside layers, and finally to the outer layer – the crust. Heat is also generated by the decay of naturally occurring radioactive materials beneath the surface. There are estimates that quantity of this energy may be available for human use and these quantities, regardless of the estimation method, is enormous. For example, it has been estimated that the total heat available within the upper 5 km of the Earth's surface is about 140×10^6 EJ. If only 1% of this could be used at the current rate of world energy consumption of about 500 EJ/year, this would provide the world with all its energy for about 3,000 years. Moreover, the utilization of geothermal energy increases the regional and local net product and the economy. It relieves dependence from fossil fuels and helps to conserve the valuable chemical resources for the future. Deep geothermal resources provide thermal and electrical energy, being a reliable energy source for the future. Utilization of deep geothermal resources extracts hot fluid from thermal reservoirs, while these hot waters are reinjected to the reservoirs, thus maintaining the natural equilibrium, permitting a sustainable and sparing energy resource management. Geothermal installations and power plants are characterized by small land uses, insignificant visual, and landscape impacts, aspects that are particularly critical in densely populated areas. Electrical energy from geothermal resources can provide an important contribution to the base-load electrical energy supply and may replace large-scale power plants fired with fossil fuels. It is worthwhile to note that the utilization of geothermal energy from shallow resources for the production of energy at low temperature for heating, hot water and cooling applications made tremendous progress over the past decades.

4.3   WIND ENERGY

Wind energy is a special form of kinetic energy in air as it flows. Wind energy can be either converted into electrical energy by power converting machines or directly used for pumping water, sailing ships or grinding gain. Wind energy is not a constant source of energy. It varies continuously and gives energy in sudden bursts. About 50% of the entire energy is given out in just 15% of the operating time. Wind strengths vary and thus cannot guarantee continuous power. It is best used in the context of a system that has significant reserve capacity such as hydro-power, compressed air storage or reserve load, such as a microgrid, desalination plant, to mitigate the economic effects of resource variability. Electricity generation from wind can be economically achieved only where a significant wind resource exists. For securing maximum power output, wind energy resource assessment at a prospective site is critical. Due to wind speed and turbine power relationship, accurate knowledge of the wind characteristics is critical to all wind energy exploitation aspects, from the site identification and predictions of the economic viability of wind energy projects through the wind turbine design and understanding the effects on electricity networks.

The most striking characteristic of the wind is its spatio-temporal variability, persisting over a very wide range of spatio-time scales. Accurate measurements of wind speed distributions are an important factor in wind energy potential analysis and assessment. There are three phases involved with wind power project planning and operations. Prospecting that uses historical data, retrospective forecasts and statistical methods to identify potential sites for wind power projects; site assessment that determines the optimum placement of a wind power project. An operations phase that is using wind forecasting and prediction to determine available power output for hour-ahead and day-ahead periods. The most critical is the first one: identifying and characterizing the wind resources. Appropriate statistical and modeling methods to compute the wind speed probability density function (PDF) are critical in wind resource analysis. The power available in the wind varies with the cube of the wind speed, and depending also on the air density and rotor cross-sectional area, being expressed as:

$$P_{wind} = 0.5\rho A v^3 \tag{4.1}$$

Here, ρ is the air density (kg/m³), A is the cross-sectional area (m²) and v is the wind speed (m/s). Equation (4.1) reveals that in order to obtain a higher wind power, it requires a higher wind speed, a longer length of blades for gaining a larger swept area and a higher air density. Because the wind power output is proportional to the cubic power of the mean wind speed, a small variation in wind speed can result in a large change in wind power. A common unit of measurement is the wind power density, or the power per unit of area normal to the wind direction from the wind is blowing:

$$p_w = \frac{P_{wind}}{A} = 0.5\rho v^3 \tag{4.2}$$

Here, p_w is wind power density (W/m²). The ultimate wind energy project goal is to extract the wind energy and not just producing power, an important parameter in site selection is the mean wind power density, expressed as if the wind frequency distribution $f_{PDF}(v)$ is known as:

$$\bar{p}_w = 0.5\rho v^3 f_{PDF}(v) \tag{4.3}$$

Example 4.1   For an average wind speed of 10 mph, a small wind turbine produces 100 W/m². What is the power density for a speed of 40 mph?

SOLUTION

Form Equation (4.2) the power density is proportional to cube of the wind speed so:

$$\text{Speed Ratio} = \frac{40}{10} = 4$$

$$P_{40} = 4^3 P_{10} = 64 \times 100 = 6.4 \, \text{kW}$$

Weibull or Rayleigh probability distribution is the most used in wind energy assessment, characterization and analysis. Instead of integration of the mathematical Weibull function, the mean value of the third power of the wind speeds in appropriate time intervals can also be used.

Example 4.2 **The diameter of a large offshore wind turbine is 120 m, assuming air density, 1.2 kg/m³, compute the available wind power 5, 10, 15 m/s What is the available wind power density for wind speed of 7.5 m/s?**

SOLUTION

By using Equation (4.1) the available power densities are:

$$P_5 = 0.5 \cdot 1.2 \left(\pi \frac{120^2}{4} \right)(5)^3 = 847.8\,\text{kW}$$

$$P_{10} = 0.5 \cdot 1.2 \left(\pi \frac{120^2}{4} \right)(10)^3 = 6.7824\,\text{MW}$$

$$P_{15} = 0.5 \cdot 1.2 \left(\pi \frac{120^2}{4} \right)(15)^3 = 22.8906\,\text{MW}$$

The power density for a wind speed of 7.5 m/s is then calculated as:

$$p_w = 0.5 \cdot 1.2(7.5)^3 = 253.12\,\text{W/m}^2$$

Wind results from the movement of air due to atmospheric pressure gradients. Wind flows from regions of higher pressure to regions of lower pressure. The larger the atmospheric pressure gradient, the higher the wind speed and thus, the greater the wind power that can be captured from the wind by means of wind energy converting machinery. The wind regime is determined due to a number of factors, the most important factors being uneven solar heating, the Coriolis force due to the Earth's rotation and local geographical conditions, e.g. the surface roughness results from both natural geography and manmade structures. Frictional drags and obstructions near the Earth's surface retard the wind speed, inducing a phenomenon known as wind shear. The rate at which wind speed increases with height varies on the basis of local conditions of the topography, terrain and climate, with the greatest rates of increases observed over the roughest terrain. A reliable approximation is that wind speed increases about 10% with each doubling of height. In addition, some geographic structures can strongly enhance the wind intensity. Air masses move because of the different thermal conditions, and the motion of air masses can be a global phenomenon (i.e., the jet stream) or a regional or local phenomenon. The local phenomena are determined by orography (e.g. the surface structure of the area) in connection with global phenomena. Wind energy resources rely on the incident wind speed and direction, both of which vary in time and space due to changes in large-scale and small-scale circulations, surface energy fluxes and topography. The wind energy production viability is governed by such factors as: the potential for large scale energy production, the predictability of the power to be supplied to the grid and the expected return on investment. The various wind energy uncertainties impact the reliable determination of these viability factors. An optimal wind farm planning include: (a) site optimal selection based on the quality of the local wind resources; (b) maximization of the annual energy production and/or minimization of the generated energy costs; and (c) maximization of the reliability of the predicted energy output. The most important activity in site selection is to determine the wind resource potential, consisting in the estimated local wind probability density function. Another important activity is to determine the turbulence levels and the resulting wind loads at the concerned site, promoting better decision-making in selecting the most suitable wind turbines for that site and in optimum life cycle cost prediction; higher wind loads generally result in higher costs. Site selection criteria include: (1) local topography, (2) distance to electric grid, (3) vegetation, (4) land acquisition issues and (5) site accessibility

TABLE 4.1

Classes of Wind Power Density at 10 m and 50 m

Wind Power Class	Wind Power Density (W/m²) – at 10 m Level	Wind Power Density (W/m²) – at 50 m Level
1	< 100	< 200
2	100 - 150	200 – 300
3	150 - 200	300 – 400
4	200 - 250	400 – 500
5	250 - 300	500 – 600
6	300 – 350	600 – 700
7	➤400	➤800

for turbine transport and maintenance. It is also important to have information about wind regime characteristics, air density and turbulence intensity. Wind power density is a comprehensive index in evaluating the wind resource at a particular site. It is the available wind power in airflow through a perpendicular cross-sectional unit area in a unit period. The classes of wind power density at two standard wind measurement heights are listed in Table 4.1.

The use of wind power classes to describe the magnitude of the wind resource was first defined in conjunction with the preparation of the 1987 US Department of Energy Wind Energy Resource Atlas. The atlas is currently available through the American Wind Energy Association and is an excellent source of regional wind resource estimates for the United States and its territories. The atlas wind resource magnitude is expressed in terms of the seven wind power classes, as well as the wind velocity. The wind power classes range from class 1, for winds containing the least energy, to class 7, for winds containing the greatest energy (Table 4.1). Mean wind speed estimates hare are based on Rayleigh wind speed probability distribution of equivalent mean wind power density for standard sea-level conditions, and to maintain the same power density, speeds are increased by 3%/1000 m (5%/5000 ft.) elevation. Wind resource assessment is the most important step in planning a wind project because it is the basis for determining initial feasibility and cash flow projections, being vital for financing. Assessment and project progress through several stages: (1) initial assessment; (2) detailed site characterization; (3) long-term data validation; and (4) detailed cash flow projection and financing. Prediction of wind energy resources is crucial in the development of a commercial (large-scale) wind energy installation. The single most important characteristic to any wind development is the wind velocity. The performance and wind farm power output are very sensitive to uncertainties and errors in wind velocity estimates, so the wind resource assessment must be extremely accurate in order to procure funding and accurately estimate the project economics. Commercial wind resource assessment performed by wind developers uses both numerical and meteorological data. Wind speed and direction measurements are collected by permanent or semi-permanent meteorological towers designed to measure wind velocity using a variety of wind sensors, e.g. anemometers, LIDARs, sodars. An important aspect is to gain an understanding of the wind profile both spatially across the location of interest and in elevation above terrain level. The main factors affecting the wind flows are orography, surface roughness and the atmospheric stability.

4.3.1 AIR DENSITY AND TURBULENCE EFFECTS, WIND SHEAR AND WIND PROFILES

Since wind speed generally increases with height, higher elevation sites potentially offer greater wind resources than comparable lower ones, being advantageous to site wind turbines at higher elevations and taking advantage of higher wind speeds. However, the decrease of air density with height can make an impact on the output power, wind power density being proportional to air density, so a given wind speed therefore produces less power from a particular turbine at higher elevations, because the air density is less. Output power and the power curve depend on the air density. For example, the air density values encountered at measurement sites in western Nevada are mostly between 0.936 kg/m³ and 1.025 kg/m³ with a multi-annual mean value of 0.982 kg/m³, significantly lower than the standard air density of 1.225 kg/m³. Power curves for various air density effects must be accounted for to improve the power output estimate accuracy. Air density is usually computed from temperature and pressure data, as expressed by:

$$\rho = \rho_0 \left(\frac{T}{T_0} \right)^{-(g/cR+1)} \text{ or } \rho = \rho_0 \left(1 + \frac{c \cdot z}{T_0} \right)^{-(g/cR+1)} \tag{4.4}$$

Where T is the local air temperature (°K), T_0 is the air temperature at the ground (°K), z is the elevation in m, $c = \mathrm{d}T/\mathrm{d}z$ is the atmosphere thermal gradient (~4.80°C/km), R is the gas constant (287 J/kg-K for air). Alternate relationships to estimate the air density dependence on the elevation are:

$$\rho = 1.229 \frac{P - VP}{760} \frac{273}{T} \text{ kg/m}^3 \tag{4.5a}$$

$$\rho = \frac{353.049}{T} \exp \left(-\frac{0.034 \cdot z}{T} \right) \tag{4.5b}$$

Here, the atmospheric pressure, P is expressed in mm Hg, VP is the vapor pressure in mm Hg, and T is the local absolute temperature in Kelvin degrees. This relationship yields to a value of 1.225 kg/m³ for dry atmosphere in standard atmospheric conditions. The vapor pressure represents a small correction, around 1%, and can be neglected. High temperatures and low pressures reduce the air density, which reduces the wind power. A major factor for air density change is the pressure change with elevation. If only elevation is known, air density can be estimated by using:

$$\rho = 1.225 - 1.194 \times 10^{-4} z \tag{4.6}$$

Depending on the turbine's method of control, either the power or velocity is normalized for use in power density calculations, as here where the velocity is normalized with the reference air density ρ_0:

$$v_{norm} = \bar{v} \left(\frac{\bar{\rho}}{\rho_0} \right)^{1/3} \tag{4.7}$$

Example 4.3   For a pressure of 750 mmHg and local air temperature equal to 21.5°C, estimate the air density and the normalized value of a wind speed of 10 m/s. Assume VP = 0.

SOLUTION

The air density, by using Equation (4.4), is 1.124 kg/m³, and by using Equation (4.6) the normalized value of the wind speed (10 m/s) is equal to 4.18 m/s.

At today's usual hub-heights at 80 m or even higher, the turbine rotors encounter large vertical gradients of wind speed and turbulence. Wind turbine rotors are susceptible to fatigue damage that results from turbulence. Wind turbulence represents the fluctuation in wind speed in short time scales, especially for the horizontal velocity component. The wind speed $v(t)$ at any instant time t can be in two components: the mean wind speed V_{mn}, and the instantaneous speed fluctuation $v'(t)$, i.e.:

$$v(t) = V_{mn} + v'(t)$$

Wind turbulence has a strong impact on the wind turbine power output fluctuations. Heavy turbulence may generate large turbine dynamic fatigue loads, reducing the expected turbine lifetime or resulting in turbine failure. In selection of wind farm sites, the knowledge of wind turbulence intensity is crucial for the stability of wind power production, turbine control and design. Quantification of the turbulence effects on wind turbine is done by computing an equivalent fatigue load, as function of the wind fluctuation amplitudes within an averaging period, blade material properties, number of averaging bins and a total number of samples. Turbulent fluctuations are the main source of the blade fatigue. The turbulence intensity (TI), a measure of the overall turbulence level, is defined as:

$$TI = \frac{\sigma_v}{v} \tag{4.8}$$

where σ_v is the wind speed standard deviation (m/s), usually at the nacelle height over a specified averaging period (e.g. 10 min). There are also differences in the output power standard deviations. In the wind speed range 4 m/s–15 m/s, the standard deviation of certain turbulence intensity classes (4–8% and 10–15%) differ up to about 50% with the standard deviation for all turbulence intensities. TI is affected by atmospheric stability, so the theoretical wind turbine power curves. A turbulence intensity correction factor can be expressed as:

$$V_{corr} = V_{norm}(1 + 3(TI)^2)^{1/3} \tag{4.9}$$

Example 4.4 **If the standard deviations for the following wind speeds 6.5 m/s, 10 m/s and 13.5 m/s are 0.90 m/s, 1.05 m/s and 1.15 m/s, respectively. What are the turbulence-intensity-corrected wind speeds?**

SOLUTION

Form the Equation (4.7) the turbulence intensity (TI) levels for these data are:

$$TI_{6.5} = \frac{0.90}{6.5} = 0.1385$$

$$TI_{10.0} = \frac{1.05}{10.0} = 0.1050$$

$$TI_{13.5} = \frac{1.15}{13.5} = 0.0852$$

By using modified Equation (4.8), the corrected wind speeds are:

$$V_{corr} = 6.5(1 + 3(0.1385)^2)^{1/3} = 6.624 \, m/s$$

$$V_{corr} = 10.0(1 + 3(0.105)^2)^{1/3} = 10.11 \, m/s$$

$$V_{corr} = 13.5(1 + 3(0.0852)^2)^{1/3} = 13.60 \, m/s$$

Notice that these are turbulence corrected wind speeds, while the wind turbine's power output is depended of cube of the wind speed.

Vertical wind shear is important as wind turbines become larger and larger. It is therefore questionable how well representative the hub height wind speed is. Various methods exist concerning the extrapolation of wind speed to the wind turbine hub height. There are several theoretical relationships for determining the wind speed profile. Obstacles can cause the displacement of the boundary layer, affecting the wind velocity. The roughness length (z_0) is the height at which the wind is zero, meaning that surfaces with large roughness lengths have larger effects on the wind. It ranges from 0.0002 m for open sea, 0.005–0.03 m for open land, 0.03–0.1 m for agricultural land, 0.5–2 m for very rough terrain or urban areas. Winds are usually recorded at 10 m, the standard meteorological height, while wind turbines have hub heights of 60 m, 80 m or higher. In cases, which lack elevated measurements, the wind velocity is estimated by extrapolations of the surface measurements. There exist several wind speed extrapolation methods. The wind speed $v(z)$ at a height z can be calculated directly from the wind speed $v(z_{ref})$ at reference height z_{ref} (the standard measurement level) by using the logarithmic law (Hellmann exponential law) expressed by:

$$\frac{v(z)}{v_0} = \left(\frac{z}{z_{ref}} \right)^{\alpha} \tag{4.10}$$

where, $v(z)$ is the wind speed at height z, v_0 is the speed at z_{ref} (usually 10 m height, the standard meteorological wind measurement level) and α is the friction coefficient or power low index. This coefficient is a function of the surface roughness at a specific site and the thermal stability of the Prandtl layer. It is frequently assumed to be 1/7 for open land. For 10 m and $z_0 = 0.01$ m, the parameter $\alpha = 1/7$, which is consistent with the value of 0.147 used in the wind turbine design standards (IEC standard, 61400-3, 2005) to represent the change of wind speeds in the lowest levels of the atmosphere. However, this parameter can vary diurnally and seasonally as well as spatially. It was found that a single power law is insufficient to adequately project the power available from the wind at a given site, especially during nighttime and also in presence of the low-level jets. However, there are significant discrepancies of values for α, especially for arid and dry regions, ranging from 0.09 to 0.120, quite smaller comparing to the standard 0.147 value. Moreover, α can vary from one place to other, during the day and year. Another formula, the logarithmic wind profile law and widely used across Europe, is the following:

$$\frac{v}{v_0} = \frac{\ln\left(\dfrac{z}{z_0} \right)}{\ln\left(\dfrac{z_{ref}}{z_0} \right)} \tag{4.11}$$

where, z_0 is called the roughness coefficient length and is expressed in meters; it depends basically on the land type, spacing and height of the roughness factor (water, grass, etc.) and it ranges from 0.0002 up to 1.6 or more. These values can be found in the common literature.

Example 4.5   A meteorological tower is located close to a town limit. If the wind speed is 7 m/s at 10 m height (standard measurement level), what is the wind speed at 50 m, using Equation (4.9)? If you are using Equation (4.10) and select z_0 =1.2, what is the wind speed at 50 m level?

SOLUTION

By using Equation (4.9) with $\alpha = 0.147$, the wind speed at 50 m is:

$$v = 7\left(\frac{50}{10} \right)^{0.147} = 8.7\,m/s$$

By using Equation (4.10), the wind speed at 50 m is:

$$v = 7 \frac{\ln\left(\dfrac{50}{1.2}\right)}{\ln\left(\dfrac{10}{1.2}\right)} = 12.3 \text{m/s}$$

This compares to 8.7 m/s using the power law with a shear exponent $\alpha = 0.147$.

In addition to the land roughness, these values depend on several factors, varying during the day and at night, and even during the year. Aside from ground level to hub height shear, wind shear over the rotor disc area can also be significant. The standard procedure for power curve measurements is given by the IEC standard (IEC Standard, 6-1400-12-1, 2005) where the wind speed at hub height is considered to be representative of the wind over the whole turbine rotor area. This assumption can lead to considerable wind power estimate inaccuracies, since inflow is often non-uniform and unsteady over the rotor-swept area. In most studies about the effect of wind shear on power performance, the wind speed shear is described by the shear exponent, obtained from the assumption of a power law profile. By integrating the wind profile over the rotor span, the corrected wind speed at the turbine nacelle can be obtained:

$$U_{avrg} = \frac{1}{2R} \int_{H+\frac{D}{2}}^{H-\frac{D}{2}} v(z)dz = v(H) \cdot \frac{1}{\alpha+1} \cdot \left(\left(\frac{3}{2}\right)^{\alpha+1} - \left(\frac{1}{2}\right)^{\alpha+1}\right) \tag{4.12}$$

where, H is the nacelle height and D is the rotor diameter. From (4.12), it is obvious that the hub height wind speed $z(H)$ is α corrected based on the profile it is experiencing.

TABLE 4.2

Sample Frequency Distribution of Monthly Wind Velocity

Wind Speed Range (m/s)	Hours per Month	Cumulative Hours
0–1	13	13
1–2	37	50
2–3	50	100
3–4	62	162
4–5	78	240
5–6	87	327
6–7	90	417
7–8	77	494
8–9	65	559
9–10	54	613
10–11	40	653
11–12	31	684
12–13	21	705
13–14	14	719
14–15	9	728
15–16	6	734
16–17	5	739
17–18	4	743
18–19	2	745
19–20	1	746
>20	1	747

4.3.2 WIND VELOCITY STATISTICS

Wind speed and direction are the most critical parameters needed to assess the power potential of a candidate site, due to the cubic dependence of the wind power density. The weather systems, the terrain and the height above the ground influence the wind. Wind speed varies by the minute, hour, day, season and even by the year. Therefore, the annual mean speed needs to be averaged over several years. In this subsection the most common wind speed probability distributions used by the wind energy community are discussed in some details. Usually, the time series of wind speeds and directions are rather large, differences among parameter estimation methods are not as important as differences among distributions. There are several PDF parameters' estimators, such as the Moment Method, Maximum Likelihood, Least-Square and Percentile Estimators Methods. These estimators are unbiased, so there is no reason to give preference to any of them. Once the wind probability distribution function is obtained, the mean power available can be deduced. The goal of any wind energy assessment and analysis is to obtain expressions allowing in giving responses to questions about statistical distribution of the maximum power obtainable from the wind, regardless of the WT chosen.

In order to predict the power generated on a yearly basis, statistical models of the wind velocity frequency of occurrence are needed. It has been found that Weibull and Rayleigh probability distributions can be used to describe wind variations with acceptable accuracy. The advantage of using well-known analytic distributions like these is that the probability functions are already formulated plenty of information and papers are available. The Weibull density distribution is a commonly applied statistical distribution to model wind speed distributions. The Weibull curve is a probability density function and indicates both the frequency and magnitude of a given wind speed over a period of time. It has been established that the Weibull distribution can be used to characterize wind speed regimes in terms of its probability density and cumulative distribution functions, and it is commonly used to estimate and to assess wind energy potential. Weibull distribution is well accepted and widely used for wind data analysis and is given by:

$$f_{WB} = k \frac{v^{k-1}}{c^k} \exp\left(-\left(\frac{v}{c}\right)^k\right) \tag{4.13}$$

The Weibull distribution is a function of two parameters, k, the shape parameter, and c, the scale factor, defining the shape or steepness of the curve and the mean value of the distribution. For wind analysis or modeling, typical k values range from 1 to 2.5 and can vary drastically form site to site, as well as during years and/or seasons. The scale parameter, c, corresponds to the average wind speed for the site. The main inaccuracy of the Weibull distribution is that it always has a zero probability of zero wind speed, which is not the case, since there are frequently times in which no wind is blowing. However, most turbines are not operational in speeds below 3 m/s, and the distribution is more accurate within the turbine operation range of 4 m/s–25 m/s. The higher the k value, the sharper the increasing part of the curve is. The higher c values correspond to a shorter and fatter distribution, with a higher mean value. In all of these statistics, c and k are Weibull coefficients that are dependent on the elevation and location. In general, frequency data would be accumulated for a particular site and wind turbine hub height elevation being considered. The data would then be fitted to a Weibull distribution to find the best c and k. The availability of high-quality fitted wind speed distributions is crucial to accurately assess the site wind energy potential. Once the distribution is found, important parameters to characterize wind regime can be computed. The average wind speed, by using Gamma function, is then:

$$V_m = c \int_0^\infty e^{-x} x^{1/k} dx = c\Gamma\left(1 + \frac{1}{k}\right) \tag{4.14}$$

The cumulative distribution function, $F(v)$, can be used to estimate the time over which the wind speed is between some interval, V_1 and V_2, such as:

$$F(V_1 \leq v \leq V_2) = F(V_2) - F(V_1) = \exp\left(-\left(\frac{V_2}{c}\right)^k\right) - \exp\left(-\left(\frac{V_1}{c}\right)^k\right) \tag{4.15}$$

Example 4.6 **A wind turbine with a cut-in velocity of 4 m/s and a cut-out velocity of 21 m/s is installed at a site where the Weibull coefficients are $k = 2.0$ and $c = 7.85$ m/s. How many hours in a 24-hour period will the wind turbine generate power?**

SOLUTION

Applying Equation (4.15) the probability that the wind speed is between cut-in and cut-off wind turbine speeds is:

$$F(4 \leq v \leq 21) = \exp\left(-\left(\frac{21}{7.85}\right)^{2.0}\right) - \exp\left(-\left(\frac{3}{7.85}\right)^{2.0}\right) = 0.86959 - 0.02125 = 0.84834$$

Therefore, the number of hours in a 24-hour period (one day) where the wind speed is between 4 and 21 m/s is: H = (24)(0.84834) = 20 h and 22 min. The Weibull shape and scale parameters are also height-dependent. Suggested corrections to Weibull coefficients k and c to account for different altitudes, z, are:

$$k = k_{ref} \frac{1 - 0.088 \cdot \ln\left(\frac{z_{ref}}{10}\right)}{1 - 0.088 \cdot \ln\left(\frac{z}{10}\right)}$$

$$c = c_{ref}\left(\frac{z}{z_{ref}}\right)^{\beta}$$

$$\beta = \frac{0.037 - 0.088 \cdot \ln(c_{ref})}{1 - 0.088 \cdot \ln\left(\frac{z_{ref}}{10}\right)}$$

The most used methods for estimating the best k and c for a Weibull distribution include: graphical method, standard deviation method, moment method, maximum likelihood method and energy pattern factor method. Graphical method is based on the use of Weibull cumulative distribution function and consist of plotting $ln[-ln(1 - F(v))]$ versus $ln(v_i)$ for the velocity samples v_i, for $i = 1,..., N$, the slope of the best-fit straight line represents the Weibull coefficient, k, and the y-intercept represents $-k \cdot ln(c)$, from which the Weibull scale factor, c, can be found. Alternatively, one can perform a least-square curve fit of the linear function to find the slope and intercept. Another method involves the wind speed sample mean and standard deviation. The shape and scale parameter, k and c, can be found by using:

$$k \simeq \left(\frac{s}{V_m}\right)^{-1.09} \tag{4.16a}$$

$$c = \frac{V_m}{\Gamma\left(1 + \frac{1}{k}\right)} \tag{4.16b}$$

Energy pattern method is based on the energy pattern factor, EPF, which is the ratio of the total power available in the wind and the power corresponding to the cube of the mean wind speed:

$$EPF = \frac{\frac{1}{N}\sum_{i=1}^{N} v_i^3}{\left[\frac{1}{N}\sum_{i=1}^{N} v_i\right]^3} \tag{4.17}$$

Having found the energy pattern from the wind velocity data, the approximate value of k is found from:

$$k = 3.957 \times (EPF)^{-0.898} \tag{4.18}$$

The value for c can be found using Equation (4.16b), for example. For the sake of brevity, moment method and maximum likelihood method are not discussed here. Interested readers are directed to the references at the end of chapter for details and implementation of these methods, or elsewhere in the literature.

Another commonly used probability distribution in wind energy analysis and assessment is the Rayleigh distribution, which is a special case of the Weibull distribution where $k = 2$. The Rayleigh distribution depends only on the mean wind speed, and is given by:

$$f_{RL}(v) = \frac{\pi}{2}\frac{v}{c^2} \exp\left[-\frac{\pi}{4}\left(\frac{v}{c}\right)^2\right] \tag{4.19}$$

These two probability distribution functions are the most commonly used for wind energy analysis and assessment. The simpler of the two is the Rayleigh distribution which has a single parameter c. The Rayleigh distribution is actually a special case of the Weibull distribution with $k = 2$. Setting $k = 2$ in the Weibull distribution gives the Rayleigh distribution. For both distributions, $V_{min} = 0$ and $V_{max} = \infty$. Setting $k = 2$ in this result gives the cumulative Rayleigh distribution.

$$F(v) = 1 - \exp\left(-\left(\frac{v}{c}\right)^2\right) \tag{4.20}$$

For the Rayleigh distribution the single parameter, c, relates the following the property:

$$c = V_m\sqrt{2} = \frac{2\mu}{\sqrt{\pi}} = \sigma\sqrt{\frac{4}{8-\pi}} \tag{4.21}$$

The Rayleigh distribution can be written using V_m or the mean velocity, μ. The minimum-least-squares-error (MLE) estimate of the mean of the normal distribution is the arithmetic mean. The MLE estimate of the variance is also familiar. The parameter c in the Rayleigh distribution is evaluated from a set of N wind velocity, v_i, expressed as:

$$\hat{c} = \sqrt{\frac{1}{N}\sum_{i=1}^{N} v_i^2} \tag{4.22}$$

Example 4.7:   For the wind speed data of Table 4.2 compute the average wind speed, standard deviation, the Weibull scale and shape parameters and the wind power density.

SOLUTION

From Equations 4.14a and 4.14b the average wind speed and standard deviation are: 9.10 m/s and 4.32 m/s. The Weibull shape and scale parameters are then computed by using:

$$k = \left(\frac{4.32}{9.10}\right)^{-1.09} = 2.246$$

And

$$c = \frac{9.10}{\Gamma\left(1+\dfrac{1}{2.246}\right)} = 10.28 \; m/s$$

Note: If we calculate the scale parameter from another relationship relating Weibull scale parameter the average wind speed, $c = 1.12 \times \bar{v} = 1.12'\ 9.10 = 10.19 \; m/s$ $c = 1.12 \cdot 9.10 = 10.19 \; m/s$, almost the same as the previous estimated scale parameter value.

4.3.3   WIND DIRECTION

Wind direction is one of the main wind characteristics. Statistical data of wind directions over a long period of time is very important in the site selection of wind farm and the layout of wind turbines in the wind farm. Changes in wind direction are due to the general atmospheric circulation, on an annual basis (seasonal) to the mesoscale (up to 5 days) or even smaller scale, such as hours. The seasonal changes of prevailing wind direction could be as little as 30° in trade wind regions to as high as 180° in temperate regions. In the US plains, the predominant directions of the winds are from the south to southwest in the spring and summer and from the north in the winter. Traditionally, the wind rose diagram (Figure 4.2a) illustrates wind directions. The wind rose diagram is a useful tool of analyzing wind data that are related to wind directions at a particular location over a specific time period (year, season, week, etc.). This circular diagram displays the wind direction relative frequencies in 8 or 16 principal directions.

To ensure the most effective wind turbine use, it should be exposed to the most energetic wind. The wind may blow frequently from some predominant directions, so more wind energy may come from directions with stronger winds. The wind direction distribution is crucially important for the evaluation of utilizing wind power, being given by wind roses or histograms of wind directions. The wind rose diagrams and wind direction frequency histograms provide useful information on the prevailing wind direction and availability in different wind speed bin. Notice that a wind vane points toward the source of the wind. Wind direction is reported as the direction from which the wind blows, not the direction toward which the wind moves. A north wind blows from the north toward the south. The wind direction varies due to the local features (topography, altitude, orientation, distance from the shore, vegetation, etc.). The wind direction can also be analyzed using continuous variable probability models to represent distributions of directional wind speeds, such as von Mises circular statistics. The model usually comprised of a finite mixture of the von Mises distributions (Figure 4.2b).

4.3.4   WIND ENERGY ESTIMATION

The ultimate estimate objective to be made in selecting a site for a wind turbine, wind power plant or wind farm is the energy that is available in the wind at that specific site. This involves calculating the wind energy density, E_{den}, for a wind turbine unit rotor area and unit time, which is a function of the wind speed and its temporal distribution at the site. Three important wind speed parameters used for

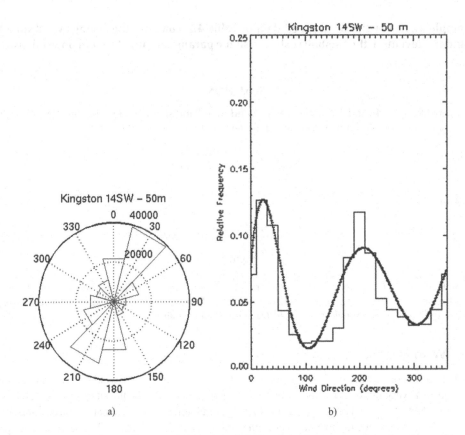

FIGURE 4.2 (a) A wind rose diagram and (b) the wind direction histogram and the fitted von Mise wind direction probability distribution.

wind energy assessment or in turbine design are the most probable (frequent) wind speed, V_{mP}; the wind speed carrying the maximum energy, V_{MaxE}; and the most frequent speed, V_{MF}. The parameters are estimated, once the wind probability distribution (e.g., Weibull or Rayleigh) is determined. The most frequent wind velocity corresponds to the maximum of the probability distribution, $f(v)$. As a result, the power generated scales as the wind velocity cube, the maximum energy usually corresponds to velocities that are higher than the most frequent. Horizontal wind turbines are usually designed to operate most efficiently at its design power wind speed, V_d. Therefore, it is advantageous if V_d and V_{MaxE} at the site are made to be as close as possible. Once V_{MaxE} is computed for the selected site, it is then possible to match the characteristics of the wind turbine to be most efficient at the site conditions. The wind power density, p_w, available in a wind stream of velocity, V, is given by the Equation (4.2). For a given velocity, V, the unit amount of time that velocity is present is $1 \times f(V)$, $f(v)$ is the wind speed probability distribution. The total energy for all possible wind velocities at a site is therefore computed as:

$$E_{den} = \int_0^\infty p_w f(v) dv \tag{4.23}$$

In the case of Weibull distribution function, after some mathematical manipulations, the energy density is expressed as:

$$E_{den} = \frac{\rho \cdot k}{2c^k} \int_0^\infty v^{k+2} \exp\left[-\left(\frac{v}{c}\right)^k\right] dv = \frac{\rho \cdot c^3}{2} \Gamma\left(\frac{3}{k}+1\right) \tag{4.24}$$

Applying the general reduction formula for a Gamma function, the following form for the energy density is obtained:

$$E_{den} = \frac{3\rho \cdot c^3}{2k} \Gamma\left(\frac{3}{k}\right)$$ (4.25)

With E_{den} estimated for a site or location, the density energy that is available over a period of time, T, is then given by:

$$E_T = E_{den} \cdot T = \frac{3\rho \cdot T \cdot c^3}{2k} \Gamma\left(\frac{3}{k}\right)$$ (4.26)

The most probable wind speed (v_{MP}), denoting the most frequent wind speed for Weibull distribution, is given by:

$$v_{MP} = c \cdot \left(\frac{k-1}{k}\right)^{1/k}$$ (4.27)

The wind speed carrying the maximum energy (v_{MaxE}) is expressed by:

$$V_{MaxE} = c \cdot \left(\frac{k+2}{k}\right)^{1/k}$$ (4.28)

The most frequent wind speed (v_{MF}) is defined by the maximum relative frequency:

$$V_{MF} = c \cdot \left(1 - \frac{1}{k}\right)^{1/k}$$ (4.29)

When considering a Rayleigh wind speed probability distribution, the wind energy density, with the mean wind speed, V_m, by using Equation (4.23) is given by:

$$E_{den} = \int_0^\infty \frac{\pi\rho}{4V_m^2} v^4 \exp\left(\frac{\pi}{4}\left(\frac{v}{V_m}\right)^2\right) dv = \frac{3}{\pi}\rho V_m^3$$ (4.30)

The energy density available at a site, having Rayleigh distribution of the wind speed, over a period of time, T, is then expressed as:

$$E_T = E_{den} \cdot T = \frac{3}{\pi}\rho \cdot T \cdot V_m^3$$ (4.31)

The most frequent wind velocity in the case of a Rayleigh wind speed probability distribution is given by:

$$V_{MF} = \sqrt{\frac{2}{\pi}}V_m$$ (4.32)

The velocity that maximizes the energy density for a Rayleigh wind distribution is computed with:

$$V_{MaxE} = \sqrt{\frac{8}{\pi}}V_m$$ (4.33)

Example 4.8 The following monthly mean wind velocity data (m/s) at a location is given in the following table. Calculate the wind energy density, the monthly energy density availability, the most frequent wind velocity and the wind velocity corresponding to the maximum energy, assuming a Rayleigh wind speed distribution and the air density 1.20 kg/m³.

SOLUTION

The energy density and the monthly energy density available are calculated with Equations (4.30 and 4.40), respectively, while the most frequent wind speed and the wind speed maximizing the energy density are computed using Equations (4.32) and (4.33). The results are summarized in the table below:

Month	E_{den} (W/m²)	E_T (MW/m²/Month)	V_{MF} (m/s)	V_{MaxE} (m/s)
January	907.0	2.4292	7.38	14.76
April	445.8	1.4940	5.82	11.65
July	1742.9	4.4680	9.18	18.35
October	368.3	0.9865	5.47	10.93

4.3.5 WIND ENERGY CONVERSION SYSTEMS

A wind turbine is a rotating mechanical machine which converts the wind kinetic energy into mechanical energy. If the mechanical energy is then converted to electricity, the machine is called a wind generator, wind turbine, wind power unit (WPU), wind energy converter (WEC) or aerogenerator. Wind turbines can be separated into two types based on the axis in which the turbine rotates. Turbines that rotate around a horizontal axis are more common. Vertical-axis turbines are less frequently used. Horizontal-axis wind turbines (HAWT) have the main rotor shaft and electrical generator at the top of a tower and must be pointed into the wind. Most have a gearbox, which turns the slow rotation of the blades into a quicker rotation that is more suitable to drive an electrical generator. Since a tower produces turbulence behind it, the turbine is usually pointed upwind of the tower. Turbine blades are made stiff to prevent the blades from being pushed into the tower by high winds. Additionally, the blades are placed at a considerable distance in front of the tower and are sometimes tilted up a small amount. The horizontal-axis wind turbines are the most used today for electricity generation, operate on the aerodynamic forces, not on trust forces, that develop when wind flows around a blade of aero-foil design. Actually, the windmills that work on thrust forces are less efficient. The wind stream at the top of the aero-foil has to traverse a longer path than that at the bottom, leading to a difference in velocities, giving rise to a difference in pressure (Bernoulli's principle), and a lift force is produced. There is also a drag force that tries to push the aero-foil back in the direction of the wind. The aggregate force is determined by the resultant of these forces.

Vertical-axis wind turbines (or VAWTs) have the main rotor shaft arranged vertically. Key advantages of this arrangement are that the turbine does not need to be pointed into the wind to be effective. This is an advantage on sites where the wind direction is highly variable. VAWTs can utilize winds from varying directions. With a vertical axis, the generator and gearbox can be placed near the ground, so the tower doesn't need to support it, and it is more accessible for maintenance. Drawbacks are that some designs produce pulsating torque. Drag may be created when the blade rotates into the wind. Savonius rotors are very simple vertical-axis wind energy turbines. The basic equipment is a drum cut into two halves vertically. The two parts are attached to the two opposite sides of a vertical shaft. The wind blowing into the assembly meets two different surfaces – convex and concave – and different forces are exerted on them, giving torque to the rotor. Providing a certain overlap between drums increases the torque because wind blowing on the concave side turns around and pushes the inner surface of the other drum, which partly cancels the wind thrust on the convex side. In a Darrieus wind turbine, two or more flexible blades are attached to a vertical shaft. The blades bow outward taking the shape of a parabola and are of a symmetrical aero-foil section. When the rotor is stationary and no torque is produced, it must be started by some external means. HAWT is based on the wind rotor configuration with respect to the wind direction that are classified as upwind and downwind wind turbines. The majority of horizontal-axis wind turbines used today are upwind turbines, in which the wind rotors face the wind. The main advantage of upwind designs is to avoid the distortion of the flow field as the wind passes though the wind tower and nacelle. For a downwind turbine, wind blows first through the nacelle and tower and then the rotor blades. This configuration enables the rotor blades to be made more flexible without considering tower strike. However, because of the influence of the distorted unstable wakes behind the tower and nacelle, the wind power output generated from a downwind turbine fluctuates greatly. In addition, the unstable flow field may result in more aerodynamic losses and introduce more fatigue loads on the turbine. Furthermore, the blades in a downwind wind turbine may produce higher impulsive or thumping noise. Downwind machines have been built, despite the problem of turbulence, because they do not need additional mechanism to keep them in line with the wind, and because in high winds the blades can be allowed to bend which reduces their swept area and thus their wind resistance.

The conversion of wind energy to electricity or mechanical power involves two steps: first, the wind kinetic energy is converted by the turbine rotor, and second, the drivetrain transfers the mechanical power to a load, such as an electric generator. The amount of power of any rotating mechanical device is the product of the torque and angular velocity, being the power available at its shaft. Most operations of transferring shaft power try to have a large angular velocity because of structural considerations. The power coefficient is the power delivered by the device divided by the power available in the wind. Since the area cancels out, the power coefficient, C_P, in the case of wind turbine, considering the input power as specified in Equation (4.1), the power available in the wind is then expressed as:

$$C_p = \frac{\text{Power Out}}{\text{Power In}} = \frac{\text{Power Out}}{0.5 \cdot \rho \cdot A \cdot v^3} \qquad (4.34)$$

A maximum value of the power coefficient is defined by the Betz limit, stating that a wind turbine can extract up to 59.3% of the power from an air stream. However, wind turbine rotors have maximum Cp values in the range 25–45%. Applying the fluid mechanic principals, conservation of energy and momentum equation can be determined that the output power of a wind turbine is expressed as:

$$P_{WT} = 0.5\rho A v^3 a(1-a)^2 \qquad (4.35)$$

Here, a is the fractional decrease in the wind velocity once it has reached the rotor due to a change in pressure (depending on how much energy the rotor captured to slow the wind). We can define the performance power coefficient, C_P, as the ratio of the power in the rotor to the power in the wind:

$$C_P = 4a(1-a)^2 \qquad (4.36)$$

The power coefficient indicates the efficiency of the turbine based solely on the stream tube concept, without accounting for non-ideal conditions and inevitable losses from the blades, the mechanics, turbine generator, control and wind conversion system electronics. Taking the derivative of the power coefficient, Equation (4.36) with respect to a, setting it equal to zero yields the axial induction factor of 1/3 which maximizes the efficiency. At this value of a, the power coefficient equals $16/27 \approx$ 0.59, the maximum extractable raw incoming wind kinetic energy (the Betz's Limit). The wind turbine power curve displays the power output as a function of the mean wind speed. Common ways to characterize wind turbine performances is expressing them through non-dimensional characteristics performance curve. Power curves are usually determined from field measurements. The tip-speed ratio, λ, is a variable relating the peripheral blade speed and wind speed, computed as:

$$\lambda = \frac{\omega R}{v} \tag{4.37}$$

The tip-speed ratio, λ, and the power coefficient, C_P, are dimensionless and are used to describe the performance of a wind turbine rotor. Then, the power extracted by wind turbine is expressed as:

$$P_{WT} = 0.5\rho \cdot A \cdot C_P(\lambda) \cdot v^3 \tag{4.38}$$

Figure 4.3a shows that the maximum power coefficient is only achieved at a single tip-speed ratio and for a fixed rotational speed of the wind turbine, this only occurs at a single wind speed. Hence, one argument for wind turbine operating at variable rotational speeds is that it is possible to operate at maximum Cp over a wind speed range. The wind turbine power output at various wind speeds is described by a power curve, giving the steady-state power output, function of the wind speed at the hub height, as shown in Figure 4.3b. The power curve has three key points on the velocity scale:

- Cut-in wind speed, the minimum wind speed at which the machine will deliver useful power.
- Rated wind speed, the wind speed at which rated power is obtained (the maximum power output of the electrical generator).
- Cut-out wind speed, the maximum wind speed at which the turbine is allowed to operate, limited by safety constraints.

As mentioned in above paragraph, the mechanical energy captured by the wind turbine blades is further converted in electrical energy via turbine electric generator. In this stage the converted efficiency is determined by several parameters, such as: gearbox efficiency, η_{gear}, generator efficiency, η_{gen}, and electrical transmission efficiency η_{ele}, counting for all losses in power electronics, converter, switches, control and cables. The overall power conversion efficiency, η_{tot} is then expressed as:

$$\eta_{tot} = C_P(\lambda) \cdot \eta_{gear} \cdot \eta_{gen} \cdot \eta_{ele}$$

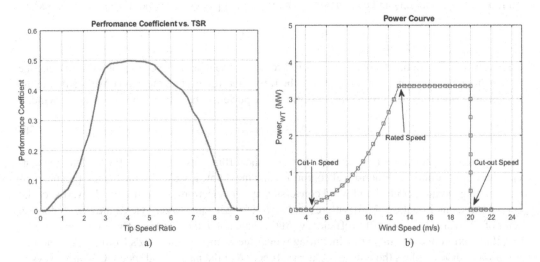

FIGURE 4.3 (a) Power coefficient vs. TSR and (b) typical large wind turbine power curve.

The effective wind turbine power output becomes:

$$P_{eff} = \eta_{tot} \cdot 0.5\rho A v^3 \tag{4.39}$$

Example 4.9 An 8 m/s wind enters a wind turbine rotor with a diameter of 36 m. Assume the air density 1.2 kg/m³. Calculate: (a) the power of incoming wind, (b) the theoretical maximum power extracted by the wind turbine and (c) if the gearbox, generator and electrical transmission and processing efficiencies are 0.85, 0.96 and 0.95, respectively, what is the turbine effective converted power.

SOLUTION

a. The incoming wind power is

$$P_{wind} = 0.5\rho A v^3 = 0.5 \cdot 1.2 \cdot \left(\frac{\pi}{4} 36^2\right) \cdot 8^3 = 312533 \text{ W}$$

b. The maximum extracted power, the Betz limit is:

$$P_{max} = 0.59 \cdot P_{wind} = 0.59 \times 312533 = 184394.5 \text{ W}$$

c. The effective power is then

$$P_{eff} = 0.85 \cdot 0.96 \cdot 0.95 \cdot P_{max} = 142942.6 \text{ W}$$

Critical components of WTG system include the rotor, gearbox (not all WT has one, but most of them), anemometer, generator, electric transmission, control system, tower and foundation. The HAWT are either rotor-upwind, facing the wind or rotor-downwind types, enabling the wind to pass the tower and nacelle before entering the rotor. The rotor diameter, number and twist angle of blades, tower height, rated electrical power and control strategy are the main factors considered in design. HAWTs usually have two or three blades. In order to improve the power output performance, a selection of ratio between the rotor diameter and the hub height need to be considered. Two-blade rotors are faster and cheaper, but three-blade ones operate more smoothly, with less flickers and higher efficiency. The tower height is also an important parameter regarding the WT performance. Higher towers offer usually more wind, larger rotors, higher power output on the trade-off of increased overall cost and installation. There are various types of towers for various classes of wind turbines. Foundation is important aspect of WT installation, requiring careful assessment of structural loads, materials, construction, geotechnical parameters, tower flange dimensions and serviceability requirement. The rotor is one the most critical element of a wind turbine. Downwind and upwind rotors are best suited for high-capacity wind turbines, operating at higher tip-speed ratio. To achieve dynamic stability, safe and reliable operation under various wind conditions, a complex control system is required. Pitch control, is an expensive subsystem, located inside the hub where it rotates around radial axis as wind velocity changes. It changes the attack angle, by pitching the blades for almost optimal adjustment for every wind speed, improving the dynamic stability and increasing the wind power capture. There are two types of pitch control: hydraulic and electromechanical. A yaw control is also needed to maintain the system dynamic stability in turbulent environments. The nacelle is rotating with respect to the tower not with respect to the rotor, and this rotation is provided by the yaw system. This is necessary because the wind direction is not fixed. The yaw system directs the rotor in respect to the wind and consists of yaw bearing, motor and drive. The yaw stem is either active or passive, determined by the rotor type, upwind or downwind. Gearbox, another heavy turbine component is used to increase the rotational speed from a low-speed rotor to high-speed electric generator. Proper gearbox maintenance is required in order to reduce the operation cost, being expensive in repair or replace it. Its lifetime is strongly affected by the wind regime. Wind turbines can run efficiently with minimum maintenance if the wind is smooth and there are less turbulence effects. The nacelle is the housing located at the tower top and serves as protection for some wind energy system components, such as generator, gear box and control and must be strong enough to handle the loads. Wind turbines must also be able to stop in case of failure of critical components or if the wind speed is higher than critical limit. There are two types of breaking systems: aerodynamic and mechanical breaking. Electrical generators convert mechanical power into electrical power with the most common are induction and synchronous generators, first type being the most used in wind energy conversion. Induction generators are reliable and not very expensive, and having also mechanical properties that suit wind turbines. The most common type of induction generator rotor is squirrel-cage system. The generator speed changes according to the rotor speed, putting less stress on the tower, gearbox and other components in the transmission lines or lower peak torque, and being important reasons for choosing induction generators rather than synchronous type. The synchronous generator is dependent on the speed of rotation, so the stator is connected to a DC-link converter system. This is increasing the system complexity and cost. It can operate at a wide range speed.

4.4   SOLAR ENERGY

The potential solar resource reflects the ubiquitous nature of sunlight and yields a huge technical resource. Solar resource is exploited through solar photovoltaic and solar thermal panels and by heat pumps. While the potential solar resource is large, solar energy is currently relatively costly to extract and deployment rates depend on economics. The practical resource for solar PV and thermal technologies is usually associated with buildings with suitable south-facing roofs and is therefore a function of the built environment rather than the resource itself. The choice of whether solar PV or solar thermal technology is installed (or a combination thereof) is a decision taken building-by-building. Solar radiation amount received by a given surface is controlled at the global scale by the geometry of the Earth, atmospheric transmittance and the relative Sun location. At the local scale, solar radiation is controlled by surface slope, aspect and elevation. Clear sky solar radiation estimates for sloped surfaces are very important in renewable energy, civil engineering and agricultural applications, which need accurate estimate of total energy striking a given surface. Table 4.3 provides a list of day numbers for the first day of each month of the year, needed in solar radiation calculations.

Solar energy is in the form of electromagnetic radiation with wavelengths ranging from approximately 0.3 μm (10^{-6} m) to over 3 μm, which correspond to ultraviolet (less than 0.4 μm), visible (0.4 and 0.7 μm) and infrared (over 0.7 μm). Most of this energy is concentrated in the visible and the near-infrared wavelength range. The Sun-generated energy divided by the Sun surface area gives the Sun specific emission, 63.11 MW/m^2, the radiant power per square meter. A sphere with the radius equal with the average Sun-Earth distance (1.5×10^8 km) receives the same total radiant power as

TABLE 4.3
Day Number of the First Day of Each Month

Month	N	Month	N
January	1	July	182
February	32	August	213
March	60	September	244
April	91	October	274
May	121	November	305
June	152	December	335

the Sun's surface. This value determines the extraterrestrial radiance at the top of the Earth atmosphere. The extraterrestrial solar radiation also varies due to the Earth's elliptical orbit around the Sun. The total energy flux incident on the Earth surface is obtained by multiplying SC by πR^2 (Earth disk are), where R is the Earth radius. The average flux incident on a unit surface area is then:

$$\frac{SC}{4} = 342 \, W/m^2 \tag{4.40}$$

4.4.1 SOLAR RESOURCE

The ultimate energy source for the vast majority of the energy systems is the Sun. Knowledge of the quantity and quality of solar energy available at a specific location is of prime importance for the design of any solar energy system. Although the solar radiation (*insolation*) is relatively constant outside the Earth's atmosphere, local climate influences can cause wide variations in available insolation on the Earth's surface from site to site. There are regions on the Earth of high or very high insolation where solar energy conversion systems are expected to produce the maximum amount of energy from a specific collector field size or type. It is the primary task of the solar energy system designer to determine the amount, quality and timing of the solar energy available at the site selected for installing a solar energy conversion system. Just outside the Earth's atmosphere, the sun's energy is continuously available at the rate of 1,367 W/m^2. Due to the Earth's rotation, asymmetric orbit about the sun and the contents of its atmosphere, a large fraction of this energy does not reach the ground. Earth's axis is tilted 23.5° with respect to the plan of its orbit around the Sun. This tilting results in longer days in the northern hemisphere from the spring equinox (approximately March 23) to the autumnal equinox (approximately September 22) and longer days in the southern hemisphere are during the other six months. On the equinoxes, the Sun is directly over the equator, both poles are equidistant from the Sun and the Earth experiences 12-hour daylight and 12-hour darkness. In the temperate latitude regions ranging from 23.450 to 66.50 north and south, variations in insolation are large. Sun's position in the sky is a function of time and latitude, being defined by its solar altitude and solar azimuth angles. Sun's position relative to a location is determined by the location's latitude, L, a location's hour angle, W, and the sun's declination angle. Latitude is the angular distance north or south of the Earth's equator, measured in degrees along a meridian. The hour angle is measured in the equatorial plane. It is the angle between the projection of a line drawn from the location to the Earth's center and the projection of a line drawn from the center of the Earth to the sun's center. Thus, at solar noon, the hour angle is zero. At a specific location, the hour angle expresses the time of day with respect to solar noon, with one hour of time equal to 15 degrees angle. By convention, the westward direction from solar noon is positive. The Sun's declination is the angle between projection of the line connecting the center of the Earth with the center of the sun and the Earth's equatorial plane. Declination varies from −23.45° on the winter solstice (December 21), to +23.45° on the summer solstice (June 22). Approximate estimates of declination angle, used in practical application are given by the following relationships:

$$\delta = 23.45 \cdot sin\left[360 \cdot \frac{284 + n}{365}\right] \tag{4.41}$$

And

$$\delta = arcsin\left[0.4 \cdot sin\left(\frac{360}{365}(n - 81)\right)\right] \tag{4.42}$$

Or

$$\delta = 23.45 \cdot sin\left[\frac{360}{365}(n-81)\right] \qquad (4.43)$$

Solar declination as a function of Julian day number is shown in Figure 4.4. Solar altitude angle (α) defines the elevation of the sun above the location horizon. In the following, the term zenith refers to an axis drawn directly overhead at a site. The solar altitude is related to the solar zenith angle (θ_z), the angle between the Sun rays and the vertical, being calculated by using the following relationships:

$$\theta_z + \alpha = \frac{\pi}{2} = 90^{o}$$

$$\cos(\theta_z) = \sin(\alpha_S) = \sin(L) \cdot \sin\delta + \cos(L) \cdot \cos(H_S) \cdot \cos\delta \qquad (4.44)$$

Here, L is the local latitude, δ is the declination angle (Equations (4.42) or (4.43)) and H_S is the solar hour angle (i.e., the angular distance between the sun and the local meridian line). In other words, this is the difference between the local meridian and the Sun meridian, with positive values occurring in the morning before the Sun is crossing the local meridian and negative values in the afternoon. Solar azimuth angle, the angle between the Sun and true north is then given by:

$$\sin\alpha_S = \frac{\cos(\delta) \cdot \sin(H_S)}{\cos(\alpha)} \qquad (4.45)$$

As, at the solar noon, by definition, the Sun is exactly on the meridian (north-south line), and consequently, the azimuth angle is 0°. Therefore, the noon latitude angle (also known as the altitude angle), α_n is then given by:

$$\alpha_n = 90^{o} - L + \delta \qquad (4.46)$$

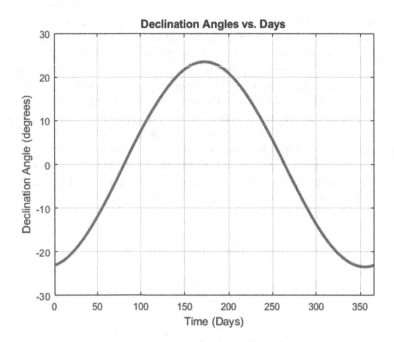

FIGURE 4.4 Declination angle vs. Julian date.

During an equinox, at solar noon, the Sun is directly over the local meridian (line of longitude), the solar rays are striking a solar collector at the best possible angle, perpendicular to the collector face. At other times of the year, the Sun is a little high or a little low for normal incidence. However, on the average it seems to be a good tilt angle. Solar noon is an important reference point for almost all solar calculations. In the Northern Hemisphere, at latitudes above the Tropic of Cancer, solar noon occurs when the Sun is due south of the observer. South of the Tropic of Capricorn, the opposite, it is when the Sun is due north, while in the tropics, the Sun may be either due north, due south or directly overhead at solar noon. On average, facing a collector toward the equator (facing it south in the Northern Hemisphere) and tilting it up at an angle equal to the local latitude is a good rule-of-thumb for better annual performances. The tilt angle that would make the Sun's rays perpendicular to the module at noon is given by:

$$Tilt = 90 - \alpha_n \qquad (4.47)$$

Example 4.10   Find the optimum tilt angle for a south-facing photovoltaic module at latitude 32.3° at solar noon on May 1.

SOLUTION

From Table 4.3 for May 1, $n = 121$ and the declination angle (by using Equation (4.43)), is then:

$$\delta = 23.45 \sin\left[\frac{360}{365}(121 - 81)\right] = 14.9°$$

Using Equations (4.46) and (4.47) the tilt angle of the photovoltaic panel, facing south is:

$$\alpha_N = 90° - 32.3° + 14.9° = 72.6°$$
$$Tilt = 90° - \alpha_N = 17.4°$$

4.4.2 Photovoltaics

Photovoltaic (PV) cells or solar cells are used to directly convert the solar energy (radiation) into electricity through the photoelectric effect. Solar electric energy conversion systems, or PV systems, are cost-effective and viable solutions to supply electricity for locations not connected to the conventional electrical grid or for special applications. PV systems are utilized almost everywhere, for terrestrial and space applications, and from Tropical to Polar Regions. However, the still higher PV capital cost means it is most economical to employ them for remote sites or applications where other, more conventional power generation options are not competitive. Solar cells are made from a variety of semiconductor materials and coated with special additives. The most widely used material for the various types of fabrication is crystalline silicon. A typical silicon cell, with a diameter of 4 in, can produce more than 1 W of direct current (DC) electrical power in full Sun (1000 W/m² solar radiation intensity). Individual solar cells can be connected in series and parallel to obtain desired voltages and currents. Silicon PV cells manufactured today can provide over 40 years of useful service life, with the average lifecycle of PV modules of about 25 years. Large scale PV applications for power generation, either on the house rooftops or in large fields connected to the utility grid are promising electricity generation option, clean, reliable, safe and strategically sound alternatives to current methods of electricity generation.

Solar cell is the component responsible for converting solar radiation into electricity. Some materials, silicon being the most common, can produce a PV effect, consisting of freeing electrons, when sunlight is striking the cell material. A PV module is composed of interconnected solar cells, encapsulated between a glass cover and weatherproof backing. The modules are typically framed in aluminum frames suitable for mounting and protection. PV modules are connected in series and parallel to form PV arrays, thus increasing total available power output to the needed voltage and current for a particular application. PV modules are rated by their total power output (W). A peak Watt is the amount of power output a PV module produces at standard test condition (STC): 25°C operating temperature and full noon time sunshine (irradiance) of 1,000 W/m². However, PV modules often operate at temperatures higher than 25°C in all but cold climates, thus reducing crystalline module operating voltage and power by about 0.5% for every 1°C above STC. Therefore, a 100 W module operating at 45°C (20° hotter than STC, yielding a 10% power drop), produces about 90 W. Amorphous PV modules do not have this effect. PV cells have been made with silicon (Si), gallium arsenide (GaAs), copper indium diselenide (CIS), cadmium telluride (CdTe) and a few other materials.

The common denominator of PV cells is that a *p-n* junction, or the equivalent, is needed to enable the photovoltaic effect. Understanding the *p-n* junction is thus critical for understanding how a PV cell converts sunlight into electricity and how a PV system operates. The main parameters used to characterize the performance of solar cells are: the peak power P_{max}, the short-circuit current density J_{SC} or short-circuit current I_{SC}, the open-circuit voltage V_{OC} and the fill factor FF. These parameters are determined from the illuminated *I-V* characteristic, as shown in Figure 4.5. The conversion efficiency, η_{pv}, is determined from these parameters. Short-circuit current, I_{SC}, is the current that flows through the external circuit when the electrodes of the solar cell are short-circuited. The short-circuit current of solar cells depends on the photon flux density incident on the solar cell, determined by the spectrum of the incident light. For standard solar cell measurements, the solar spectrum is standardized to the AM1.5 spectrum. The I_{SC} depends on the area of the solar cell. In order to remove the dependence of the solar cell area on I_{SC}, often the short-circuit current density is used to describe the maximum current delivered by a solar cell. The maximum current that the solar cell can deliver strongly depends on the optical properties of the solar cell, such as absorption in the absorber layer and reflection. The

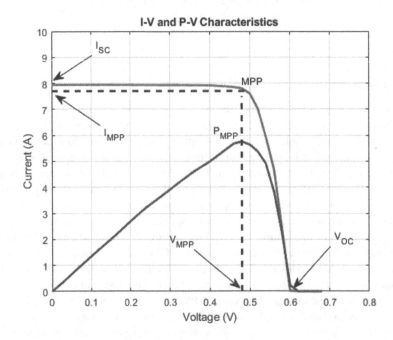

FIGURE 4.5 Solar cell I-V characteristics and output power vs. voltage.

gross current generated by a solar cell, I_L (the light current), since it occurs when the cell is illuminated is calculated taking into account the losses occurring in the cell. When a solar cell is connected to an external circuit, the photo-generated current then flows from the p-type semiconductor-metal contact, through the conductor loop, powers the load, until it reaches the n-type semiconductor-metal contact. Under a certain sunlight illumination, the current passed to the load from a solar cell depends on the external voltage applied to the solar cell normally through a power electronic converter for a grid-connected PV system. If the applied external voltage is low, only a low photo-generated voltage is needed to make the current flow from the solar cell to the external system. Nevertheless, if the external voltage is high, a high photo-generated voltage must be built up to push the current flowing from the solar cell to the external system. This high voltage also increases the diffusion current, so that the net output current of the solar cell is reduced. The I-V characteristic curve for the p–n junction diode is described by the (Shockley) diode equation, diode current, I_D, expressed as:

$$I_D = I_0\left(\exp\left(\frac{qV_D}{mkT}\right) - 1\right) \qquad (4.48)$$

Here, I_0 is the reverse saturation current (A), and V_D is the forward diode voltage, T is the junction temperature (K), $k = 1.381\times10^{-23}$ J/K, Boltzmann constant, $q = 1.605\times10^{-19}$ C the elementary (electron) charge, and m, the diode factor depends on the voltage at which the cell is operating. The diode factor m is equal to 1 for an ideal diode; however, a diode factor between 1 and 5 allows a better description of PV cell characteristics. The so-called thermal voltage $V_T = kT/q$ has a value of 25.7 mV at 25°C (STC) and the magnitude of the saturation current I_0 is of the order of 10^{-10} to 10^{-5} A. For standard temperature of the junction, 25°C Equation (4.72), and ideal diode (m = 1) has the form:

$$I_D = I_0(\exp(38.9\cdot V_D) - 1) \qquad (4.49)$$

The dark current, I_D, is flowing in the opposite direction of the photovoltaic (light) current, I_L, so the net diode current is computed as:

$$I = I_L - I_D = I_L - \left[\exp\left(\frac{qV_D}{mkT}\right) - 1\right] \qquad (4.50)$$

Plotting current I vs voltage V, using Equation (4.74) for the representative cell parameters and the insulation level the I-V diagram (Figure 4.5) is obtained. I-V curve typically passes through the two end points: the short-circuit current, I_{SC}, and the open-circuit voltage, V_{OC}. I_{SC} is the current produced with the positive and negative terminals of the cell shorted, the voltage between the terminals is zero, corresponding to zero load resistance. The V_{OC} is the voltage across the positive and negative terminals under open-circuit conditions with no current, corresponding to infinite load resistance, and the peak power point is located on the farthest upper right corner of where the rectangular area is greatest under the curve. At zero voltage, the amount of current produced is the *short-circuit current, I_{SC}*, which is equal to the light current since the dark current is making no contribution. The *open-circuit voltage V_{OC}*, the voltage at which no current flows (due to the exponential nature of I_D) through the external circuit is the maximum voltage that a solar cell can deliver, depending on the photo-generated current density. V_{OC} is calculated, assuming that the net current is zero, Equation (4.50):

$$V_{OC} = \frac{mkT}{q}\ln\left(\frac{I_L}{I_0} + 1\right) \approx \frac{mkT}{q}\ln\left(\frac{I_L}{I_0}\right) \qquad (4.51)$$

For 25°C, the standard junction temperature the above equation becomes:

$$Voc = 0.0257 \ln\left(\frac{I_L}{I_0} + 1\right) \qquad (4.52)$$

Equation (4.51) shows that V_{OC} depends on the saturation (reverse diode) current of the solar cell, I_0, and the photo-generated current, I_L. The photo-generated current density J_{ph}, typically has a small variation, key effect being the saturation current, since it may vary by orders of magnitude. The saturation current density, J_0, depends on the recombination processes in the solar cell, so V_{OC} is a measure of the amount of recombination in the device. Laboratory crystalline silicon solar cells have a V_{OC} of up to 720 mV under the standard $AM1.5$ conditions, while commercial solar cells typically have V_{OC} exceeding 0.6 V, depending on the recombination in the solar cell.

Example 4.11   An ideal PV cell has a saturation current of 10^{-8} A and is operation at 35°C. Find the open-circuit voltage, assuming that light current is 600 mA.

SOLUTION

The short-circuit current is equal to the light current at zero voltage:

$$I_{SC} = I_L = 600 \, mA$$

Assuming the ideal diode (m=1) in Equation (4.51), the open-circuit voltage is:

$$Voc = \frac{1.38 \times 10^{-23}(35 + 273.15)}{1.602 \times 10^{-19}} \ln\left(\frac{600 \times 10^{-3}}{10^{-8}} + 1\right) = 0.4755 \, V$$

The net current can be written in terms of the fixed cell parameters I_{SC} and V_{OC} and independent variable, V, the diode voltage. We are now able to determine the values V_{mp} and I_{mp} (the so-called Maximum Power Point, MPP on the I-V curve, as shown in Figure 4.6) that maximize the solar cell power output, P_{max}. The PV cell may be operated over a wide range of voltages and currents, by varying the load resistance from zero (a short circuit) to infinity (an open circuit), it is possible to determine the highest efficiency as the point where the cell delivers maximum power. Because power is the product of voltage and current, the maximum-power point (P_{max}) occurs on the I-V curve where the product of current (I_{mp}) and voltage (V_{mp}) is a maximum. No power is produced at the short-circuit current or at open-circuit voltage, so maximum power generation is expected to be between these points. Keep in mind that the power a PV cell delivers to a load

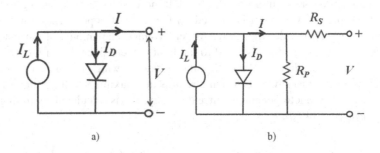

a)   b)

FIGURE 4.6   Solar cell equivalent circuit: (a) simple equivalent circuit; (b) extended one diode equivalent circuit model.

depends also on the load resistance. The optimum operating point occurs at V_{mp} and I_{mp}. The relationship between the maximum power $P_{max} = V_{mp} \cdot I_{mp}$ and the product of open-circuit voltage and short-circuit current is referred as the fill factor, FF, expressed as:

$$FF = \frac{P_{max}}{V_{OC} \cdot I_{SC}} \tag{4.53}$$

Assuming that the solar cell behaves as an ideal diode, the fill factor can be expressed as a function of open-circuit voltage V_{OC}, as given by:

$$FF = \frac{v_{OC} - \ln(v_{OC} + 0.73)}{v_{OC} + 1} \tag{4.54}$$

Here, $v_{OC} = \dfrac{q \cdot V_{OC}}{kT}$ is the normalized voltage. Equation (4.54) is a good approximation for normalized voltage values higher than 10. However, FF does not change drastically with a change in V_{OC}, because large variations in V_{OC} are not common. For example, at standard illumination condition, a typical commercial solar cell's maximal FF is about 0.85. The conversion efficiency is calculated as the ratio between the maximal generated power and the incident power. The irradiance value P_{in} of 1000 W/m² for the AM1.5 spectrum has become a standard for measuring the conversion efficiency of solar cells, is:

$$\eta_{pv} = \frac{P_{max}}{P_{IN}} = \frac{V_{OC} I_{SC} \cdot FF}{P_{IN}} \tag{4.55}$$

Typical conversion efficiency lies in the range of 15–20% for commercial solar cells. An ideal photovoltaic cell can be described by a current source in parallel with diode. This simple equivalent circuit is well suited to describe the behavior of an irradiate solar cell. This simple equivalent circuit (as shown in Figure 4.6a) is sufficient in many applications. The current source generates the light (photocurrent), I_L, which depends on the irradiance (solar radiation intensity), E and a coefficient, C_0 as:

$$I_L = C_0 \cdot E \tag{4.56}$$

An ideal solar cell can be modeled by a current source, representing the photo-generated current I_L, in parallel with a diode, representing the ideal p-n junction of a solar cell. In a real solar cell, there exist other effects not accounted for by the ideal model. The differences between calculated and measured characteristics of the solar cells are in the range of few percent. However, only the extended solar cell equivalent circuit can describe its behavior over an extended range of operating conditions. Charge carriers in an actual solar cell are experiencing voltage drop on their way through the junction to external contacts. A series resistance, R_S, expresses this voltage drop, and in addition a parallel resistance, R_P, is included to describe the leakage currents at the cell edges. These two extrinsic effects are summarized as: (1) current leaks proportional to the terminal voltage of a solar cell and (2) losses of semiconductor itself and of the metal contacts with the semiconductor. The first is characterized by a parallel resistance R_P accounting for current leakage through the cell, around the edge of the device and between contacts of different polarity (see Figure 4.6b). The second is characterized by a series resistance R_S, which causes an extra voltage drop between the junction voltage and the terminal voltage of the solar cell for the same flow of current. The series resistance of real cells is in the range of milliohms (mΩ), while the parallel resistance is usually higher than 10 Ω. The mathematical model of a solar cell is described by the following equations:

$$I = I_L - I_0 \left(\exp\left(\frac{qV_D}{mkT} \right) - 1 \right) - \frac{V_D}{R_P} \tag{4.57a}$$

And

$$V_C = V_D - R_S \cdot I \tag{4.57b}$$

where I_L is proportional to the sunlight intensity, m is the diode ideality factor ($m = 1$ for an ideal diode), the diode reverse saturation current I_0 depends on temperature. At 25°C and standard insolation testing condition, Equation (4.80a) becomes:

$$I = I_L - I_0\left(\exp\left(38.9 \cdot V_D\right) - 1\right) - \frac{V_D}{R_P} \tag{4.58}$$

Important solar cell characteristics are the output current, power and output voltage. Several factors affect the PV cell, causing variations from the theoretical behavior. The most important factors are the temperature and the solar radiation. Increasing the solar irradiance increases the magnitude of the light current and consequently increases the short-circuit current and the one-circuit voltage, so the cell output power. Cell temperature affects linearly the thermal voltage V_T, while the saturation current, I_0 and the light current, I_L have nonlinear temperature dependence. The net result is that the open-circuit voltage is reduced when the temperature increases. However, cell performances vary in temperature not only because ambient temperatures change but also because insolation on the cells changes. Even the cell current increases with the temperature, V_{OC} falls with temperature, leading to lower power output because the voltage dominates. An indicator of the temperature effects on the solar cell is the nominal operating cell temperature (NOCT), the cell temperature at 20°C, solar irradiance of 800 W/m² and wind of 1 m/s. To account for other ambient conditions, the following expression may be used:

$$T_{PV\ Cell} = T_{amb} + \frac{NOCT - 20\ °C}{800} \cdot S \tag{4.59}$$

where $T_{PV\ cell}$ is cell temperature (°C), T_{amb} is ambient temperature (°C) and S is solar insolation (W/m2). For Si solar cell the open-circuit voltage, V_{OC} drops by 0.37% per Celsius degree increases, while the output power drops about 0.5% per degree.

Example 4.12   Estimate cell temperature, open-circuit voltage and maximum power output for a 150 W PV module operating at 32°C and 1000 W/m² insolation. The NOCT of this module is 48°C and the open circuit voltage is 43.5 V.

SOLUTION

The cell operating temperature is:

$$T_{PV\ Cell} = 32 + \frac{48 - 20}{800} \cdot 1000 = 67\ °C$$

The open circuit voltage and power are computed as:

$$V_{OC} = 43.5 \cdot (1 - 0.0037(67 - 25)) = 37.74\,V$$

And

$$P_{OUT} = 150 \cdot (1 - 0.005 \cdot (67 - 25)) = 118.5\,W$$

4.4.2.1 PV Cell Manufacturing Technologies

The most common material for the production of solar cells is silicon. Silicon is obtained from sand and is one of the most common elements in the Earth's crust, so there is no limit to the availability of raw materials. Current solar cell manufacturing technologies are: monocrystalline, polycrystalline, bar-crystalline silicon and thin-film technology. Cells made from crystal silicon (Si) are made of a thinly sliced piece (wafer), a crystal of silicon (mono-crystalline) or a whole block of silicon crystals (multi-crystalline); their efficiency ranges between 12% and 19%. Mono-crystalline Si cells have conversion efficiency for this type of cells ranges from 13% to 17% and can generally be said to be in wide commercial use. In good light conditions, it is the most efficient photovoltaic cell. This type of cell can convert solar radiation of 1.000 W/m^2 to 140 W of electricity with the cell surface of 1 m^2. The production of monocrystalline Si cells requires an absolutely pure semiconducting material. Monocrystalline rods are extracted from the molten silicon and sliced into thin chips (wafer). Such type of production enables a relatively high degree of usability. Expected lifespan of these cells is typically 25–30 years and, of course, as well as for all photovoltaic cells, the output degrades somewhat over the years. Multi-crystalline Si cells are converting solar radiation of 1.000 W/m^2 to 130 W of electricity with the cell surface of 1 m^2. The production of these cells is economically more efficient compared to monocrystalline. Liquid silicon is poured into blocks, which are then cut into slabs. During the solidification of materials, crystal structures of various sizes are being created, at whose borders some defects may emerge, making the solar cell to have a somewhat lower efficiency, ranging from 10% to 14%. The expected lifespan is up to 25 years. Ribbon silicon has the advantage of not needing a wafer cutting (which results in loss of up to 50% of the material in the process of cutting). However, the quality and the possibility of production of this technology is not making it a leader in the near future. Their efficiency is around 11%.

In the thin-film technology the modules are manufactured by piling extremely thin layers of photosensitive materials on a cheap substrate such as glass, stainless steel or plastic. The process of generating modules in thin-film technology has resulted in reduced production costs compared to crystalline silicon technology, which is somewhat more intense. Today's price advantage in the production of a thin-film is balanced with the crystalline silicon due to lower efficiency of the thin-film, which ranges from 5% to 13%. The share of thin-film technology in the market is 15% and constantly increasing. It is also expected an increase in years to come and thus reduce the adverse market ratio in relation to the photovoltaic module of crystalline silicon. Lifespan is around 15–20 years. There are four types of thin-film modules (depending on the active material) that are now in commercial use. Amorphous Si Cells, with efficiency is around 6%, a cell surface of 1 m^2 can convert 1.000 W/m^2 of solar radiation to about 50 W of electric energy. Progresses in research of this type of module have been made and it is expected a greater efficiency in the future. If a thin film of silicon is put on a glass or another substrate, it is called amorphous or thin layer cell. The layer thickness is less than 1 μm, therefore the lower production costs are in line with the low cost of materials. However, the efficiency of amorphous cells is much lower compared to other cell types. It is primarily used in equipment where low power is needed or, more recently, as an element in building facades. Cadmium tellurium (CdTe) cells, with an efficiency is around 18%, a cell surface of 1 m^2 can convert solar radiation of 1.000 W/m^2 to 160 W of electricity in laboratory conditions. Cadmium teleurid is a fusion of metal cadmium and tellurium semimetal. It is suitable for use in thin photovoltaic modules due to the physical properties and low-technology manufacturing. However, it is not widely used due to cadmium toxicity and suspected carcinogenicity. Copper indium gallium selenide (CIS, CIGS) cells have the highest efficiency among the thin-film cells, which is about 20%. This cell type can convert solar radiation of 1.000 W/m^2 to 160 W of electricity in laboratory conditions. Thermo-sensitive solar cells and other organ cells (DSC) are still in the development stage, since it is still testing and it is not increasingly commercialized. Cell efficiency is around 10%. The tests are going in the direction of using the facade integrated systems, which has proven to be high-quality solutions in all light radiation and all temperature conditions. Also, a great potential of this technology is its low cost compared to silicon cells. There are other types of photovoltaic technologies that are still developing or about to be commercialized. Regardless of the lifespan, the warranty period of most common commercial PV modules is 10 years at 90% power output and 25 years at 80% power output. The period of energy depreciation of PV cells is period that must pass using a photovoltaic system to return the energy that has been invested in the construction of all parts of the system, as well as the energy required for the breakdown after the lifetime of a PV system. Of course, the energy depreciation time is different for different locations at which the system is located, thus it is a lot shorter on locations with a large amount of irradiated solar energy, up to 10 or more times shorter than its lifetime.

Since an individual cell produces only about 0.5–0.7 V and about 1 W power output, there are rare applications for which just a single cell is needed to provide the required power. To increase the power ratings, the PV cells are connected in series and parallel configurations. The series connection increases the overall output voltage, while the parallel connection increases the overall output current. The interconnected PV cells are called PV module or panel, the basic building block for PV applications. A typical PV module consists of a number of pre-wired PV cells in series, all encased in tough, weather-resistant packages. A typical PV module has 36 cells in series, designated as a "12-V PV module", even though it is capable of delivering higher voltages than that. Some 12 V modules have only 33 cells, which, as will be seen later, may be desirable in certain very simple battery charging systems. Large 72-cell modules are now quite common, some of which have all of the cells wired in series, in which case they are referred to as 24 V modules. Some 72-cell modules can be field-wired to act either as 24 V modules with all 72 cells in series or as 12 V modules with two parallel strings having 36 series cells in each. Multiple modules can be wired in series to increase voltage and in parallel to increase current, to provide the required power. The interconnected PV modules form a PV array, while several PV arrays forms a PV system. An important element in PV system design is deciding how many modules should be connected in series and how many in parallel to deliver whatever energy is needed. Such combinations of modules are referred to as a PV array. Figure 4.7 shows this distinction between PV cells, PV modules and PV arrays. Several of these PV arrays form a PV system. In order to maximize the power output of a PV system tracking devices to follow the Sun throughout the day are mounted, they are tilting the PV arrays to maximize the solar cell exposure to the solar radiation, thus increasing the system power output.

A PV module *I-V* curve has the same set of operation points as solar cells, which are critical in order to properly install and troubleshoot PV power systems: short-circuit current (I_{SC}), the maximum current generated by a PV module and is measured when no load (resistance) is connected (i.e., the module is shorted). Its value depends on the cell surface area and the solar radiation

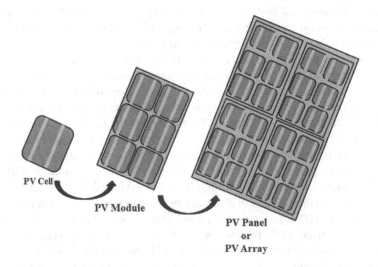

FIGURE 4.7 Diagrams of PV cells, modules and arrays.

TABLE 4.4

Sample of PV Module Specifications

Parameters	PV Module (36 cells)	PV Module (72 cells)
Operating Point	Model BP VLX-53	NE-Q5E2U
P_{mp}	53 Wp (peak W)	165 Wp
V_{mp}	17.2 V	34.6 V
I_{mp}	3.08 A	4.77 A
V_{OC}	21.5 V	43.1 V
I_{SC}	3.5 A	5.46 A
Standard test conditions (STCs)	1,000 W/m^2, 25°C	1,000 W/m^2, 25°C

incident upon the surface. I_{SC} is used for all electrical ampacity design calculations. Nameplate current production is given for a PV cell or module at standard reporting condition (SRC). The SRC commonly used by the PV industry is for a solar irradiance of 1,000 W/m^2, a PV cell temperature of 25°C and a standardized solar spectrum referred to as an air mass 1.5 spectrum (AM = 1.5), which is the standard test condition (STC). However, in reality, unless one is using PV in a relatively cold climate, the modules are operating at higher temperatures (often 50°C or more), which reduces their power performances. As module operating temperature increases, module voltage drops while current essentially holds steady. PV module operating voltage is reduced on average for crystalline modules approximately 0.5% for every degree Celsius above STC (i.e., 25°). In general, when sizing terrestrial PV systems, we expect a 15–20% drop in module power from STC. This is important to remember when calculating daily actual energy production. Open-circuit voltage (V_{OC}) is the maximum voltage generated by the module, measured when no external circuit is connected to the PV module, similar as for the PV cells. Similarly, the rated maximum power voltage (V_{mp}) corresponds to the maximum power point on the module I-V curve. Maximum power (P_{max}) is the maximum power available from a PV module, occurring at the maximum power point on the I-V curve, the product of the PV current (I_{mp}) and voltage (V_{mp}). If a PV module operates outside its maximum power value, the amount of power delivered is reduced and represents needless energy losses. Thus, this is the desired point of operation for any PV module or system. Manufacturers are providing PV module specifications, such as ones shown in Table 4.4.

The PV cell model developed before can be used to compute the values of the PV module, array and system if the cell parameters and the environmental conditions are known. When the solar cells are connected in series and parallel, we are making the assumption that the cell parameters and the environmental conditions are the same for every cell in the module, array or system. Modules must be fabricated so the PV cells and interconnects are protected from moisture and are resistant to degradation from the ultraviolet radiation. Since the modules are usually exposed to a wide range of temperatures, they must be designed so that thermal stresses are not causing delamination. Modules must also be resistant to blowing sand, salt, hailstones, acid rain and other unfriendly environmental conditions and must be electrically safe over the long period. The electrical characteristics of the PV array are the same as the individual modules, with the power, current and voltage modified according to the number of modules connected in series, parallel or series-parallel configuration. However, the module or array efficiencies are usually less than the constituting cells or modules, unless the cells or modules are perfectly similar. For modules in series, *I-V* curves are simply added along the voltage axis. For modules in parallel, the same voltage is across each PV module and the total current is the sum of the individual currents. When high power is needed, the array usually consists of a combination of series and parallel PV modules for which the total I-V curve is the sum of the individual module I-V curves. There are two ways to imagine wiring a series/parallel combination of modules: (a) the series modules are wired as strings and the strings wired in parallel and (b) the parallel modules are wired together first and those units combined in series. The total *I-V* curve is the sum of the individual module curves, being the same in either case when everything is working. However, the wiring of strings in parallel is preferred for the reason that if an entire string is removed from service, the array is still delivering the needed voltage to the load, though the current is diminished, which is not the case when a parallel group of modules is removed. When photovoltaics are wired in series, they all carry the same current, and at any given current their voltages add. For a PV module having n cells in series, the current is simply calculated using Equations (4.57a) or (4.58), while the voltage, by using Equation (4.57b), is then:

$$V_{Module} = n(V_D - R_S \cdot I) \tag{4.60}$$

Example 4.13 **A PV module consist of 36 identical cells, all wired in series. Calculate the module voltage, current and the delivered power considering 1000 W/m² insolation. Each cell has short-circuit current I_{SC} of 3.5 A, its reverse saturation current $I_0 = 6 \times 10^{-10}$ A at 25°C, junction voltage 0.5 V, a parallel resistance R_P equal to 7.50 Ω and series resistance R_S equal to 0.005 Ω.**

SOLUTION

by using Equation (4.57b) the current is:

$$I = 3.5 - 6 \times 10^{-10} \cdot (\exp(38.9 \times 0.5) - 1) - \frac{0.5}{7.5} = 3.265 \, A$$

Then the module voltage is calculated by using Equation (4.83), as:

$$V_{Module} = 36 \cdot (0.5 - 0.005 \cdot 3.265) = 17.4 \, V$$

The power delivered by this module is:

$$P_{Module} = V_{Module} \times I = 17.4 \times 3.265 = 56.86 \, W$$

4.5 GEOTHERMAL ENERGY

Geothermal resources span a wide range of Earth's heat sources, which include easily developed, currently economic hydrothermal resources, the Earth's deeper, stored thermal energy, that is present everywhere. Conventional hydrothermal resources are used effectively for electric and/ or nonelectric applications; however, they are limited to their location and ultimate potential for supplying electricity. Earth's geothermal resources are theoretically more than adequate to supply world energy needs; however, only a very small fraction may or can be economically exploited. Geothermal energy is derived from: (1) steam trapped deep into the Earth, brought to the surface, used to drive steam turbine-generator units to produce electricity and (2) water pumped and heated through deep hot rocks, to provide heat or steam for buildings or industrial processes. Geothermal energy, in the form of natural steam and hot water, has been exploited for long time for space heating and industrial processes or to generate electricity. The Earth is giving the impression that it is dependably constant, because over the human life time scale, little seems to change. However, the Earth is quite a dynamic entity, with time scales spanning for seconds for the earthquakes, a few years that volcanoes appear and grow, over millennia that landscapes slowly are evolving and to over millions of years the continents rearrange themselves on the planet's surface. The energy source driving these processes is heat, with a constant flux from every square meter of the Earth's surface. The average heat flux for the Earth is 87 mW/m², or for the Earth surface of $5.1 \times 108 \, km^2$

is equivalent to about 4.5×10^{13} W. For comparison, it is estimated that the total annual world power demand is approximately 1.6×10^{13} W. Clearly, the geothermal energy has the huge potential to significantly contribute to the human energy needs. The geothermal energy, contained in the rock and fluid into the globe layers, is linked with the Earth's internal structure and composition and associated physical processes. Despite the fact it is present in huge, inexhaustible quantities into the Earth's crust or deeper layers, it is unevenly distributed, seldom concentrated, often at depths too great to be economically or even technological possible for exploitation. There are almost 4,000 miles from the Earth's surface to its center and the deeper it is the hotter it gets. The outer layer of the Earth, the Crust, about 35 miles thick, insulates the surface from the hot interior. The inner generated heat flows towards the surface where it dissipates, the Earth temperature increases with depth, a geothermal gradient of about 30°C/km of depth exists. During the last century, many countries started to use geothermal energy, as it becomes economically competitive with other energy sources. Moreover, the geothermal energy is in some regions, the only energy source available locally. Geothermal energy is coming from two main sources:

1. Heat that flows upward and outward across the entire Earth' surface from the very deep, mantle and core radioactive decay of uranium, thorium and potassium. However, usually this energy flux is too small to be commercially useful for any application.
2. The localized heat resulting from the movement of magma into the crust. In some areas, this localized heat, with higher temperatures and heat fluxes, can be found between the surface and about 3500 m (about 10,000 ft.) depth. Where these heat fluxes meet the requisite conditions, geothermal energy can be used for multiple purposes, power generation, providing heat or hot water for buildings or industrial processes.

Example 4.14 Estimate the available power for two 2000 km² areas, one having the average geothermal flux and the second one (an active geothermal area) has the geothermal flux of 200 mW/m².

SOLUTION

The available power for the average geothermal flux area and for the active area are:

$$P_{active} = 2 \times 10^9 \times 200 \times 10^{-3} = 400\,MW_t$$
$$P_{average} = 2 \times 10^9 \times 87 \times 10^{-3} = 164\,MW_t$$

Geothermal energy can provide heat and hot water for homes, greenhouses or industrial processes, dry vegetables or generate electricity. Some of these applications can be pursued anywhere, while others require special circumstances, being restricted to specific areas. In order to use these energy resources in a way that is both economical and environmentally sound requires that their characteristics to be known, through the description of the Earth compositional and physical structure. Earth is compositionally inhomogeneous, consisting of an iron-nickel core, a dense rocky mantle and a thin, low-density rocky crust. The Earth's radius is about 6370 km. Extending outward from Earth's center, significant systematic changes occur in both composition and rheological behavior (material physical or mechanical properties). The Earth's structure and its interior are shown in Figure 4.8, consisting of several layers, *the crust*, a relatively thin region of low-density silicates; *the mantle*, a thick region of higher-density iron-rich silicates; and *the core*, a *central region* of iron mixed with various impurities, being usually depicted as concentric spheres, in ultra-simplified schematics. However, the interfaces are likely so irregular and the boundaries so fuzzy that such a representation is misleading. The crust has continental regions, made of even lower-density aluminum-rich

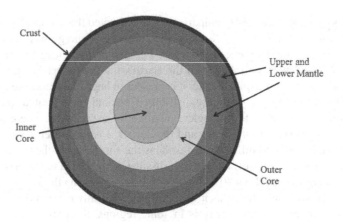

Crust

Upper and
Lower Mantle

Inner
Core

Outer
Core

FIGURE 4.8 Earth's internal structure and layers.

silicates and oceanic regions, made of denser iron-rich silicates. The mantle is divided into upper mantle within which the iron-rich silicates are gradually compressed from lower-density more open mineral structures, to higher-density more compact mineral structures, the lower mantle, where the mineral structures are compacted to their densest forms. The core, extending from the center to a depth of about 2900 km, with the temperature of about 6000°C consists of a molten (liquid) outer core layer, primarily of iron, which is lowering its melting temperature, and a solid inner core, consisting of almost certainly of a crystalline mixture of iron and nickel. Overlying the core is the mantle, made of partly rock and partly magma, which extends from a depth of about 2900 km to less than 100 km. Its volume makes up the largest part of Earth's interior, and its temperature decreases upward from about 5000°C to less than 1500°C. The last layer is the Earth's crust, consisting of a thin shell, varying from 70 to 80 km thick under the continents to less than a few kilometers thick under the ocean floor. The denser, oceanic crust is made of basalt, whereas the continental crust is often referred to as being largely granite. The crust and the uppermost mantle layer are relatively rigid and brittle. The lithosphere is divided into lithospheric plates, large blocks at continental scale or bigger. The lithosphere is about 70 km thick beneath the oceans and up to 125 km thick beneath the continents. The lithosphere seems to be in continual movement, likely as a result of the underlying mantle convection, and brittle lithospheric plates seems to move easily over the asthenosphere.

First experimental power generation was installed in Larderello, Italy, on July 4 1904, about 40 years later than the commercial electricity uses, and the first commercial geothermal power plant (250 kW) in 1913, and the first large-power installation in 1938 (69 MW). It would be 20 years before the next large geothermal power installation was built in Wairakei, New Zealand, commissioned in 1958, that grew to 193 MW of installed capacity by 1963. In the United States, the installation of the first unit of 11 MW at Geysers, in Sonoma, California in 1960, eventually became later the world's largest geothermal power complex, with a capacity of 1890 MW. Plant retirements and declining steam supply have since reduced its generation capacity to an annual average of 1020 MW from 1421 MW of installed capacity, still the largest geothermal plant. Since those early efforts, a total of 2564 MW of geothermal power generation capacity is currently installed in the United States, generating approximately 2000 MW each year. In 2005, annual worldwide geothermal power was estimated at 56,875 GWh from 8,932 MW of installed capacity. Geothermal energy is also utilized in direct-heat uses for space heating, recreation and bathing, and industrial and agricultural uses. Geothermal energy in direct-use applications is estimated to have an installed capacity of 12,100 MW in thermal capacity, with annual average energy usage of 48511 GWh, excluding the ground-coupled heat pumps (GCHPs). In the United States, the first district geothermal heating was installed in Boise, Idaho, in 1892, still in operation today. GCHP units are reported to have 15,721 MW of installed capacity and 24,111 GWh, representing 56.5% worldwide direct use, respectively.

4.5.1 Geothermal Energy Origins and Resources

Geothermal energy originates from the planet formation and from radioactive decay of materials, roughly in equal proportions. The Earth's heat flow, measured in mW/m^2 varies on its surface and with the time at any particular place. This heat flow originates from the primordial heat, generated during the Earth's formation, or due to the decay of radioactive isotopes. The average heat flow through the continental crust is 57 mW/m^2, through the oceanic crust, is 99 mW/m^2, while the global average heat flow is 82 mW/m^2, while the total global output is over 4×10^{13} W, four times than the present world energy consumption. Continental heat flow seems to originate mainly from the radiogenic decay within the upper crust, the heat generated in the most recent magmatic episode and the heat from the mantle. In the oceanic crust, the concentration of radioactive isotopes is very low, so the heat flow is largely derived from heat from the mantle. The geothermal gradient, due to the core-surface temperature differences, drives continuous thermal energy conduction in the form of heat to the surface. The crust base temperature is about 1100°C, the temperature gradient between the surface (~20°C) and the crust bottom is 31.1°C/km (the normal temperature gradient). Good geothermal sources occur where the thermal gradient is several times greater than the normal one. The rate of natural heat flow per unit area, the normal heat flux, is roughly 1.2 x 10^6 cal/cm^2·s, in non-thermal Earth areas. The Earth conductive heat flow is the product of the geothermal gradient and the thermal conductivity of rocks. The geothermal gradient is measured in wells, while the conductivity of rocks is measured in laboratory, on samples taken from the well where the gradient was measured. The two heat transfer forms occurring within the Earth are conduction and convection, the former being more efficient in heat transfer. Earth's thermal behavior studies imply the determination of temperature variations with depth, and how such temperature variations may have changed throughout geological time. The thermal gradient values as lower as 10°C/km are found in ancient continental crust, while very high values of about 100°C/km are found in active volcanic areas. Once the gradient is measured, it can be used to determine the upward heat rate through a particular area. As the heat moves upwards through solid impermeable rock, the principal heat transfer mechanism is conduction. The heat flow rate, proportional with the geothermal gradient and thermal conductivity of rocks is defined as the amount of heat conducted per second through unit area, for a temperature gradient 1°C/m perpendicular to that area. If the gradient is expressed in °C/km and conductivity in W/(m°C), then the heat flow rate is in mW/m^2. Earth's thermal energy is distributed between the constituent host rock and the natural fluids, contained in hot rock fractures and pores at temperatures above ambient levels, mostly water with varying dissolved salts, being present as a liquid phase or sometimes a saturated, liquid-vapor mixture or superheated steam vapor. Notice that the amounts of hot rock and contained fluids are larger and more widely distributed in comparison to oil and natural gas contained in sedimentary rock formations.

The Earth's heat flux is estimated to be equivalent to 42 million megawatts, far greater than the coal, oil, gas and nuclear energy combined. It is estimated that a recovery of even a small fraction of this heat would supply the world's energy needs for centuries. Geothermal energy is also permanently available, unlike the solar and wind energy sources, which are dependent upon factors, like weather variations and daily and seasonal fluctuations. Electricity from geothermal energy is more consistently available, once the resource is tapped. Several different types of geothermal systems can be exploited: (a) convective or hydro-thermal systems, (b) enhanced geothermal systems (EGS), (c) conductive sedimentary systems, (d) hot water produced from oil and gas fields, (e) geo-pressured systems and (f) magma bodies. Convective hydro-thermal systems have seen several decades of commercial exploitation for electric generation in several countries to date, but with limited distribution. There are two basic convective system classes, depending on the thermal energy source type: volcanic and non-volcanic. A volcanic convective system drives its thermal energy from the convecting magma, while a non-convective system drives its thermal energy from meteoric water that has heated up by deep circulation in Earth's high heat flow areas, with no associated magmatic body. The installed power capacity that exploits such systems totals about 10 GW worldwide and

3,000 MW in the United States only, with a reserve base only in the United States of about 20 GW. It has been suggested that a positive correlation exists between the geothermal resource potential available from volcanic convective systems and the number of active volcanoes in the country. However, an exploitable geothermal resource base may exist in the form of non-volcanic convective systems. The heat moves from the Earth's interior towards the surface where it dissipates, although this is generally not noticed, being aware of its existence as the depth increased. There are areas which are accessible by drilling, and where the gradients are well above the average. This occurs when a few kilometers below surface, there are magma bodies undergoing cooling, in a fluid state or solidification process, releasing heat. In other areas, where there is no magmatic activity, the heat accumulation is due to particular crust geological conditions, such that the geothermal gradient reaches anomalously high values.

The extraction and utilization of large heat quantities require a suitable carrier to transfer it to accessible depths beneath the surface. The heat is transferred from depth to sub-surface regions by conduction and convection processes, with geothermal fluids acting as the carrier in the former. These fluids are rain water that has penetrated into the crust from the recharge areas, being heated by the contact with the hot rocks and accumulated into aquifers, occasionally at high pressures and temperatures (above 300°C). These aquifers (reservoirs) are the essential parts of many geothermal fields, usually covered with impermeable rocks, preventing the hot fluids to easily reach the surface and keeping the fluids under pressure. To obtain industrial superheated steam, steam mixed with water, or only hot water depends on the local hydro-geology and the temperature of the rocks. Wells are drilled into the reservoir to extract the hot fluids, and the usage depends on the fluid temperature and pressure. Electricity generation requires higher temperatures, while space heating and industrial processes are often run at the lower temperature range. Geothermal fields are usually systems with continuous heat and fluid circulations, where fluid enters the reservoir from the recharge zones, leaving through discharge areas, hot springs or wells. During the exploitation, the fluids are recharged to the reservoir by reinjecting through wells the fluids from the utilization plants. Reinjection process may also compensate, at least part of the extracted fluid, prolonging to a certain limit the field lifetime. Geothermal energy is therefore to a large extent a renewable energy source; hot fluid production rates tend, however, to be larger than recharge rates. Enhanced geothermal systems imply a man-made reservoir created by hydro-fracturing impermeable or very "tight" rock through wells. By injecting into wells normal temperature water, in such artificially fractured reservoir and extracting heated water through other wells for industrial uses. The EGSs represent conductive systems that have been enhanced their flow and storage capacity by hydro-fracturing, and in theory can be developed anywhere by drilling deep enough to encounter attractive temperature levels. However, this technology is still experimental and posing technical challenges, such as: (a) creating a pervasively fractured large rock volume, (b) securing commercially productivity, (c) minimizing heated water cooling rate, (d) minimizing the losses and (e) minimizing any induced micro-seismicity. Another geothermal energy resource type, considered for exploitation is the heat contained in the water produced from deep oil and gas wells, and co-produced with petroleum or from the abandoned oil or gas wells. While there are no significant challenges to exploiting this resource, the energy cost may not be always attractive due to relatively water low temperature and production rates. Other geothermal energy resource types of quite restricted distributions worldwide are the "geo-pressured" systems. These are confined sedimentary reservoirs with pressures much higher than the local hydrostatic pressure, allowing the exploitation of the kinetic energy of the produced water in addition to its thermal energy. Furthermore, because of its high pressure, such a system may contain methane gas dissolved in the water that can be used to generate electricity or for other uses. Such geo-pressured wells can provide thermal, kinetic and gas-derived energy production. However, there are technical challenges to make such energy systems commercially viable.

Geothermal energy can be utilized as direct use (heat) not only for electricity generation. Most of the geothermal energy resources are inaccessible because of the depths and other area characteristics. However, along the plate boundaries, geothermal activity is close enough to the surface to be accessible, while the zones with the high earthquake activity are the most suitable for geo-thermal power generation. Geothermal resources, characterized by the thermal and compositional characteristics, are divided into four categories: *hydrothermal (geo-hydrothermal) resources, geo-pressurized, molted rocks (magma)* and *enhanced (hot, dry rock) geothermal systems.* Hydrothermal resources are the most limited type among the four classes. However, they are the easiest to harvest. In these resources, water is heated and/or evaporated by direct contact with hot porous rock or per-meable rock, and bounded with low permeability rock. Water flows through the porous rocks, heated (perhaps evaporated) and discharged to the surface. Hydrothermal systems producing steam only are called vapor-dominated, and if they are producing hot water and steam mixture, they are called liquid-dominated. Geo-pressurized energy resources include sediment-filled and hot water confined under pressure reservoirs. The fluid temperature range is 150–180°C, and the pressures up to 600 bars. In many of these systems the fluid contains methane called *geothermal brine,* a highly cor-rosive mixture. Magma or molten rock systems, under active volcanoes at accessible depths, have temperatures in excess of 650°C. Hot dry rock (HDR) has the temperature in the excess of 200°C, and as the name implies, contains small liquid amount. The method for harvesting this resource is through EGS, by directing water under the rock and rejecting the heated water back for various uses.

Further, the geothermal systems are classified as: convective, liquid- and vapor-dominated hydro-thermal reservoirs, lower temperature aquifers and conductive, hot rock and magma over a wide range of temperatures. Lower temperature aquifers contain deeply circulating fluids in porous media or fracture zones, with no localized heat source, being further sub-divided into systems at hydrostatic pressure and systems at pressure higher than geo-pressured systems. Resource utilization technolo-gies are grouped under types for electrical power generation and for direct heat uses. GHPs are a subset of the direct use, and EGS, where the fluid pathways are engineered by rock fracturing, are a subset under both types. A geothermal system requires heat, permeability and water, the EGS tech-niques make up for reservoir deficiencies in any of these areas. EGS technologies enhance existing rock fracture networks, introduce water or another working fluid or otherwise build on a geothermal reservoir that would be difficult or impossible to derive energy from by using conventional technolo-gies. Currently, the most widely exploited geothermal systems for power generation are hydrothermal systems of continental subtype. In areas of magmatic intrusions, temperatures above 1000°C often occur at less than 10 km depth. Magma typically involves mineralized fluids and gases, mixed with deeply circulating groundwater. Typically, a hydrothermal convective system is established whereby local surface heat-flow is significantly enhanced. Such shallow systems can last hundreds of thou-sands of years, and the gradually cooling magmatic heat sources can be replenished periodically with fresh intrusions from a deeper magma. Finally, geothermal fields with temperatures as low as 10°C are also used for direct heat pumps. Subsurface temperatures increase with depth according to the local geothermal gradient, and if hot rocks within drillable depth can be stimulated to improve per-meability, using hydraulic fracturing, chemical or thermal stimulation methods, they form a potential EGS that can be used for power generation and/or direct use applications.

Example 4.15 Calculate the geothermal power potential of a site that covers 30 km² with a thermal crust of 2 km, where the temperature gradient is 240°C. At this depth the specific heat of the rocks is 2.5 MJ/m³, and the mean surface temperature is 10°C. Assuming that only 2% of the available thermal energy could be used for electricity generation, how much takes to produce $5 \cdot 10^4$ MWh?

SOLUTION

From the estimated slab volume, its stored heat is then:

$$Q = VC\Delta T = 2 \times 30 \times 10^9 \times 2.5 \times 10^6 (240 - 10) = 34.5 \times 10^{18} \, \text{J}$$

The total capacity to generate power is:

$$W = 0.02 \times Q = 69 \times 10^{16} \, J$$

$$Time = \frac{69 \times 10^{16} \, J}{5 \times 10^{10} \times 3600 \, J} = 3833.3 \, s \, or \, 1.0648 \, hours$$

4.5.2   SURFACE GEOTHERMAL TECHNOLOGY AND RESERVOIR CHARACTERISTICS

Once a reservoir is found and characterized, the surface technology, power plant and related infrastructure is designed and the equipment selected to optimize the resource use and sustainability. The goal is to construct an energy efficient, low cost, minimal environmental impact power plant. Geothermal fluid, a hot, mineral-rich liquid or vapor, is the carrier that brings geothermal energy up through wells from deeper layers to the surface. This hot water and/or steam extracted from an underground reservoir, isolated during production, flowing up through wells is converted into electricity at a geothermal power plant or is used in direct use systems for heating, cooling or providing hot water to buildings and industrial facilities. Once used, the water and the condensed steam are reinjected into the geothermal reservoir to be reheated. It is separated from groundwater by encased pipes, making the facility virtually water-pollution-free. Such resources, using the existing appropriate hot water and steam accumulations of appropriate are the *hydrothermal* resources. While other geothermal resources exist, all US geothermal power is using hydrothermal resources. The use of natural steam for electricity generation is not the only possible geothermal energy application. Hot waters, which are present in large continental areas, can be exploited, for space heating and industrial processes. The geothermal energy distribution for nonelectric applications is: (a) 42% for bathing and swimming pool heating, (b) 23% for space heating, (c) 12% for geothermal heat pumps, (d) 9% for greenhouse heating, (e) 5% for industrial applications and (f) 9% for fish farm pond, agricultural drying, snow melting, air conditioning or other uses.

Direct use of geothermal energy is one of the oldest, most versatile and common forms of utilizing geothermal energy. Unlike geothermal power generation, in which heat energy is converted to electricity, direct-use applications use heat energy directly to accomplish a broad range of uses. The temperature range of these applications is from about 10°C to about 150°C. Given the ubiquity of this temperature range in the shallow subsurface, these types of applications of geothermal energy have the potential to be installed almost anywhere that has sufficient fluid available. Approximately 5.4×10^{27} J of thermal energy is available worldwide, of which nearly a quarter is available at depths less than 10 km. For direct use, the heat must be significantly above ambient surface temperatures and transferrable efficiently. Such conditions are satisfied in areas where hot springs emerge at the surface or in locations where high thermal gradients allow shallow drilling to access heated waters. Such sites are quite restricted in their distribution, being concentrated in volcanic activity area or where continent rifting has occurred. For such reasons, a relatively small fraction of this large amount of continental heat contained can be economically employed for geothermal direct-use applications. The fraction of the heat readily available is not well known because thorough assessment efforts to quantitatively map the distribution of such resources have thus far been limited. As drilling technology improves and fluid circulation to support heat harvesting at depth improves, the continental thermal resource that can be accessed will significantly expand. As of 2010, approximately 122 TWh/yr of thermal energy was used for direct-use purposes worldwide, which was derived from an installed capacity of 50,583 MW. For comparison, global consumption of electricity in 2006 was 16,378 TWh/yr. The growth in installed capacity of direct-use applications reflects a rapid growth in international development of this type of system. In 1985, 11 countries reported using more than 100 MW of direct-use geothermal energy, while by 2010, the number increased to

78. The global distribution of such systems reflects the diversity of applications for which they have been engineered.

A geothermal system that can be developed for beneficial uses requires heat, permeability and water. When hot water or steam is trapped in cracks and pores under impermeable rock layers, a geothermal reservoir is formed. The exploration of a geothermal reservoir for potential developments includes exploratory drilling and testing for satisfactory conditions to produce useable energy, particularly the resource temperature and flows, water being a critical system component. The hot water, which comes from the geothermal system, is reinjected then into the reservoir to maintain reservoir pressure and to prevent reservoir depletion. However, rainwater and snowmelt usually continue to feed underground thermal aquifers, naturally replenishing geothermal reservoirs. Reinjection keeps the mineral-rich, saline water found in geothermal systems separate from ground water and fresh water sources to avoid cross-contamination. Injection wells are encased by thick borehole pipe and are surrounded by cement. Once the water is returned to the geothermal reservoir, it is reheated by the Earth's hot rocks and can be used over and over again to produce electricity or to provide heat. The key ingredients for geothermal energy production can be summarized by the following equation:

$$P_{conv} = C_P \cdot (T_{Rsvr} - T_{Rjec}) \cdot F_{rate} \cdot \eta - P_{Loss} = C_P \cdot \Delta T \cdot F_{rate} \cdot \eta - P_{Loss} \qquad (4.61)$$

where C_P is the specific heat of the working fluid, F_{rate} is the flow rate from the production well (in Kg/s), ΔT is the sensible heat that can be extracted from the fluid produced by the production hole ($T_{reservoir} - T_{rejection}$), η is the efficiency with which the heat energy can be used and P_{Loss} represents the fluid transfer and conversion losses. The goals of geothermal system developments are to optimize these parameters to increase electrical or heat output relative to the investment capital costs of the development. Based on the current experience in power generation from convective hydrothermal resources, the minimum amount of net energy produced by a well is about 4 MW. For most of the geothermal systems, the working fluid is water with varying salinity or other dissolved materials. The specific heat is almost constant for all types of geothermal resources. The ΔT is often in the order of 50°C to 150°C, and the efficiency of current power cycles is about 10%. Based on these numbers and ignoring parasitic losses, a well needs to flow at a minimum of 70 kg/s to be viable. This rate is in orders of magnitude higher than average flows in the US oil industry, and at the upper end of production rates for water wells, particularly at the depths needed to access high temperatures. The flow problem is not as significant in convective hydrothermal resources as these typically produce steam rather than water. Although the specific heat and density of steam is lower than water, high flow rates are achieved because of steam's low viscosity and density, allowing wells to produce without pumping. Exploration targets are the two components of Equation (10.1), ΔT and F_{rate}. High temperature differences (ΔT) are targeted by looking for areas with very high thermal gradients. Flows are targeted by looking for areas with high natural permeability or with characteristics that are suitable for EGS techniques. The aims are to reduce the risks and therefore costs of discovering a geothermal resource. However, the exploration for any natural resource is inherently a high risk one. The risks can be reduced by the selection of suitable exploration targets. Targets need to be developed at all scales to enable the appropriate basin selection, selection of tenements within these basins, the development of prospects within tenements and the locating exploration wells within the prospects.

4.5.3 ELECTRICITY FROM GEOTHERMAL ENERGY SOURCES

Geothermal power plants use the natural hot water and/or steam to turn turbine-generator units for electricity generation. Unlike fossil fuel-based power plants, no fuel is burned in these plants. Geothermal power plant's byproduct is water vapors, with no smoky emissions. Geothermal power plants are for the base load power as well as the peak load demand units. Geothermal electricity has become competitive with conventional electricity generation in many world regions. Main types of geothermal power plants are listed and discussed here. *Dry steam power plants* are the simplest and most economical technologies, and therefore are widespread. It is suitable for sites where the geothermal steam is not mixed with water. Geothermal wells are drilled down to the aquifer, and the superheated and pressurized steam (180°C–350°C) is brought to the surface and passed through a steam turbine to generate electricity. In simple power plants, the low-pressure steam output from the turbine is vented to the atmosphere. However, usually after passing the turbine-generator unit, the steam condensates, resulting in almost pure water, which is reinjected into the aquifer or used for other purposes. This can improve the overall plant efficiency, while avoiding the environmental problems associated with the direct steam release into the atmosphere. Italy and United States have the largest dry steam geothermal resources. This type of resource is also found in Indonesia, Japan and Mexico. In *single flash steam technology*, hydrothermal resource is in a liquid form. The fluid is sprayed into a flash tank, held at a much lower pressure than the fluid, causing it to vaporize (flash) rapidly to steam. The steam is then passed through a turbine coupled to a generator. To prevent the geothermal fluid flashing inside the well, the well is kept under high pressure. Flash steam plant generators range from 10 MW to 55 MW, with a standard power size of 20 MW is used in several countries. *Binary cycle power* plants are used where the geothermal resource is insufficiently hot to produce steam, or where the resource contains too many chemical impurities to allow flashing. In addition, the fluid remaining in the tank of flash steam plants can be utilized in binary cycle plants (e.g. Kawerau in New Zealand). In the binary cycle process, the geothermal fluid is passed through a heat exchanger. The secondary fluid (e.g. isobutene or pentane) which has a lower boiling point than water is vaporized and expanded through a turbine to generate electricity. The working fluid is condensed and recycled for another cycle. The geothermal fluid is then reinjected into the ground in a closed-cycle system. Binary cycle power plants can achieve higher efficiencies than flash steam plants and allow the utilization of lower temperature resources. In addition, corrosion problems are also avoided.

The world geothermal electrical capacity installed in the year 2000 was about 8 GWe with the generation in that year of 49.3 billion kWh, while in 2015 was about 12 GWe, generating about 80 TWh/year. In the industrialized countries, where the installed electrical capacity reaches very high levels, in the range of tens or even hundreds of thousands of MWe, the geothermal energy is unlikely to account for more than 2%, at most, of the total in the near future. On the other side, in the developing countries, with quite limited electrical consumption but good geothermal prospects, geothermal electricity generation could make quite a significant contribution to the total, with estimates of about or over 15% in countries like Philippines, El Salvador, Nicaragua or Costa Rica. The efficiency of the generation of electricity from geothermal steam ranges from 10% to 20%, about three times lower than the efficiency of nuclear or fossil-fueled plants. Geothermal power plants have the lower efficiencies due to the low steam temperature, 250°C or lower. Furthermore, geothermal steam has a different chemical composition than the pure water vapor, containing usually non-condensible gases, that reduces the overall system efficiency. The simplest and cheapest of the geothermal cycles used to generate electricity is the direct-intake non-condensing cycle. Steam from the geothermal well is passed through a turbine and exhausted to the atmosphere, with no condensers at the outlet of the turbine. Such cycles consume about 20 kg of steam per kWh. Non-condensing systems can be used if the content of non-condensible gases in the steam is very high, greater than 50% in weight, and are used in preference to the condensing cycles for gas contents exceeding 15%, because the high energy required extracting these gases. In power plants where electricity is produced from dry or superheated steam, vapor-dominated reservoirs, steam is piped directly from the wells to the turbine. This is a well-developed, commercially available technology, with typical turbine-size units in the 20–120 MWe capacity range. Recently, a new trend of installing modular standard generating units of 20 MWe has been adopted in Italy. Vapor-dominated systems are less common in the world, and steam from these fields has the highest enthalpy (energy content), generally close to 670 kcal/kg (2800 kJ/kg). At present these systems have been found only in Indonesia, Italy, Japan and the United States. These fields produce about half of the geothermal electrical energy of the world. Water-dominated fields are much more common. Flash steam plants are used to produce energy from these fields that are not hot enough to flash a large proportion of the water to steam in surface equipment, either at one or two pressure stages.

If the geothermal well produces hot water instead of steam, electricity can still be generated, provided the water temperature is above 85°C, by means of binary cycle plants. These geothermal plants operate with a secondary, low boiling-point working fluid (Freon, Isobutane, Ammonia, etc.) in an organic Rankine cycle. The working fluid is vaporized by the geothermal heat in the vaporizer, and then passes through the organic vapor turbine, coupled to the generator. The exhaust vapor is then condensed in a condenser and is recycled to the vaporizer by a fluid cycle pump. The efficiency of these cycles is quite low, up to 6%. Typical unit sizes are from 1 MWe to 3 MWe. However, the binary power plant technology has emerged as the most cost-effective and reliable way to convert large amounts of low temperature geothermal resources into electricity; such large low-temperature reservoirs at accessible depths exist in almost any world areas. The power rating of geothermal turbine-generator units tends to be smaller than in conventional thermal power stations, with common power levels of 55 MWe, 30 MWe, 15 MWe and 5 MWe or even smaller. One of the main advantages of geothermal power plants is that they can be built economically in relatively much smaller units than, e.g. hydropower stations. In developing countries with a small electricity market, geothermal power plants with units from 15 to 30 MWe can be more easily adjusted to the annual increase in electricity demand than larger hydropower or fossil fuel power plants. The reliability of geothermal power plants is very good, the annual load factor and availability factor are commonly about 90% and geothermal fields are not affected by annual or monthly fluctuations in rainfall or weather, since the essentially meteoric water has a long residence time in geothermal reservoirs. There are basically three types of geothermal power plants, which are determined primarily by the nature of the geothermal resource at the site. The first ones are direct steam geothermal plant, used where the geothermal resource produces steam directly from the well. These are the earliest types of plants developed in Italy and in the United States. However, such steam resources are one of less common of all the geothermal resources, existing in only a few places. Obviously, steam plants are improper to the low-temperature resources. Flash steam plants are employed in cases where the geothermal resource produces high-temperature hot water or a combination of steam and hot water. The fluid from the well is delivered to a flash tank where a part of the water flashes to steam and is directed to the turbine. The remaining water is directed to the disposal. Depending on the resource temperatures it may be possible to use two stages of flash tanks, in which the water separated at the first stage tank is directed to a second stage flash tank where more (but lower pressure) steam is separated. Remaining water from the second stage tank is then directed to disposal. The *double flash* plant delivers steam at two different pressures to the turbine. Again, this type of plant cannot be applied to low-temperature resources. The third type of geothermal power plant is the binary geothermal power plant. The name derives from the fact that a second fluid in a closed cycle is used to operate the turbine rather than geothermal steam. Geothermal fluid passes through a heat exchanger (boiler or vaporizer), in some plants, two heat exchangers in series are used, the first a preheater and the second a vaporizer where the heat in the geothermal fluid is transferred to the working fluid causing it to boil. Past working fluids in low temperature binary plants were CFC (Freon type) refrigerants, while the new ones use hydrocarbons (isobutane, pentane, etc.) as refrigerants with the specific fluid chosen to match the geothermal resource temperature. The working fluid vapor passes to the turbine where its energy is converted to mechanical energy, delivered to the generator. The vapor exits the turbine to the condenser where it is converted back to a liquid. In most plants, cooling water is circulated between the condenser and a cooling tower to reject this heat to the atmosphere. An alternative is to use so called "dry coolers" or air-cooled condensers which reject heat directly to the air without the need for cooling water. This design essentially eliminates any consumptive use of water by the plant for cooling. Dry cooling is operating at higher temperatures (in the summer season) than cooling towers does result in lower plant efficiency. Liquid working fluid from the condenser is pumped back to the higher-pressure pre-heater/vaporizer by the feed pump to repeat the cycle. The binary cycle is the type of plant which would be used for low temperature geothermal applications, while the processes in binary geothermal plant are very similar to ones in steam turbine-based power generation. Currently, the binary equipment is available in modules of 200 to 1,000 kW.

The process of generating electricity from a low-temperature geothermal heat source (or from steam in a conventional power plant) involves a process engineers refer to as a Rankine Cycle. A conventional power plant includes a boiler, turbine, generator, condenser, feed water pump, cooling tower and cooling water pump. Steam is generated in the boiler by burning a fuel (coal, oil, gas or uranium). The steam is passed to the turbine where, in expanding against the turbine blades, the heat energy in the steam is converted to mechanical energy causing rotation of the turbine. This mechanical motion is transferred through a shaft to the generator where it is converted to electrical energy. After passing through the turbine, the steam is converted back to liquid water in the condenser of the power plant. Through the process of condensation, heat not used by the turbine is released to the cooling water. The cooling water is delivered to the cooling tower where the "waste heat" from the cycle is often rejected to the atmosphere. However, the modern power plants are using advanced heat recovering technologies and combined heat and power generation in order to increase the overall system efficiency. Steam condensate is delivered to the boiler by the feed pump to repeat the process. In summary, a power plant is simply a cycle that facilitates the conversion of energy from one form to another. In this case the chemical energy in the fuel is converted to heat (at the boiler) and then to mechanical energy (in the turbine) and finally to electrical energy (in the generator). EGS systems are referring to the creation of the artificial conditions at a site or a location where a reservoir has the potential to produce geothermal energy. A geothermal system requires heat, permeability and water, so EGS techniques make up for reservoir deficiencies in any of these areas. These systems involve injecting water into the source and circulating it through the dry rocks. Because of the low thermal conductivity of the rocks large surface areas are necessary for such systems.

4.6 CHAPTER SUMMARY

Sustainable energy systems need to use renewable energy resources that are environmentally friendly and socially acceptable. Modern society also requires an increasingly reliable energy infrastructure as more and more loads are connected. Generation before the smart grids has been dominated by large, centralized power plants, often located far from load centers to gain economy scale and power system resilience and stability. Moreover, any power system must operate in prescribed voltage and frequency limits, and if these limits are exceeded the insulation of the power system components and consumer equipment may be damaged, leading to short-circuit faults. The power distribution capacity is also limited by the voltage variations between times of maximum and minimum loads. Such issues lead to the interests in connecting distributed generation and dispersed energy resources to the medium and low voltage sections of the power grid. Renewable energy resources, such as wind power, solar PV systems, biomass, geothermal energy, together with energy storage are strong candidates for distributed power generation and replacing the conventional fossil fuel-based generation options. Notice that the stability, reliability and cost implications of renewable energy resources in the development and penetration into the smart grid are critical and vital. All the Earth's renewable energy sources are generated from solar radiation, which can be converted directly or indirectly to energy using various technologies. This radiation is perceived as white light since it spans over a wide spectrum of wavelengths, from the short-wave infrared to ultraviolet. Such radiation plays a major role in generating electricity either producing high temperature heat to power an engine mechanical energy which in turn drives an electrical generator or by directly converting it to electricity by means of the photovoltaic (PV) effect. Wind and solar energy have the potential to play an important role in future energy supply and generation mix in many areas and countries of the world. Wind regime is ultimately a consequence of the Sun's energy being determined by the global synoptic circulation and by the local flows and topography. The most important characteristics of wind are its variability and intermittency on a broad range of spatio-temporal scales. Wind regime knowledge and characteristics are important for assessment and analysis of wind energy potential for an area or location, exploitation of wind energy, design, management or operation of wind energy conversion systems, as well as in other engineering and technology branches. Wind regime is ultimately a consequence of the Sun's energy being determined by the global synoptic circulation and by the local flows and topography. Geothermal energy has tremendous potential to provide many areas and regions of the world with reliable, base-load, dispatchable and clean renewable energy for centuries to come. Photovoltaic devices are rugged and simple in design, requiring very little maintenance and their biggest advantage being their construction as standalone systems to give outputs from microwatts to megawatts. Many modern products incorporate PV cells or modules in order to operate independently of other electrical supplies. Electricity produced from photovoltaic (PV) systems has a far smaller impact on the environment than traditional methods of electrical generation. During their operation, PV cells need no fuel, give off no atmospheric or water pollutants and require no cooling water. The use of PV systems is not constrained by material or land shortages and the sun is a virtually endless energy source. Geothermal growth and development of electricity generation has increased significantly over the past 30 years, with higher rates in the early part of this period and lower rates in the last 10 years due to lower overall economic increasing in many countries and the low price of competing fuels. Geothermal heat can be used for different applications, not only electricity generation, based primarily on its temperature range. The other key aspect is the type of geothermal resource.

4.7 QUESTIONS AND PROBLEMS

1. What is the air density difference between sea level and a height of 1,000 m and 2,500 m?
2. Solar power potential is around 1 kW/m². What wind speed gives the same power potential?
3. The area of New Mexico is 121,666 square miles. The average annual insolation (hours of equivalent full sunlight on a horizontal surface) is approximately 2000 h. If one-third of the area of New Mexico is covered with solar panels of 10% efficiency, how much electricity can be generated per year? What percentage of US energy needs can be satisfied?
4. How much geothermal energy is used in the United States? How much is the potential of using geothermal resources in the United States?
5. What geological areas are most likely to have higher heat flows?
6. On the map of your country or of an US state, select the most suitable site for a geothermal energy development; describe the tectonic setting and likely types of rocks and geologic structures that would be found. Justify your conclusions.
7. What regions in your country are best suited for geothermal power facilities? Why? What regions are the least suited?
8. What is the global average heat flow at the surface of the Earth? What is its range? What controls heat flow?
9. A wind power plant has 15 wind turbines, each one rated for 1.0 MW. The capacity factor is 37.5%. What is the wind power plant's annual energy yield?
10. Calculate the power, in kilowatts, across the following areas for wind speeds of 5, 15 and 25 m/s. Use the turbine diameters of 5, 10, 50 and 100 m for the rotor area. Assume the standard air density of 1.225 kg/m³.

11. A wind turbine is rated at 300 kW in a 10 m/s wind speed in air at standard atmospheric conditions. If we assume that the power output is directly proportional to air density, what is the power output of the turbine in a 10 m/s wind speed at elevation 1500 m above sea level at a temperature of 20°C?

12. Calculate the factor for the increase in wind speed if the original wind speed was taken at a height of 10 m. New heights are 30 m, 60 m and 90 m. Use the power law with an exponent equal to: (a) 0.1 and (b) 0.20.

13. Calculate the wind speed distribution using the Rayleigh distribution for an average wind speed of 6.3 m/s. Use 2 m/s bin widths. Calculate the wind energy density, the monthly energy density availability, the most frequent wind velocity and the wind velocity corresponding to the maximum energy, for Rayleigh wind speed distribution of previous problem, assuming and air density 1.225 kg/m³.

14. Compute the wind power density for a site located at an elevation of about 500 m when the air temperature is 30°C and the wind speed is equal to 10 m/s, 12.5 m/s and 15 m/s.

15. Calculate the wind speed distribution for a Weibull distribution for c = 7.2 m/s and k = 1.8. Use 1 m/s bin widths. How many hours per day the wind speed is between 5 m/s and 12 m/s, the cut-in speed and the rated speed of a medium-size wind turbine?

16. For a wind turbine having a rotor diameter of 60 m, for incoming winds of 6 m/s, 9 m/s and 12 m/s, calculate: (a) the power of incoming wind, (b) the theoretical maximum power extracted by the wind turbine and (c) if the gearbox, generator and electrical transmission and processing efficiencies are 0.85, 0.96 and 0.95, respectively, what is the turbine effective converted power. Assume standard air density.

17. What is the solar time in El Paso, Texas (latitude: 31.8° N, and longitude: 106.4° W) at 10 AM, and 3 PM, Mountain Standard Time on March 15, July 15 and October 1st?

18. Find the altitude angle and azimuth angle of the sun at the following (solar) times and places:
    a. March 1st at 10:00 A.M. in New Orleans, latitude 30°N.
    b. August 10the at 2:00 P.M. in London, Ontario, Canada, latitude 43°N
    c. July 1st at 5:00 P.M. in San Francisco, latitude 38°N.
    d. December 21st at 11 A.M. at latitude 68°N.

19. Find the solar altitude and azimuth angles at solar noon in Gainesville, Florida, on February 28 and August 8. Also find the sunrise and sunset times in Gainesville on that day. Repeat the calculations for Sydney, Australia.

20. Find the solar altitude and azimuth angles in San Juan, Puerto Rico on (a) June 1 at 7 a.m. and (b) December 1 at 2 p.m. Also find the sunrise and sunset times on these days.

21. Determine the solar altitude and azimuth angles at 11:00 A.M. local time at Bucharest, Romania, on July 15.

22. Current output from a solar module is proportional to what variable? What kind of current produces a PV module?

23. Given an I-V curve (select a typical one from any manufacturer), find I_{mp}, V_{mp}, I_{SC} and V_{OC}. Calculate the fill factor (FF). Given P_{in}, 1000 W/m², calculate the conversion efficiency.

24. Estimate the cell temperature and power delivered by a 120 W PV module, assuming 0.5% per degree power loss, a NOCT = of 56°C, ambient temperature of 25°C, insolation of 1kW/m².

25. Using the 36 cell PV module specifications, as found from a manufacturer's specifications, determine the nominal current and voltage outputs for a PV system that is wired 4 strings in parallel, each having 5 PV modules in series.

26. How does a conventional geothermal power plant work? How does geothermal energy benefit local economies?

27. What factors influence the cost of a geothermal power plant? Identify locations in the United States that are suitable for geothermal energy.

28. If we are assuming that the US geothermal energy sources generated some 15 billion kWh of electricity. How many conventional 1000 MW power plants are needed to produce this electricity if they operate 9000 hours per year?

29. It is accepted that geothermal fluids must about or exceed 200°C in order to be practical for electricity generation. Assuming a surface temperature of 25°C, how deep must one drill to hit this temperature at a location having a normal temperature gradient?

30. The average geothermal heat flux is about 0.050208 J/m²s in the continental regions. Assuming that three-fourths of this is attributed to the crust, by using the average thermal conductivity for rock, calculate the temperature at 5000 m.

5 Energy Storage in Future Power Systems

5.1 NEW POWER SYSTEM PARADIGM

Economic, technological and environmental incentives and issues are changing the face of electricity generation, transmission, distribution and uses. Centralized energy generation structures, using complex power networks to produce and transfer electricity to the consumers, have high investment, operation and maintenance costs, while the overall system efficiency is low due to the large system losses. While overall electricity dependency has intensified, smaller generation using a broad generation mix, combined with energy storage has emerged as competitive alternatives to large central generation. The pressure of improving the power system efficiency, power quality, supply security, stability and environmental impacts has determined significant power system changes. There are significant increases in the electricity demands, while the DG and RES seems a way to cope with energy demands and grid issues. DER and RES integration into the existing energy networks can result in several benefits, in addition to the reduced grid losses and environmental impacts, e.g. relieved transmission and distribution congestion, peak demand shavings, voltage support, reduced price fluctuations and the deferred investments to upgrade existing systems. The energy storage systems can harness the excess of the produced electricity during the off-peak, low demand periods, to be used during the peak periods or when needed, reducing the needs for the high-cost peak-load generators. The energy storage systems can provide dispatchability of the renewable energy sources, having no own dispatchability. Power systems generate and supplies electricity through complex processes, consisting of electricity generation in large power plants, usually located close to the primary energy sources (e.g. coal mines, water reservoir), far from large consumer (load) centers, delivering the electricity by a complex distribution infrastructure, involving high-voltage (HV), medium-voltage (MV) and low-voltage (LV) electric networks, in which the power is flowing in one direction, from HV levels down to customers. New technologies allow the electricity to be generated in smaller size units, located closer to user, in the MV and LV grid sections, as depicted in Figure 5.1. Moreover, the increasing RES uses to reduce the environmental impacts and to diversify the supply lead to a new electricity grid. In this new paradigm, part of the power generation is shifted to the MV and LV networks, with part of the energy supplied by the centralized generation and part produced by the DG units (see diagram of Figure 5.1). Large-scale DG integration is a main trend in modern power systems.

Energy storage technology has been in existence for a long time and has been utilized in many forms and applications from a flashlight to spacecraft systems. Energy storage systems (ESS) become critical in enabling renewable energy uses, making the non-dispatchable renewable energy resources dispatchable. In order to match power demand, in the renewable energy intermittent and non-dispatchable context and the fairly predictable electrical demand, energy storage is critical, allowing de-coupling of generation from usage, reducing the needs for monitoring and demand predictions. Energy storage also provides economic benefits by allowing a plant generation to meet average demands rather than peak demands. There are several drivers that are spurring the growth in the energy storage, such as: the renewable energy usage, increasingly strained grid infrastructure as new lines lags well behind demands or the microgrid emergence as part of smart grid architecture, demands for higher energy supply reliability, security or efficiency. However, some of the issues regarding the optimal integration (operational, technical and market) of energy storage into the electric grid are still in research. The ESSs integration must be based on the existing

DOI: 10.1201/9780429174803-5

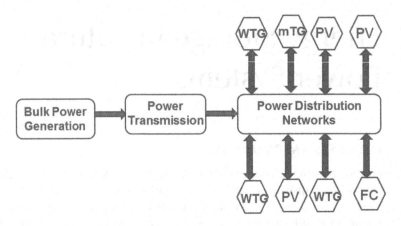

FIGURE 5.1   Distributed energy resources on the power distribution network.

electric infrastructure, requiring an optimal integration of the energy storage systems. This chapter describes the main energy storage technologies currently available and parameters used to describe and characterize an energy storage device. It also contains the analysis of each energy storage technology currently available indicating: cost, advantages and disadvantages and applications. The functions and objectives of the energy storage system in a smart grid can be classified into five categories: (i) bulk energy services, (ii) grid ancillary services, (iii) power transmission infrastructure services, (iv) power distribution infrastructure services and (v) customer energy management services. In grids with high penetration of renewable energy, the functions and goals of energy storage are primarily the first three items. Note that most types of energy storage devices can support multiple sub-goals or sub-functions.

5.2   ENERGY STORAGE IN SMART GRIDS

Energy storage has become a critical factor that can solve several power system problems and issues. A renewable energy system with its corresponding energy storage system can behave as a conventional power plant, at least for time intervals in the order of half an hour up to a day, depending on the storage capacity. Such renewable energy sources are usually not providing immediate response to an energy demand, as they do not deliver a supply easily adjustable to consumption needs. Thus, the growth of this decentralized generation means greater network load stability problems and requires energy storage, as one of the potential solution. The energy storage is also a crucial element in the energy management from renewable energy sources, allowing energy to be released into the grid during peak hours when it is more valuable. There are several well-established electricity storage technologies, as well as a large number in process of development offering significant application potential. Economically viable energy storage requires efficient conversion of electricity into other energy form that is converted back to electricity when needed. All energy storage methods need to be feasible, efficient and environmentally safe. Energy storage systems can be separated in four major classes: mechanical, electrical, thermal and chemical-based systems. Mechanical energy storage includes pumped hydropower storage, compressed air energy storage and flywheels. Electro-chemical energy storage includes batteries, hydrogen-based energy storage and fuel cells. Here also is included thermochemical energy storage, such as solar-hydrogen, solar-metal techniques, etc., not discussed in this chapter. Electromagnetic energy storage includes super-capacitors and superconducting magnetic energy storage. Thermal energy storage includes two broad categories: low-temperature and high-temperature energy storage. Figure 5.2 summarizes the most common energy storage technologies. Energy storage

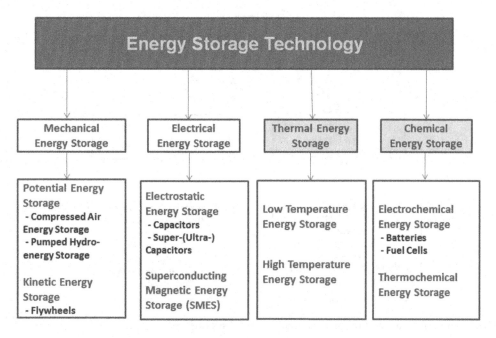

FIGURE 5.2 Classification of the major energy storage technologies.

systems can increase the stability and reliability of the intermittent energy sources, facilitating their grid integration and uses. There also are opportunities for significant improvement power quality, load shifting and energy management through energy storage applications. In a weak power grid, the RES integration at remote connection points may generate unacceptable voltage variations due to power fluctuations. Upgrading the power transmission line to mitigate this problem is expensive, while the EES inclusion for power smoothing and voltage regulation at the point of connection allows power utilization, offering an economic alternative to costly upgrading transmission options.

Even if there are several drivers that are spurring the growth in energy storage applications, there are still issues regarding the optimal active integration (operational, technical and market) of energy storage technologies into the electric grid, not fully developed, tested and standardized. The ESSs integration and further development of energy converting units including renewable energies must be based on the existing electric infrastructure, requiring optimal integration of energy storage systems. Renewable energy systems with optimum energy storage can behave as conventional power plants, at least for short-time intervals of order of half an hour to a day, depending on the storage capacity. Electricity generated from renewable sources can rarely provide immediate response to demands as these sources do not deliver a supply easily adjustable to consumption, being in this regard low-inertia systems. The major energy storage system applications are summarized in Figure 5.3, where these systems are classified according to their applications. Energy storage technologies for electric applications have achieved various levels of technical and economic maturity in the marketplace. For grid energy storage, challenges include roundtrip efficiencies, representing a tradeoff between the increased cycled electricity cost through the energy storage system, and the value of greater dispatchability or other grid services. The capital cost of grid energy storage units is still relatively higher than conventional alternatives, such as gas-fired power plants, which can be constructed quickly, being perceived as low-risk investments. Understanding the potential of energy storage in electric applications is complicated by a number of factors. First, the wide range of energy storage technologies either commercially available, in development or being researched, making it

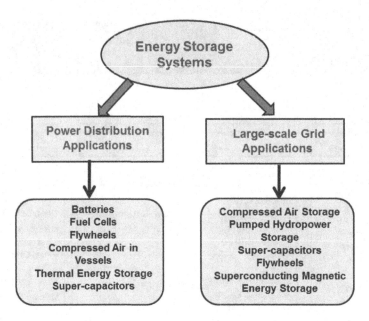

FIGURE 5.3   Applications of the principal energy storage systems.

difficult to have a balanced understanding of the fundamental capabilities, costs and comparative advantages of these different energy storage options. Second, there are multiple applications of energy storage, each with distinct operational requirements. Certain energy storage technologies may suit certain applications better than others. Finally, there are many aspects of market structure and economic regulation that affect energy storage deployment. While there is a consensus that storage technology improvements are needed, there are multiple potential pathways to such improvements that cut across scientific disciplines. A number of obstacles have hampered the energy storage use, such as: (1) lack of design experience, (2) inconclusive benefits, (3) high capital costs or (4) who should pay for energy storage, utilities or RES developers? However, as renewable energy resources and power quality become increasingly important, costs and concerns regarding energy storage are expected to decline.

5.2.1   ENERGY STORAGE FUNCTIONS AND APPLICATIONS

Electrical energy can be stored in the forms of kinetic, potential, electro-chemical or electromagnetic energy, and transferred back into electrical energy when required or needed. The conversion of electrical energy to different forms and back to electrical energy is done by a specific conversion process. The electricity generated during off-peak periods can be stored and used to meet the loads during peak periods when the energy is more expensive, improving the power system economics and operation. Compared to conventional generators, the storage systems have faster ramping rates to respond to the load fluctuations. Therefore, the EESs are a perfect spinning reserve, providing a fast load following and reduces the need for conventional and more expensive spinning reserves. EESs were initially used only for load leveling applications, while now are seen as tools to improve the power quality, stability, to ensure a reliable and secure power supply and to black start the power systems. Breakthroughs are reducing dramatically the EES costs and are driving significant changes into the power system design, structure and operation. Peak load problems could be reduced, stability improved, power quality issues reduced or even

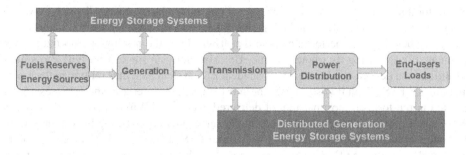

FIGURE 5.4 New chain of electricity, with energy storage as the sixth dimension.

eliminated. Storage can be applied at the power plant level, in support of the transmission system, at various power distribution system points and on particular equipment on the customer side. Figure 5.4 shows how the new electricity value chain is changing supported by the integration of energy storage systems. EESs in combination with advanced power electronics can have a great technical role and can lead to many benefits. EESs are also suitable for particular applications, primarily due to their potential energy storage capacities. Therefore, to provide a fair comparison between the various energy storage technologies, they are grouped based on the size of power and energy storage capacity. There are four categories: devices with large power (>50 MW) and energy storage (>100 MWh) capacities, medium power (1–50 MW) and storage capacities (5–100 MWh), medium power or medium storage capacities but not both, and small scale energy storage systems, with power less than 1 MW and energy capacity less than 5 MWh. Storage systems such as pumped-hydro-electric energy storage (PHES) have been in use since 1929 for daily load leveling. Today, with the smart grid advent, energy storage is a realistic option for: (1) electricity market restructuring; (2) integrating renewable energy resources; (3) improving power quality; (4) shifting towards distributed generation; and (5) helping network operate under more stringent environmental requirements.

Energy storage systems are needed by the electricity industry, because unlike other commodities, the conventional electricity industries have very limited energy storage facility. The electricity transmission and distribution systems are operated through a simple one-way transportation from large power plants to the consumers, so the electricity must always be used precisely when produced. However, the electricity demand varies considerably emergently, daily, weekly and seasonally, while the maximum energy demand may only last for short periods, leading to inefficient, overdesigned and expensive peak power plants. ESS allows energy production to be de-coupled from the supply, self-generated or purchased. Having large-scale electricity storage capacities available over any time, system planners need to build only sufficient generating capacity to meet average electricity demands rather than peak demands. This is particularly important to large utility generation systems, e.g. nuclear power plants, which must operate near full capacity for economic reasons. Therefore, ESS can provide substantial benefits including load following, peaking power and standby reserve, higher overall efficiencies of thermal power plants, reducing the harmful pollutant emissions. Furthermore, ESS is regarded as an imperative technology for dispersed and distributed energy systems. The traditional electricity value chain has been considered to consist of five links: energy (fuels) sources, generation, transmission, distribution and customer-side energy service as shown in Figure 5.4. Supplying power when and where is needed, ESS is on the brink of becoming the *sixth link* by integrating the existing segments and creating a more responsive market. Storage technologies are various and covering a full spectrum from larger scale, generation and transmission, to those related to power distribution and even *beyond the meter*, into the end-user sites.

ESS performances include cycle efficiency, cost per-unit capacity, energy density, power capacity, lifecycle, environmental effect including end-of-life disposal cost. An ideal ESS system is one exhibiting the best possible performances so that it will have the minimum amortized (dollar or environmental) cost during its whole lifetime. Unfortunately, no single ESS type can simultaneously fulfill all the desired characteristics of an ideal ESS system, and thus minimize the amortized lifetime cost of ESS. Capital cost, one of the most important criteria in the ESS design, can be represented in the forms of cost per unit of delivered energy ($/kWh) or per unit of output power ($/kW). Capital cost is an especially important concern when constructing hybrid energy storage (HES) systems. Such systems often consist of several ESS elements with relatively low unit cost (e.g. lead-acid batteries) and ESS elements with relatively high unit cost (e.g. super-capacitors). The overall HES system cost is minimized by allocating the appropriate mix of low-cost vs. high-cost ESS elements while meeting other constraints such as cycle efficiency, total storage capacity or peak output power rate. The ESS cycle efficiency is defined by the "roundtrip" efficiency, i.e. the energy efficiency for charging and then discharging. The cycle efficiency is the product of charging efficiency and discharging efficiency, where charging efficiency is the ratio of electrical energy stored in an ESS element to the total energy supplied to that element during the entire charging process, and discharging efficiency is the ratio of energy derived from an ESS element during the discharging process to the total energy stored in it. Charging-discharging efficiency is significantly affected by the charging/discharging profiles and the ambient conditions. The ESS state of health (SOH) is a measure of its age, reflecting the ESS general condition and its ability to store and deliver energy compared to its initial state. During the ESS lifetime, its capacity or "health" tends to deteriorate due to irreversible physical and chemical changes which taking place. Term "replacement" implies the discharge, meaning the use, until the ESS is no longer usable (its end-of-life). To indicate the rate at which SOH is deteriorating, the lifecycle may be defined as the number of ESS cycles performed before its capacity drops to a specific capacity threshold, being one of the key parameters and gives an indication of the expected ESS working lifetime. The lifecycle is closely related to the replacement period and full ESS cost. The self-discharge rate is a measure of how quickly a storage element loses its energy when it simply sits on the shelf, being determined by the inner structure and chemistry, ambient temperature and humidity and significantly affect the sustainable energy storage period of the given storage element.

Deferred from the conventional power system, which has large centralized units, DERs are installed at the distribution level, close to the consumers and generate lower power typically in the range of a few kW to a few MW. The electric grid is undergoing the change to be a mixture of centralized and distributed subsystems with higher and higher DER penetration. However, more drastic load fluctuations and emergent voltage drops are anticipated due to smaller capacity and higher line fault probability than in conventional power system. ESS is a key solution to compensate the power flexibility and provide a more secure power supply, being also critically important to the integration of intermittent renewable energy. The renewable energy penetration can displace significant amounts of energy produced by large conventional power plants. A suitable ESS could provide an important (even crucial) approach to dealing with the inherent RES intermittency and unpredictability as the energy surplus is stored during the periods when generation exceeds the demand and then used to cover periods when the load demands are greater than the generation. Future development of the RES technologies is believed to drive the energy storage cost down. Nonetheless, the widespread deployments are facing the fundamental difficulty of intermittent supplies, requiring demand flexibility, backup power sources and enough energy storage for significant time. For example, the EES applications to enhance wind energy generation are: (i) transmission curtailment, mitigating the power delivery constraint due to insufficient transmission capacity; (ii) time-shifting, firming and shaping of wind generated energy by storing it during the off-peak interval (supplemented by power from the grid when wind generation is inadequate) and discharging during the peak interval; (iii) forecast hedge mitigating the errors in wind energy bids prior to required delivery, reducing the price volatility and mitigating consumer risk exposure to this volatility; (iv) grid frequency support through the energy storage during sudden, large decreases in wind generation over a short discharge interval; and (v) fluctuation suppression through stabilizing the wind farm generation frequency by suppressing fluctuations (absorbing and discharging energy during short duration variations in output). The key energy storage functions are equally relevant to intermittent renewable energy source, such as:

1. **Grid voltage support,** meaning the additional power provided to the electrical distribution grid to maintain voltages within the acceptable range.
2. **Grid frequency support,** meaning real power is provided to the grid to reduce any sudden, large load-generation imbalance to keep the frequency within the permissible tolerance for up to 30-minute periods.
3. **Grid angular (transient) stability,** meaning the reduction of the power oscillations (due to rapid events) by real power injection and absorption.
4. **Load leveling (peak shaving),** consisting of rescheduling certain loads to lower power demands, and/or the energy generation during off-peak periods for storage and use during peak demand periods.
5. **Spinning reserve,** defined as the amount of generation capacity that is used to produce active power over a given period which has not yet been committed to the production of energy during this period.
6. **Power quality improvement,** which is basically related to the changes in magnitude and shape of voltage and current, energy storage can help to mitigate the power quality problems.
7. **Power reliability,** defined as the percentage/ratio of interruption in delivery of electric power (may include exceeding the threshold and not only complete loss of power) versus total uptime. Distributed energy storage systems (DESSs) can help provide reliable electric service to consumers.
8. **Ride through support,** meaning the electric unit staying connected during system disturbance (voltage sag), ESSs have the potential of providing energy and support to ride-through.
9. **Unbalanced load compensation,** which is done by injecting and absorbing power individually at each phase to supply unbalanced loads.

The advanced electric energy storage technologies, when utilized properly, would have an environmental, economic and energy diversity advantages to the power systems. These include:

1. *Matching electricity supply to load demand:* energy is stored during periods when production exceeds consumption (at lower cost possible) and the stored energy is utilized at periods when consumption exceeds production (at higher cost level), so the electricity production need not be scaled up and down to meet demand variations, instead the production is maintained at a more constant and economic level. This has the advantage that fuel-based power plants (i.e. coal, oil, gas) are operated more efficiently and easily at constant production levels, while maintaining a continuous power to the customer without fluctuations.
2. *Reducing the risks of power blackouts:* energy storage technologies have the ability to provide power to smooth out short-term fluctuations caused by interruptions and sudden load changes. If applied properly, real long-term energy storage can also provide power to the grid during longer blackouts.
3. *Enabling renewable energy generation:* solar and wind energy systems are intermittent sources, expected to produce about 20% of the future electricity, generating during off-peak, when the energy has a low financial value, and the energy storage can smooth out their variability, allowing the electricity the dispatch at a later time, during peak periods, make them cost-effective and more reliable options.

TABLE 5.1

Energy Storage Applications in Power Systems

Applications	Matching Supply and Demand	Providing Backup Power	Enabling Renewable Technology	Power Quality
Discharged power	1 MW–100 MW	1 MW–200 MW	20 KW–100 MW	1 KW–20 MW
Response time	<10 min.	<10 ms (quick) <10 min. (conventional)	< 1 s	
Energy capacity	1–1000 MWh	1–1000 MWh	10 KWh–200 MWh	50–500 kWh
Efficiency needed	High	Medium	High	Low
Life time needed	High	High	High	Low

4. *Power Quality:* it may cause poor operations or failures of end-user equipment. Distribution network, sensitive loads and critical operations suffer from outages and service interruptions, lading to financial losses to utility and consumers. Energy storage, when properly engineered and implemented, can provide electricity to the customer without any fluctuations, overcoming the power quality problems such as swells/sags or spikes. A summary of energy storage applications and requirements are given in Table 5.1.

The electricity supply chain is deregulated with clear divisions between generation, transmission system operators, distribution network operators and supply companies. Energy storage applications to power distribution can benefit the customer, supply company and generation operator (conventional and DG) in several ways. Major areas where energy storage systems can be applied can be summarized as follow:

1. Voltage control, supporting a heavily loaded feeder, providing power factor correction, reducing the need to constrain DG, minimizing on-load tap changer operations and mitigating flicker, sags and swells.
2. Power flow management, by redirecting power flows, delay network reinforcement, reduce reverse power flows and minimizing losses.
3. Restoration, by assisting voltage control and power flow management in a post-fault reconfigured network.
4. Energy market, through the arbitrage, balancing market, reduced DG variability, increased DG yield from non-firm connections, replacing the spinning reserve.
5. Commercial/regulatory functionality, by assisting the grid compliance with energy security standard, reducing customer minutes lost, while reducing generator curtailment.
6. Network management, by assisting islanded networks, support black starts, switching ESS between alternative feeders at a normally open point.

It is evident that developing a compelling EES installation at power distribution level in today's electricity market with present technology costs is difficult if value is accrued from only a single benefit. The importance of understanding the interactions between various objectives and quantifying the individual benefits brought is a critical issue in ESS potential evaluation. EES systems can contribute significantly to meeting the needs for more efficient and environmentally benign energy use in buildings, industry, transportation, and utilities. Overall, the ESS uses often results in significant benefits as: reduced energy costs and consumption, improved air and water quality, increased operation flexibility, and reduced initial and maintenance costs, reduced size, more efficient and effective utilization of the equipment, fuel conservation, by facilitating more efficient energy use and/or fuel substitution, and reduced pollutant emissions. EESs have significant potential to increase the effectiveness of energy-conversion equipment use and for facilitating large-scale fuel substitutions. EESs are complex and cannot be evaluated properly without a detailed understanding of energy supplies and end-use considerations.

5.3 ENERGY STORAGE TECHNOLOGIES

The energy storage technologies can be classified in four categories depending on the type of energy stored, mechanical, electrical, thermal and chemical energy storage technologies each offering different opportunities, but also consisting of own disadvantages. In the following, each method of electricity storage is assessed and the characteristics of each technology, including overall storage capacity, energy density (energy stored per kilogram), power density (energy transfer time rate per kilogram) and round trip efficiency of energy conversion are compared. Energy storage can be defined as the conversion of electrical energy from a power network into a form in which it can be stored until converted back to electrical energy. The energy storage systems are currently characterized by: (a) disagreement on the role and design of energy storage systems; (b) common energy storage uses; (c) new available technologies are still under demonstration and illustration; (d) no recognized planning tools/models to aid understanding of storage devices; (e) system integration including power electronics must be improved; and (f) it seems small-scale storage will have great importance in the future. However, several energy storage systems are available nowadays with different characteristics, capabilities and applications. In the next chapter's subsections, a quite comprehensive presentation of various energy storage technologies is presented.

5.3.1 PUMPED-HYDRO-ELECTRIC ENERGY STORAGE (PHES)

PHES is the most mature and largest energy storage technique available. It consists of two large reservoirs located at different elevations and a number of pump/turbine units (as shown in Figure 5.5). During off-peak electrical demand, water is pumped from the lower reservoir to the higher reservoir and stored until it is needed. Once required the upper reservoir water is released through penstocks and the turbines, connected to generators producing electricity. Therefore, during generation a PHES operates similarly to a conventional hydro-electric system. The efficiency of modern PHES facilities is in the range of 70–85%. However, variable speed machines are now being used

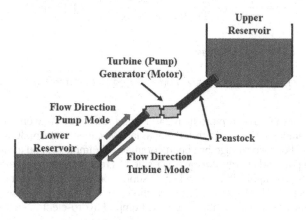

FIGURE 5.5 Typical pumped hydro-electric energy storage.

to improve it. The efficiency is limited by the pump/turbine unit efficiency used. Until recently, PHES units have always used fresh water as the storage medium. A typical PHES facility has 300 m of hydraulic head (the vertical distance between the upper and lower reservoir). The power capacity (kW) is a function of the flow rate and the hydraulic head, whilst the energy stored (kWh) is a function of the reservoir volume and hydraulic head. To calculate the mass power output of a PHES facility, the following relationship can be used:

$$P_C = \rho g Q H \eta \qquad (5.1)$$

where P_C is the power capacity in W, ρ is the mass density of water in kg/m³, g is the acceleration due to gravity in m/s², Q is the discharge through the turbines in m³/s, H is the effective head in m, and η is the efficiency. And to evaluate the storage capacity of the PHES the following must be used:

$$S_C = \frac{\rho g H V \eta}{3.6 \times 10^9} \qquad (5.2)$$

where S_C is the storage capacity in megawatt-hours (MWh), V is the volume of water that is drained and filled each day in m³. It is evident that the power and storage capacities are both dependent on the head and the volume of the reservoirs. However, facilities should be designed with the greatest hydraulic head possible rather than largest upper reservoir possible, being cheaper to construct a facility with a large hydraulic head and small reservoirs, than to construct a facility of equal capacity with a small hydraulic head and large reservoirs because: (1) less materials are removed for the required reservoirs, (2) smaller piping is necessary, hence, smaller boreholes during drilling and (3) the turbine is physically smaller.

Example 5.1   A pumped storage facility has a head of 450 m and an efficiency of 93%. What is the flow rate needed to generate 100 MW for 3.5 hour per day? Assume the water density is 10³ kg/m³ and the acceleration due to gravity 9.80 m/s. What is the required working volume per day?

SOLUTION

From Equation (5.1) the flow rate is:

$$Q = \frac{P_C}{\eta \rho g H} = \frac{10^8}{0.93 \times 9.806 \times 10^3 \times 450} = 24.383\,\text{m}^3/\text{s}$$

Working volume can be estimated as:

$$V = Q \cdot \Delta t = 24.383 \times 3.5 \times 3600 = 307{,}225.8\,\text{or} \approx 307.226 \cdot 10^3\,\text{m}^3$$

Today, there is over 90 GW in more than 240 PHES facilities in the world, roughly 3% of the world's global generating capacity. Each individual facility can store from 30 MW to 4,000 MW of electrical energy. Pumped storage has been commercially implemented for load balancing for over 80 years. The pumped energy storage is classified as real long-term response energy storage and is typically used for applications needing to supply power for periods between hours and days (power outages). In the United States only, there are 38 pumped energy storage facilities, providing a total power capacity of about 19 GW. Pumped hydro-electric energy storage usually comprises the following parts: an upper reservoir, waterways, a pump, a turbine, a motor, a generator and a lower reservoir, as shown in Figure 5.5. As in any system, including hydraulic ones, in a hydro-electric energy storage system there are losses during operation, such as frictional losses, turbulence and viscous drag, and the turbine itself is not 100% efficient. However, the efficiency of large-scale water-driven turbines can be quite high, even over 95%, while the efficiency of the dual-cycle reversible hydropower storage system typically is about 80%. There also are other losses, such as water evaporation from the reservoirs and leakage around the turbine. The water retains some kinetic energy even when it enters the tailrace. For the final conversion of hydropower to electricity, the turbine-generator losses need to be taken into account. Therefore, the overall PHES efficiency is as the ratio of the energy supplied, while generating, E_{Gen} and the energy consumed while pumping, E_{Pump}, depending on the pumping and generation efficiencies:

$$\eta_{PHES} = \frac{E_{Gen}}{E_{Pump}} = \eta_{Gen} \times \eta_{Pump} \tag{5.3}$$

The energy used for pumping a volume V of water up to height h with a pumping efficiency, η_{Pump}, and the energy supplied to the grid or load while generating with generating efficiency η_{Gen} are given by:

$$E_{Pump} = \frac{\rho g h V}{\eta_{Pump}}$$
$$E_{Gen} = \rho g h V \cdot \eta_{Gen}$$

The volumetric energy density for a pumped hydro-electric energy storage system will therefore depend on height h and is given by:

$$W_{PHES} = \frac{E_{Pump}}{V} = \frac{\rho g h}{\eta_{Pump}} \tag{5.4}$$

At first, pumped hydro storage stations usually copied conventional hydro-electric design in having the power transformation (extraction) system located outdoors close to the lower reservoir. The increase in power capacity ratings and pumping heads, combined with the higher rotational speeds of turbines, has required the hydraulic unit to be set at considerable depths below the minimum tail-water level in order to avoid cavitation. To meet these requirements, a massive concrete construction is needed to withstand the external water pressure and resist hydrostatic uplift, so the system has become increasingly expensive. A variant of pumped hydro-power storage are the underground PHES (UPHES) facilities, having the same operating principle as PHES system: two reservoirs with a large head between them. The only major difference between the two designs is the reservoir locations. In conventional PHES, suitable geological formations must be identified to build the facility, while in UPHES facilities have the upper reservoir at ground level and the lower reservoir deep below the earth's surface. The depth depends on the amount of hydraulic head required for a specific application. UPHES has the same disadvantages as PHES (large-scale required, high capital costs, etc.), with one major exception. As stated previously, the most significant problem with PHES is their geological dependence. As the lower reservoir is obtained by drilling into the ground and the upper reservoir is at ground level, so no stringent geological dependences. The major disadvantage for UPHES is its commercial youth. To date there are a very few, if any, UPHES facilities in operation, making it difficult to analyze and to trust the performance of this technology. However, the UPHES has a very bright future if cost-effective excavation techniques can be identified for its construction. Its relatively large-scale storage capacities, combined with location independence, provide storage with unique characteristics. The cost, design, power and storage capacities and environmental impacts need to be investigated in order to prove that UPHES systems are a viable option.

5.3.2 Compressed Air Energy Storage (CAES)

CAES is an established energy storage technology, being in the grid operation since late 1970s. The energy stored mechanically by compressing the air, being released to the grid when the air is expanded. Compressed air energy storage is achieved at high pressures (70 atm) and near the ambient temperatures, which means less volume and smaller reservoirs. Large caverns made of high-quality rock deep in the ground, ancient salt mines or underground natural gas storage caves are the best CAES options, as they benefit from geostatic pressure, which facilitates the containment of the air mass (see Figure 5.5). Studies have shown that the air could be compressed and stored in underground, high-pressure piping (20 to 100 bars). A CAES facility may consist of a power train motor that drives a compressor, compressing the air into the cavern, high-pressure and low-pressure turbines and a generator. In a gas turbine (GT), 66% of the energy is used is required to compress the air (Figure 5.6). Therefore, the CAES is pre-compressing the air using off-peak electrical power, taken from the grid to drive a motor (rather than using GTs) and stores it in large storage reservoirs. CAES may use the energy peaks generated by renewable energy plants to run compressors that are compressing the air into underground or surface reservoirs. The compressed air is used into turbine-generator units to generate electricity during peak demand. The energy storage capacity depends on the compressed air volume and storage pressure. When the grid is producing electricity during peak hours, the compressed air, stored by using cheaper off-peak electricity from the storage facility, is used instead of using more expensive natural gas. However, when the air is released from the reservoir it is mixed with small amounts of gas before entering the turbine, to avoid air temperature and pressure issues. If the pressure using air alone is high enough to achieve a significant power output, the air temperature would be far too low for the materials and connections to tolerate. The required gas amount is very small so a GT working with CAES can produce three times more electricity than operating alone, using the same natural gas amount. The reservoir can be man-made – the expensive choice – or by using suitable natural geological formations, such as salt-caverns, hard-rock caverns, depleted gas fields or an aquifer, selected to suit specific requirements. In a salt-cavern, fresh water is pumped into the cavern and left until the salt dissolves and saturates the fresh water, then transferred at surface to remove salt, and the cycle is repeated until the required cavern volume is created. This process is expensive and can take up to two years to complete. Hard-rock caverns are even more expensive, about 60% higher than salt-caverns. Finally, the aquifers cannot store the air at higher pressures, having relatively lower energy capacities. CAES efficiency is difficult to estimate, especially when is

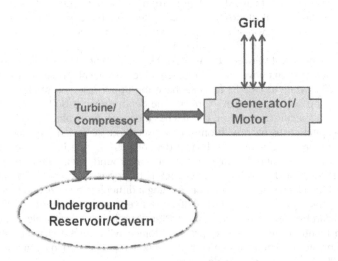

FIGURE 5.6 Diagram of typical compressed air energy storage.

using both electrical energy and natural gas, with estimated efficiencies based on the compression and expansion cycles in the range of 68–75%. CAES systems with typical capacities between 50 MW and 300 MW are used for large- and medium-scale applications. Their lifetimes are far longer than existing gas turbines and the charge/discharge ratio is dependent on the compressor and reservoir size and pressure. With assumption of ideal gas and isothermal process, the energy stored by compressing m amount of gas at constant temperature from initial pressure, P_i to final pressure, P_f is given by:

$$E_{fi} = -\int_{P_i}^{P_f} VdP = mR_g \ln\left(\frac{P_i}{P_f}\right) = P_f V_f \ln\left(\frac{P_i}{P_f}\right) = P_f V_f \ln\left(\frac{V_i}{V_f}\right)$$ (5.5)

Here, air is assumed to be an ideal gas whose specific heat is constant, R_g is the ideal gas constant (8.31447 J·K^{-1}mol^{-1}), V_i, and V_f are the initial and final volumes of the compressed air. The compression power, P_C, depends on the air flow rate, Q (the volume per unit time) and the compression ratio (P_f/P_i), expressed as:

$$P_C = \frac{\gamma}{\gamma-1} P_i \cdot Q\left[\left(\frac{P_f}{P_i}\right)^{\frac{\gamma}{\gamma-1}} - 1\right]$$ (5.6)

where γ is the ratio of air specific heat coefficients ($\gamma \approx 1.4$), P_i, P_f are the initial and final pressure, the atmospheric and compressed state, respectively. A 290 MW CAES (300000 m^3 and 48000 Pa), Hundorf, Germany was built in 1978, and a 110 MW CAES (540000 m^3 and 53000 Pa) was built in 1991 at McIntosh, Alabama. Other CAES projects are implemented or in process to be built in Canada, U.S. and E.U., using different technologies, geological structures and approaches at various power capacities.

Example 5.2   A CAES has a volume of 450000 m³, and the compressed air pressure range is from 75 bars to atmospheric pressure. Assuming isothermal process and an efficiency of 30% estimate the energy and power for a three-hour discharge period.

SOLUTION

Form Equation (5.5), assuming the atmospheric pressure 1 bar the energy is:

$$E = 45 \cdot 10^4 \times 75 \cdot 10^5 \ln\left(\frac{75}{1}\right) \approx 1457.2 \cdot 10^4 \text{ MJ}$$

The average power output is then

$$P = \frac{\eta \cdot E}{\Delta t} = \frac{0.3 \times 1457.2 \cdot 10^{10}}{2 \times 3600} \approx 404.76 \text{ MW}$$

Compression of a fluid/air in CAES generates heat, while after the decompression the air is colder. If the heat generated during compression can be stored and used again during decompression, the overall system efficiency improves considerable. CAES is achieved through *adiabatic, diabatic* and *isothermal* processes. An adiabatic storage has basically no heat exchange during the compressions-expansion cycle, heat being stored in fluids, such as oil or molten salt solutions, while in diabatic storage the heat is dissipated with intercoolers. In an isothermal process, the operating temperature is maintained constant (or rather quasi-constant) through heat exchange with the environment, being practical only for low power levels and some heat losses are unavoidable, compression process not being truly isothermal. For an isothermal processes, the maximum energy that can be stored and released is given by:

$$E_{fi} = nRT \ln\left(\frac{V_i}{V_f}\right)$$ (5.7)

Here, T is the absolute temperature (K), and n is the number of moles of the air in the reservoir.

Example 5.3   Determine the maximum stored energy if a mass of 2900 kg of air is compressed isothermally at 300 K from 100 kPa to 1500 kPa, assuming a heat loss of 33 MJ and ideal gas.

SOLUTION

With the ideal gas assumption, molar mass for air 29 kg/Kmol and R 8.314 kJ/kmol·K, the number of moles of air, is:

$$n = \frac{2900}{29} = 100 \, \text{kmol}$$

The mechanical energy stored, by using Equation (5.7) is then:

$$E_{fi} = -nRT \ln\left(\frac{V_i}{V_f}\right) = -100 \times 8.314 \times 300 \cdot \ln\left(\frac{100}{1500}\right) = 675,441.9 \text{kJ}$$

Net stored energy is then:

$$E_{net} = E_{fi} - \text{Heat Losses} = 675441.9 - 33000 = 642,441.9 \text{kJ}$$

CAES systems, the only very large grid-scale energy storage other than PHES, have fast reaction time, usually able to go from 0% to 100% in less than ten minutes, 10% to 100% in approximately four minutes and from 50% to 100% in less than fifteen seconds. As a result, CAES is ideal for acting as a large bulk energy sink, being also able to undertake frequent start-ups and shut-downs. CAES do not suffer from excessive heat when operating on partial load, as traditional gas turbines. These flexibilities mean that CAES can be used for ancillary services such as load following, frequency regulation and voltage control. As a result, CAES has become a serious contender in the wind energy industry. A number of possibilities being considered such as integrating a CAES facility with several wind farms within the same area, so the excess off-peak power from the wind farms is stored into CAES facilities. CAES advantages are high energy and power capacity, long lifetime, while the major disadvantages are low efficiencies, some adverse environmental impacts, and difficulty of the siting. CAES coupled with natural gas storage is based on the idea of coupling underground natural gas storage with electricity storage. The pressure difference between high-pressure gas storage in reservoirs deep underground and the gas injected into the conduits leads to the energy consumption for compression that can be released in the form of electricity during decompression. The natural gas and air liquefaction requires a large amount of energy that can be uses through two storage reservoirs for liquefied natural gas and liquid air, regenerative heat exchangers, compressors, and gas turbines for energy storage. Such systems are still under development.

5.3.3 Electro-Chemical Energy Storage

Chemical energy storage is classified into electro-chemical and thermochemical energy storage. The electro-chemical energy storage refers to conventional batteries, such as lead-acid (LA), nickel-metal hydride and lithium-ion (Li-ion) and flow batteries (zinc/bromine (Zn/Br) and vanadium redox) and metal-air batteries. Electro-chemical energy storage is also achieved through fuel cells (FCs), most commonly hydrogen fuel cells, but also include direct-methanol, molten carbonate and solid oxide fuel cells. Batteries and fuel cells are common energy storage devices used in power systems and in several other applications. In batteries and fuel cells, electrical energy is generated through redox reactions at the anode and cathode. The difference between batteries and fuel cells is the locations of energy storage and conversion. Batteries are closed systems, with the anode and cathode being the charge-transfer medium, taking an active role in the redox reaction as *active masses*, the energy storage and conversion occurring in the same compartment. Fuel cells are open systems where the anode and cathode are just charge-transfer media and the *active masses* undergoing the redox reaction are delivered from outside the cell, either from the environment, e.g. oxygen from air, or from a tank, e.g. hydrogen or hydrocarbons. Energy storage (in the tank) and energy conversion (in the fuel cell) are thus separated. Battery configuration consists of two electrodes and an electrolyte, placed together in a container and connected to an external device (source or load). The battery cells, consisting of positive and negative electrodes joined by an electrolyte are converting chemical energy into electrical energy, through a chemical reaction between the electrodes and the electrolyte, generating DC electricity. In the case of secondary (rechargeable) batteries, the chemical reaction is reversed by reversing the current and the battery returned to a charged state. Battery types are of two forms, *disposable* or *primary batteries* and *rechargeable* or *secondary batteries*. A primary battery is a cell or group of cells intended to be used until exhausted and then discarded, and are assembled in the charged state, while the discharge is the process during operation. A secondary battery is a cell or group of cells for the generation of electrical energy in which each cell, after being discharged, may be restored to its original charged condition by an electric current flowing in the opposite direction than during discharge. Such batteries are assembled in the discharged state, being first charged before the use through secondary processes. Mature secondary battery chemistries are: (a) lead-acid (LA), (b) nickel-cadmium (Ni-Cd), (c) nickel-metal hydride (Ni-MH), (d) lithium-ion (Li-ion), (e) lithium-polymer/lithium metal (Li-polymer), (f) sodium-sulfur (Na-S), (g) sodium-nickel chloride and (h) lithium-iron phosphate. With the increased demand growing from electric vehicles and portable consumer products, significant funds are spent by companies on the research of new battery technologies, such as zinc-based chemistries and silicon as a material for improving battery properties and performances. Higher energy density and life cycle, environmental friendliness and safer operation are among the general design research targets for secondary batteries. Primary batteries are a reasonably mature technology, in terms of chemistry, but still there is research to increase the energy density, reduce self-discharge rate, increase the battery life, or to improve the usable temperature range.

Chemical energy storage is usually achieved through accumulators and batteries, characterized by a double function of storage and release of the electricity by alternating the charge-discharge phases. They transform chemical energy through electro-chemical reactions into electrical energy and vice versa, without almost any harmful emissions or noise, while requiring little maintenance. There is a wide range of battery technologies in use, and their main assets are their energy densities (up to 2000 Wh/kg for lithium-based types) and technological maturity. Chemical energy storage devices and electro-chemical capacitors (ECs) are among the leading EES technologies today. Both technologies are based on electrochemistry, the fundamental difference being that batteries store energy in chemical reactants capable of generating charges, whereas electro-chemical capacitors store energy directly as electric charges. Although the electro-chemical capacitor is a promising technology for energy storage, especially considering its high power capabilities, the energy density is too low for large-scale energy storage. The most common rechargeable battery technologies are lead-acid, sodium-sulfur, vanadium redox and lithium ion types. During discharge, electro-chemical reactions at the two electrodes generate an electron flow through an external circuit. Vanadium redox batteries, having good prospects because they can be scaled up to much larger storage capacities, are showing great potential for longer lifetimes and lower per-cycle costs than conventional batteries requiring refurbishment of electrodes. Li-ion batteries are also displaying high potential for large-scale energy storage. A battery consists of one or more electro-chemical cells, connected in series, in parallel or series-parallel configuration to provide the desired voltage, current and power; the anode (the electronegative electrode) from which the electrons are generated to do the external work; the cathode is the electropositive electrode to which positive ions migrate inside the cell and the electrons migrate through the battery external electrical circuit; the electrolyte allows the ions and electrons to flow from one electrode to another and is commonly a liquid solution containing a dissolved salt and must be stable in the presence of both electrodes. The current collectors allow the transport of electrons to and from the electrodes, typically are made of metals and must not react with the electrode or electrolyte materials. The cell voltage is determined by the chemical reaction energy occurring inside the cell. The anode and cathode are, in practice, complex composites, containing, besides the active material, polymeric binders to hold together the powder structure and conductive diluents such as carbon black to give the whole structure electric conductivity so that electrons can be transported to the active material. In addition, these components are combined to ensure sufficient porosity to allow the liquid electrolyte to penetrate the powder structure and permit the ions to reach the reacting sites. During the charging process, the electro-chemical reactions are reversed via the application of an external voltage across the electrodes.

The battery technologies range from the mature, long-established lead-acid type through to various more recent and emerging systems and technologies. Newest technologies are attracting an increased interest for possible use in power systems, having achieved market acceptance and uptake in consumer electronics in the so-called 3Cs sector (cameras, cellphones and computers). Batteries have the potential to span a broad range of energy storage applications, in part due to their portability, ease of use, large power storage capacity (from 100 W up to 20 MW) and ease to be connected in series-parallel combinations to increase their power capacity for specific applications. Major battery advantages include: standalone operation, no need to be connected to an electrical system, easy to expand and reconfigure, while the disadvantages are: cost, limited lifecycle and maintenance. These systems could be located in any place, buildings or industrial facilities, near the demand point, can be rapidly installed, less environmental impacts of other ESS technologies. Grid-connected BESS uses an inverter to convert the battery DC voltage into AC grid-compatible voltage. These units present fast dynamics with response times near 20 ms and efficiencies ranging from 60% to 80%. The battery temperature change during charge and discharge cycles must be controlled because it affects its life expectancy. Depending on how the battery and cycle are, the BESS can require multiple charges and discharges per day. The battery cycle is normal while the discharge depth is small, but if the discharge depth is high the battery cycle duration could be degraded. The expected useful life of a Ni-Cd battery is 20,000 cycles if the discharge depth is limited to 15%. Example of large scale BESSs installed today are: 10 MW (40 MWh) Chion system, California and 20 MW (5 MWh) Puerto Rico. Their main inconvenience, however, is their relatively low durability for large-amplitude cycling. They are often used in emergency back-up, renewable-energy system storage, etc. The minimum discharge period of the electro-chemical accumulators rarely reaches less than 15 minutes. However, for some applications, power up to 100 W/kg, even a few kW/kg, can be reached within a few seconds or minutes. As opposed to capacitors, their voltage remains stable as a function of charge level. Nevertheless, between a high-power recharging operation at near-maximum charge level and nearing full discharge, the voltage can easily vary by a ratio of two.

5.3.3.1 Battery Operation Principles and Battery Types

Batteries convert the chemical energy contained in its active materials into electrical energy through an electro-chemical oxidation-reduction reversible reaction. Battery fundamental principles and operation, regardless of the battery type, can be explained by using the so-called galvanic element or electro-chemical cell. A galvanic cell consists of three main components: the *anode*, the *cathode* and the *electrolyte*. In this case of a galvanic cell, the electrons, needed for conduction, are produced by a chemical reaction. From thermodynamics we know that the work done by the electro-chemical cell comes at a cost, which in turn implies that the chemical reactions taking place within the cell must lead to a decrease in free energy. In fact for a reversible process at constant temperature and pressure the maximum work done by the system, W, is equal the free energy (Gibbs free energy) change $-\Delta G$. The work performed when transporting an electric charge e (in C) through a potential difference E (in V) is simply the product of eE. Here, of interest is to express this work in a per-mole basis. The total charge carried by one mole of positively charged ions of valence +1 is 96,487 C and this number, denoted by F, the Faraday's constant. Thus, the work produced by the electro-chemical cell is:

$$W = -\Delta G = \xi F \cdot E \Rightarrow E_{max} = \frac{-\Delta G^0}{\xi F} \tag{5.8}$$

where ξ is the valence of the ions produced in the chemical reaction. The electric potential difference across the cell electrodes, E, is the electromotive force, or EMF, of the galvanic cell. It is clear from Equation (5.8) that in order to have higher work and potential, we need to find reactions with the highest driving force, the free energy change $-\Delta G$.

Example 5.4 Estimate the maximum output voltage that a Zn-Cu electro-chemical cell can generate.

SOLUTION

The Zn-Cu reaction taking place in this electro-chemical cell is:

$$Zn + Cu^{2+} \rightarrow Cu + Zn^{2+}$$

By using equation 5.91 and from the data tables of the book appendixes (for Zn-Cu cell ξ is equal to 2, and ΔG^0 is equal to 216,160 kJ/kmole), the maximum voltage is:

$$E_{max} = \frac{-\Delta G^0}{\xi F} = \frac{216,160 \text{ kJ/kmole}}{2 \times 96,500,.0} = 1.12V$$

To understand the role of each of the electro-chemical cell components and its operation, it is best to refer to specific examples. If, e.g. a galvanic element, made of a zinc (Zn) electrode, a copper (Cu) electrode and an electrolyte ($CuSO4$, for example), the two electrodes are electrically connected (see diagram of Figure 5.7), the Zn^{2+} ions flow from Zn electrode, through electrolyte, while the electrons migrate from Zn, and eventually combine with Cu^{2+} ions, residing into the electrolyte and from copper atoms, increasing the Cu electrode volume. The direction of electrons is determined by the potential difference between metal and electrolyte, $\Delta\phi_{metal-eletrolyte,}$ and the electron transport is in such way that the metal (Zn, here) separates the electrons and ions because it has a lower metal-electrolyte potential difference than that of the Cu electrode against electrolyte. The metal (and electrode) with higher potential is serving as positive electrode (cell terminal) of a galvanic element. The metals are usually arranged in voltaic sequences, in a way that all metals (e.g. Fe, Cd, Ni, Pb or Cu) in the right side of a certain metal (e.g. Zn) for a positive pole in a combination of electrodes, chosen based of the metal-electrolyte potential difference values

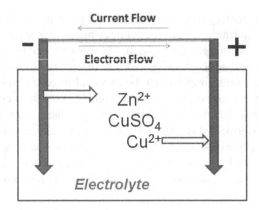

FIGURE 5.7 Galvanic element diagram (consisting of Zn and Cu electrodes, and CuSO$_4$ electrolyte). Current flows are also shown here.

of Table 5.2. For example, Zn is a negative pole with respect of Fe, Cd, Ag or Au electrodes, while Li electrode form a negative pole with respect to K, Na, Mg, Zn or Fe electrodes. The potential difference (external voltage) between the cell (galvanic element) terminals is the voltage difference, for this galvanic element, existing between Zn electrode and electrolyte (CuSO$_4$) and between Cu electrode and electrolyte. Notice that is not possible to directly measure the potential difference between the metal electrode and electrolyte, only the metal-metal potential difference can be measured.

Example 5.5 **Determine the potential differences between Li-Au, Li-Ag and Pb-H$_2$, by using the values in Table 5.2.**

SOLUTION

The required potential differences are:

$$\Delta\phi_{Li-Au} = -3.02 - 1.50 = -4.52\,V$$
$$\Delta\phi_{Li-Au} = -3.02 - 0.86 = -3.88\,V$$
And
$$\Delta\phi_{Ni-Cu} = -0.25 - 0.345 = -0.595\,V$$

TABLE 5.2

Metal Voltaic Series

Metal Electrode	Li	K	Na	Mg	Zn	Fe	Cd
$\Delta\phi_{metal\text{-}eletrolyte}$ (V)	−3.02	−2.92	−2.71	−2.35	−0.762	−0.44	−0.402
Metal Electrode	Ni	Pb	H$_2$	Cu	Ag	Hg	Au
$\Delta\phi_{metal\text{-}eletrolyte}$ (V)	−0.25	−0.126	0.0	+0.345	+0.80	+0.86	+1.50

5.3.3.2 Battery Fundamentals, Parameters and Electric Circuit Models

Battery capacity $(CAP(t)$ represents the quantity of electrical energy that can be delivered by a cell or battery. The battery capacity is proportional to the size of the battery for a given battery chemistry or technology. It is computed by the integral of current, i(t), over a defined time period as expressed:

$$CAP(t) = \int_0^t i(t)dt \tag{5.9}$$

The above relationship applies to either battery charge or discharge, meaning the capacity added or capacity removed from a battery or cell, respectively. The capacity of a battery or cell is measured in milliampere-hours (mAh) or ampere-hours (Ah). This basic definition is simple and straight; however, several different forms of capacity relationship are used in the battery industry. The distinctions between them reflect differences in the conditions under which the battery capacity is measured. Standard capacity measures the total capacity that a relatively new, but stabilized production cell or battery can store and discharge under a defined standard set of application condition, assuming that the cell or battery is fully formed, that it is charged at standard temperature at the specification rate and that it is discharged at the same standard temperature at a specified standard discharge rate to a standard end-of-discharge voltage (EODV). The standard EODV is subject to variation depending on discharge rate. When the application conditions differ from standard ones, the cell or battery capacity changes, so the term actual capacity includes all nonstandard conditions that alter the amount of capacity the fully charged new cell or battery is capable of delivering when fully discharged to a standard EODV. Examples of such situations might include subjecting the cell or battery to a cold discharge or a high-rate discharge. The portion of actual capacity that is delivered by the fully charged new cell or battery to some nonstandard EODV is called available capacity. Thus, if the standard EODV is 1.6 V/cell, the available capacity to an EODV of 1.8 V/cell would be less than the actual capacity. Rated capacity is defined as the minimum expected capacity when a new, fully formed cell is measured under standard conditions. This is the basis for C rate (defined later) and depends on the standard conditions used which may vary depending on the manufacturers and the battery types. If a battery is stored for a period of time following a full charge, some of its charge will dissipate. The capacity which remains that can be discharged is called retained capacity. In most of the practical engineering applications, the battery capacity, C, for a constant discharge rate of I (in A) as:

$$C = I \times t \tag{5.10}$$

It is clear from above equation that the capacity of a battery is reduced if the current is drawn more quickly. Drawing 1 A for 10 hours does not take the same charge from a battery as running it at 10 A for 1 hour. Notice that the relationship between battery capacity and discharge current is not linear, and less energy is recovered at faster discharge rates. This phenomenon is particularly important for electric vehicles, as in this application the currents are generally higher, with the result that the capacity might be less than is expected. It is important to be able to predict the effect of current on capacity, both when designing electric vehicles and when are designing and making instruments to measure the charge left in a battery, the so-called battery fuel gauges. The best way to do this is by using the Peukert model of battery behavior, although this model is not very accurate at low currents, as for higher currents where it models battery behavior well enough. The starting point of this model is that there is a capacity, called the Peukert capacity, which is constant and is given by the equation:

$$C_{Pkt} = I^k \cdot t \tag{5.11}$$

where C_{Pkt} is the amp-hour capacity at a 1 A discharge rate, I is the discharge current in Amperes, t is the discharge time, in hours, and k is the Peukert coefficient, typically with values from 1.1 to 1.3.

Example 5.6   A lead-acid battery has a nominal capacity of 50 Ah at a rate of 5 h, and a Puekert coefficient of 1.2. Estimate the battery Peukert capacity.

SOLUTION

This battery of a capacity of 50 Ah if discharged at a current of:

$$I = \frac{50 \text{ Ah}}{5 \text{ h}} = 10A$$

If the Peukert Coefficient is 1.2, then the Peukert Capacity is:

$$C_{Pkt} = 10^{1.2} \cdot 5 = 79.3 \text{ Ah}$$

The capacity of a battery, sometimes referred to as C_{load} or simply C, is an inaccurate measure of how much charge a battery can deliver to a load. It is an imprecise value because it depends on temperature, age of the cells, state of the charge and rate of discharge. It has been observed that two identical, fully charged batteries, under the same circumstances, will deliver different charges to a load depending on the current drawn by the load. In other words, C is not constant and the value of C for a fully charged battery is not an adequate description of the characteristic of the battery unless it is accompanied by an additional information, *rated time of discharge*, with the assumption that the discharge occurs under a constant current regime. Usually, lead-acid batteries are selected as energy storage for the building of DC microgrids because of relatively low cost and mature technology. The energy storage systems are usually operated by current closed-loop control, while the storage power is controlled by supervision unit which calculates the corresponding power reference. The storage state of charge (SOC) must be respected to its upper (maximum) and lower (minimum) SOC limits, SOC_{max} and SOC_{min}, respectively, to protect the battery from over-charging and over-discharging, as given in the Equation (5.16), below. SOC is then calculated with Equation (5.17), where SOC_0 is the initial SOC at t_0 (initial time), CREF is the storage nominal capacity (Ah) and V_s is the storage voltage.

$$SOC_{min} \leq SOC(t) \leq SOC_{max} \tag{5.12}$$

And

$$SOC(t) = SOC_0 + \frac{1}{3600 \times V_S \times CREF} \int_{t_0}^{t} (P_{SC} - P_{SD}(t))dt \tag{5.13}$$

Accurate battery models are required for the simulation, analysis and design of energy consumption of electric vehicles, portable devices or renewable energy and power system applications. The major challenge in modeling a battery are the nonlinear characteristics of the equivalent circuit parameters, which depend on the battery state of charge and are requiring complete and complex experimental and/or numerical procedures. The battery itself has internal parameters, which need to be taken care of for modeling purposes, such as internal voltage and resistance. All electric cells in a battery have nominal voltages which gives the approximate voltage when the cell is delivering electrical power. The cells can be connected in series to give the overall voltage required by a specific application.

Example 5.7   A battery bank consists of several cells, connected in series. Assuming that the cell internal resistance is 0.012 Ω and the cell electro-chemical voltage is 1.25 V. if the battery bank needs to delivery 12.5 A at 120 V to a load determine the number of cells.

SOLUTION

With cells in series, the number of cell can be estimated as:

$$N(cells) = \frac{V_L}{V_{OC}(cell)} = \frac{120}{1.25 - 12.5 \times 0.012} = 109.09$$

We are choosing 110 cells, round off to upper integer, meaning a higher terminal voltage than 120 V. However, the terminal voltage is decreasing, over the battery bank lifetime.

There are three basic battery models mostly used in engineering applications: the ideal, linear and Thevenin models (diagrams Figure 5.8). The battery ideal model, a very simple one, is made up only by a voltage source (Figure 5.8a), while ignoring the internal cell or battery impedance. The battery linear model (Figure 5.8b), a widely used battery model in applications, consists of an ideal battery with open-circuit voltage, V_0, and an equivalent series resistance, R_S, while V_{Out} represents the battery terminal (output) voltage. This terminal voltage is obtained from the open-circuit tests as well as from load tests conducted on a fully charged battery. Although this model is quite widely used, it still does not consider the varying characteristics of the internal impedance of the battery with the varying state of charge (SOC) and electrolyte concentration. Any battery, regardless of the type in the first approximation (for steady-state operation), works as constant voltage source with an internal (source/battery) resistance by considering battery linear model as shown in Figure 5.8b. These parameters, the internal voltage (V_{OC}) and resistance (R_S), are dependent of the discharged energy (Ah) as:

$$V_{OC} = V_0 - k_1 \times DoD \tag{5.14a}$$

And

$$R_S = R_0 + k_2 \times DoD \tag{5.14b}$$

Here, V_{OC}, known also as the open-circuit or electro-chemical voltage decreases linearly with depth of discharge, DoD, while the internal resistance, R_S, increases linearly with DoD. V_0 and R_0 are the values of the electro-chemical (internal) voltage and resistance, respectively, when the battery is fully charged, DoD is 0 and when fully discharged DoD is 1.0. The constant, k_1 and k_2, are determined form the battery test data, through curve fitting or other numerical procedures. The depth of discharge (DoD) is defined from the battery state of charge (SoC), as:

$$DoD = \frac{Ah \text{ drained form battery}}{Battery \text{ rated Ah capacity}} = 1 - SoC \tag{5.15}$$

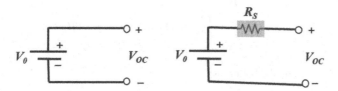

FIGURE 5.8   Battery electric diagrams, steady-state conditions, (a) ideal model and (b) linear model.

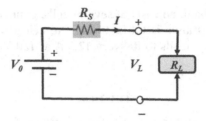

FIGURE 5.9 Electric diagram of battery-load, by using the steady-state battery linear model.

SoC (as defined in Equation 5.13) is computed for practical application by a much simpler relationship that can be estimated from the battery monitoring (test data) as:

$$SoC = 1 - DoD = \frac{\text{Ah remaining in the battery}}{\text{Rated Ah battery capacity}} \tag{5.16}$$

Notice that the battery terminal voltage is lower and the internal resistance is higher in a partially battery discharge state (i.e. any time when $DoD > 0$). All electric battery cells heave nominal voltages which gives the approximate voltage when the battery cell is delivering electrical power. Cells can be connected in series to give the required voltage. The terminal voltage, V_L of a partially discharged battery, with notation of Figure 5.9 is expressed as:

$$V_L = V_0 - I \cdot R_S = V_0 - k_1 \times DoD - I \cdot R_S \tag{5.17}$$

The load delivered power is $I^2 R_L$, the battery internal loss is $I^2 R_S$, dissipated as heat inside the battery. In consequence, as the battery discharge internal resistance, R_L increases, and more heat are generated.

Example 5.8 A 12 V lead-acid car battery has a measured voltage of 11.2 V when delivers 40 A to a load. What are the load and the internal battery resistances? Determine its instantaneous power and the rate of sulfuric acid consumption.

SOLUTION

For the battery equivalent circuit of Figure 5.9, the voltage across the load is:

$$V_L = V_0 - I \cdot R_S$$

And the internal battery and load resistances are:

$$R_S = \frac{12.0 - 11.2}{45} = 0.02\,\Omega$$

$$R_L = \frac{V_0}{I} - R_S = \frac{12}{40} - 0.02 = 0.3 - 0.02 = 0.28\,\Omega$$

The power delivered by the battery is:

$$P = V_L \times I = 11.2 \times 40 = 448\ \text{W}$$

5.3.3.3 Summary of Battery Parameters

1. **Cell and battery voltage**: all battery cells have a nominal voltage which gives the approximate voltage when delivering power. Cells are connected in series to give the overall battery voltage required. Figure 5.8 is showing one of the equivalent battery electric circuits.

2. **Battery charge capacity (Ah)**: the most critical parameter is the electric charge that a battery can supply and is expressed in ampere-hour (Ah). For example, if the capacity of a battery is 100 Ah, the battery can supply 1 A for 100 hours. The storage capacity, a measure of the total electric charge of the battery, and is an indication of the capability of a battery to deliver a particular current value for a given duration. Capacity is also sometimes indicated as energy storage capacity, in watt-hours (Wh).

3. **Cold-cranking-amperage (CCA)** for a battery composed of nominal 2 V cells is the highest current (A) that the battery can deliver for 30 seconds at a temperature of 0°F and still maintain a voltage of 1.2 V per cell. The CCA is commonly used in automotive applications, where the higher engine starting resistance, in cold winter conditions is compounded by reduced battery performance. At 0°F the cranking resistance of a car engine may be increased more than a factor of two over its starting power requirement at 80°F, while the battery output at the lower temperature is reduced to 40% of its normal output. The battery output reduction is due to the decrease of chemical reaction rates with temperature decreases.

4. **The storage capacity** is a measure of the total electric charge of the battery, is usually quoted in ampere-hours rather than in coulombs (1 Ah = 3600 C), being an indication of the capability of a battery to deliver a particular current value for a given duration. Thus a battery that can discharge at a rate of 5 A for 20 hours has a capacity of 100 Ah. Capacity is also sometimes indicated as energy storage capacity, in watt-hours. The energy stored in a battery depends on the *battery voltage*, and the *charge* stored. The SI unit for energy is Joule (J), however this is an inconveniently small unit, and in practical application Watt-hour (Wh) is used instead. The energy is expressed in Wh as:

$$\text{Energy (Wh)} = V \times A \cdot \text{hr} \tag{5.18}$$

5. **Specific energy** is the amount of electrical energy stored in a battery per unit battery mass (kg), and in practical applications is expressed in Wh/kg.

6. **Energy density** is the amount of electrical energy per unit of battery volume, expressed in practical applications and engineering in Wh/m^3.

7. **Charge (Ahr) efficiency** of actual batteries is less than 100%, and depends on the battery type, temperature, rate of charge and varies with the battery state of charge. An ideal battery will return the entire stored charge to a load, so its charge efficiency is 100%.

8. **Energy efficiency** is another important parameter, defined as the ratio of the electric energy supplied by the battery to the electric energy required to return the battery at its state before discharge. In other words, it is the ratio of the energy delivered by a fully charged battery to the recharging energy required to restore it to its original state of the charge.

9. **Self-discharge rate** refers to the fact that most of the battery types are discharging when left unused. The self-discharge, an important battery characteristic, means that some batteries must be recharged after longer periods. The self-discharge rate varies with battery type, temperature and storage conditions.

10. **Battery temperature, heating and cooling**, either most of the battery types are running at ambient temperatures, there are battery types that need heating at start and then cooling when in use. For some battery types, performances vary with temperature. The

temperature effects, cooling and heating are important parameters that designers need to take in consideration.

11. Most of the rechargeable batteries have limited **number of deep cycles** of 20% of the battery charge, in the range of hundreds or thousands cycles, which is also determine the **battery life**. The number of deep cycles depends on the battery type, design details and the ways and conditions that a battery is used. This is a very important parameter in battery specifications, reflecting the battery lifetime and its cost.

5.3.4 Fuel Cells and Hydrogen Energy

Fuel cells are an interesting alternative for power generation technologies because they have higher efficiencies and quite very low environmental effects. A fuel cell is an electro-chemical conversion device that has a continuous supply of fuel such as hydrogen, natural gas or methanol and an oxidant such as oxygen, air or hydrogen peroxide. It can have auxiliary parts to feed the device with reactants as well as a battery to supply energy for start-up. In conventional power generation systems, fuel is combusted to generate heat and then heat is converted to mechanical energy before it can be used to produce electrical energy. The maximum efficiency that a thermal engine can achieve is when it operates at the Carnot cycle, which is related to the ratio of the heat source and sink absolute temperatures. Fuel cell operation is based on electro-chemical reactions and not fuel combustion, by avoiding the chemical energy conversion into mechanical energy, through thermal phase enables fuel cells to achieve higher efficiency than that of conventional power generation technologies. A fuel cell can be considered as a "cross-over" of a battery and a thermal engine, resembling an engine because theoretically it can operate as long as it is fed with fuel. However, its operation is based on electro-chemical reactions, resembling batteries providing significant advantages for fuel cells. On the other hand, batteries are devices that when their chemical energy is depleted, they must be replaced or recharged, whereas fuel cells can generate electricity as long are fueled. However, fuel cells resemble rechargeable batteries, while their theoretical open-circuit voltage is given by Equation (11.7). The open-circuit voltage of single fuel cell is about 1.2 V.

5.3.4.1 Hydrogen Storage and Economy

Hydrogen has been advocated for quite a long time as environmentally friendly and a powerful energy storage medium and fuel. Among the most important advantages of using the hydrogen as energy storage medium are the lightest element, very stable compound, on volumetric basis can store several times more energy than compressed air, reacting easily with oxygen to generate energy, forming water, harmless to the environment, can be easily used in fuel cells and has a long industrial application history. However, the hydrogen has a few disadvantages as energy storage medium: it is flammable and explosive, requiring special containers and transportation, highly diffusive due to its specific energy content and being the lightest element has high pressure and large containers must be used for significant mass storage. Hydrogen (H_2) can be produced with electrolysis, consisting of an electric current applied to water, which separates it into components O_2 and H_2. The oxygen has no inherent energy value, but the higher heat value (HHV) of the resulting hydrogen can contain up to 90% of the applied electric energy. This hydrogen can then be stored and later combusted to provide heat or work or to power fuel cells. Compression to a storage pressure of 350 bar, the value usually assumed for automotive technologies, consumes up to 12% of the hydrogen's HHV if performed adiabatically, although the loss approaches a lower limit of 5% in a quasi-isothermal compression. Alternatively, the hydrogen can be stored in liquid form, a process that costs about 40% of HHV, using current technology and that at best would consume about 25%. Liquid storage is not possible for automotive applications because mandatory boil-off from the storage container cannot be safely released in closed spaces. Hydrogen can also be bonded into metal hydrides using an absorption process. The storage energy penalty may be lower for this process, which requires pressurization to only 30 bars. However, the density of the metal hydride can be up to 100 times the density of the stored hydrogen. Carbon nanotubes have also received attention as a potential hydrogen storage medium. A major issue of the use of hydrogen as energy medium is the hydrogen embrittlement (grooving) consisting of hydrogen diffusion through metal matrices that can lead to small cracks compromising the hydrogen storage container quality. However, despite its disadvantages, its higher storage capacity, abundance and relatively easy way to be produced through electrolysis from water make the hydrogen a strong candidate for energy storage medium. For these reasons, some scientists are suggesting that the widespread use of hydrogen may transform our economy into the *hydrogen economy.*

In the hydrogen economy context, the hydrogen is an energy carrier rather than a primary energy source. It may be generated through electrolysis or other industrial processes, using the energy harnessed by wind or solar power conversion systems, or chemical methods and used as fuel with almost no harmful environmental impacts. However, the establishment of the hydrogen economy requires that the issue of hydrogen storage and transportation are solved and suitable materials for storage are available. A hydrogen economy may lead to widespread of the use of renewable energy for power generation, avoiding the use of expensive and environmentally harmful fossil fuels. Whether a hydrogen economy evolves in the near or far future is strongly dependent on the technological advances in hydrogen-based storage and transportation. Proponents of the hydrogen economy are making the case based on the fact that hydrogen is the cleanest end-user energy source, especially for transportation and one of the most abundant natural elements. Almost every country can become energy independent in this scenario. Critics of this transition to the hydrogen economy are arguing that the cost is prohibitive, and a transition in intermediate steps may be more economically viable, with the transition focusing on more locally than regionally, entire country or globally. A future and extensive hydrogen infrastructure. It is believed be established in the same ways as the conventional energy distribution networks were established during the 20th century. Moreover, the options and technologies exist for the storage of hydrogen: compressed gas, storage based on chemical compounds, liquid hydrogen storage, and metallic hydrides.

5.3.4.2 Fuel Cell Principles and Operation

Fuel cells are electro-chemical devices that are producing electricity form paired oxidation or reduction reactions, being in some way batteries with flows/supplies of reactants in and products out. Fuel cells are hardly a new idea, being invented in about 1840, but they are really making their mark as a power source for electric vehicles, space applications and consumer electronics, only in the last part of 20th century and their time is about to come. A battery has all of its chemicals stored inside, and it converts those chemicals into electricity too. This means that a battery eventually "goes dead" and you either throw it away or recharge it. With a fuel cell, chemicals constantly flow into the cell so it never goes dead – as long as there is a flow of chemicals into the cell, the electricity flows out of the cell. Fuel cells, similar to batteries, exhibit higher efficiency at partial load than at full load and with less variation over the entire operating range, having good load following characteristics. Fuel cells are modular in construction with consistent efficiency regardless of size. Reformers, however, perform less efficiently at part load so that overall system efficiency suffers when used in conjunction with fuel cells. Fuel cells, like batteries, are devices that react chemically and instantly to changes in load. However, fuel cell systems are comprised of predominantly mechanical devices, each of which have their own response time to changes in load demand. Nonetheless, fuel cell systems that operate on pure hydrogen tend to have excellent overall response. Fuel cell systems operate on reformate using an on-board reformer, which can be sluggish, particularly if steam reforming techniques are used. The fuel cells are distinguished form the secondary rechargeable batteries by their external fuel storage and extended lifetime. Fuel cells have the advantages of high efficiency, low emissions, quite operations, good reliability and fewer moving parts (only pumps and fans to circulate coolant and reactant gases) over other energy generation systems. Their generation efficiency is very high in fuel cells, higher power density, lower vibration characteristics. A fuel cell is a DC voltage source, operating at about 1 V level. However, this might be set to change over the next 20–30 years. The basic principle of the fuel cell is that it uses hydrogen fuel to produce electricity in a battery-like device, as discussed later. The fuel cell basic chemical reaction is:

$$2H_2 + O_2 \rightarrow 2H_2O + \text{Energy} \qquad (5.19)$$

The reaction products are thus water and energy; the sole reaction product of a hydrogen-oxygen fuel cell is water, an ideal product from a pollution standpoint. In addition to electrons, heat is also a reaction product. This heat must be continuously removed as it is generated, in order to keep the cell reaction isothermal. There are quite a few problems and challenges for fuel cells to overcome before they become a commercial reality as a vehicle power source or other power applications, such as cost, thermal management and catalyst degradation. Hydrogen is the preferred fuel for fuel cells, but hydrogen is very difficult to store and transport. However, there is great hope that these problems can be overcome, and fuel cells can be the basis of less environmentally damaging transport. There are for main parts of a fuel cell: *the anode, the cathode, the catalyst and the proton exchange membrane (PEM)*. The catalyst is a special material that facilitates the reaction of oxygen and hydrogen, made usually of platinum powder very thinly coated onto carbon paper or cloth. The catalyst is rough and porous so that the maximum surface area of the platinum can be exposed to the hydrogen or oxygen. The platinum-coated side of the catalyst faces the PEM. The electrolyte is the proton exchange membrane, being a specially treated material that only conducts positively charged ions, and the membrane blocks the electrons. In order to understand the fuel cell operation, the separate reactions taking place at each electrode must be considered. These important details vary for different types of fuel cell, but if we start with a cell based on an acid electrolyte, we shall consider the simplest and the most common type. At the anode of an acid electrolyte fuel cell, the hydrogen gas ionizes, releasing electrons and creating H+ ions:

$$2H_2 \rightarrow 4H^+ + 4e^- \qquad (5.20)$$

During this reaction energy is released. At the cathode, oxygen reacts with electrons taken from the electrode, and H+ ions from the electrolyte, to form water.

$$O_2 + 4H^+ + 4e^- \rightarrow 2H_2O \qquad (5.21)$$

In order to make sure that these reactions proceed continuously, electrons produced at the anode must pass through an electrical circuit to the cathode. Also, H^+ ions must pass through the electrolyte, so an acid, having free H^+ ions, serves this purpose very well. Certain polymers can also contain mobile H^+ ions. These reactions may seem simpler, but they are not in normal circumstances. Also, the fact that hydrogen has to be used as a fuel is a disadvantage. To solve these and other problems, several fuel cell types have been researched. The different types are usually distinguished by the used electrolyte, though there are other important differences as well. In fuel cells, similar to batteries, the electrode reactions are surface phenomena, occurring at a liquid-solid or gas-solid interface and therefore proceed at a rate proportional to the exposed solid areas. For this reason, porous electrode materials are used, often porous carbon impregnated or coated with a catalyst to speed the reactions. Because of the reaction rate-area relation, fuel cell current and power output increase with increased cell area. The surface power density (W/m^2) is an important parameter in comparing fuel cell designs, and the fuel cell power output can be scaled up by increasing its surface area. The electrolyte acts as an ion transport medium between electrodes. The theoretical maximum energy of an isothermal fuel cell (or other isothermal reversible control volume) is the difference in free energy (Gibbs) functions of the cell reactants and the products. The drop in free energy is mainly associated with the Equation (5.31), expressing free energy of the chemical potential, e.g.

of the H^+ ions dissolved in the electrolyte, which related to the maximum (open-circuit) voltage of a fuel cell. The maximum voltage of a fuel cell is about 1.23 V, as estimated in the example below. The maximum work output of which a fuel cell is able to perform is given by the decrease in its free energy, ΔG. The fuel cell conversion efficiency, η_{fc}, is defined as the electrical energy output per unit mass (or mole) of fuel to the corresponding heating value of the fuel consumed, the total energy drawn from the fuel, given by the maximum energy available from the fuel in an adiabatic steady-flow process, which the difference in the inlet and exit enthalpies, ΔH. The thermal fuel cell efficiency is expressed as:

$$\eta_{fc}(thermal) = \frac{\Delta G}{\Delta H} \tag{5.22}$$

5.3.4.3   Fuel Cell Types and Applications

An important element of fuel cell design is that, similar to the large battery systems, it is built from a large number of identical units or cells. Each has an open-circuit voltage on the order of 1 V, depending on the oxidation-reduction reactions taking place in the cell. The fuel cells are usually built in sandwich-style assemblies called *stacks*, while the fuel and oxidant crossflow through a portion of the stack. A fuel cell stack can be configured with many groups of cells in series and parallel connections to further tailor the voltage, current and power produced. The number of individual cells contained within one stack is typically greater than 50 and varies significantly with stack design. The basic components that comprise a fuel cell stack include the electrodes and electrolyte with additional components required for electrical connections and/or insulation and the flow of fuel and oxidant through the stack. These key components include current collectors and separator plates. The current collectors conduct electrons from the anode to the separator plate. The separator plates provide the electrical series connections between cells and physically separate the oxidant flow of one cell from the fuel flow of an adjacent cell. The channels in the current collectors serve as the distribution pathways for the fuel and oxidant. Often, the two current collectors and the separator plate are combined into a single unit called a bipolar plate. Electrically conducting bipolar separator plates serve as direct current transmission paths between successive stack cells. This modular type of construction allows research and development of individual cells and engineering of fuel cell systems to proceed in parallel. The preferred fuel for most fuel cell types is the hydrogen. Hydrogen is not readily available; however, the infrastructure for the reliable extraction, transport or distribution, refining and/or purification of hydrocarbon fuels is well established. Thus, fuel cell systems that have been developed for practical applications to date have been designed to operate on hydrocarbon fuels. In addition to the fuel cell system requirement of a fuel processor for operation on hydrocarbon fuels, a power conditioning and for grid connection to supply an AC load, an inverter is also needed. There are five major types of fuel cells known or used in the stationary and mobile applications. All these fuel cell types have the same basic design as mentioned above, but with different chemicals used as the electrolyte. These fuel cells are:

1. Alkaline Fuel Cell (AFC);
2. Phosphoric Acid Fuel Cell (PAFC);
3. Molten Carbonate Fuel Cell (MCFC);
4. Solid Oxide Fuel Cell (SOFC); and
5. Proton Exchange Membrane Fuel Cell (PEMFC)

All these fuel cells, regardless of the type, require fairly pure hydrogen fuel to run. However, large amount of hydrogen gas is difficult to transport and store. Therefore, a reformer is normally equipped inside these fuel cells to generate hydrogen gas from liquid fuels such as gasoline or methanol. Among these five types of fuel cell, PEMFC has the highest potential for widespread use. PEMFC is getting cheaper to manufacture and easier to handle. It operates at relatively low temperature when compared with other types of fuel cell. AFC systems have the highest efficiency and are therefore being used to generate electricity in spacecraft systems for more than 30 years. However, it requires very pure hydrogen and oxygen to operate and thus the running cost is very expensive. As a result, AFCs are unlikely to be used extensively for general purposes, such as in vehicles and in our homes. In contrast, the MCFC and the SOFC are specially designed to be used in power stations to generate electricity in large-scale power systems. Nevertheless, there are still a lot of technical and safety problems associated with the use of these fuel cells (MCFC and SOFC) in the long-term applications. Apart from these fuel cell types, a new type of fuel cell, the direct methanol fuel cell (DMFC), being under vigorous on-going research, is coming on the market. This type of fuel cell has the same operating mechanism as PEMFC, but instead of using pure hydrogen, it is able to use methanol directly as the basic fuel. A reformer is therefore not essential in this fuel cell system to reform complex hydrocarbons into pure hydrogen.

5.3.5 Flywheel Energy Storage (FES)

A flywheel energy storage system converts electrical energy supplied from DC or three-phase AC power source into kinetic energy of a spinning mass or converts kinetic energy of a spinning mass into electrical energy (as shown in Figure 5.10). Basically, a flywheel is a disk with a certain amount of mass that can spin, holding kinetic energy. Modern high-tech flywheels are built with the disk attached to a rotor in upright position to prevent gravity influence. They are charged by a simple electric motor that simultaneously acts as a generator in the process of discharging. When dealing with efficiency, however, it gets more complicated, as stated by the rules of physics, they will eventually have to deal with friction during operation. Therefore, the challenge to increase that efficiency is to minimize friction. This is mainly accomplished by two measures: the first one is to let the disk spin in a vacuum, so there will be no air friction; and the second one is to bear the spinning rotor on permanent and electromagnetic bearings so it basically floats. The spinning speed for a modern single flywheel reaches up to 16.000 RPM and offers a capacity up to 25 kWh, which can be absorbed and injected almost instantly. These devices are comprised of a massive or composite flywheel coupled with a motor-generator and special brackets (often magnetic), set inside a housing at very low pressure to reduce self-discharge losses. They have a great cycling capacity (a few 10,000 to a few 100,000 cycles) determined by fatigue design. For power system applications, high-capacity energy flywheels are needed. Friction losses of a 200 tons flywheel are estimated at about 200 kW. Using this hypothesis and an efficiency of about 85%, the overall efficiency would drop to 78% after 5 h and 45% after one day. Long-term energy storage with flywheels is therefore not foreseeable. A FES device is made up of a shaft, holding a rotor, rotating on two magnetic bearings to reduce friction. These are all contained within a vacuum to reduce aerodynamic drag losses. Flywheels store energy by accelerating the rotor/flywheel to a very high speed and maintaining the

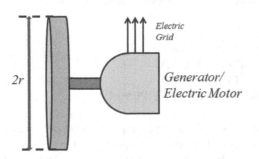

FIGURE 5.10 Flywheel and generator-motor configuration.

energy in the system as kinetic energy, and release the energy by reversing the charging process so that the motor is used as a generator. As the flywheel discharges, the rotor slows down until eventually coming to a complete stop. The rotor dictates the amount of energy that the flywheel is capable of storing. Due to their simplicity, flywheel energy storage systems have been widely used in small power units (about 3 kWh) in the range of 1 kW per 3 hours to 100 kW per 3 seconds. Energy is stored as kinetic energy using a rotor:

$$E = \frac{1}{2} J \omega^2 \tag{5.23}$$

where J is the momentum of inertia and ω is the angular velocity. The moment of inertia is given by the volume integral taken over the product of mass density ρ, and squared distance r^2 of mass elements with respect to the axis of rotation:

$$J = \int \rho r^2 dV$$

However, in the case of regular geometries, the momentum of inertia is given by:

$$J = kmr^2 \tag{5.24}$$

Here, m is the flywheel total mass, and r the outer radius of the disk. Equation (5.23) then became:

$$E = \frac{1}{2} kmr^2 \omega^2 = \frac{1}{2} k(\rho \Delta V) r^2 \omega^2 \tag{5.25}$$

Here, ΔV is the increment of the volume. The inertial constant depends on the shape of the rotating object. For thin rings, $k = 1$, while for a solid uniform disk, $k = 0.5$. Flywheel rotor is usually a hollow cylinder, and has magnetic bearings to minimize the friction. The rotor is located in a vacuum pipe to decrease the friction even more. The rotor is integrated into a motor/generator machine that allows the energy flow in both directions. The energy storage capacity depends on the mass and shape of the rotor and on the maximum available angular velocity.

Example 5.9 A flywheel is a uniform circular disk of diameter of 2.90 m, 1500 kg and is rotating at 5000 RPM. Calculate the flywheel kinetic energy.

SOLUTION

In Equation (5.25), $k = 0.5$, so:

$$E = \frac{1}{2} kmr^2 \omega^2 = \frac{1}{2} \times 0.5 \times 1500 \times \left(\frac{2.90}{2}\right)^2 \times \left(\frac{2\pi 5000}{60}\right)^2 = 2.16142 \times 10^8 J = 216.142 MJ$$

There are two topologies, *slow* flywheels (with angular velocity below 6,000 rpm) based on steel rotors and *fast* flywheels (below 60,000 rpm), using advanced material rotors (carbon fiber or glass fiber) that present higher energy and power densities than steel rotors. The flywheel designs are modular and systems of 10 MW are possible, the efficiency of 80–85%, with a useful life of 20 years. The advances on the rotor technology have permitted high dynamics and a durability of tenths of thousands of cycles, making them suitable for power quality applications: frequency deviations, temporary interruptions, voltage sags and voltage swells. Usually, the trend is that when FES systems are applied to the renewable energy systems is to combine them with other energy storage technologies, like micro-CAES or thermal energy storage. Flywheels store power in direct relation to the mass of the rotor, but to the square of its surface speed. The amount of energy which can be stored by a flywheel is determined by the material design stress, material density and total mass, as well as flywheel shape factor K. It is not directly dependent on size or angular speed since one of these can be chosen independently to achieve the required design stress. Material properties also govern flywheel design and therefore allowable K values. In order to take maximum advantage of the best properties of highly anisotropic materials, the flywheel shape is such that lower K values have to be accepted, compared to those normally associated with flywheels made from isotropic material. However, at first sight from Equations (5.33) to (5.35), we are tempted to maximize spatial energy density by increasing the product of the three factors ω^2, ρ and the squared distance r^2 of the relevant mass elements with respect to the axis of rotation. This approach, however, cannot be extended to arbitrarily high spatial energy densities because the centrifugal stresses caused by the rotation also roughly increase with the product of these three factors. To avoid fragmentation of the flywheel, a certain material dependent tensile stress level must not be exceeded. Consequently, the most efficient way to store energy in a flywheel is to make it spin faster, not by making it heavier. The energy density within a flywheel is defined as the energy per unit mass:

$$\frac{W_{KIN}}{m_{FW}} = 0.5 \cdot v_l^2 = \frac{\sigma}{\rho} \tag{5.26}$$

where W_{KIN} is the total kinetic energy in Joules (J), m_{FW} is the mass of the flywheel in kg, v_l is the linear velocity of the flywheel in m/s, σ is the specific strength of the material in Nm/kg and ρ is the density of the material in kg/m^3. For a rotating thin ring, therefore, the maximum energy density is dependent on the specific strength of the material and not on the mass. The energy density of a flywheel is normally the first criterion for the selection of a material. Regarding specific strength, composite materials have significant advantages compared to metallic materials. Table 5.3 lists some flywheel materials and their properties. The burst behavior is a deciding factor for choosing a flywheel material. However, not all the stored energy, during charging phase can be used during discharging phase. The useful energy per mass unit (the energy released during the discharge) is expressed as:

$$\frac{E}{m} = (1 - s^2)K\frac{\sigma}{\rho} \tag{5.27}$$

where s is the ratio of minimum to maximum operating speed, usually taken to be 0.2.

Example 5.10   Compute the energy density for a steel flywheel.

SOLUTION

Using Equation (5.26) and the values of the density and specific strength for steel, from Table 5.3, the flywheel energy density is:

$$\frac{W_{KIN}}{m_{FW}} = \frac{0.22 \times 10^6}{7800} = 28.21 J/kg$$

The power and energy capacities are decoupled in flywheels. In order to obtain the required power capacity, you must optimize the motor/generator and the power electronics. These systems, the *low-speed flywheels*, have relatively low rotational speeds, approximately 10,000 RPM and a heavy rotor made form steel. They can provide up to 1650 kW for very short times

TABLE 5.3

Typical Materials Used for Flywheels and Their Properties

Material	Density (kg/m³)	Strength (MN/m²)	Specific Strength (MNm/kg)
Steel	7800	1800	0.22
Alloy (AlMnMg)	2700	600	0.22
Titanium	4500	1200	0.27
GRFP[a]	2000	1600	0.80
CFRP[a]	1500	2400	1.60

[a] GFRP, glass fiber reinforced polymer; CFRP, carbon fiber-reinforced polymer

(up to 120 s). To optimize the flywheel storage capacities, the rotor speed must be increased. These high-speed flywheels spin on a lighter rotor at much higher speeds, with prototype composite flywheels claiming to reach speeds in excess of 100,000 RPM. However, the fastest flywheels commercially available spin at about 80,000 RPM. They can provide energy up to an hour, with a maximum power of 750 kW. Over the past years, the flywheel efficiency has improved up to 80%, although some sources claim efficiencies as high as 90%. Flywheels have an extremely fast dynamic response, a long life (~20 years), require little maintenance and are environmentally friendly. As the storage medium used in flywheels is mechanical, the unit can be discharged repeatedly and fully without any damage to the device. Flywheels are used for power quality enhancements, uninterruptable power supply, capturing waste energy that is very useful in electric vehicle applications and finally, to dampen frequency variation, making FES very useful to smooth the irregular electrical output from wind turbines. The stored energy in flywheels has a significant destructive potential when released uncontrolled, safety being a major disadvantage. Efforts are being made to design rotors such that, in the case of a failure, many thin and long fragments, having little trans-lateral energy, are released, so the rotor burst can be relatively benign. However, even with careful design, a composite rotor still can fail dangerously. The safety of a flywheel system is not related only to the rotor. The housing enclosure, and all components and materials within it can influence the result of a burst significantly. To facilitate mechanical ESS by flywheel, low-loss and long-life bearings and suitable flywheel materials need to be developed. Some new materials are steel wire, vinyl-impregnated fiberglass and carbon fiber. However, the major advantages of the flywheels include: high power density, nonpolluting, high efficiency, long life (over 20 years) and independent operation from extreme weather conditions. Their major disadvantages are: safety, noise and high-speed operations leading to wear, vibration and fatigue.

5.3.6 SUPERCONDUCTING MAGNETIC ENERGY STORAGE

Superconducting magnetic energy storage (SMES) exploits advances in materials and power electronics technologies to achieve novel energy storage based on three principles of physics: (a) superconductors carry current with no resistive losses, (b) electric currents induce magnetic fields and (c) magnetic fields are energy forms that can be stored. These principles provide the potential for the highly efficient electricity storage in superconducting coils. Operationally, SMES is different from other storage technologies in that a continuously circulating current within the superconducting coil produces the stored energy, the only conversion process in the SMES system is from AC to DC power conversion, i.e. there are no thermodynamic losses inherent in this conversion. Basically, SMESs store energy in the magnetic field created by the flow of direct current in a superconducting coil which has been cryogenically cooled to a temperature below its superconducting critical temperature. The idea is to store energy in the form of an electromagnetic field surrounding the coil, which operating at very low temperatures, to become superconducting, which made the system a superconductor. SMES makes use of this phenomenon and – in theory – stores energy without almost any energy loss (practically about 95% efficiency). SMES was originally proposed for large-scale load leveling; however, because of its rapid discharge capabilities, it has been implemented on electric power systems for pulsed-power and system-stability applications.

The power and stored energy in a SMES system are determined by application and site-specific requirements. Once these values are set, a system can be designed with adequate margin to provide the required energy on demand. SMES units have been proposed over a wide range of power $(1–1000\ MW_{AC})$ and energy storage ratings (0.3–1000 MWh). Independent of size, all SMES systems include three parts: superconducting coil, power conditioning system (power electronics and control) and cryogenically cooled refrigerator. Once the superconducting coil is charged, the current is not decaying and magnetic energy can be stored indefinitely. The stored energy can be released back to the network by discharging the coil. SMES loses the least electricity amount in the energy storage process compared to any other methods of storing energy. There are several reasons for using superconducting magnetic energy storage instead of other energy storage methods. The most important advantage of SMES is that the time delay during charge and discharge is quite short as compared with other energy storage technologies. Power is available almost instantaneously and very high power output can be provided for a brief period of time. Thus, if demand is immediate, SMESs are a viable option. Another advantage is that the loss of power is less than other storage methods because electric currents encounter almost no resistance. Additionally, the main parts in a SMES are motionless, resulting in high reliability. The magnetic energy stored by a carrying current coil is given by:

$$E_{SMES} = \frac{1}{2}LI^2 \qquad (5.28)$$

where E is the energy, expressed in J, L is the coil inductance measured in H and I is the current. The total stored energy, or the level of charge, can be found from the above equation and the current in the coil. Alternatively, the SMES magnetic energy density per unit volume, for a magnetic flux density B (in T) and magnetic permeability of free space, μ_0 $(= 4\pi \times 10^{-7}$ H/m) is expressed as:

$$E_{SMES} = \frac{B^2}{2\mu_0} \approx 4 \times 10^5 B^2\ \mathrm{J \cdot m^{-3}} \qquad (5.29)$$

Example 5.12   Find the energy density for a SMES that has a magnetic flux density of 4.5 T.

SOLUTION

Applying Equation (4.29) the volume energy density is:

$$E_{SMES} \approx 4 \times 10^5 (4.5)^2 = 8.1\mathrm{MJ \cdot m^{-3}}$$

The maximum practical stored energy, however, is determined by two factors: the size and geometry of the coil, which determine the inductance. The characteristics of the conductor determine the maximum current. Superconductors carry substantial currents in high magnetic fields. For example, at 5 T, which is 100,000 times greater than the earth's field, practical superconductors can carry currents of 300,000 A/cm². For example, for a cylindrical coil with conductors of a rectangular cross-section, with mean radius of coil R, a and b are width and depth of the conductor, f called the form function determined by the coil shapes and geometries, ξ and δ are two parameters to characterize the dimensions of the stored energy is a function of coil dimensions, shape, geometry, number of turns and carrying current, given by:

$$E = \frac{1}{2}RN^2I^2f(\xi,\delta) \tag{5.30}$$

where, I is the current, $f(\xi,\delta)$ is the form function $(J/A\text{-}m)$ and N is the number of turns of coil.

Example 5.13   If the current flowing in a SMES coil having 1000 turns, a radius of 0.75 m is 250 A, compute the energy stored in this coil. Assume a form function of 1.35.

SOLUTION

For the SEMS characteristics the stored energy is:

$$E = 0.5 \times 0.75 \times (10^3)^2 (250)^2 \times 1.35 = 31.64063 \times 10^9 \, J$$

The superconductor is one of the major costs of a superconducting coil, so one design goal is to store the maximum amount of energy per quantity of superconductor. A primary consideration in the design of a SMES coil is the maximum allowable current in the conductor. It depends on conductor size, the superconducting materials used, the resulting magnetic field and the operating temperature. The magnetic forces can be significant in large coils, a containment structure within or around the coil is needed. The superconducting SMES coil must be maintained at a temperature sufficiently low to sustain the conductor superconducting state, about 4.5 K (–269°C, or –452°F). This thermal operating regime is maintained by a special cryogenic refrigerator that uses helium as the refrigerant, being the only material that is not a solid at these temperatures. Thermodynamic analyses have shown that power required to remove heat from the coil increases with decreasing temperature. Including inefficiencies within the refrigerator itself, 200–1000 W of electric power is required for each watt that is removed from the 4.5 K environment. As a result, design of SMES and other cryogenic systems places a high priority on reducing losses within the superconducting coils and minimizing the flow of heat into the cold environment. Both the power requirements and the physical dimensions of the refrigerator depend on the amount of heat that must be removed from the superconducting coil. However, small SMES coils and modern MRI magnets are designed to have such low losses that very small refrigerators are adequate.

Charging and discharging a SMES coil is different from that of other energy storage technologies, because it carries a current at any state of charge. Since the current always flows in one direction, the power conversion system (PCS) must produce a positive voltage across the coil when energy is to be stored, while for discharge, the power electronic system is adjusted to make it appear as a load across the coil, producing a negative voltage, causing the coil to discharge. The applied voltage times the instantaneous current determines the power. SMES manufacturers design their systems so that both the coil current and the allowable voltage include safety and performance margins. The PCS power capacity determines the rated SMES unit capacity. The control system establishes the link between grid power demands and power flow to and from the SMES coil. It receives dispatch signals from the power grid and status information from the SMES coil, and the integration of the dispatch request and charge level determines the response of the SMES unit. The control system also measures the condition of the SMES coil, the refrigerator and other equipment, maintaining system safety and sends system status information to the operator. The power of a SMES system is established to meet the requirements of the application, e.g. power quality or power system stability. In general, the maximum power is the smaller of two quantities the PCS power rating and the product of the peak coil current and the maximum coil withstand voltage. The physical size of a SMES system is the combined sizes of the coil, the refrigerator and the PCS, each depending on a variety of factors. The overall efficiency of a SMES plant depends on many factors. In principle, it can be as high as 95% in very large systems. For small power systems, used in power quality applications, the overall system efficiency is lower. Fortunately, in these applications, efficiency is usually not a critical factor. The SMES coil stores energy with absolutely no loss while the current is constant. There are, however, losses associated with changing current during charging and discharging, and the resulting magnetic field changes. In general, these losses referred to as eddy current and hysteresis losses are also small. Major losses are in the PCS and especially in the refrigerator system. However, power quality and system stability applications do not require high efficiency because the cost of maintenance power is much less than the potential losses to the user due to a power outage.

5.3.7 SUPERCAPACITORS

Supercapacitors are very-high-capacity electrolytic devices that store energy in the form of electrostatic charge. They are composed of two electrodes with a very thin separator. Energy storage capacity increases as the surface area of the electrodes increases. Energy is stored as a DC electric field in the supercapacitor, while the systems uses power electronics to both charge and discharge the supercapacitors. Supercapacitors can have very high discharge rates and could handle fast load changes in a power system or a microgrid, for example. In electro-chemical capacitors (or supercapacitors), energy may not be delivered via redox reactions and, thus the use of the terms anode and cathode may not be appropriate but are in common usage. By orientation of electrolyte ions at the electrolyte/electrolyte interface, so-called electrical double layers (EDLs) are formed and released, which results in a parallel movement of electrons in the external wire, that is, in the energy-delivering process. Capacitors consist of two conducting plates are separated by an insulator, as shown in Figure 5.11. A DC voltage is connected across the capacitor, with one plate being positive and the other negative. The opposite charges on the plates attract and hence store energy. The electric charge $Q(C)$ stored in a capacitor of capacitance C (in F) at a voltage of V (V) is given by the equation:

$$Q = C \cdot V \tag{5.31}$$

Similar to the flywheels, the capacitors can provide large energy storage capabilities, and they are usually used in small size configurations as components in electronic circuits and systems. The large energy storing capacitors with very large plate areas are the so-called supercapacitors or ultracapacitors. The energy stored in a capacitor is given by the equation:

$$E = 0.5C \cdot V^2 \tag{5.32}$$

where $E(J)$. The capacitance C of a capacitor in Farads will be given by the equation:

$$C = \frac{\varepsilon A}{d} \tag{5.33}$$

Here, ε is the electric permittivity of the material between the plates, A is the plate area and d is the separation of the plates, all expressed in standard units. The key technology of the

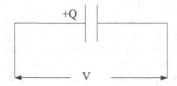

FIGURE 5.11 Capacitor symbol and diagram.

supercapacitors is that the separation of the plates is very small while the plate area is very large. In contrast to the conventional capacitors, the electric double layer capacitors do not have any dielectrics in general, but rather utilize the phenomenon typically referred to as the electric double layer. In the double layer, the effective "dielectric" thickness is exceedingly thin, and because of the porous nature of the carbon the surface area is extremely large, which translates to a very high capacitance. Generally, when two different phases come in contact with each other, positive and negative charges are set in array at the boundary. At every interface an array of charged particles and induced charges exist. This array is known as *Electric Double Layer.* The high capacitance of an EDLC arises from the charge stored at the interface by changing electric field between anode and cathodes. The very large capacitances arise from the formation on the electrode surface of a layer of electrolytic ions (the double layer), having huge surface areas, leading the capacitances of tens, hundreds or even thousands of *Farads,* with the capacitor fitted into a small size container.

However, the problem with this technology is that the voltage across the capacitor can only be very low, usually lower than 3 V. Equation (5.40) severely limits the energy that can be stored for a given capacity and so the voltage. In order to store charge at a reasonable voltage, several capacitors are usually connected in series. This not only increases the cost, but putting capacitors in series the total capacitance is reduced, as well as the charge equalization problem. In a string of capacitors in series, the charge on each one should be the same, as the same current flows through any series circuit. However, the problem is that there is always a certain amount of self-discharge in each one because of the fact that the insulation between the plates of the capacitors is not perfect. This self-discharge is not equal in all the capacitors, which is an issue that needs to be corrected, otherwise there may be a relative charge build-up on some of the capacitors, and this will result in a higher voltage on those capacitors. The solution to this issue, being essential in systems of more than about six capacitors in series, is to have charge equalization circuits. They have relatively high specific power and relatively low specific energy. Supercapacitors are inherently safer than flywheels as they avoid the problems of mechanical breakdown and gyroscopic effects. Power electronics are needed to step voltages up and down as required.

5.3.8 THERMAL ENERGY STORAGE

Thermal energy storage (TES) technologies store thermal energy for later use as required, rather than at the time of production. They are, therefore, important counterparts to intermittent renewable energy generation and also provide a way to use the waste process heat and reduce the energy demand of buildings, facilities and industrial processes. A variety of TES techniques have been developed, including building thermal mass utilization, phase change materials (PCM), underground thermal energy storage, heat pumps and energy storage tanks. TES systems can either be centralized or distributed systems. Centralized TES systems are used in district heating or cooling systems, large industrial plants, large CHP units or renewable energy plants. Solar thermal systems are often applied in residential and commercial buildings to capture solar energy for water and space heating or cooling. In such cases, TES systems can reduce the energy demand during peak times. The economic performance depends significantly on the specific application and operational needs and the number and frequency of storage cycles. TES systems for cooling or heating are used where there is a time mismatch between the demand and the economically most favorable energy supply. TES can provide short-term storage for peak shaving as well as long-term storage for the introduction of renewable and natural energy sources. Sustainable buildings need to take advantage of renewable and waste energy to approach ultra-low-energy buildings. TES systems are methods that enable the collection and preservation of excess heat for later utilization. Practical situations where TES systems are often installed are solar energy systems, geothermal systems, DG units and other energy conversion systems where heat availability and peak use periods do not coincide. The three basic types of TES systems are *sensible heat storage, latent heat storage and thermochemical heat storage.* Energy storage by causing a material temperature to change is the sensible heat storage, for which the system performance depends on the storage material specific heat, and if the volume is important, on density. Sensible heat storage systems usually use rocks, ground, oil or water as the storage medium. Latent heat storage systems store energy in PCMs, with the thermal energy stored when the material changes phase, e.g. from a solid to a liquid. Latent heat storage systems use the fusion heat, needed or released when a storage medium changes phase by melting, solidifying, liquefaction or vaporization. Thermochemical energy storage is based on chemical reactions in inorganic substances. The TES choice depends on the required storage time period, e.g. day-to-day or seasonal, and outer operating conditions. The specific heat of solidification or vaporization and the temperature at which the phase change occurs are design criteria. Both sensible and latent heat types may occur in the same storage material.

The need of thermal energy storage is often linked to the cases, where there is a mismatch between thermal energy supply and energy demand, when intermittent energy sources are utilized and for compensation of the solar fluctuations in solar heating systems. Possible technical solutions to overcome the thermal storage needs may be the following: building production over-capacity, using a mix of different supply options, adding back-up/auxiliary energy systems, only summer-time utilization of solar energy and short/long-term thermal energy storage. In traditional energy systems, the need for thermal storage is often short-term one and therefore the technical solutions for thermal energy storage may be quite simple, and for most cases the water is used as storage medium. Large-volume sensible heat systems are promising technologies with low heat losses and attractive prices. Sensible heat-based thermal energy storage is based on the temperature change in the material and the unit storage capacity (J/kg) is equal to the medium heat capacitance times the temperature change. Phase-change-based energy storage involves the material changes its phase at a certain temperature while heating the substance, then heat is stored in the phase change. Reversing the process, the heat is dissipated when at the phase change temperature it is cooled back. The storage capacity of the phase change materials is equal to the phase change enthalpy at the phase change temperature plus the sensible heat stored over the whole energy storage temperature range. The sorption or thermo-chemical reactions provide thermal storage capacity, based on the endothermic chemical reaction (when a chemical reaction absorbs more energy than it releases) principle:

$$AB + \text{Heat} \Leftrightarrow A + B \tag{5.34}$$

The heat used for a compound AB to be broken into components can be stored separately, and later by bringing the A and B compounds together to form again the AB compound, the heat is released.

The energy storage capacity is the reaction heat or the reaction free energy. TESs based on chemical reactions have negligible losses whereas the sensible heat storage systems dissipate the stored heat to the environment and need to be isolated to perform efficiently. Materials are the key issues for thermal energy storage. There are a large range of different materials that can be used for thermal storage as shown by Table 5.4. One of the most common storage medium is the water. TES in various solid and liquid media or materials is used for solar water heating, space heating and cooling as well as high-temperature applications such as solar-thermal power generation. Important parameters in a storage system include the storage duration, energy density (or specific energy) and the charging and discharging cycle (storage and retrieval) characteristics. However, the energy density is a critical factor for the size and application of any energy storage system. The rate of charging and discharging depends on thermo-physical properties such as thermal conductivity and design of the thermal energy storage systems. TES systems are dealing with the storage of energy by material cooling, heating, melting, solidifying or vaporizing; the thermal energy becomes available when the initial process is reversed. The materials that store heat are typically well-insulated. Primary

TABLE 5.4

Examples of Materials Suitable for Thermal Energy Storage

Thermal Storage Methods	Material	Thermal Storage Methods	Material	Thermal Storage Methods	Material
Sensible Heat	Water Ground Rocks Ceramics	Phase Change	Inorganic Salts Organic and Inorganic Compounds Paraffin	Thermo-chemical Reactions	*Working Fluid*: Water, ammonia, hydrogen, carbon dioxide, alcohols *Sorption Materials*: Hydroxides, hydrates, ammoniates, metal hydrides, carbonates, alcoholates

disadvantage of a thermal energy storage system is the large initial investments required to build the energy storage infrastructure. However, it has two primary advantages: (1) the energy-system efficiency is improved with the implementation of a thermal energy storage system (CHP has approximately 85–90% efficient while conventional power plants are only 40% efficient or lower), and (2) these techniques have already been implemented with good results. On the negative side, thermal energy storage does not improve flexibility within the transportation sector, like in the hydrogen energy storage system this not being a critical issue. TES does have disadvantages, but these are small when compared to its advantages. Due to the efficiency improvements and maturity of these systems, it is likely that they are becoming more prominent, enabling the utilization of intermittent renewable energy, but also to maximize the fuel use within power plants. These systems already put into practice with promising results. Therefore, it is evident this technology can play a crucial role in future energy and power systems. TES operation and system characteristics are based on thermodynamics and heat transfer principles and laws. There are two major TES types for storing thermal energy, sensible heat storage and latent heat storage. First consists of changing the temperature of a liquid or solid, without changing its phase. Thermal energy quantities differ in temperature, and the energy required E to heat a volume V of a substance from a temperature T_1 to a temperature T_2 is expressed by the well-known relationship:

$$E = m \int_{T_1}^{T_2} CdT = mC(T_2 - T_1) = \rho VC(T_2 - T_1) \tag{5.35}$$

where C is the specific heat of the substance, m is the mass and ρ is its density. The energy released by a material as its temperature is reduced, or absorbed by a material as its temperature is increased, is called the sensible heat. Second type of energy storage implies the phase change. The ability to store sensible heat for a given material is strongly dependent of the value of ρC. For high-temperature sensible heat TES (i.e. in the range of hundred Celsius degrees), iron and iron oxide have good characteristics comparable to water, low oxidization in high-temperature liquid or air flow, with moderate costs. Rocks are unexpansive sensible heat TES materials; however, the volumetric thermal capacity is half that of water. Some common TES materials and their characteristics are listed in Table 5.5. Latent heat is associated with the changes of material state or phase changes, e.g. from solid to liquid. The amount of energy stored (E) in this case depends upon the mass (m) and latent heat of fusion (λ) of the material:

$$E = m \cdot \lambda \tag{5.36}$$

TABLE 5.5

Thermal Capacities at 20°C for Some Common TES Materials

Material	Density (kg/m³)	Specific Heat (J/kg·K)
Aluminum	2710	896
Brick	2200	837
Clay	1460	879
Concrete	2000	880
Glass	2710	837
Iron	7900	452
Magnetite	5177	752
Sandstone	2200	712
Water	1000	4182
Wood	700	2390

The storage operates isothermally at the melting point of the material. If isothermal operation at the phase change temperature is difficult to achieve, the system operates over a range of temperatures T_1 to T_2 that includes the melting point. The sensible heat contributions have to be considered in the top of latent heat, and the amount of energy stored is given by:

$$E = m \left[\int_{T_1}^{T_{melt}} C_{Sd} dT + \lambda + \int_{T_{melt}}^{T_2} C_{Lq} dT \right] \tag{5.37}$$

Here, C_{Sd} and C_{Lq} represent the specific heats of the solid and liquid phases and T_{melt} is the melting point. It is relatively straightforward to determine the value of the sensible heat for solids and liquids, being more complicated for gases. If a gas restricted to a certain volume is heated, both the temperature and the pressure increases. The specific heat here is the specific heat at constant volume, C_v. If instead the volume is allowed to vary and the pressure is fixed, the specific heat at constant pressure, C_p, is obtained. The ratio $\gamma = C_p/C_v$ and the fraction of the heat produced during compression can be saved, affecting the energy storage system efficiency. TES specific applications determine the used method. Considerations include, among others: storage temperature range; storage capacity, having a significant effect on the system operation; storage heat losses, especially for long-term storage; charging and discharging rate; initial and operation costs. Other considerations include the suitability of container materials, the means adopted for transferring the heat to and from the storage and the power requirements for these purposes. A figure of merit that is used occasionally for describing the performance of a TES unit is its efficiency, which is defined by Equation (5.46). The time period over which this ratio is calculated would depend upon the nature of the storage unit. For a short-term storage unit, the time period would be a few days, while for a long-term storage unit it could be a few months or even one year. For a well-designed short-term storage unit, the value of the efficiency should generally exceed 80%.

$$\eta = \frac{T_{max} - T_{min}}{T_{charging} - T_{min}} \tag{5.38}$$

where T_{min}, T_{max} are the maximum and minimum temperatures of the storage during discharging respectively, and $T_{charging}$ is the maximum temperature at the end of the charging period. Heat losses to environment between the discharging end and the charging beginning periods, as well as during these processes, are usually neglected. Two particular problems of thermal energy storage systems are the heat exchanger design and in the case of phase change materials, the method of encapsulation. The heat exchanger should be designed to operate with as low a temperature difference as possible to avoid inefficiencies. In the case of sensible heat storage systems, energy is stored or extracted by heating or cooling a liquid or a solid, which does not change its phase during this process. A variety of substances have been used in such systems, such as: (a) liquids (water, molten salt, liquid metals or organic liquids) and (b) solids (metals, minerals or ceramics). In the case of solids, the material is invariable in porous form and the heat is stored or extracted a goas or a liquid flowing through the pores or voids. For incompressible type of thermal storage, e.g. the ones using heavy oils or rocks, the maximum work that can be produced is given in terms of specific heat capacity, C and mass:

$$W_{\max} = m\left[C(T_{str} - T_{amb}) + CT_{amb} \ln\left(\frac{T_{str}}{T_{amb}} \right) \right] = \rho V\left[C(T_{str} - T_{amb}) + CT_{amb} \ln\left(\frac{T_{str}}{T_{amb}} \right) \right] \quad (5.39)$$

Here, T_{str} and T_{amb} are the storage material temperature and ambient temperature (in Kelvin degrees), respectively, and m is the storage material mass, ρ the storage material density and V the volume. The storage materials (water, steam, molten salt, heavy oil or solid rocks) are at temperatures significantly higher than the ambient one, so the heat is continuously lost from the thermal storage, regardless of the insulation quality. Given enough time, the stored energy, if not used, is dissipated. For this reason, TES are suitable for short-term or intermediate period applications rather than long-term ones. The total rate of heat transfer, q, from a TES reservoir depends on the overall heat transfer coefficient, C_{trsf} the reservoir instant temperature, T, the ambient temperature and the reservoir total surface, A_{tot}, expressed as:

$$q = C_{trsf} \cdot A_{tot} \cdot (T - T_{amb}) \quad (5.40)$$

A number of TES applications are used to provide building or facility heating and cooling including *aquifer thermal storage* (ATS) and *duct thermal storage* (DTS). However, these are heat generation techniques rather than energy storage techniques. An aquifer is a ground water reservoir, consisting of highly water permeable materials such as clay or rocks, having large volumes and high thermal storage capacities. When heat extraction and charging performances are good, high heating and cooling powers are achieved by such systems. The energy that can be stored in an aquifer depends on the local conditions (allowable temperature changes, thermal conductivity and groundwater flows). An aquifer storage system is used for short-to-medium storage periods – daily, weekly, seasonal or mixed cycles. In terms of storing energy, there are two primary thermal energy storage options. One option is a technology used to supplement building air conditioning. The thermal energy storage can also be used very effectively to increase the energy system capabilities to facilitate the penetrations of renewable energy sources can be increased. Unlike other energy storage systems, which enabled interactions between the electricity, heat and transport sectors, thermal energy storage only combines the electricity and heat sectors with one another. By introducing district heating into an energy system, electricity and heat can be provided from the same facility to the energy system using CHP plants. This brings additional flexibility to the system, enabling larger renewable energy penetrations.

Example 5.14 **A residence requires 72 kWh of heat on a winter day to maintain a constant indoor temperature of 21°C. (a) How much solar collector surface area is needed for an all-solar heating system that has 20% efficiency? (b) How large does the storage tank have to be to provide this much energy? Assume the average solar energy per square meter and per day for the area is 6.0 kWh/m²/day.**

SOLUTION

a. Daily thermal energy per unit of area converted into thermal energy is:

$$\text{Thermal Energy} = \frac{6.0 \times 0.20}{1.0} = 1.20 \, \text{kWh/m}^2/\text{day}$$

The minimum converter area is then:

$$A_{Collector} = \frac{72}{1.2} = 60 \, \text{m}^2$$

b. If we are assuming the storage medium, water, the most common storage medium in residential applications, heat capacity of water is 1 kcal/kg/°C, and the temperature difference is that between the hot fluid and the cold water going into the storage tank is about 40°C. Therefore, the required mass of water for a day's worth of heat is:

$$\text{Tank Mass} = \frac{72}{1.116 \times 10^{-3} \times 40} = 1612.9 \text{ or } \approx 1613 \text{ kg}$$

5.4   ENERGY STORAGE IN SMART GRID OPERATION

The development and use of renewable energy has experienced rapid growth over the past few years. In the next 20–30 years, all energy systems will be based on the rational use of conventional resources and greater renewable energy use. Decentralized and renewable energy-based electricity production yields a more assured supply for consumers with fewer environmental impacts. However, the unpredictable character of these energy sources requires that network provisioning and usage regulations be established for optimal system operation. *The criteria to identify the most suitable EES are the following for each: How it works, advantages, applications, cost, disadvantages and future.* A brief comparison indicates the broad range of operating characteristics available for energy storage technologies. For both utility and renewable energy integration, energy storage capacity, power output and life cycle are key performance criteria. The need for long life cycle has motivated the use of storage systems from reversible physics such as CAES or pumped-hydro as an alternative to electro-chemical batteries that present problems of aging and are difficult to recycle. In transportation applications, portability, scalability and energy and power density are key performance criteria. Therefore, due to their modularity and portability, and in spite of the numerous issues, including limited life, batteries are still considered the most viable option for transport applications. Energy storage is a well-established concept, yet still relatively unexplored. Storage systems such as pumped-hydro-electric energy storage (PHES) have been in use since 1929, primarily to level the daily load on the network between night and day. As the electricity sector is undergoing a lot of change, energy storage is starting to become a realistic option for: (1) restructuring the electricity market; (2) integrating renewable resources; (3) improving power quality; (4) aiding shift towards distributed energy; and (5) helping network operate under more stringent environmental requirements. It is possible to divide grid storage applications into two broad categories based on the length of time a storage device needs to provide service: high power applications where the device must respond rapidly and be able to discharge for only short-term periods (up to about one hour), and energy-management-related applications where the device may respond more slowly but must be able to discharge for several hours or more. Ideally, all energy storage devices would be able to provide all services, but some technologies are technically restricted to provide only short-term services. However, many of these services have very high value in the grid, so short-term storage can still provide considerable benefits.

Energy storage systems are increasing their impact on the utility grids and even more applications into the smart grids as solution to the stability problems, increasing renewable energy penetration, reducing the overall operation costs, improving grid resilience, service restoration and power quality. The main ESS advantages are to contribute to the quality of the grid by maintaining the power constant and reducing operation cost. The main role of these ESS units is to increase the RES penetration to level load curve, to contribute to the frequency control, to upgrade the transmission line capability, to mitigate the voltage fluctuations and to increase the power quality and reliability. Drawn from recent literature surveys, the major grid areas where the energy storage systems can be applied can be summarized as:

1. With the widespread use of sensitive electrical equipment and devices, the power quality is becoming an increasingly important topic. Energy storage units (e.g. batteries), commercially available and cost effective are used in uninterruptible power supplies (UPSs) to mitigate short-term loss of power and power fluctuations. Energy storage can also be used to mitigate voltage fluctuations and improve other power quality issues such as harmonics. The alternative to energy storage for power quality applications is to make the control systems of the sensitive equipment more robust and more expensive to operate.

2. Intermittent supply and lack of controllability are inherent characteristics of renewable energy generation, challenging for the secure operation of the power system. Energy storage could support both the power system and renewable energy sources by smoothing their output, matching contract positions (or enabling scheduled dispatch) and time-shifting the generation. Energy storage could mitigate the output forecast errors of renewable energy generators by supplying the energy deficit or absorbing the excess. The output of a renewable energy sources has no direct correlation to the electrical demand, energy storage can store the energy when the production is greater than the demand and supply the load when demand is greater than the supply, particularly useful when low demand conditions coincide with high wind or solar power conditions. The alternatives to compensate for the variable output of renewable energy generation are fast response gas turbine generators, or interconnection of the renewable energy generation over a wide geographic area to smooth the output and demand-side integration.

3. Electrical energy time-shifting involves storing energy during periods when demand is low or at times when the price is low, and discharging the energy when demand is high or at times when the price is high, enabling efficient energy uses. It also supports distribution networks by relieving congestion during peak demand periods, providing the energy storage unit is positioned correctly within the distribution system.

4. Energy storage could provide benefits to end-users who are on time-of-use tariff through electrical energy time shifting or who have micro-generation. The main alternative to energy storage for energy time shifting and end-use energy management is demand-side integration.

5. Under normal operating conditions, system voltage is maintained within prescribed limits, usually achieved by transformer tap changers and/or reactive power flows. Distributed energy storage may be attractive, due to the ability to provide active and reactive power, voltage control, while reducing the reactive power flows in the network.
6. The reserve support, an ancillary service that is maintained to ensure system stability under unexpected connection/disconnection of load/generation. A fraction of the reserve is spinning and supplied by part-loaded large generators. These part-loaded generators operate at reduced efficiency. Energy storage, replacing spinning reserve, does not discharge on a regular basis but is available to discharge if needed. In addition to spinning reserve, the balancing task can be supported by a so-called standing reserve, which is supplied by higher fuel cost plant, such as open-cycle gas turbines and energy storage. The advantage of energy storage lies in its ability to store surpluses in generation during periods of high wind and low demand, and subsequently make a part of this energy available, and hence reduce costs. Further, storage can store and release energy while the OCGT plant can only provide energy to the power system.
7. Load following, an ancillary service purchased by utilities to follow frequently changing power demand, can be performed by energy storage devices, operating at partial output levels with high efficiencies and their response is quick, energy storage could be an ideal candidate for this service.
8. The increase uses in distributed generation causes the power distribution lines to be overloaded, thus demanding distribution circuit upgrades. Energy storage could be used to relieve the congestion of power distribution circuits, deferring circuit reinforcement, and reducing the overall costs.

In summary, energy storage systems have been recognized as viable solutions for implementing the smart grid paradigm, but are also creating challenges for load leveling, integrating renewable and intermittent energy sources, voltage and frequency regulation, grid resiliency, improving power quality and reliability, reducing energy import during peak demand periods and so on. In particular, distributed energy storage systems can address a wide range of the above potential issues and problems, and they are gaining attention from customers, utilities, professionals and regulators. Distributed energy storage systems have considerable potential for reducing operation costs and improving the quality of the electric services. However, installation costs and lifespan are ones of the main drawbacks to the wide diffusion of this technology. In this context, a serious challenge is the adoption of new techniques and strategies for the optimal planning, control and management of grids that include distributed-energy storage devices. Regulatory guidance and proactive policies are urgently needed to ensure a smooth rollout of this technology. The electricity system expansion and future developments can be accelerated by the widespread deployment and extended applications of energy storage units, since storage can be a critical component of grid stability, resiliency and operation costs. The future for energy storage need to address the following issues: *energy storage technologies should be cost competitive with other power technologies providing similar services, energy storage should be recognized for its value in providing multiple benefits simultaneously and finally, the energy storage technology need to be seamlessly integrated with existing power systems and sub-systems leading to its ubiquitous deployment.* In reviewing the barriers and challenges, and the future for energy storage, a strategy that would address these issues should comprise three broad outcome-oriented goals: energy storage should be a broadly deployable asset for enhancing renewable energy penetration, specifically to enable the energy storage deployment at high levels of the new renewable energy-based power generation; energy storage should be available to industry and regulators as an effective option to resolve issues of grid resiliency and reliability; and energy storage should be a well-accepted contributor to realization of smart grid benefits, specifically enabling confident deployment of electric transportation and optimal utilization of demand-side assets. To realize these outcomes, the principal challenges to focus on are on the four major factors.

1. *Cost competitive energy storage technology,* requiring attention to factors such as life-cycle cost and performance (round-trip efficiency, energy density, life-cycle cost, capacity fade, etc.) for energy storage technology as deployed. It is expected that early deployments are in high-value applications, but long-term success requires cost reduction and capacity to realize revenue for all grid services.
2. *Validated reliability, safety and performance* of any energy storage technology are essential and critical for the user confidence and for industry acceptance.
3. *Equitable regulatory environment*, meaning that grid energy storage values depend on reducing institutional and regulatory hurdles to levels comparable with other grid resources.
4. *Industry acceptance and adoption* are requiring that the utilities have confidence that the energy storage systems are deployed as expected, and are delivering as predicted and promised.

5.4.1 ENERGY STORAGE FOR ELECTRIC GRID AND RENEWABLE ENERGY APPLICATIONS

Energy storage can optimize the existing generation and transmission infrastructures while also preventing expensive upgrades. Power fluctuations from renewable resources will prevent their large-scale penetration into the network. However, energy storage devices can manage these irregularities and thus aid the amalgamation of renewable technologies. In relation to conventional power production, energy storage devices can improve overall power quality and reliability, which is becoming more important for modern commercial applications. Finally, energy storage devices can reduce emissions by aiding the transition to newer, cleaner technologies such as renewable resources and the hydrogen economy. The concept of having energy storage for an electric grid provides all the benefits of conventional generation, such as enhanced grid stability, optimized transmission infrastructure, high power quality, excellent renewable energy penetration and increased wind farm capacity. However, energy storage technologies produce no carbon emissions and do not rely on imported fossil fuels. As a result, energy storage is an attractive option for increasing wind penetration onto the electric grid when it is needed. The rapid advances in energy storage technology have permitted such devices of reasonable size to be designed and commissioned successfully aiming at balancing any instantaneous mismatch in active power during abnormal operation of the power grid. Thus, fast-acting generation reserve is provided to the microgrid so that the dynamic security can be significantly enhanced. High power and energy density with outstanding conversion efficiency, and fast and independent power response in four quadrants make the selected DESs capable of providing significant benefits to many potential microgrid applications. Most of these advanced DES technology applications are described below:

- **Distributed energy storage**: the selected advanced DES units could provide the potential for energy storage of up to several MWh with a high return efficiency (higher than 95% for SMES and SCES, while 85% for FES systems) and a rapid response time for dynamic change of power flow (half millisecond for SMES and SCES devices, up to some milliseconds for FES systems), making them ideal for energy management with large variations in energy requirements, as well as for backup power supply in case of loss of the utility main power supply or as a replacement of major generating unit trips in the microgrid.
- **Spinning reserve**: in case of contingencies, such as failures of generating units or other microgrid components, a certain amount of short-term generation must be kept unloaded as spinning reserve. This reserve must be appropriately activated by means of the primary frequency control (PFC). The selected advanced EESs can be effectively used in order to store excess energy during off-peak periods for substituting the generation reserve during the action of the PFC, enhancing the grid dynamic security.

- **Load following**: the selected advanced DES devices have the ability to follow system load changes almost instantaneously which allows for conventional generating units to operate at roughly constant or slowly changing output power.
- **Grid stability**: the selected advanced DES units have the capability to damp out low-frequency power oscillations and to stabilize the system frequency as a result of system transients. Since the considered advanced DES systems are capable of controlling both the active and reactive powers simultaneously, they can act as a good device in order to stabilize the microgrid with high level of penetration of renewable energy sources, such as photovoltaic or particularly wind generation.
- **Automatic Generation Control (AGC)**: the advanced DES systems can be used as a controlling function in an AGC system to support a minimum of area control error (ACE).
- **Tie line power flow stabilization and control**: A schedule of power between various microgrids or control areas inside the same microgrid requires that actual net power matches closely with the scheduled power. Unfortunately, generators with highly fluctuating active power profile in one microgrid produce an error in the actual power delivered respect to the scheduled one, which can result in inefficient use of generation and system components. Advanced DES systems can be designed with appropriate controls to provide power in order to nearly eliminate this error and to ensure that generation is efficiently used and power schedules are met.
- **Power quality improvement**: SMES can provide ride through capabilities and smooth out disturbances on the microgrid that would otherwise interrupt sensitive customer loads. All these devices have very fast response and can inject active power in less than one power cycle; thus providing premium power supply to critical customers and preventing from losing power.
- **Reactive power flow control and power factor correction**: The selected advanced DES units are capable of controlling the generated reactive power simultaneously and independently of the active power, which enables the correction of the power factor.
- **Voltage control**: The selected advanced DES systems have shown to be effective for providing voltage support and regulation by locally generating reactive power.

ESSs are increasing their impact on the utility grid as a solution to stability problems. The main advantage of a storage plant is to contribute to the quality of the grid by maintaining the power constant. The main role of these ESSs is to increase the RES penetration, to level load curve, contribute to the frequency control, to upgrade the transmission line capability, to mitigate the voltage fluctuations, and to increase the power quality and reliability.

1. *Increasing RES Penetration:* although RESs are environmentally beneficial, the intermittent nature of two fast growing energies, wind and solar, causes voltage and frequency fluctuations on the grid. That represents a significant barrier to widespread penetration and replacement of fossil fuel source base-load generation, because integrating renewable sources introduces some new issues on the operation of the power system, such as potential unbalancing between generation and demand. However, the intermittent RESs, such as solar and wind, needs to be supported with other conventional utility power plants. It is estimated that, for every 10% wind penetration, a balancing power from other generation sources equivalent to 2–4% of the installed wind capacity is always required for a stable power system operation. Thus, with more penetration of intermittent renewable energy like wind power, the system operation will be more complex, and it will require additional balancing power. This is critical in countries with a large penetration of solar and wind systems, such as Denmark or Spain, where it is estimated that approximately 20% and 10% of the electricity generation come from wind power, respectively. A large storage capacity will allow a high percentage of wind, photovoltaic, and other power plants in the electrical

mix contributing to fulfil the objectives for a more sustainable future. In order to integrate RESs, it is necessary to propose a suitable storage system that offers capacities of several hours and power level from 1 to 100 MW. Nowadays, high-temperature thermo-solar power plants are including a TEES, and it is expected that other storage systems will be included in the new generation of RES and the distributed generation sources in general.

Recently, the concept of vehicle-to-grid (V2G) has been introduced. It describes a system in which electric or plugin hybrid vehicles communicate with the power grid to sell demand respond services by either delivering electricity into the grid or throttling their charging rate. When coupled to an electricity network, EVs can act as a controllable load and energy storage in power systems with high penetration of RESs. The reliability of the renewable electricity will be enhanced with the vast untapped storage of EV fleets when connected to the grid. It is estimated that the market area for V2G represents over one million vehicles with 20–50 kWh capacity, where 10% of this capacity is available for utility applications, including integration of RES. The benefits of energy storage applications in RESs have been deeply studied in the bibliography.

2. *Load Leveling:* load leveling refers to the use of electricity stored during times of low demand to supply peak electricity demand, which reduces the need to draw on electricity from peaking power plants or increase the grid infrastructure. To deliver more power to the load, there are two possibilities: increase the infrastructure and the generator capacity or install an ESS. The ESS allows one to postpone a large infrastructure investment in transmission and distribution network. New technologies, which are not restricted by their geographic limitations, have been proposed as more suitable for load leveling such as TESS and BSS. Rechargeable battery technologies like sodium sulfur (NaS) technology are attractive candidates for use in many utility scale energy storage applications. These advanced battery systems can be utilized with existing infrastructure, helping energy providers to meet peak demands and critical load.

3. *Energy Arbitrage:* energy arbitrage refers to earning a profit by charging ESS with cheap electricity when the demand is low and selling the stored energy at a higher price when the demand is high. This activity can also be used to influence in the demand side, such as using higher peak prices to induce a reduction in peak demand through demand charges, real-time pricing or other market measures. This function has been traditionally performed by pumped-hydro storage (PHS). PHS is appropriate for energy arbitrage because it can be constructed at large capacities over 100 MW range and discharged over periods of time from 100 to 1000 minutes. These installations allow storage when the demand is low and the energy is cheap. This ESS is the most widely used energy storage technology at utility scale (100 GW installed worldwide). CAES is also appropriate for energy arbitrage because it can be constructed in capacities of a few hundreds of megawatts and can be discharged over long periods of time. A new trend for this application is to use the ancillary services that offer the battery of electrical V2G. The large quantity of this V2G expected in a next future could contribute to a new concept of the energy marker.

4. *Primary Frequency Regulation:* the technical application of ESS includes transient and permanent grid frequency stability support. To contribute to the frequency stabilization during transient, called grid angular stability (GAS), low- and medium-capacity ESSs are needed. This low-energy storage requirement is because GAS operation consists of injection and absorption of real power during short periods of time, 1–2 s. Modern variable-speed wind turbines and large photovoltaic power plants connected to the utility grid do not contribute to the frequency stability as the synchronous generators of the conventional gas or steam turbine do. This creates a new application of ESS that is to be used to emulate the inertia of these steam turbine generators to complement this angular stability deficit. Another solution is to use the power electronic converter of variable-speed wind turbines to emulate the steam turbine inertia using the inertial energy storage of the rotors of these

wind turbines. SMESs are getting increasing acceptance in variation applications of damping frequency oscillations because of their higher efficiency and faster response. EDLC, FESS and BSS are also very suitable for this application.

5. *End-User Peak Shaving:* there are several undesired grid voltage effects at the end-user level, depending on the duration and variability. Typical voltage effects are long-period interruptions (blackouts), short-period interruptions (voltage sags), voltage peaks and variable fluctuation (flicker).

To perform peak shaving and prevent against blackouts, the typical approach involves installing UPSs. If an online UPS is installed in series, this isolates the load from the grid, and fluctuations produced by the utility have no effect on the users. However, this solution may not be optimal for all applications.

One solution, presented as grid voltage stability in, involves the mitigating against degraded voltage by providing additional reactive power and injecting real power for durations of up to 2 s. The energy storage needed to protect the load against this voltage degradation is low. The energy storage demanded is lower in applications with ride-through capability, where the electric load or the generator stays connected during the system disturbance, because part of the energy is from the grid during the under-voltage period. The voltage flicker is caused by rapid changes of RES units or loads, such as electric arc furnaces, rolling mills, welding equipment and pumps operating periodically. An ESS can help to reduce voltage fluctuations at the point of common coupling produced by these transitory generators and loads.

5.5 CHAPTER SUMMARY

The present chapter identifies the characteristics, possible applications, strengths and weaknesses of the different energy storage concepts and technologies. ESSs are the key enabling technologies for transportation, building energy systems, conventional and alternative energy systems for industrial processes and utility applications. In particular, the extended applications of energy storage are enabling the integration and dispatch of renewable energy generation and are facilitating the emergence of smarter grids with less reliance on inefficient peak power plants. Energy storage systems are playing critical roles in an efficient and renewable energy uses; much more so than it does in today's fossil-based energy economy. Major power and energy storage systems, e.g. compressed air storage, fuel cells, pumped hydropower energy storage thermal energy storage are reviewed and discussed in this chapter. It is concluded that EESs can contribute significantly to the design and optimal operation of power generation and smart grid systems, as well as the improved energy security and power quality, while reducing the overall energy costs. There is a range of options available to store intermittent energy until it is needed for electricity production. In the transportation sector, the emergence of viable onboard electric energy storage devices such as high-power and high-energy Li-ion batteries will enable the widespread adoption of plug-in and HEVs, which will also interact with the smart grids of the future. Mature energy storage technologies can be used in several applications, but in other situations, these technologies cannot fulfill the application requirements. Thus, new energy storage systems have appeared, passing new challenges that have to be solved by the research community. They may have many attractive features including independent sizing of power and energy capacity, longer lifetimes, high efficiency, fast response and relatively low costs, e.g. low initial investment and/or reduced operational and maintenance expenditures.

5.6 QUESTIONS AND PROBLEMS

1. What are the benefits of the energy storage systems?
2. What are the major problems for using: a) pumped water energy storage; b) compressed air energy storage; and c) the flywheels?

3. List the major potential applications of energy storage in smart grids.
4. List the benefits of electricity energy storage for power grid operation.
5. Classify and briefly explain each compressed-air storage systems.
6. List the essential criteria for comparing energy storage systems.
7. List the by-products that may have been produced by fuel cells?
8. List the major advantages and disadvantages of fuel cells.
9. How much energy can be delivered form a pumped storage facility of 8 million m³, a head of 500 m and the overall efficiency of 85%?
10. What energy storage options are available for solar energy applications?
11. List the key measures of merit for an energy storage system.
12. How much kinetic energy does a flywheel (steel disk of diameter 6 in, thickness 1 in), speed of rotation 30,000 RPM have?
13. A water-pumped energy storage facility has a level difference (head) of 500 m and a working volume of 081 km3. Estimate how much power is generated if it is required to operate 1.5 hours per day. The overall facility efficiency is 83% and the water density is 1000 kg/m³.
14. Estimate the volumetric energy density and the overall efficiency of a PHES, having the difference level between the lower and upper reservoirs 250 m, and the efficiencies of the pumped and generating phases, 0.85 and 0.90.
15. A CAES has a volume of 500000 m³, and the compressed air pressure range is from 80 bars to 1 bar. Assuming isothermal process and an efficiency of 33% estimate the energy and power for a 3-hour discharge period.
16. A hydropower pumping storage station has the following characteristics, the effective storage capacity of the of the upper reservoir 375x103 m³, the generation efficiency 0.91, the average head 200 m, the pumping efficiency 0.87. What are the overall efficiency and the volumetric energy density?
17. If the power generated by CAES of 300000 m³, in 2-hour discharge period from 66 bars to the atmospheric pressure is 360 MW, what is the system efficiency?
18. An old salt mine, having a 15,000 m³ storage capacity has been selected for pressurized air storage at 33 bar, if the temperature during the filling/charging phase is 175°C, assuming an isothermal process and an efficiency of 35% estimate the energy and average power for a 3-hour discharge period.
19. An underground cavern of volume 35,000 m³ is used to store compressed air energy isothermally at 300 K. Determine the maximum stored energy by air compression form 100 kPa to 1750 kPa, assuming heat loss of 63,000 kJ. Hint: one mol of air occupies 22.4 l.
20. Determine the maximum available stored energy if a 1450 kg of air is compressed from 100 kPa to 1600 kPa at 27°C, assuming isothermal condition ad a heat loss of 27.5 MJ.
21. A flywheel has a weight of 20 kg, 8 m diameter, an angular velocity of 1200 rad/s, a density of 3200 kg/m³ and a volume change of 0.75 m³. Evaluate, for k equal to 0.5 and 1.05, how much energy is stored in this flywheel.
22. Very-high-speed flywheels are made of composite materials. If a flywheel has the following characteristics: a) a ring with radius of 2.5 m, mass of 100 kg and speed of 25000 RPM; and b) a solid uniform disk, with same mass, radius and running at the same speed of rotation. How much energy is stored in each system?
23. If the speed of rotation in the previous problem decreases to 2500 RPM during discharge phase, what is the useful energy density for this device?
24. Estimate the energy stored in a CAES with capacity of 250000 m³, if the reservoir air presses is 60 bar, and the overall system efficiency is 33%. Estimate also the total generated electricity if the air is discharged over a period of 90 minutes.
25. A flywheel is constructed in a toroidal shape, resembling a bicycle wheel, has a mass of 300 kg. Assuming all mass is concentrated at 1.8 m, what is its RPM to provide an 800 kW for 1 min? In your opinion is this system physical possible.

26. What is the significance of depth-of-discharge (DOD), and how this parameter is affecting the battery lifecycle?
27. A battery has an internal resistance of 0.02 Ω per cell needs to deliver a current of 25 A at 150 V to a load. If the cell electro-chemical voltage is 1.8 V determine the number of cells in series.
28. A battery stack, used as an electrical energy storage system, is designed for 20 MW peak power supply for duration of 4 hours. This energy storage system uses 600 Ah batteries operating at 420 V DC. Estimate the stack minimum number of batteries and the current in each during peak operation.
29. Calculate the required time for a heat storage temperature to decrease from 36°C to 8°C, if there is no load heat removal from, and the ambient temperature is 12°C, the reservoir storage capacity is 2.0 m³, the heat storage area is 30 m², and the overall heat transfer reservoir liquid-environment is 6.5 W/m²·°C.
30. Assuming a cylindrical reservoir, made of bricks, with 10 m diameter and height of 12 m, compute the heat transfer rate, assuming the ambient temperature of 20°C degrees and the storage material temperature of 105°C.
31. Calculate the required storage medium volume for the following thermal energy storage systems in order to store 1 MWh of thermal energy for the temperature range from 350°C to 20°C, assuming the solid rocks and heavy oil storage materials.
32. Briefly describe the major applications of energy storage technologies in the smart grids
33. List the major challenges in the grid and utility application of the energy storage.

6 Smart Power Distribution, the Heart of Smart Grid

6.1 INTRODUCTION, POWER DISTRIBUTION IN TRANSITION

Electric utilities are transferring electrical energy generated at large power plants, in efficient and reliable ways, to the consumers through a complex network of transmission and distribution subsystems. Power delivery systems are designed to transfer the electricity produced in large generation centers to the final load points where consumers demand it. Power systems consist of four main subsystems: generation, transmission and sub-transmission, power distribution and end-users (loads). In a deregulated market, each subsystem is usually owned and operated by a different company, and free competition is permitted in each of them. Power distribution systems receive electricity from substations and distribute it to every consumer in a designated area. Power distribution lines are serving a critical, but limited, role within the power systems, delivering electricity to the customers within standard voltage ranges. At the basic level, the role of local power distribution is to deliver power to consumers consistent with their demands and expectations for quality and reliability. Three-phase AC systems are commonly used for transmission and in large part of the power distribution. Distribution networks are serving as links from the distribution substation to the end-users, providing safe, secure and reliable electricity to various loads throughout the service territory. Power distribution networks begin as the medium-voltage three-phase circuits, usually 30 kV through 60 kV, and terminate at lower secondary three- or single-phase circuits, usually below 1 kV at the customer's premise. Voltage levels of primary distribution are anywhere from about 5 kV to as high as 35 kV with the most common voltages in the 10 kV to 15 kV range. However, the areas served by a given voltage are proportional to the voltage level; for the same load density, a 35 kV system can serve considerably longer lines than a 12.8 kV system. Lines can be as short as a mile and as long as about 30 miles or 50 km, commonly they are about 10 miles or 15 km or shorter. Distribution feeders consist of overhead and/or underground circuits in a mix of branching laterals from the distribution substation, designed around requirements such as peak load, voltage limits, distance to customers and other local conditions, e.g. terrain, visual regulations or customer requirements. The branching laterals are mostly operated in a radial, looped or ring networked configurations. High-density urban areas are often connected in complex underground networks providing highly redundant and reliable costumer connections. Most three-phase AC systems are for large loads such as large buildings, commercial or industrial customers. The loading of a distribution feeder is inherently unbalanced because of the large number of unequal single-phase loads that must be served. An additional unbalance is introduced by the non-equilateral conductor spacings of the three-phase overhead and underground line circuit segments.

Power distribution is a complete system, structured in the most suitable architecture, according to the standards, local regulations and energy supply constraints, load types or end-user requirements and demands. Electric distributions, a major electric infrastructure section, are taking the electricity from the high-voltage (HV) transmission networks and delivering it to the end-users, via complex circuits and specific equipment and devices and are broadly divided into primary and secondary power distribution circuits. Primary distribution lines are *medium-voltage* (MV) networks, mostly in the range from 600 V to 35 kV. The first primary distribution component is the distribution substation, where the energy delivered by transmission and sub-transmission subsystems is received, and where the voltages are reduced to required levels. From the distribution substation, MV distribution lines take this energy one step closer to the customers. As

DOI: 10.1201/9780429174803-6

in the transmission system, larger loads are directly connected to the primary distribution. At distribution substations, transformers are stepping-down the transmission voltages to the primary distribution levels, which are fanning out from the substation. Close to end-users, other power distribution transformers are further lowering the primary distribution voltage levels to the LV secondary circuit levels (120/240 V or other utilization voltages). From power distribution transformers, the secondary distribution circuits connect the end-users at the service entrance. Secondary distribution consists of step-down transformers and low-voltage lines (e.g. 230 V) that are delivering the energy to low power customers, such as most of commercial and residential loads. In summary, power distribution provides the infrastructure delivering electricity from distribution substations to end-users. Typically radial in nature, primary distribution system includes feeders and connecting networks, the laterals, with typical voltages are 34.5 kV, 14.4 kV, 13.8 kV, 13.2 kV, 12.5 kV, 12 kV or lower voltages, e.g. 4.16 kV. The distribution voltages in all service territories are similar because it is easier and cost effective when the system parameters are consistent. Industrial, commercial and residential facilities are increasingly dependent on computer and automated control power electronics for processes, operation and management, and as a consequence are requiring an increase in the cleanliness and reliability of the electricity supply at the expected power quality standards.

Power distribution networks are very large and extensive systems consisting of hundreds of thousands of nodes that can generate huge data volumes. However, the existing power distribution systems as a whole are often poorly monitored and the availability of required real-time measurements is impractical due the massive system size and associated costs. In order to provide affordable, reliable and sustainable energy, a modern grid needs to become smarter and more efficient. The major characteristics of distribution assessment are: reliability, efficiency, voltage regulation, cost and safety, environmental and esthetic impacts. Moreover, power distribution design and operation present a multitude of complex and often conflicting objectives. For example, reliability, efficiency and safety must be maximized for asset protection, while minimizing the costs and customer disturbances. These conflicting objectives impact the type, choice and structure of the power distribution assets, substation topologies, locations, feeder design, feeder and substation protection, control and other significant technical decisions. Transitioning from conventional power distribution to smart power distribution requires a paradigm shift in the design, management and operation. This shift is due in part to the increased use of distribution management systems, demand side response and more significant is the burgeoning of distributed energy resources. Massive underway DG deployment is expected to increase in the near future. The smart distribution systems must be designed, so that rapid service restoration, improved safety and reliability are implemented. It is envisioned that smart distribution systems will increasingly migrate to networked configurations, especially in the high load density areas. Rapid service restoration can accomplish several objectives, including reduction of the average interruption duration and frequency, or the minimization of unserved energy to loads. A highly reliable, reconfigurable and fault-tolerant distribution system must contain multiple redundant paths. However, more important than multi-connection paths are the smart strategies for the fault detection, isolation and reconfiguration (FDIR) to manage the distribution redundancy.

6.2   POWER DISTRIBUTION IN CONVENTIONAL POWER SYSTEMS

In the United States, power companies provide electricity to medium or large loads at voltages form 4.2 kV to 63 kV, usually at 13.8 kV level. For residential customers, utilities are lowering the HV and MV voltages at substations and in power distribution networks, and the required voltage levels. From there, the electricity is fed through a meter and into the building. Power flow on a transmission line is a function of voltage and current, while the current itself being inherently bi-directional, the electricity can flow in either direction. However, other operational constraints, e.g. circuit breakers and control devices, may not be able to accommodate power flow reversal without replacement or modification. The grid operates as a three-phase AC network down to the service point level. Feeders, power lines transferring power from a distribution substation to the power distribution transformers are three-phase overhead and underground lines. Closer to the loads (many of which are single-phase), three-phase or single-phase laterals provide the customer connections. Depending on the residential customer needs, the voltage supplied can be 120 V single-phase or 120/240 V single-phase, where the 240 V distribution transformer secondary has a center tap that also provides two 120 V single-phase circuits. Larger customers utilize three-phase power, with 120/208 V or 277/480 V service and often higher voltage levels. Electricity is commonly delivered to industrial users by three-phase AC networks at medium or high voltages. The values of medium and high voltage as defined by standards are medium voltages range from 5 to 35 kV and HV range from 50 KV to 500 kV. However, higher and lower values can be found according to local conditions. Industrial users reduce the voltage levels to lower voltages inside facilities. Although the transformer efficiency is very high (about 95 %), it is worth reducing transformer losses as much as possible to lower the total site energy consumed. Transformer losses are no-load and load losses, no-load losses are almost exclusively iron losses, proportional to the squared voltage, occurring whenever the transformer is connected, being independent of the load, while the load losses are function of the squared current, or load apparent power, mainly copper losses, occurring only when a load is connected to the transformer and supplying circuit.

Figure 6.1 is showing a typical power distribution network, in which the electricity is coming from HV transmission lines, through step-down substation transformers and is eventually delivered to the end-users through radial feeders from central substation switchboards via protection and control devices. Feeders 1 through 3 are shown as line diagrams, while feeder 4 is shown as a three-phase four-wire configuration (phases A, B and C and the neutral N). The lateral feeders are taped

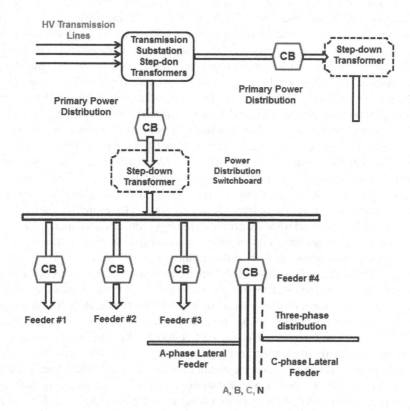

FIGURE 6.1   Typical power distribution system.

off from each phase (line-to-neutral connection) through a circuit breaker to service single-phase loads that are almost equally distributed on all system three phases, a quasi-balanced three-phase system. If the voltage drop at the end of main feeder is larger than the allowable limits, with cables meeting the ampacity (current limit) but not the voltage drop requirement. Capacitor banks or synchronous machines (condensers) may be connected to reduce the voltage drop and to improve the system power factor. To minimize the capacitor losses, they are switched off during the light load periods and switched back on during heavy load periods. Functionally, distribution circuits are those feeding loads, regardless of voltage level or configuration. Distribution infrastructure, an extensive energy system is located in cities, suburbs, rural areas and even remote places. Distribution circuits are found along the roads and streets. Urban structures are mainly underground, while rural ones are mainly overhead. Suburban structures are a mix, with a good deal of new construction underground. An urban utility may have less than 50 ft. (~15 m) of distribution circuit for each customer. A rural utility can have over 300 ft. (~90 m) of primary circuit per customer. Several entities may own distribution systems, e.g. municipalities, state and federal agencies, rural cooperatives or investor-owned utilities. In addition, large industrial facilities often have own power distribution. While there are differences, the engineering issues are similar for all entities. Due to the extensive nature, power distribution is capital-intensive businesses. Cost, simplified and standardized designs are critical for improvements. Few components or installations are individually engineered, standardized equipment and design is the norm.

6.2.1 POWER DISTRIBUTION STRUCTURE AND ELEMENTS

The electricity is delivered from the generation stations to consumers (loads) via a complex and vast structure, consisting of transmission lines, transformers, protection, sensing, monitoring and control equipment. The electricity is transferred to the load centers over long distance HV transmission lines, operating at voltages between 220 kV to about 1200 kV. Transmission lines distributing the electricity within an area are MV and LV distribution lines. There are several such transmission line categories, e.g. sub-transmission, primary and secondary power distribution networks. Power distribution substations are the interconnection element between the power distribution and the upstream electricity delivery systems. At the substation the step-down transformer reduces the sub-transmission voltage level to an appropriate value for primary power distribution lines. Different protection, switching and measurement equipment and devices are installed at the substation to ensure a safe and secure operation. The primary distribution lines, the so-called *feeders*, are spread across the consumption area served by the substation. Lateral lines (or laterals) are branching from distribution feeders, extending until they reach the step-down (MV-to-LV) distribution transformers, responsible for performing the final voltage reduction to obtain a voltage level adequate for customer uses (e.g. 400 V and 230 V). The secondary distribution lines operating at a LV level transfer the energy to the customer's interconnection point. These lines are usually single-phase circuits, but they can also be as three-phase circuits. Overhead lines are primarily used in rural areas, whereas in urban areas the distribution lines are mostly underground. In suburban areas there can be a mixture of overhead and underground circuits, depending on the specific location. Large industrial and commercial zones are usually served by dedicated circuits as they represent quite very large loads that can affect the service of other loads. The power distribution is coming in several configurations, types and circuit lengths, while sharing common characteristics. Table 6.1 shows typical distribution parameters. A main *feeder*, a power delivery circuits out of a power substation is the three-phase backbone circuit, the *mains* or *mainline*. The main feeder is often set at 400 A, allowing emergency ratings of 600 A. Branching from the mains are *laterals*, called also taps, lateral taps or branch lines. The laterals are single-, two- or three-phase circuits, have circuit breakers separating them from the mainline in the event of a fault. The most common power distribution primaries are four-wire, multi-grounded circuits, three-phase conductors and a multi-grounded neutral. Single-phase loads are serviced by power transformers, connected between one phase and the neutral, acting as a return conductor and

TABLE 6.1

Typical Distribution Circuit Parameters

Substation and Feeder Characteristics	Most Common Value	Other Common Values
Voltage	12.47 V	4.16, 4.8, 13.2, 13.8, 24.94, and 34.5 kV
Number of station transformers	2	1 to 6
Substation transformer size	21 MVA	5 to 60 MVA
Number of feeders per bus	4	1 to 8
Peak current	400 A	100 to 600 A
Peak load apparent power	7 MVA	1 to 15 MVA
Power factor	0.98 lagging	0.8 lagging to 0.95 leading
Number of customers	400	50 to 5000
Length of feeder mains	4 mi	2 to 15 mi
Length including laterals	8 mi	4 to 25 mi
Area covered	25 mi^2	0.5 to 500 mi^2
Mains wire size	500 kcmil	4/0 to 795 kcmil
Lateral tap wire size	1/0	#4 to 2/0
Lateral tap peak current	25 A	5 to 50 A
Distribution transformer size (1-ph)	25 kVA	10 to 150 kVA

as an equipment safety ground. A single-phase line has one phase conductor and the neutral, while a two-phase line has two-phases and the neutral. Some distribution primaries are three-wire circuits, with no neutral, where single-phase loads are connected phase to phase. Distribution systems are power system essential parts, distributing the electricity from the highly meshed, HV transmission networks to the end-users. In order to transfer electricity from an AC or a DC power supply to the loads, specific type of power distribution network is structured and employed.

Electrical infrastructure is a very extensive and complex infrastructure ever built in human history. Modern industrial and commercial facilities are using close to 50% of all the electrical energy, while the building sector about 40%. Power distribution systems are separated into three major components: *distribution substation*, *primary* and *secondary power distribution*. At a substation, the voltages are lowered as needed, the power is distributed to the customers, with one substation supplying power to thousands customers. The number of transmission lines in the power distribution is very large, and most customers are connected to a single-phase circuit. Therefore, the power flowing in each of these lines is different, the system being typically unbalanced. Primary distribution lines are MV networks, ranging from 600 V to 35 kV. At a distribution substation, transformers take the incoming transmission-level voltage, steps it down to several distribution primary circuits, fanning-out from substation. Close to end-users, other power distribution transformers step-down the primary voltages to LV secondary circuits at 120/240 V or other voltage levels, and the secondary distribution circuits connect to the end-users at the service entrance. Power distribution systems employ power transformers, circuit breakers, monitoring or metering devices in order to deliver a safe and reliable power. The power distribution in the United States is in the form of three-phase 60 Hz AC current. The ultimate purpose of the transmission and distribution is to supply the required electricity needed for residential and commercial uses. Primary distribution circuits are usually radial in design, unlike transmission systems where circuit designs are meshed. The radial designs have certain advantages: (1) protection is basically overcurrent, (2) lower fault currents, (3) voltage regulation and power flow control are easier to implement and (4) system design is less expensive. The radial circuit design presents several variations, such as the single feeder and open-loop configurations. In *single feeder configuration* all power demanded by laterals and secondary circuits are served by a single primary line. However, in case of failure or any other event, forcing the feeder out of service (e.g. maintenance), all loads experience a service interruption. The single feeder layout can also present a branched-configuration, where several branches stem from the original feeder in order to cover a larger area. These branches are not to be confused with laterals, having a much lower current capacity, whereas the branches have the same (or similar) capacity as the main feeder. In an *open-loop configuration* two feeders parting from the same substation are connected at their end terminals through a normally open tie-switch. Under normal conditions each feeder serves a number of lateral circuits but has the capacity to provide the necessary power to all circuits connected to both feeders. Load transfer between feeders is possible by closing the tie-switch (either manually or automatically). This configuration presents a greater reliability level than the single feeder configuration but requires that each feeder have the capacity to carry the load corresponding to both feeders; additionally, extra equipment is needed (e.g. the tie-switch).

In rural and suburban areas radial configurations are the common design for secondary circuits, while in urban circuits different configurations are used depending on the load types. The spot (single-point) configuration is used for large loads concentrated at a single location (e.g. factories and large buildings), whereas the network configuration is used to serve a large number of loads distributed over an extended area. The radial design in secondary circuits is equivalent to the configuration used in primary circuits. A secondary circuit parts from a step-down (MV/LV) distribution transformer, spreading over the area where the customers are located. Usually, due to the covered area size, the secondary circuits normally present a branched-configuration. Single-point configuration is used for loads, requiring dedicated circuits due to their high power demand, typically consisting of three or five feeders delivering load demanded power, allowing normal operation, even with the loss of one or two of the primary circuits. Each feeder arrives at step-down (MV to LV) distribution transformers that are serving part of the total load. Note that all transformers are equipped with a protection device installed on the secondary side. In the networked configuration several primary circuit lines feed the secondary network from multiple step-down power distribution transformers. The secondary circuits connected at the low-voltage side of the distribution transformers form a meshed network, from which the load power is provided. This configuration is used to serve commercial and residential loads (three- and single-phase). The three-phase power is distributed directly to large industrial and commercial facilities from transmission subsection. Substations are using massive three-phase power transformers and associated equipment (circuit breakers, HV conductors, insulators, etc.) to distribute the electricity to industrial, residential and commercial users. Facilities and buildings must provide maximum supply safety, consume fewer energy resources during construction and operation and be flexibly to adapt to any future power requirements or changes. The intelligent integration of all building and facility service installations offers an optimum for safety, efficiency, flexibility and environmental compatibility, offering maximum comfort. Buildings are responsible for about 40% of the total energy consumption within the United States and developed countries. In large facilities the distribution system choice depends on the building type, dimension, the length of supply cables and the loads.

Power distribution feeder circuits consist of overhead and underground transmission line networks in a mix of branching circuits (laterals) from the substation to the various customers. The use of overhead distribution lines is because they are less expensive than underground ones to install, operate or maintain. Where the underground power distribution is more cost effective, it should be used. When exceptions are considered, the rule is to follow the local practices. Each circuit is designed around constraints such as peak load power, voltage levels, distance to the customers and other local conditions (area or street layout, terrain configurations, visual and environmental regulations or consumer requirements). The secondary voltage in North America consists of a split single-phase service, providing the customers with 240 V and 120 V, depending on their service type and ratings. The services are supplied from a three-phase distribution feeder, usually Y-connected consisting of a neutral center and three phase conductors. In most of the world, single-phase voltages of 220 V or 230 V are supplied directly from a large power distribution transformer, providing also a secondary voltage circuit often serving hundreds of customers. The branch laterals are operated in radial or looped configurations, where two or more feeders are connected together usually through an open distribution switch. Distribution networks are usually highly redundant and reliable complex electric networks. Overhead lines are mounted on concrete, wooden or steel poles designed to support distribution transformers and other needed equipment, besides the conductors. Underground distribution networks use conduits, cables, manholes and necessary equipment installed under the street surface. The system choice depends on factors, such as safety, initial, operation and maintenance costs, flexibility, accessibility, appearance, lifetime, fault probability, location or interference with communication systems. The underground distribution systems compared to the overhead systems are more expensive, requiring higher investment, maintenance and operation costs, lower fault probabilities, being more difficult to locate and repair a fault, lower cable capacities and voltage drops. Each system has advantages and disadvantages, while the most important is the economic factors. However, non-economic factors are sometimes more important as the economic ones.

6.2.2 Load Parameters, Voltages and Frequency Characteristics

The power line frequency (grid frequency) in modern electricity infrastructure is 50 Hz, in most of the world, and 60 Hz in the United States, Canada and few other countries. The countries using the 50 Hz frequency tend to use 220–240 V, and those that are employ 60 Hz tend to use 100–127 V. Several factors influence the choice of frequency in an AC system. Electrical machines, transformers, lighting and transmission lines all have characteristics which depend on the frequency. All of these factors interact, making the frequency selection a matter of considerable importance. Loads, e.g. electrical machines, lighting and heating equipment shows different characteristics with the voltage and frequency changes. In power systems, voltage and frequency at load bus always change due to the disturbances, causing load power fluctuations, which depend on the load characteristics. For the purpose of system stability analysis, it is essential to determine the effects of the load changes due to the voltage and frequency changes. The power system operators regulate the daily average frequency so that clocks stay within few seconds of the correct time. In practice the nominal frequency is changed by a specific percentage to maintain synchronization. The average frequency is maintained at its nominal value within a few hundred parts per million. In the European Union synchronous grids, the deviation between grid phase time and UTC (set by International Atomic Time) is calculated at 08:00 each day at a Swiss control center. The frequency is then adjusted up to ±0.01 Hz (±0.02%) to ensure a long-term frequency average of exactly 50 Hz × 60 sec × 60 min × 24 hours = 4,320,000 cycles per day. In North America, whenever the error exceeds 10 s for the East, 3 s for Texas or 2 s for the West, a correction of ±0.02 Hz (0.033%) is applied. Time error corrections start and end on the hour or on the half hour. Since 2011, The North American Electric Reliability Corporation (NERC) discussed of relaxing the frequency regulation requirements for grids which is reducing the long-term clock accuracy that is using the 60 Hz grid frequency as a time base.

The primary reason for accurate frequency control is to allow that the AC power flow from multiple generators through the grid is controlled. Trends in system frequency are a measure of mismatch between demand and generation is a critical parameter for load control in interconnected systems. Grid frequency varies as the load and/or generation change. Increasing the mechanical input power to a synchronous generator will not greatly affect the system frequency but produces larger electric power from that unit. During a severe overload caused by tripping or failure of generators or transmission lines the frequency declines, due to an imbalance of load versus generation. Loss of an interconnection while exporting power causes the system frequency to rise. Automatic generation control is used to maintain scheduled frequency and interchange power flows. Power plant control systems detect changes in the network-wide frequency and adjust the mechanical power input to the generators back to the target frequency. This action takes a few tens of seconds due to the turbine-generator system inertia. Temporary frequency changes are an unavoidable consequence of changing demand. Unusual or rapidly changing frequency is often a sign that an electricity distribution network is operating near its capacity limits, sometimes being observed shortly before major outages. Frequency protective relays on the power system network sense the decline of frequency and automatically initiate load shedding or tripping of interconnection lines, to preserve the operation of at least part of the network. Small frequency deviations (i.e. 0.5 Hz on 50 Hz or 60 Hz grids) result in automatic load shedding or other control actions to restore system frequency. Smaller power systems, not extensively interconnected with many generators and loads, are not maintaining frequency with the same degree of accuracy. Where system frequency is not tightly regulated during heavy load periods, the system operators allow system frequency to rise during periods of light load, maintaining a daily average frequency of acceptable accuracy. Portable generators, not connected to the grid, need not tightly regulate their frequency, because typical loads are insensitive to small frequency deviations. Power system stability is classified as rotor angle and voltage stability. Voltage stability is an issue in power systems which are heavily loaded, disturbance or have a shortage of reactive power supply.

Power distribution networks are designed to supply electricity to loads or end-users, in agreement with their characteristics, expectations and requirements. All utilities are supplying a broad range of loads from the rural areas with load densities of 10 kVA/mi^2 to the urban areas with the load densities 300 MVA/mi^2 or higher. A utility feeds houses and buildings within specific peak loads on the circuit providing power. The feeder load is the sum of all individual customer loads, while a customer load is the sum of the power drawn by the customer equipment. Customer loads have common characteristics, with several parameters being used for characterization. Load levels vary through the day, usually peaking up in later afternoon or early evening. The *peak power (load)* is the maximum power that consumed by loads during a specific or given time interval and it is equal to the maximum actual power, generated by the power plant, minus transmission line losses. *Average load* is the average power that consumed by the loads during a certain time period, equal to the actual power that generated by the power plants during the same time period, less transmission line losses. The *load demand* is the load average over a specified time period – often 15, 20 or 30 min – being used to characterize real (active) power, reactive power, total power or current. *Peak load demand* over a time period is the most common way utilities quantify a circuit's load. In substations, it is common to track the current demand. *Load factor* is the ratio of the average load over the peak load for a certain period of time. Basically, the load factor is the ratio of the load that equipment actually draws when it is in operation to the full load. The *utilization factor* is the ratio of the time that equipment is in use to the total time that it could be in use. For a day, the load factor is a daily load factor and if the period of time is month, the load factor is monthly load factor and similarly for the yearly load factor. The *load curve* shows the load characteristics over a certain period of time (day, month or year). The curve is plotted by placing the ordinate (kW) with their proper time sequin as shown in the diagram of Figure 6.2. Peak load is normally the maximum demand but may

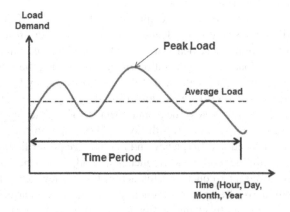

FIGURE 6.2   A typical load curve for a generic time period.

be the instantaneous peak power, being a number between zero and one. A load factor close to 1.0 indicates that the load runs almost constantly. A low load factor indicates a widely varying load. From the utility point of view, it is desired to have higher load factors. Load factor is usually found from the total energy used (kWh) as:

$$LF = \frac{E(kWh)}{d_{kW} \times hrs} \tag{6.1}$$

Here, LF is the load factor, $E(kWh)$ represents the energy use in kWh, d_{kW} is the peak load demand in kW and hrs is the number of hours during that specific time period. The load factor value used for determining the overall cost per generated unit is less than one, maximum energy demand being always higher than the average demand. Higher the load factor, the lesser is the cost per unit. Load factor does not appear on utility bills, but it affects electricity costs, indicating how efficiently the customer is using the peak demand. A high load factor means power use is relatively constant. Low load factor shows that occasionally a high demand is set. To service that peak, capacity is sitting idle for long periods, thereby imposing higher costs on the system. Electrical rates are designed so that customers with higher load factors are charged less overall per kWh. *Coincident factor* is the ratio of the peak demand of a whole system to the sum of the individual peak demands within that system. The peak demand of the whole system is referred to as the peak *diversified load* demand or the peak *coincident* demand. The individual peak demands are the *non-coincident load* demands. The coincident factor is less or equal to one, usually lower than one because individual loads are not reaching coincidentally their peaks. *Diversity* relates the rated loads of the equipment downstream of a connection point, and the connection point rated load. *Diversity factor* is the ratio of the individual peak load demands in a system to the overall system peak load demand, being greater than or equal to one and the reciprocal of coincident factor. The diversity factor is also defined as the probability that particular equipment is coming on at the time of the facility peak load. The diversity factor for all other installations is based upon a local evaluation of the loads, applied at different moments in time. The load is dependent upon time and equipment characteristics. The diversity factor shows that a load does not equal the sum of its parts due to the time interdependence (i.e. diverseness). *Responsibility factor* is the ratio of a load demand at the system peak load time to its peak load demand. A load with a responsibility factor of one is peaking at the same time as the system. It applies to the individual customers, customer classes or circuit sections. The loads of certain customer classes tend to vary in similar patterns. Commercial loads tend to run from 8 a.m. to 6 p.m., the industrial loads runs continuously, and as a class have a higher load factor, while the residential loads are peaking usually in the evening. Weather significantly changes loading levels, affecting both the load demand and peak demand. For example, on hot summer days, air conditioning increases the demand and reduces the diversity among loads.

Example 6.1   An oversized motor 20 HP drives a constant 15 HP load whenever it is ON, operating only eight hours a day and 50 weeks in a year. Estimate its load factor and the utilization factor.

SOLUTION

The motor is driving a constant load all the time when the load is ON, so the load factor is:

$$\text{Load Factor} = \frac{P_{Load}}{P_{Motor}} = \frac{15\ HP}{20\ HP} = 0.75 \text{ or } 75\%$$

The hours of operation would then be 2000 hour, and the motor use factor for a base of 8760 hour per year would be 2000/8760 = 22.83%. With a base of 2000 hour per year, the motor use factor would be 100%. The bottom line is that the use factor is applied to get the correct number of hours that the motor is in use.

Notice that any load is time dependent as well as being dependent upon equipment characteristics. When the maximum load demand of a supply is assessed, it is not sufficient to simply add together the ratings of all electrical equipment that could be connected to that supply. If this is done, a figure somewhat higher than the true maximum demand will be produced. This is because it is unlikely that all the electrical equipment on a supply will be used simultaneously. The concept of being able to *de-rate* a potential maximum load to an actual maximum load demand is known as the application of a diversity factor. For example, 70% diversity means that the device in question operates at its nominal or maximum load level 70% of the time that it is connected and turned on. If all electrical equipment were running at full load at the same time, the diversity factor is equal to *one. Notice that the greater the diversity factor, lesser is the cost of power generation.* Diversity factor in a power distribution network is the ratio of the sum of the peak demands of the individual customers to the peak demand of the network, being determined by the service type, i.e. residential, commercial, industrial or combinations. Diversity factor is used for sizing the distribution feeders and transformers. To determine the maximum peak load and diversity factor data records or demand estimates are needed. The feeder diversity factor is the sum of the maximum demands of the individual consumers divided by the overall feeder maximum demand. In the same way, diversity factor of a substation, a transmission line or a whole utility system is determined. Residential loads have the highest diversity factor, the industrial loads have lower diversity factors, about 1.4, the street lighting practically unity, while other loads vary between these limits. Diversity factors are used by utilities for distribution transformer sizing and load predictions. Demand factors are more conservative and are used by NEC for service and feeder sizing. Demand factors and diversity factors are design parameters. For example, the sum of the connected loads supplied by a feeder, multiplied by the demand factor determines the load for which the feeder is sized, which is the feeder maximum demand. The sum of the maximum demand loads for a number of sub-feeders divided by the diversity factor for the sub-feeders gives the maximum demand load to be supplied by that feeder. The utilization factor must be applied to each individual load, with particular attention to electric motors, which are rarely operated at full load.

Example 6.2   Four individual feeder-circuits (similar to diagram of Figure 6.1) **with connected loads of 250 kVA, 200 kVA, 150 kVA and 400 kVA and demand factors of 90%, 80%, 75% and 85%, respectively. By assuming a diversity factor of 1.5, estimate the transformer size, supplying these feeders.**

SOLUTION

Calculating demand for feeder-circuits

> Feeder 1: 250 kVA x 90% = 225 kVA
> Feeder 2: 200 kVA x 80% = 160 kVA
> Feeder 3: 150 kVA x 75% = 112.5 kVA
> Feeder 4: 400 kVA x 85% = 340 kVA

The sum of the individual demands is equal to 837.5 kVA. If the main feeder circuit were sized at unity diversity: kVA = 837.5 kVA ÷ 1.00 = 837.5 kVA. The main feeder circuit would have to be supplied by an *850 kVA transformer*. However, using the diversity factor of 1.5, the kVA = 837.5 kVA ÷ 1.5 = 558 kVA for the main feeder. For diversity factor of 1.5, a *600 kVA transformer* could be used.

6.2.3   POWER DISTRIBUTION SYSTEM ELEMENTS

The safe and secure operations of a power distribution system are requiring much dedicated equipment and devices. These equipment and devices are installed throughout the power distribution system and they include elements and devices such as power transformers, circuit breakers, fuse and control and monitoring devices, equipment and apparatuses. The most important elements and a brief definition are presented as follows. *Distribution lines* are responsible for transporting electrical energy between two distant points; overhead lines are typically made of bare aluminum (being ACSR a commonly used type), whereas underground lines commonly use cables with polymer-insulation, such as XLPE and EPR. Cables and conductors used for distribution lines are characterized by their current capacity and rated voltage. A *power distribution transformer* usually consists of two or more windings coupled by their electromagnetic fields; it transfers power from one winding to another without changing the frequency and is capable of performing voltage level transformation (reduction or increment). Power distribution transformers are used to perform successive voltage reductions along the power distribution system in order to adjust the voltage level to an adequate value for every system section and end-user. A *circuit breaker* is a switching device designed to open and close a circuit by non-automatic means and to open the circuit automatically on a predetermined overcurrent in order to avoid damage to itself and other equipment and devices. A *fuse* is a protective device used in distribution systems to protect laterals, secondary circuits and low power transformers; it consists of a strip of wire that melts and clears an electric circuit when an overcurrent or short-circuit current passes through it. Melting and clearing times depend on the fuse's time-current curves. The most commonly used fuses are the types *K* and *T*.

Voltage (potential) transformer (VT) is a conventional transformer with primary and secondary windings on a common core. Standard voltage transformers are single-phase units designed and constructed so that the secondary voltage maintains a fixed relationship with primary voltage, being employed to reduce the primary circuit voltage to a safe value (e.g. 120 V), to be used as an input for monitoring and protection devices. Usually, such transformers only have the capacity to serve low rating meters and relays. A current transformer transforms line current into values suitable for standard protective and monitoring devices and isolates the relays from line voltages. A *current transformer* (CT) has two windings, designated as primary and secondary, which are insulated from each other. The primary winding is connected in series with the circuit carrying the line current to be measured, and the secondary winding is connected to protective devices, instruments, meters or control devices. The secondary winding supplies a current in direct proportion and at a fixed relationship to the primary current. Voltage and current transformers are known as generic instrument transformers. A *relay* is an electronic, low-powered device used to activate a high-powered device. In power distribution systems, relays protect feeders and system equipment from damage in the event of a fault by issuing tripping commands to the corresponding circuit breakers in order to interrupt the current produced by the fault. The *automatic circuit recloser* is a protective device with the necessary intelligence to sense overcurrents and interrupt fault currents, and to re-energize the line by reclosing automatically. In case of a permanent fault, the recloser locks open after a preset number of operations (usually three or four), isolating the faulted section from the main part of the system. A *switch* is a switching device used to isolate a system element for repair or maintenance. It must be capable of carrying and breaking currents during normal operating conditions; a switch may include specified operating overload conditions and also carrying for a specified time currents under specified abnormal circuit conditions such as those of a short circuit. A switch, therefore, is not expected to break fault current, although it is normal and usual for a switch to have a fault making capacity.

A *voltage regulator* is a transformer with a 1:1 nominal transformation ratio equipped with an on-load tap changer. Such taps are allowing the transformer to vary its transformation ratio to react to voltage variations at the primary side. Voltage regulators are installed at intermediate points of long primary lines in order to compensate the voltage drop produced along the circuit; voltage control will impact the voltage profile of all loads downstream from the voltage regulator. A *capacitor bank* is a local source of reactive power, usually consisting of several capacitors connected in series and parallel. By correcting power factor it can perform voltage regulation and reduce system losses. Capacitor banks are generally three-phase and are installed within the distribution substation or at intermediate points of a primary circuit line. *SCADA (System Control and Data Acquisition)* is a communication system that allows real-time monitoring of the distribution system; it collects information from equipment installed throughout the system and stores it in a database accessible to different users and applications. The measured values reflect different time varying quantities, such as bus voltages, line currents and tap changer positions. SCADA also has the capability to remotely operate circuit breakers and switches, which provides a greater flexibility for system operation and reduces response times for switching actions. *Distributed generation (DG)* or embedded generation refers to generation applied at the power distribution level. DG units can be directly connected at the distribution substation or dispersed throughout a power distribution system. Due to their small size, distributed generators can be placed close to load consumption, typically DG present sizes of up to 5 MW (IEEE STD 1547 applies for generators under 10 MW). However, utilities can limit the rated power of the distributed generation units according to their own operation policies. The origins of distributed generation can be found in the cogeneration (combined power and heat generation) practiced by many industries. These small-size generators serve a portion of the load at these industrial facilities and inject any excess of generated power into the utility system. DG unit can also provide emergency power to the industrial facility during utility outages. This is a common practice in industries such as pulp and paper, steel mills, petrochemical facilities, as well as military bases and some university campuses that had internal generation within their electrical facilities, operating in parallel with the grid. Distributed generation concept does not make reference to a specific technology; different types of technologies are used to drive DG depending on the selected primary energy source. In recent years there has been a great effort to develop and improve technologies based on renewable energy sources; much of this effort has been oriented to work on small-size generators that can have an application in distributed generation; however, classical combustion engines remain a cost-effective option for the small-scale generation of electrical energy. Common technologies used in DG units are microturbines, fuel cells, Stirling engines, internal combustion engines, flywheels, photovoltaic systems, solar thermal units, geothermal heat pumps and medium and small size wind turbines. The connection of DG units to the power distribution system can be used for supporting voltage, reducing losses, providing backup power, providing ancillary services, service restauration, power system resilience enhancement or deferring power distribution system upgrade. Aspects to be considered when embedding DG units into a power distribution system are the great variety of generating technologies, and the intermittent nature of some of the renewable energy sources.

6.2.4 Feeder Voltage Drops, Electric Distribution Losses and Power Factor Control

Electrical energy is generated at generation stations and transferred over high-voltage transmission line and power distribution networks to the utilization points. The voltage into power distribution must be kept within a narrow deviation range from the nominal service voltage to provide satisfactory service, required power quality and to avoid damage to customers' equipment and devices. In the United States, the allowable deviations are specified by the ANSI C84.1 (1) standard to be at 5% of the nominal value. For example, for a 120 V service, acceptable rage is 114 V to 126 V. If a current flows through impedances a voltage drop occurs and for a conductor with impedance $Z = R + jX$, carrying a current, I, is given by:

$$|\Delta V| = RI \cos(\theta) + XI \sin(\theta) = RI_d + XI_q \tag{6.2}$$

The conductor impedance is characterized by its resistance, R, reactance, X, and by $cos(\theta)$, the power factor of the received apparent power. Here, I_d and I_q are the real and imaginary component of the current corresponding to the transmitted active and reactive powers. The reactance, X, is usually several times greater than R for power distribution lines. Voltage drop can be reduced by

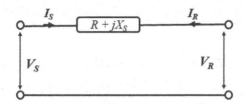

FIGURE 6.3 Cable impedance between source and load ends.

lowering the reactive current, I_q, without affecting the transmitted active power. The use of HV lines is motived by the increasing the line capacity and by loss reduction. A transmission line is modeled by series resistance, series inductance, shunt (parallel to ground) capacitance and conductance. Series resistance and inductance are the most important, parallel elements being often neglected in calculations. The term *wire* refers to one or more insulted small size conductors (solid or stranded), while *cable* refers to one or more insulated conductors, grouped in insulated jacket, often with a ground shield. In models, a cable is represented by a resistance, R in series with leakage reactance, X_L ($= \omega L$), where L is the leakage inductance between the current carrying conductors. A cable calculation requires selecting the conductor size with required ampacity at the operating temperature, meeting the voltage limitations over the feeder length under the steady-state and inrush motor or equipment current. In such calculations, per-phase current, I, power factor, resistance and inductive reactance are required. The voltage drop of the transmission line in Figure 6.3 is given by:

$$V_{drop} = I \times (R \cdot \cos\theta + X_L \cdot \sin\theta) = I \times Z_{cable} \tag{6.3}$$

Here, θ is the phase angle (PF $= \cos\theta$), and all calculations can be performed in standard units or in percent or p.u. values, taking θ positive for lagging power factor and negative for leading power factor. Equation (6.4) is convenient for the definition of the effective cable impedance because its product with the cable current is simply giving the voltage drop magnitude or the cable voltage drop in volt/amperes:

$$V_{drop/A} = Z_{eff(cable)} V/A \tag{6.4}$$

Example 6.3 A 4.8 kV three-phase, line-to-line wye-connected feeder has a per-phase resistance of 50 mΩ and 0.33 mΩ inductive reactance per 1000 ft. Calculate the feeder voltage drops for: 0.85 lagging power factor, unit power factor and 0.85 leading power factor at an apparent power of 1.2 MVA.

SOLUTION

The feeder phase voltage and the current at three-phase apparent power of 1.2 MVA are:

$$V_\phi = \frac{4800}{\sqrt{3}} = 2775.6 < 0° \text{ V}$$

$$I = \frac{1.2 \times 10^6}{\sqrt{3} \times 4800} = 144.5 \text{ A}$$

Voltage drops for 0.85 lagging power factor, unit power factor and 0.85 leading power factor, using Equation (6.2) are:

$$V_{drop} = 114.5 \times (50 \times 10^{-3} \cdot 0.85 + 0.33 \times 10^{-3} \cdot 0.527) = 6.86\,\text{V}$$

$$V_{drop} = 114.5 \times (50 \times 10^{-3} \cdot 1.0 + 0.33 \times 10^{-3} \cdot 0.0) = 7.73\,\text{V}$$

$$V_{drop} = 114.5 \times (50 \times 10^{-3} \cdot 0.85 - 0.33 \times 10^{-3} \cdot 0.527) = 2.88\,\text{V}$$

Notice that in the United States the standards and codes established two ranges for the voltage variations designated as *Range A* and *Rang B*. Range A voltage is in the brackets of ±5% of the nominal value, while Range B has an asymmetric range of +6% to −12% of the nominal value. Utilization equipment is designed to give fully satisfactory operation throughout *Range A* limits for voltages. Range B allows voltages above and below Range A limits. The cables are selected and manufactured to operate for specific requirements and conditions. Power factor has a significant effect on the cable voltage drop. The energy loss, ΔE, over a conductor is the product of the cable resistance and the current magnitude squared:

$$\Delta E = RI^2 = R\left(I_d^2 + I_q^2\right) \tag{6.5}$$

On typical power distribution systems, usually with no distributed generation, the active power must be supplied by the utility (grid), so there is very limited control over the distributed reactive power. The only available option is to use reactive power compensation to reduce the reactive current flows on the power distribution system. Usually, capacitor banks are used to provide reactive power compensation. Power distribution systems have automatic voltage regulation schemes and devices at one or usually at several control power stations, such as area substation or facility switchgears to maintain the constant voltage regardless of the load currents. Voltage drops in line are in relation to the resistance and reactance of line, length and the current drawn. For the same power handled by a transmission, the lower the voltage, the higher is the current drawn and the higher is the voltage drop. The current drawn is inversely proportional to the voltage for the same power handled, while the power loss in the line is proportional to the resistance and square of current (i.e. $P_{Loss} = I^2R$). Higher voltage transmission and distribution helps to minimize line voltage drop in the ratio of voltages, and the line power loss in the ratio of square of voltages. The feeder cables deliver power from the controlled *sending end* (power source) to the *receiving end* (load), with the voltage gradually dropping form the initial maximum value to the minimum value at the load. For this reason, the power distribution system is seized not only for the required ampacity (capacity) but also to maintain the required voltage levels during the steady-state operation and during the transients such as inrush current during motor starting. If the load current of a power factor, *PF* (usually lagging the voltage) flows from the switchboard is controlled to maintain the constant bus voltage, the magnitude of the voltage drop between the ends, Equations (6.3) can be written for the cable impedance as:

$$V_{drop} = I \times (R \cdot PF + X_L \cdot \sqrt{1 - PF^2}) \tag{6.6}$$

Often the cable manufacturers are listing the cable effective impedance at 0.85 lagging power factor, therefore the catalog cable impedance values are:

$$Z_{cable} = R \times 0.85 + X_L \times \sqrt{1 - 0.85^2} = 0.85 \cdot R + 0.527 \cdot X_L\,\Omega/\text{phase} \tag{6.7}$$

However, if the power factor value is quite different than 0.85 lagging, the Equation (3.5) must be adjusted consequently. The cable size is usually selected to meet the capacity (ampacity) with up to 30% margin, limiting the solid-state voltage drop at 3–5% from the switchboard to the load. We

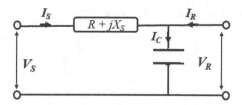

FIGURE 6.4 Voltage rise calculation circuit diagram (C is the shut capacitor).

have noticed that the feeder voltage drop depends, besides the cable impedance, also on the load power factor. Capacitors can improve the load power factor, while correcting also the feeder voltage drops by reducing or eliminating the reactive terms in Equations (6.2), (6.3), (6.6) and (6.7). The voltage boost estimation produced by the capacitors (bank) placed at the end of the feeder (load), starts with the estimate the capacitor value, so the reactive power correction and values, Q, needed to improve power factor at the desired level as:

$$Q_{Cap} = Q_{Load} - Q_{Desired} \tag{6.8}$$

Expressing the reactive power, through the apparent power factor, the apparent power, reactive power and active (real) power for a feeder, the load and the desired reactive power are expressed as:

$$Q_{Load} = \sqrt{\left(\frac{P_{Load}}{PF_{Load}}\right)^2 - P_{Load}^2}$$

$$Q_{Desired} = \sqrt{\left(\frac{P_{Load}}{PF_{Desired}}\right)^2 - P_{Load}^2} \tag{6.9}$$

Form the above relationships the needed capacitor (bank) reactive power to improve the power factor at the desired level is given by:

$$Q_{Cap} = P_{Load} \times \left[\sqrt{\frac{1}{PF_{Load}^2} - 1} - \sqrt{\frac{1}{PF_{Desired}^2} - 1} \right] \tag{6.10}$$

The pre-phase capacitor (bank) value is then computed by using the well-known relationship:

$$C = \frac{Q_{Cap}}{2\pi f \cdot V_{LL}^2} \tag{6.11}$$

Here, f is the electrical supply frequency and V_{LL} is the feeder line-to-line voltage. In addition to improving the power factor, the capacitor (bank) is reduction the feeder voltage drop and the feeder cable losses by producing a voltage rise, due to the leading capacitor current. To understand the voltage rise calculation, the circuit of Figure 6.4 is considered. The rated capacitor (bank) current is given by:

$$I_C = \frac{Q_{Cap}}{\sqrt{3} \times V_{LL}} \tag{6.12}$$

The voltage rise due to the capacitor (bank) is equal to:

$$V_{Rise} = I_C \cdot X_L = X_L \frac{Q_{Cap}}{\sqrt{3} \times V_{LL}} \tag{6.13}$$

Example 6.4 **A facility has a power demand of 1 MW at 0.75 lagging power factor. What are the capacitor bank reactive power ratings required to improve the power factor to 0.85 and 0.95? What is the voltage rise in each case, if the feeder reactance is 0.015 Ω/phase and the line-to-line voltage 460 V?**

SOLUTION

Direct application of Equation (6.10) leads to:

$$Q_{Cap-0.85} = 1000 \times \left[\sqrt{\frac{1}{0.75^2} - 1} - \sqrt{\frac{1}{0.85^2} - 1} \right] = 262.2 \text{kVAR}$$

$$Q_{Cap-0.95} = 1000 \times \left[\sqrt{\frac{1}{0.75^2} - 1} - \sqrt{\frac{1}{0.95^2} - 1} \right] = 553.2 \text{kVAR}$$

The voltage boost in this case and for the feeder characteristics (Equation 6.13) are:

$$V_{Rise-0.85} = 0.015 \times \frac{262.2 \times 10^3}{\sqrt{3} \times 460} = 4.94 \text{ V}$$

$$V_{Rise-0.95} = 0.015 \times \frac{553.2 \times 10^3}{\sqrt{3} \times 460} = 10.4 \text{ V}$$

The voltage rise (boost) due to the capacitor can be expressed in per-phase values as:

$$V_{Rise-phase} = X_L \frac{Q_{Cap-phase}}{V_\phi} \text{ V/phase and } \%V = \frac{V_{Rise-phase}}{V_\phi} \times 100 \tag{6.14}$$

Notice that in Equation (6.14) the voltage drop is zero for leading power factor that gives the $R/X_L = tan(\theta)$, therefore the reviving end voltage is constant. Typical R/X_L ratios for power distribution cables are in the range 0.2–0.3, with an average value of 0.25, for a power factor of 0.97 leading, giving a flat receiving-end voltage value, regardless the load current. Even only a small part of the electric energy transferred through the transmission lines in a distribution network (3% or less) is lost through Joule effect and other losses are still important to estimate such losses. Cable manufacturers usually provide the AC resistance and the series leakage reactance, including the skin and the proximity effects and the conduit/raceway type. However, the difference between DC and AC characteristics are often neglected in power system studies. Distribution losses are mostly due to the Joule effect, which depends on the line current and resistance. Most of electric losses occur at end-users, in electrical machines, drives, electrical heating (furnaces, boilers, etc.), microwave and lighting equipment, other loads, e.g. electrochemical equipment, monitoring, control and communication systems. The basic power relationship is expressed by:

$$P = n \times R \times I_{line}^2 \text{ (W)} \tag{6.15}$$

Here, n is the number of phase conductors, R (Ω) is the phase conductor resistance and I_{line} the RMS line current (A). In actual power distribution, due to issues with the installation, operation and maintenance costs, usually a limited number of capacitors are installed to compensate for reactive power. Moreover, due to the discrete capacitor switching nature, the compensation of the capacitor banks is not able to match the load variations throughout a day, requiring scheduling of the OFF and ON switching to achieve the best compensation results. Because the feeder loadings vary with hour, day, week, month, season and year, setting the correct reactive compensation amount is not a simple task, requiring consideration of the actual real-time feeder loading conditions, the locations and the size of the installed capacitor banks.

6.3 VOLTAGE CONTROL IN CONVENTIONAL AND ACTIVE POWER DISTRIBUTION

A power system is well designed if it provides a continuous, good quality, secure and reliable electricity supply at minimum possible costs. Good quality means that the voltage levels are maintained within the prescribed limits. Practically all the equipment and devices are designed to operate satisfactorily only when system voltage levels correspond to their rated voltages or at the variations are within narrow range, e.g. 5%. If the voltage variations are larger than specified limits, their performances are degraded and their life is reduced. When power is supplied to a load through a transmission line keeping the sending end voltage constant, the receiving end or load voltage undergoes variations depending on the load and its power factor. The higher the load with smaller power factor, the larger is the voltage variation. A node voltage variation is an indication of the unbalance between the reactive power generated and consumed. If the generated reactive power is greater than consumed, the voltage goes up and vice versa. The DG and RES intermittent nature, if present in distribution network, can significantly affect the grid voltage profile and adversely impacting the voltage control devices such as tap-changing transformers and capacitor banks. Furthermore, the growing penetration of plug-in electric vehicles (PEVs) can put additional stress on voltage control devices due to the PEV stochastic and concentrated power profiles. Such profiles may lead to high maintenance costs and reduced lifetimes for voltage control devices and limiting the PES and PEV high penetration levels. The DG integration changes the power distribution to an active distribution network (ADN), with bi-directional power flows. Voltage regulation is one of major power distribution operational challenges, accompanying high DG penetration levels. RES units can change the smart grid voltage profiles, interacting negatively with control schemes, as one based on load tap changers. In today's market environment, utilities are focused to improve operational efficiencies and service quality for higher profitability. Whenever there is a power flow through a circuit element, it results in voltage drop across the element, varying with the power flowing through it, making the voltages in a distribution network variable and hence a need to control and maintain voltages within specified ranges. An inductive circuit element also causes a phase lag between the voltage and current, degrading the power factor. Thus, to deliver same active power more current needs to flow through the system as compared to pure resistive elements. Moreover, the loads' actual power consumption could be quite different form the nominal or design value, depending on the actual voltage condition, making the voltage and reactive power control very important for power distribution operation and management. Loads, connected to the distribution networks, are characterized by the following relationship with various values of the parameter α, as:

$$P_{actual} = P_{nominal} \left(\frac{V_{actual}}{V_{nominal}} \right)^{\alpha} \tag{6.16}$$

Most common loads are the constant impedance ones ($\alpha \approx 2$), constant current load ($\alpha \approx 1$) and constant power loads ($\alpha \approx 0$). For the constant impedance loads, the power consumption varies with the voltage squared; for the constant current load model, the power varies linearly with the voltage magnitude; and for the constant power load model, the power is constant regardless the voltage magnitude. Resistive heaters are a common example of constant impedance load. If such heater is designed (rated) to use 1 kWh at rated voltage, it will consume 1.1025 kWh if the actual voltage is 1.05 of the nominal (design) value. On the other hand, the power consumption of a constant power load remains invariant, within the limits of the design voltage. When the voltage drops, such loads draw more current so the power remains the same. However, most of the loads are various combinations of constant impedance and constant power loads.

The utilities always try to improve or maintain the power quality level and to operate an efficient power distribution without considerable capital expenditure. However, part of generated electricity is lost during the transfer to the end-users. About 40% of the overall grid losses occur in power distribution, and any reductions have significant economic benefits. As the electricity demands keep growing new power plants are needed to meet peak power demands or unforeseen events. However, most of the peak demands are lasting only about 5% of the time, some hundreds of hours per year, the peak power plants being used infrequently and for very short periods. By active demand management on power distribution and through demand response, voltage, reactive power control and optimization, the peak demands are reduced, eliminating some of the expensive capital expenditure with beneficial economic impacts. Such actions are resulted in the development of technologies like Volt-VAR optimization (VVO) and conservation voltage reduction (CVR). Traditional Volt-VAR methods have been used by the power industry to reduce electric line losses and increase grid efficiency. CVR is one of the loss reduction approaches employed since the 1950s by utilities to reduce energy consumption. CVR is based on the premise that any voltage reduction leads to reduced energy consumption by the end-use loads. However, the idea that CVR can actually reduce energy usage on power systems was very controversial and did not initially gain much acceptance. This was mainly because it will be difficult to deploy CVR over wide areas without causing danger to the consumer installations and equipment. CVR methods work on the principle that energy is saved by operating in the ANSI residential voltage lower half band, i.e. 114–120V, rather than 120–126V, without causing harm to consumer appliances. It has been noticed that there is 1.2% change in energy consumption with 1% reduction in voltage. The CVR factor, a measure of how effectively voltage reductions are used to calculate energy savings or reactive power savings, can be calculated as:

$$CVR\,Factor_{Active\,Power} = \frac{\%kWh}{\%Voltage\,Reduction} \qquad (6.17a)$$

And

$$CVR\,Factor_{Reactive\,Power} = \frac{\%kVAR}{\%Voltage\,Reduction} \qquad (6.17b)$$

CVR has again been revived because of advanced metering infrastructure capabilities and enhanced two-way communications currently available in smart distribution systems. In the smart grids, CVR techniques are one of the cheapest technologies that can be used to provide better asset usage, demand reduction and efficiency improvement. Several smart grid projects have utilities deploying CVR to evaluate its benefits. CVR techniques are implemented using various strategies and approaches, such as line drop compensation, voltage control, voltage reduction methods and VVC techniques. Moreover, the VVO and CVR objectives are to maintain the voltages and phase lags within certain (standard) limits and specified values. Main VVC objective is to maintain acceptable voltages at all points along the distribution feeders, within standard limits and under all loading conditions. A second objective is to maintain the power factor closer to unity in order to minimize losses. Voltage optimization methods, by incorporating CVR methods are reducing voltage variations across the feeder, to prescribed range, reducing the load and system losses. VVC and VVO techniques are key operation methods of smart grids. Approaches refer to the optimization of reactive power and voltage regulation resources for the purpose of power factor correction, voltage correction or loss reduction as well as over- and undervoltage prevention conditions and voltage collapse. Conventional VVC and VVO techniques have downsides such as low robustness to input errors, computational complexity, erroneous state identification, etc. In summary, the most common coordinated Volt-VAR control requirements are keeping the bus voltages within standard limits, minimize active power losses and the number of tap changes, manage the reactive power source, regulate the transformer and the feeders loading and control the power factor.

Older voltage control relies on slow-acting electromechanical devices installed on primary feeders, which are not suitable for handling the changing power distribution paradigm towards smart grid. The voltage regulation (VR) of power distribution is customarily provided by load tap changer (LTC) transformers installed in the substation complemented by voltage regulators on the feeders and reactive compensation. Some voltage regulators compensate for the voltage drop on the lines to control the voltage at a certain distance. These control devices cannot, however, react fast enough in emergency conditions. Furthermore, in a distribution network with DGs, the settings of these traditional devices are different from those of conventional VR systems without DGs. ANSI C84.1 standard specifies the recommended service voltage for "range A" as ±5% of the nominal value (i.e. 114 V to 126 V for a nominal service voltage of 120 V). Under heavy load conditions, without VVC, a distribution system may experience undervoltage violation. Since distribution system voltage regulation is a fundamental requirement for all utilities, the primary goal of VVC is to maintain the voltages along a distribution feeder within an appropriate range under all operating conditions. Since distribution system voltage regulation is a fundamental requirement for all utilities, the primary goal of VVC is to maintain the voltages along a distribution feeder within an appropriate range under all operating conditions. The means to achieve VVC falls into the two categories: voltage regulation by LTC or VR, and VAR compensation by using smart inverters, capacitors (CAPs) or synchronous machines (condensers). In conventional distribution systems, voltage regulators, load tap changers and capacitor banks are the choice devices used for VVC control. With the increasing DG and RES penetration on distribution system, PV with smart inverters and solid-state transformers (SSTs) are considered as VAR control devices. In traditional VVC, VRs and LTCs are controlled based on local measurements, and they are coordinated by differentiating the time-delay settings. In this local control mode, the user can configure the voltage set point of a VR or LTC, and then controller gives command of increasing/decreasing the tap position to regulate the voltage to the set point by changing the turns ratio. Capacitors' control is realized manually at the substation or by automatic control, including time-switched, voltage-controlled, voltage-time-controlled or voltage current-controlled schemes. Advanced VVC methods like smart inverters and distributed secondary-side VR devices are key technologies to meet smart grid requirements, such as effective voltage control and ability for seamless DER integration. Despite power distribution's complex nature and limited number of regulating devices, conventional VVC approaches provided satisfactory performances. However, the growing DER and fast charging electric vehicle penetration is posing new challenges on VVC devices, relying on slow acting control assets. Additionally, the smart grid parading is making the operating voltage limits an even more stringent requirement. The sparsely located electromechanical VRs, installed on the primary distribution, cannot handle the new dynamics arising from variabilities of DG units and smart loads, to provide SG desired performances. However, with the extension of supervisory control and data acquisition (SCADA) to power distribution, VVC methods can be improved to achieve more benefits while keeping the voltages acceptable. Advanced controllers with remote control function enable VR, LTC and CAPs to receive command from SCADA system. In remote control mode, VR and LTC receive command of tap increasing/decreasing from the control center, and CAPs are receiving command of ON/OFF.

Conservation voltage reduction and Volt-VAR optimization are potentially cost-effective ways to deliver energy efficiency benefits to customers without the need to recruit and involve participants.

Increasingly, utility regulators are allowing associated energy savings to count toward voluntary energy efficiency goals or mandatory energy efficiency standards, a significant advantage. In addition, grid modernization efforts are improving the tools that distribution system operators can use to optimize the voltage. At the power system level, the savings potential from the VVO or CVR operations is driven by a variety of factors, including the customer class distribution, distribution line density and percent loading. A key component in estimating potential customer savings is the CVR factor, which is a commonly used indicator of the relationship between energy savings and changes in voltage from CVR operations. As mentioned, CVR methods attempt to maintain load voltage level into ANSI standard limits, in order to achieve energy conservation without changes into customer's behavior. This can reduce the distribution network energy consumption and peak demands, conserving the energy by lowering the ending points' voltage levels. Therefore, there is a need for effective VVC techniques that are actively controlling the feeder voltage profiles to get more capacity out of existing networks, meeting CVR objectives, while enabling the DER units to participate in the active voltage control, improving voltage profiles in the active distribution networks. Advanced VVC devices are now commercially available to provide VR and secondary-side reactive power compensation. Conventional VVC uses electromechanical devices, e.g. substation load tap changer transformers, switched capacitor banks and voltage regulators. Autonomous controllers typically mounted on the VR device take the control decisions based on local voltage and current measurements. The response of these controllers improves the network voltage. VVC key features summarized in the literature include: (1) acceptable voltage level and profile must be maintained at all feeder points within all loading conditions; (2) acceptable power factor for each feeder must be maintained for all loading conditions; (3) operator override is allowed and proper monitoring alert system must be available in the event of the control failure; (4) feeder reconfiguration actions must be properly handled by the distribution operators; (5) optimal coordinated control of all available VR devices must be provided; and (6) reactive power control from connected DERs must be properly handled.

The increased presence of DG units, electric vehicles and the smart grid deployment have made voltage control more complex, leading to additional VVC objectives that are contributing to the overall grid performance and reliability. Advanced VVC systems have broader objectives, such as: *maintain acceptable voltage, improve efficiency, minimize consumption, enable higher renewable energy penetration, coordinate all network devices, self-monitor capabilities, allowing operator override, support self-healing and feeder reconfiguration capabilities and functions and allow selectable objectives.* However, maintaining acceptable voltage levels is the primary function of any voltage control system, with or without optimization and is expected to be met at all times under normal operation. Regardless of load, voltage should remain within the ±5% range at all feeder points. This being the objective that is most at risk by higher RES penetrations, by often increasing the insertion point far outside the voltage limits or, conversely once compensated for, by dropping below the minimum voltage as for example, the loss of wind or solar generated power if the wind stops or a cloud passes over. Any system efficiency is improved by reducing its losses. Any non-unity power factor increases losses. While a unit power of factor is ideal, a realistic system should be able to operate at lower power factors. In recent years, CVR methods, based on the idea that lower voltages are reducing power consumption, are considered better approaches of the efficiency improvements based on maintaining the voltage in the lower half of the acceptable range. Conventional voltage control methods can perform poorly in the presence of significant power generation on a distribution feeder, being based on the premise that voltage usually decreases with the distance from the feeder. An equally important assumption which is also violated by the RES presence is that the load variations are the only cause of changes in current so that historical load data combined with ambient temperature are good predictors of current flow through the system. As a result, the system's ability to control voltage is one of the factors limiting RES penetration on a feeder. Even within the system's ability to maintain voltage, the variations in the apparent load caused by the wind and solar variability may cause more frequent cycling of electromechanical devices used for voltage control, reducing the operating life and maintenance intervals. A more ideal VVOC system would include the ability to compensate for voltage variations created by variable DER power contributions.

The CVR principle is that energy consumption savings and minimizing the system losses are achieved by a voltage level reduction without affecting the performance of the customer's devices. This means that total power demand can be reduced by operating in the lower allowable voltage range. Resistive loads such as incandescent lights increase power consumption proportional to the voltage square and constant current loads consume power proportional to voltage. Therefore, reducing the voltage is expected to achieve some level of demand reduction and lower the distribution circuit losses. It is assumed that a voltage reduction has a negative impact on customers. CVR, however, is not reducing the consumption in all loads. A constant power load increases the current and hence distribution system losses as the voltage decreases, even though the device itself draws constant power regardless of voltage. Inverter-based power supplies and compact fluorescents fall act like constant power loads. In heavily loaded motors, lower voltage can actually increase power consumption causing current to rise at a faster rate than voltage decreases. Notice that conventional voltage control methods can perform poorly in the presence of significant generation on a distribution feeder since they are based on the premise that voltage naturally decreases with increasing distance from the feeder. An equally important assumption which is violated by the RES presence is that load variations are the only cause of current changes so that historical load data combined with ambient temperature is a good system current flow predictor. As a result, the system's ability to adequately control voltage is one of the factors limiting RES penetration on a feeder. Even within the system's ability to maintain voltage, the variations in apparent load caused by the variability of wind and solar may cause more frequent cycling of electromechanical devices used for voltage control and shorten both operating life and maintenance intervals. A variety of devices can be used to control voltage and reactive power, but the bulk of voltage regulation today is provided by three basic types of devices: load tap changers, voltage regulators and capacitor banks. Advanced inverter based devices, including most distribution renewable generation and storage devices are also potentially capable of providing reactive power and thereby contribute to VAR control. Operating restrictions imposed by IEEE 1547 have restricted the use of this capability, but restrictions are expected to be relaxed in the near future as IEEE 1547 proposed changes are currently under review. In addition to the elements mentioned above, which are expected to constitute the primary infrastructure for voltage control, two other types of elements are finding applications in particular circumstances especially related to the DER impacts.

In order to operate properly, a power system requires combinations of real and reactive power, due to the existence of the phase lag between voltage and current. The currents caused by the reactive power flows are contributing to system resistive losses. Utilities are trying to reduce such losses by locally injecting reactive power into the system near the loads, reducing the line currents due to reactive power flow, method referred to as VAR control. The concept of optimizing the system losses by reducing reactive power flow, which in turn affects the system voltages, is referred to as the VVO method. The shortcomings and limitations of the conventional VVC methods, as described previously, lead the utilities to explore advanced VVC techniques suitable for smart grids. Such emerging methods are providing VVC on the secondary side of transformers, having the potential for addressing such limitations. Interconnections of small- and large-scale wind and solar energy systems and other DG units into the power distribution are keeping increasing. However, DG penetration levels are often limited by the adverse impacts on various power distribution operating parameters, such as: steady-state and transient voltage rise. Inverters used for solar PV and wind farms can be used to absorb or inject reactive power to mitigate the impacts of their real power injection on the power distribution voltages. Such active voltage regulation by the DER was not allowed by the standards in the past. The recent amendment of IEEE 1547 standard permits the voltage regulation by using DER controlled reactive power injection. The inverter-based DER can be operated in the VAR control modes. The most common reactive power control approach followed in North America for both small- and large-scale PV systems is the *fixed unity power factor*, meaning the reactive power Q is zero. In this case, the DER inverter is operated at constant unit power factor, with very limited or no reactive power injection into distribution network. However, such real power injection without any VR control can create unacceptable steady-state voltages during light load periods and undesirable voltage variations due to DER variabilities. In *fixed power factor, Q(P)* mode, the DER inverters, specifically designed, are operated at constant power factor such that the voltage rise caused by the real power injection is mitigated by reactive power absorption. The power factor value is decided during the system studies before the DER installations, allowing DER full power output, without causing voltage problems. In the variable *power factor method*, in which $Q(P, R/X)$ is controlled, DER reactive power injection depends on the active power injection and the reactance-resistance (X/R) ratio at the point of common coupling (PCC), varying according to the following relationship with zero voltage variation:

$$\Delta V \approx \frac{P_{Load} \times R + Q_{Load} \times (X_{Line} - X_{Cap})}{V^2} \approx 0 \tag{6.18}$$

The X/R ratios at PCC and the associated control settings for DER reactive power injections are determined during the system interconnection studies. The *Volt-VAR control* approach is effective for systems with high X/R ratio i.e., when the resistance at the PCC is small compared to the reactance. In such control approach, the utility specifies custom *Volt-VAR characteristic*, the inverter follows the reactive power injection as a function of its available reactive power injection capacity and the terminal voltage at a given time. Different *Volt-VAR* characteristics can be designed for different inverters, and depending upon the requirements, the utility transmits the control command to the inverters to determine which *Volt-VAR* characteristic to follow for the reactive power injection. The *Volt-Watt control* approach is most effective when the X/R ratio at PCC is small, the resistance and reactance being comparable. In this mode, *Volt-Watt* curve is defined for the groups of inverters by the utility. The inverter then reduces the DER active power output according to the terminal voltage and define Volt-Watt curve. Since this method reduces DER active power output, it should be used only when the existing voltage regulations are unable to prevent higher voltages. Over a distribution line, represented by its resistance R_{Line} and its reactance X_{Line}, a current causes a voltage drop that is greater the further away the location is from the feeding point. By connecting to the line series capacitors, a capacitive voltage drop is obtained. Since both the series capacitor and the line impedance voltage drops are proportional to the load current, the voltage control is instantaneous and self-regulating, meaning that the voltage can be kept constant even in the event of rapid load variations. The voltage drop ΔV along the line can be calculated by using Equations (6.2), (6.3) or (6.6). Usually, the active voltage drop ($R_{Line} \times P$) is larger than the reactive voltage drop ($\Delta X \times Q$) in a distribution line. In case of overcompensation ($X_{Cap}/X_{Line} > 1$), the value ($X \times Q$) is negative, so the resistive voltage drop can be compensated. The series capacitor is bypassed (set out of service) in the overcurrent case, when short-circuit current flows into the power line. Connecting series capacitors in a distribution line with inductive load (common case) reduces the transmission losses. If the load point voltage increases, the current is reduced in case of constantly transmitted power and the line losses are reduced as well. A distribution line loss is proportional to the line current squared. Adding a series capacitor in a line with inductive load, the load end voltage is increased. For constant load, the line current is reduced according to this relationship, $I = S_{Load}/V_{Load}$, where S_{Load} and V_{Load} are the load power and voltage. This affects the line losses, defined by: $P_{Losses} = R_{Line} \cdot I^2$. In the case of lateral radial feeder, each lateral can be considered as a load point on the main feeder with a lumped load equal to the sum of all its loads. The location of series capacitor is determined at each load point to keep its voltage at a desired value V_{dsr} (e.g., $V_{dsr} = 98.5\% \ V_{ref}$ where V_{ref} is the rated voltage) or not less than the allowable limit ($V_{limt} = 95\% \ V_{ref}$). The calculated voltage at a load point is considered as the sending voltage to the next section of the main feeder, each section being treated individually. For the laterals, each one is treated with the same method used for the main feeder. The calculated voltage at the compensated node of the lateral connection with the main feeder is considered as a source voltage of that lateral.

Example 6.5   A distribution system includes a radial overhead transmission line at 22 kV, 50 Hz, with a total length of 12 km. The overhead transmission line is of aluminum with cross-sectional area (CSA) 240/40 mm². Its resistance and reactance per kilometer are 0.1329 and 0.32 Ω, respectively. The feeder is supplying 12 MVA. There are three load points on the feeder line, with 1 MVA, 2.5 MVA and 1.5 MVA of power consumption at each load point, respectively. The load points are separated by 4 km lines. Assuming 0.8 power factor estimate the voltage drops of each transmission line section.

SOLUTION

From the above date, it is found that the V_{ref} is 22 kV, V_{limt} is 20.9 kV and V_{dsr} is 21.68 kV. The calculation of voltages at load points without compensation is as below. The voltage drop at first load point without compensating capacitor is then:

$$\Delta V = (12 \times 4 \times 0.1329 \times 0.8 + 12 \times 4 \times 0.32 \times 0.6) \cdot 100 / 22^2 = 7.48\%$$

The voltage at the first load point, which is also the sending voltage to the next line section, is:

$$V_1 = (1 - 0.0748) \times 22 = 20.35 kV$$

The power supplied to the second line section is 11 MVA and the voltage is 20.35 kV. Repeating the procedure, the voltage drop across the second line section is:

$$\Delta V = (11 \times 4 \times 0.1329 \times 0.8 + 11 \times 4 \times 0.32 \times 0.6) \cdot 100 / 20.35^2 = 8.01\%$$

The voltage at the second load point, which is also the sending voltage to the next line section, is:

$$V_1 = (1 - 0.0801) \times 20.35 = 18.72 kV$$

The power supplied to the second line section is 8.5 MVA and the voltage is 18.72 kV. Repeating the procedure, the voltage drop across the last line section is:

$$\Delta V = (8.5 \times 4 \times 0.1329 \times 0.8 + 8.5 \times 4 \times 0.32 \times 0.6) \cdot 100/18.72^2 = 7.32\%$$

The voltage at the second load point, which is also the sending voltage to the next line section, if any, is:

$$V_1 = (1 - 0.0732) \times 18.72 = 17.35 kV$$

Keeping in mind that if the voltage at each load point is out of the required limits (5%), violating the limits, series compensations are needed. The capacitor size is calculated by using Equations (6.10) and (6.11), replacing the line reactance X_{Line} by the net reactance $(X_{Line} - X_{Cap})$, maintaining the voltage drop at desired values. VVC solutions include the following components: capacitor banks, controllers, switches, substation transformer on load tap changers, voltage regulators, sensors and monitoring equipment for current, voltage and power flow, communication infrastructures, control decision computing equipment, applications and packages for control decisions, supporting IT infrastructures, e.g. supervisory control and data acquisition system (SCADA), distribution management system (DMS) and supporting applications, e.g. network models, load forecasting models, state estimation algorithms, applications for power flows, short circuit calculation methods, etc. There are several VVC methods, and on the choice of the VVC components, some of them may not be required and be present in all systems. Historically and in VVC conventional approaches, the voltage and VAR control devices and solutions are regulated with local information. A capacitor bank is switched ON and OFF based on time, temperature and on their correlation with daily load variations or on the voltage or current measurements at the capacitor bank location. This approach is simple to install, not requiring communication infrastructure. On a feeder with several VVC compensation devices, each device is controlled independently, without regard for the impact on the operation of other devices. Such approach is producing sensible control actions from each device perspective, but unstable and conflicting results for the entire feeder or substation.

6.4   SMART POWER DISTRIBUTION NETWORKS

Efficient, secure and reliable electricity delivery is a crucial and paramount issue for power distribution. Power systems are experiencing significant changes with increased DG, RES and ESS penetration; smart and responsive loads; introduction of electric vehicles; demand side management; new energy market paradigm; and so on. These changes are pushing a paradigm shift, one of the most important challenges in power system design, planning and operation, the management must shift from conventional planning and manual intervention to intelligent control, management and operation, the full *smartization* of MV and LV power networks. Peculiarities and criticalities of smart power distribution originate from the system complexities, e.g. network physical and cyber aspects, e.g. data elaboration, communication, supervision and control or fully integrated advanced monitoring, fostering the transition towards network automation. The design and development of such networks require distinct expertise in the power and information engineering fields. The increasing importance of power distribution reliability, resilience, demand side and costumer participation are changing the way power distribution is planned, managed and operated. For example, to achieve power distribution self-healing against outages, emerging technologies, such as remote-controlled switches (RCSs), two-way communication or smart metering, sensing and monitoring are employed. Higher automation levels are transforming power distribution into the smart distribution system (SDS). The SDS data availability and remote control capabilities provide distribution operators with an opportunity to optimize the operation and control. On the other hand, opening of the electricity market and new regulations cause unbundling of the vertically organized utilities into different companies, in a competitive market (e.g. generation, energy trade or local utilities) or in further regulated (e.g. power transmission and power distribution networks). However, the regulated power transmission and power distribution companies (RPC) are facing reduction of their revenue and significant increasing of their responsibility and penalties by regulation. In the new economic and technical environment, RPCs are forced to significantly increase efficiency and reduce operation costs, in order to be able to survive and operate positively. Smart grid solutions can provide such affects and are widely applied by the RPCs.

The DER massive penetration is leading to new approaches in distribution network (DN) design, operation and development. Until recently, the distribution network has been regarded as a passive termination of the transmission subsystems, with the objective of supplying reliably and efficiently electricity to the end-users. Accordingly, the DNs are mostly radial, with unidirectional power flows and with simple and efficient protection schemes. Greater DER and DG penetration is changing disruptively such well-consolidated environment and structure, DNs are no longer passive systems and a gradual, but ineluctable, transformation towards active networks has been foreseen and is in full deployment. The bi-directional power flows require changes in distribution control and protection strategies, enhanced distribution automation, enforcement of DN infrastructure and greater degrees of energy and information management according to the SG paradigm. Higher DG penetration introduces critical issues in the short-term DN planning, operation and management, requiring the following:

1. Novel DN structures, as well as new frameworks for energy management and electricity market;
2. Advanced, accurate and reliable monitoring systems, able to observe and estimate DN status at greater space-time resolution, in order to support DN control and management;
3. Appropriate management and control systems, able to optimize DN bi-directional power flows, energy saving and economic benefits according to the SG paradigm; and
4. Communication and data processing systems, which have to guarantee seamless interactions among all DN components and proper level of cyber-security.

Smart distribution networks (SDNs) are implementing the SG paradigm, namely intelligence assets integrating the actions of all actors connected, e.g. DER units, energy suppliers, aggregators and consumers to efficiently deliver sustainable, economical and secure electricity. To achieve these goals, a large integration of sensors and measurement devices for collecting data across the grid is needed. In addition, SDN exploits information and communication technologies to improve the reliability, security and efficiency of the power systems. By using such measurement and communication systems, SDN is able to integrate DERs at any network point, including prosumers. Thus, SDN structure creates the opportunity for consumer to become a power supplier and take part in energy trading mechanism. This is the key for consumer to be converted into an active energy citizen. Increasing grid resilience and faster service restauration are also smart grid critical challenges. Extreme weather events and natural hazards may further challenge the reliable operation of distribution systems. To minimize the impact of these events on reliable power supply, utilities are making an effort toward implementation of the smart power distribution systems. Traditionally, limited information is acquired along distribution feeders with few deployed sensors and limited communication. The lack of remote monitoring and control capability limits distribution operator ability to monitor system operations and take fast control actions to the extreme events. It takes often hours to determine the fault locations through field crews and calls from affected customers. The power distribution observability and controllability are enhanced by adopting emerging intelligent devices and SG applications. With the ongoing smart grid development, smart meters, sensing and remote control switches are extensively deployed into power distribution. Such SG technologies enhance the power distribution system ability to withstand extreme low probability, high impact events and restore power supply, faster and efficiently to interrupted customers (loads) after any major outage.

In a power distribution system, a fault and the activation of protective devices can lead to a power outage. Detection of a fault and identifying the activated protective devices to isolate the fault are key functions of an Outage Management System (OMS). Once the faulted line section and activated protective devices are identified, the fault clearing is set. Fault isolation and service restoration can be performed with remote or manual switches. An advanced OMS using outage reports from smart meters helps distribution operators to determine the faulted line section promptly, facilitating the service restoration. However, power distribution systems have few monitoring and data acquisition points beyond the substation. When a power outage occurs, distribution operators rely on calls from interrupted customers to determine the outage area. Once a customer outage report is received, the OMS matches the phone number of the call to the customer location, identifying the distribution transformer and protective devices on the feeder. When sufficient trouble calls are collected, the OMS predicts the actuated protective devices and faulted line section(s). However, this approach may take hours to collect enough data, prolonging the identification of outage scenarios, leading to longer durations for affected customers. Automatic meter reading (AMR) provides another way to access customer and fault information. When an outage occurs, AMR meters are polled by the distribution operators, and AMR responses determine if the customers are experiencing a power outage. Data from AMR meters, trouble calls and SCADA systems are put together for accurate outage management. In smart power distribution, monitoring, control, sensing and data acquisition of the electricity network extend further down to the distribution pole-top transformer and even to individual customers through the substation communications network, by means of separate feeder communication networks or tied into the advanced metering infrastructure (AMI). A more granular field data helps increase the operational efficiency and provide data for other smart grid applications. Higher speed and increased bandwidth communications for data acquisition and control will be needed. Power losses are reported as the difference of the energy measured on network entry points and energy measured on consumption points (customers), in a specific time period. Notice also that technical power losses are related to network energization and power flow, while non-technical losses are related to the measurements and energy theft. Power losses are direct commercial losses, since they have to be compensated by additional energy purchases or generation, significantly contributing to the environmental pollution.

6.4.1 DISTRIBUTED GENERATION IMPACTS ON VOLTAGE CONTROL AND PROFILES

Power distribution systems are planned and designed considering transmission line parameters, loads and assigned energy supply capabilities. When DG units are introduced into the power distribution without any consideration and preventive measures, they might have significant impacts on the power flow and voltage levels and profiles. Distributed generation affects the distribution network operation in several ways. Power flows and fault currents are altered and problems related to voltage quality, protection and increasing fault levels can occur. The distribution networks are expected to fulfil certain power quality requirements, set by standards and codes to avoid harmful effects on the network components or customer equipment. Deviations from designated tolerances may result in customer equipment malfunction or breakage of network components or customer devices. The DG effects on the network reliability and stability need to be analyzed in DG interconnection studies. Distribution voltage quality consists of features, such as frequency, voltage magnitude and variations, rapid voltage changes, voltage dips, interruptions, voltage unbalance and harmonics. Distributed generation affects many of the voltage quality characteristics: DG alters the network voltage levels, can induce rapid voltage changes and dips, can increase the harmonic distortion and the unbalance of the network voltage, as well as the power distribution fault level, having effect on the voltage quality. Depending on the DG unit location and size, the effect on the load profile can be both positive and negative. High DG, RES and EV penetration in power distribution networks can lead to several issues that are needed to be mitigated by the smart power distribution. Voltage rise, a problem well known by the distribution system operators (DSOs), consists of the voltage increases when DG unit generation is large, particularly at the far end of the LV feeders, which can limit the DG installations (e.g. the size of PV plants or wind farms). Higher R/X ratios in the LV systems make the voltage magnitude more sensitive to the active power injection than to the reactive power injection. This is opposite to the HV networks for which the R/X ratio is low and the grid operator uses reactive power injection to regulate the network voltage. As shown in the diagram of Figure 6.5, DG generation unit, for example, a PV power plant is injecting the apparent power to the grid (power system) through a transmission line, represented, as series impedance, set by the electric distance between the PV plant and the main grid. In actual networks, this impedance is certainly not constant, depending on the power flow level in the transmission line, network configuration, ambient weather conditions, etc. Notice that such analysis is an illustration of the voltage characteristics in various networks, being applicable to all DG types. Assuming that the receiving end voltage V_T is at the

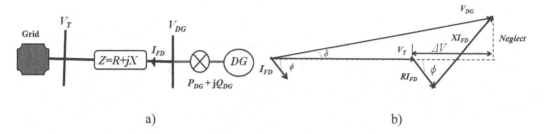

FIGURE 6.5 (a) Simplified line section with a DG unit model and (b) its phasor diagram.

standard condition state (the voltage angle is near zero), the voltage drop across the series impedance (see Figure 6.5) is given by:

$$\Delta V = \frac{PR+QX}{V_T} + j\frac{PX-QX}{V_T} \qquad (6.19)$$

In networks with low R/X ratios, such as the HV circuits, the voltage drop from the Equation
(6.18) can be expressed by neglecting the resistance effect, by the following relationship:

$$\Delta V \approx \frac{PR}{V_T} + j\frac{PX}{V_T} \qquad (6.20)$$

Here, the magnitude of the voltage drop can be approximated by ignoring the imaginary component in the above relationship. In the high R/X ratio networks, the effect of resistance is significant
and cannot be neglected. Instead, the magnitude of the voltage drops in the LV system may be
approximated by:

$$|\Delta V| = \frac{PR+QX}{V_T} \qquad (6.21)$$

**Example 6.6   A power distribution section, equipped with a PV plant, has a resistance
of 0.63 Ω and an inductive reactance of 0.12 Ω. If the line terminal voltage is 4.2 kV
estimate and the distributed generation unit is injecting 200 kW active power and 45
kVAR reactive power, estimate the voltage rise.**

SOLUTION

The R/X ration of this line is 5.25, so we can use Equation (6.26) to estimate the voltage rise for this
power distribution network section in the presence of DG unit.

$$|\Delta V| = \frac{200 \times 0.63 + 45 \times 0.12}{4.2} = 31.3\,V$$

Notice the voltage rise is about 0.7% well into the standard prescribed range.

Usually, power distribution networks have a unidirectional power flow from the distribution substation to the customers. This model leads to a descending voltage profile which may only suffer
from an undervoltage near the load center, and such issue is usually addressed by using OLTC and

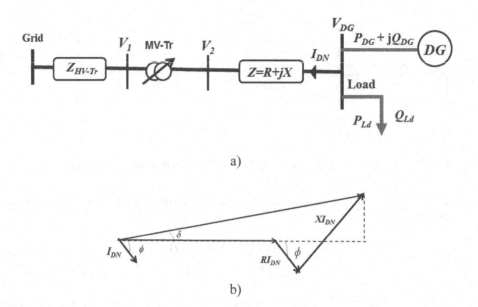

FIGURE 6.6 A distribution network section with a DG unit: (a) single-line diagram and (b) phasor diagram.

capacitor banks. On the other hand, the DG integration into power distribution networks is making the power flow bi-directional, thus overvoltage problems may also occur. To understand and assess the impact of DGs on the power distribution system voltage, the simplified distribution network model of Figure 6.6a is employed, where a DG unit is connected at a load bus. Here, R and X are the feeder resistance and reactance, respectively, I_{FD} is the feeder current, V_1 and V_2 are the primary and secondary voltage of the distribution transformer, respectively and V_{DG} is the DG unit voltage. The phasor diagram of the simplified distribution network (see Figure 6.6b), in which δ is the power angle and ϕ is the phase shift between V_{DG} and I_{FD}. Using this phasor diagram, the relationship between, V_{DG} (DG voltage) and V_2, assuming the small power angle (a reasonable assumption in many cases) can be expressed as:

$$V_{DG} \simeq V_2 + I_{DN}(R \cdot \cos\phi + jX \cdot \sin\phi) \tag{6.22}$$

Therefore, the voltage rise caused by the DG, ΔV_{DG}, is given by the difference between V_{DG} and V_2:

$$\Delta V_{DG} = V_{DG} - V_2 \simeq I_{DN}(R \cdot \cos\phi + jX \cdot \sin\phi) = \frac{(P_{DG} - P_{Ld})R + (Q_{DG} - Q_{Ld})X}{V_{DG}} \tag{6.23}$$

Here, P_{DG} and Q_{DG} are the DG unit output active and reactive powers, respectively, while P_{Ld} and Q_{Ld} are the load active and reactive powers, respectively. It can be seen from Equation (6.22) that the highest overvoltage happens when the DG unit generates its maximum power during a light load condition. This problem is mainly associated with the excessive reverse power flow caused by the DG unit presence. Due to the intermittency of RES generation and charging of the PEVs, the conventional control schemes for OLTC and DGs are failing to provide proper voltage regulation. This shortcoming can be compensated using communication-assisted voltage regulation schemes. In the literature, the communication-assisted schemes are divided into two approaches: distributed and centralized control schemes. Both methods involve investment in communication links and remote terminal units. The distributed (intelligent) approach is an expert-based control or

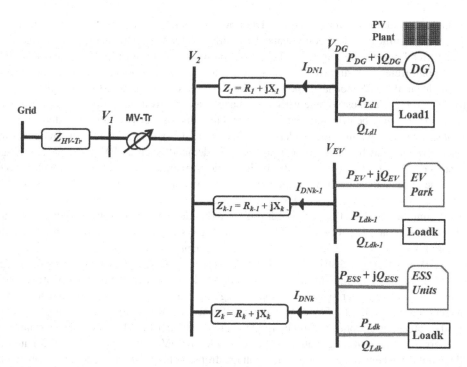

FIGURE 6.7 Simplified distribution network section with an EV park, a PV plant (DG) and EES unit.

model-free approach, which coordinates a variety of voltage control devices with the goal of providing effective and non-optimal voltage regulation with fewer communication requirements. On the other hand, the centralized approach relies on a central point that monitors the system status and optimizes the operation of voltage control equipment. Typically, a centralized control optimization problem is solved to dispatch the reactive power of different voltage control equipment based on: (i) load forecasting and (ii) generation monitoring. Several solutions and control methods have been proposed to provide optimal reactive power dispatch for DG units. Figure 6.7 represents a simplified multi-feeder power distribution network connected to a substation through an OLTC. The network has a photovoltaic PV-based DG, an EV parking lot and an energy storage unit, which are connected at different feeder terminals. Following previous derivation, the per-unit voltage deviation for DG unit, EV and ESS buses can be approximated by the following relationship:

$$\Delta V_{DG} \approx (P_{DG} - P_{Ld1})R_1 + (Q_{DG} - Q_{Ld1})X_1 \tag{6.24a}$$

$$\Delta V_{EV} \approx -(P_{EV} - P_{Ldk-1})R_{k-1} - (Q_{EV} - Q_{Ldk-1})X_{k-1} \tag{6.24b}$$

And

$$\Delta V_{ESS} \approx \pm(P_{EV} - P_{Ldk-1})R_{k-1} \pm (Q_{EV} - Q_{Ldk-1})X_{k-1} \tag{6.24c}$$

Here, P_{DG}, P_{EV}, P_{ESS}, P_{Ld1}, P_{Ldk-1} and P_{Ldk} are the DG, EV, EES and load active powers, respectively, and Q_{DG}, Q_{EV}, Q_{ESS}, Q_{Ld1}, Q_{Ldk-1} and Q_{Ldk} are the DG, PEV and load reactive powers, respectively. Equations (6.24a, b and c) are showing that two worst-case scenarios may occur: (i) the overvoltage occurs when the DG generates its maximum power during light loads and (ii) the undervoltage occurs during a peak load demand and low DG power output. The integration of DGs changes the voltage profile significantly and complicates the voltage regulation. This is due to two reasons: (i) the voltage trend not descending from the substation to the feeder terminal, thereby invalidating the target point (reference) and (ii) the voltage estimation, based on local measurements becomes inaccurate due to the RES stochastic power natures, and the EV charging. Moreover, the EV charging stochastic nature makes voltage estimates inaccurate, aggravating the undervoltage problem. Therefore, OLTCs may be affected by the wear and tear due to excessive and repeated operations. This problem worsens when some feeders are affected by overvoltage due to high DG penetration and others are affected by undervoltage during high demand, such as EV and ESS charging. In such instances, the OLTC has two contradicting actions: increase transformer secondary voltage mitigates the undervoltage problem at the expense of the system overvoltage and vice versa. Since the power profiles of parking lots and PV-based DGs naturally coincide, there is a good chance that the system simultaneously suffers from overvoltage and undervoltage. A problem solution is to use a centralized-based controller for the OLTC, exploiting the system maximum and minimum voltages.

Substation voltage is usually constant, and any increment of active power supplied from the DG unit (e.g. a PV plant) to the distribution substation increases the sending end voltage. By applying negative reactive power increment, this voltage magnitude is reduced. To regulate voltage, the PV plant can reduce active power injection or apply a negative reactive power injection (a reactive power sink) as required by the IEEE 1547-2018 standard. If reactive power cannot sufficiently regulate the voltage, the active power control could be implemented with controllable loads such as smart loads, EVs or energy storage units, e.g. batteries. For example, the EV charging points, coordinated with small DG units, can control and mitigate the voltage drops. Voltage drop is a classical problem dealt with by DSO operators. Moreover, the extended use of the EVs is adding more challenges, particularly overloads and excessive voltage drops at peak loads. Such problems are aggravated if there are no incentives or information for EV owners to schedule charging for optimized energy uses. Such process requires controlling charging rates and time of connections with a fair distribution among EV owners. Another issue is the impact from the reversed power flow due to the DG, energy storage and RES units. Conventional distribution systems are designed to operate for unidirectional power flow. Transmission line voltage regulators (LVRs) usually correct the load-side voltage drops. Voltage is sensed on the LVR load-side, taps are adjusted to correct the load-side voltage. When a DER unit is back-feeding a voltage regulator, it changes the taps to correct the voltage. In addition to the feeding grid, DER units can also control the local voltage by adjusting current. If LVR operates in constant voltage mode, it decreases the load-side voltage if DER increases it. However, DER connection point voltage may still be too high if the number of taps is not adequate, or if the LVR set-point is determined assuming unidirectional power flow and there is large distance between LVR and DER unit. If the LVR operates in line compensation mode, DER can cause significant problems since it increases the voltage at its secondary-side under reverse power flow, worsening the situation if there is a voltage rise caused by the DER unit. Even if the LVRs are bi-directional, they can only control the local voltage. When there is a voltage rise due to reversed power flow, only the DER and VR devices installed near the DER unit provide full voltage corrections. Large voltage fluctuations due to solar radiation and wind variations affecting the DER power outputs lead to poor power quality and voltages out the prescribed ranges across the entire feeder. Further, in the presence of smart inverters, ensuring system stability, the Volt-VAR curves have higher dead-bands (3%) and larger slope (3%). These factors reduce the upper and lower voltage margins in reference to the standard limits for performing CVR and VVC functions. Larger presence of DER units and EVs, and the overall push to a smart grid have complicated voltage control, leading to additional objectives for a Volt-VAR optimized control (VVOC) system that contribute to the overall smart grid performance and reliability. Advanced VVOC systems have a broader set of objectives including: maintain acceptable voltage, improve efficiency and minimize consumption, enable renewable energy penetration, coordinate all network devices, self-monitoring capabilities, allow operator override, support self-healing and feeder reconfiguration and allow selectable objectives.

Voltage control in the active and smart power distribution is classified into four types: local, decentralized, distributed and centralized control. Proper selection and trade-off of different control schemes is one major challenge for voltage control approach. In local voltage control, DER control decisions are made based on local voltage and current measurements at the point-of-interconnection. Voltage control is provided through reactive power control of DG units, DG generation active power curtailment and smart charging of EVs, which can reduce the impact of renewable energy generation and minimize the needs of auxiliary VR equipment. Although local controls do not require communication, its inability to align with the utility control strategy is a main drawback. For a simple two-bus system, the reactive power required to maintain constant voltage for an increase in active power P is approximated as $Q = \Delta P \cdot R/X$, where R and X are the branch resistance and reactance. The problem is that in MV and LV lines the R/X ratio tends to be high, resulting in higher reactive power requirements, leading to inverter high power rating requirements and can also increase circuit losses. Active power curtailment (APC) is another viable method to prevent overvoltage, in which the output power of inverters, typically operating at maximum power point, is curtailed. Adaptive techniques such as online adjustment of APC droop values have been proposed. Other methods incorporate ESS units and use a mixture of reactive and active power control strategies. Decentralized voltage control aims to enhance the local control using low-form communication systems, including coordination among various system components in an automated manner without system operator regulation (although some strategies may optionally communicate with system operators) to optimize local network operation. A global distress signal is used under overvoltage conditions, in which the controller is implemented as a finite-state machine, prioritizing reactive power support over active power curtailment. Some decentralized robust energy management methods for distribution networks with DG units are based on the alternating direction method of multiplier algorithms. The EV agents communicate messages to substation agents. The EVs are charged at minimum cost, without affecting the network, keeping customer welfare at maximum at all times. Compared with centralized schemes, decentralized control schemes provide flexible, efficient and robust regulation for smart distribution. Distributed voltage control is a control scheme without a central controller, and implemented with node-local computations using only local measurements augmented with limited information from the neighboring nodes through communication systems. Centralized voltage control, also known as active network management, utilizes sophisticated communication networks to regulate voltage. State estimation is used to estimate voltage profile, based on which DG and other components are dispatched. Through a coordinated control of on-load tap changer (OLTC), DG and voltage regulators, a centralized control approach allows optimized operation of the entire region of the grid under the system operator.

6.4.2 SMART SUBSTATIONS

Electrical substations are the focal point of power systems, from electricity generation, transmission, subtransmission and power distribution subsystems. In a substation the voltages are transformed from high to low levels or reverse and distributed to the feeders and eventually to end-users (loads), by using power transformers, monitoring, metering, protection and control equipment. Substations are critical power system units, being strategically important for power system design, management and operation. Substations are providing various access points to the power system parameters, enabling monitoring, sensing, control and secure operation of the geographically widespread power system. They are equipped with a broad range of instrumentations, sensors, equipment, communication and data acquisition to facilitate various operation functions. In any electric grid the energy flows through several substations between generating stations and consumers, with several voltage changes and corrections. There are different substation types, transmission, power distribution, collector and switching substations. Substation's major functions include: voltage transformations, connection point for transmission and distribution power lines, switchyard for electrical transmission or power distribution configurations, a monitoring point for control center, protection of power lines, equipment, devices and apparatus and communication with other substations and regional control centers. Substations and feeders are essential real-time data sources for utility's efficient and safe operation. Real-time (operational) data are the instantaneous values of power system status data and information, e.g. voltage (V), current (A), active power (MW), reactive power (MVAR), circuit breaker status and switch positions. These are time critical data, used to protect, monitor and control the power system field equipment and devices. There is also a wealth of operational (non-real-time) data available from the field devices. Non-operational data consist of files and waveforms such as event summaries, oscillographic or sequential event records and reports, in addition to the SCADA-like points (e.g. status and analog points) that have a logical state or a numerical value. Non-operational data are not needed by the SCADA dispatchers for monitoring and control, but such data can help to make power system operation and management more efficient and reliable.

SCADA refers usually to a combination of systems that collects data from various sensors at a plant, facility or in other remote locations and then sends these data to a central computer system, which then manages and controls the data and remotely controls devices in the field. Power industry has a specific set of SCADA requirements. The primary purpose of an electric utility SCADA system is to acquire real-time data from the field devices located at the power plants, transmission and distribution substations, distribution feeders, etc., providing the control of the field equipment, presenting the information and data to the operating personnel. Real-time monitoring and control of substations and feeders is typically in the time range of 1–5 s. SCADA systems are globally accepted as a means of real-time monitoring and control of electric power systems, particularly in generation and transmission. RTUs (remote terminal units) are collecting the analog and status telemetry data from field devices, and communicate control commands to the field devices. Installed at centralized locations, e.g. utility control center, SCADA systems are the frontend data acquisition equipment, operator of graphical user interface (GUI), engineering applications that act on the data, software and other components. New trends in SCADA include providing increased situational awareness through improved GUIs and presentation of data and information, intelligent alarm processing, utilization web-based clients, improved integration with other engineering and business systems and enhanced security features. Usually, system control and data acquisition equipment has at least one master station, several RTUs and communication subsystems. The utility master station is usually located at an energy control center, while RTUs are installed at the power plants, transmission and distribution substations, feeder equipment, etc. Smart grid substations moved beyond basic protection and automation schemes to bring complexity around distributed functional and communication architectures, local analytics, automation, *smartness* and massive data management. There are significant intelligence migration from the centralized functions and decisions at the energy and distribution management system level to substations and feeders, enhancing system responsiveness. To realize the smart grids, the legacy of substations is retrofitted by substation automation system packages, allowing the robust monitoring, sensing, control and communication tasks and functions.

The electrical substation is of paramount and critical importance to the electrical generation, transmission and distribution system. According to the literature, there are four main types of electric substations:

1. *Switchyard substation* at a generating station connects the electric generators to the utility grid, providing offsite power to the power plant. Generator switchyards are large installations, usually subject to significant planning, finance and construction efforts different from those of routine substations.
2. *Customer substation* functions as the main electric power supply for one particular customer. Its technical and business characteristics are set by the customer's requirements rather the utility needs.
3. *Power system substation* involves the transfer of bulk power across the electric network. Some of system substations are providing only switching facilities, no power transformers, whereas others are performing voltage conversion as well. Such substations typically serve

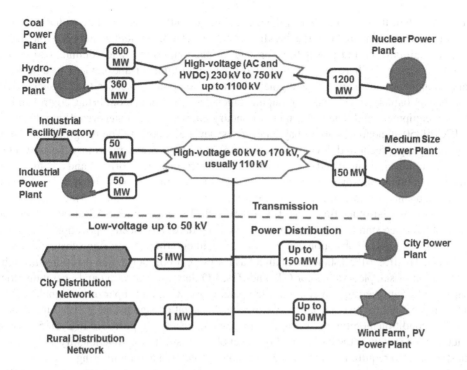

FIGURE 6.8 Substations used in electric grids.

as the end points for transmission lines originating from generator switchyards, providing the electricity for circuits, feeding the transformer stations. They are integral component to the long-term power system reliability and integrity, enabling large electric energy amounts to be transferred from the power plants to the load centers. System substations are strategic facilities and usually very expensive to construct, operate and maintain.

4. *Distribution substations* are the most common power system facilities, providing electricity to the power distribution circuits that directly supply most customers, typically located close to the load centers, the neighborhoods that they supply, being the stations most likely to be encountered by the customers.

A pictorial diagram showing how the electrical substations are used within an electric grid is given in Figure 6.8. The substation's roles clearly indicate that they are critical electric infrastructure components, especially for substations in the transmission sections of the grids, interconnecting several power system components. Smart substation are using the advanced, reliable, integrated and environmentally friendly intelligent equipment and devices to automatically achieve information collection, sensing, measurement, monitoring, control, protection, metering, sensing, monitoring and other basic functions, and support a series of advanced functions such as sequence control, intelligent alarm and analysis, real-time fault information comprehensive analysis, intelligent operation order system, source-end maintenance, device status visualization, substation area control and so on. Compared with conventional substations, smart substations achieve the intelligent management of substation equipment and devices by advanced data analysis processing methods based on advanced sensor and communication technologies. Substation automation systems (SAS) are designed to increase the efficiency of the control, management and communication schemes installed in substations. A SAS is designed to cast a local computerized network at a substation to enhance the response time for any unpredictable event as fast as possible. Substation automation (SA) consists of several functions and subfunctions, appropriately interfaced with equipment and devices. SA system uses any number of devices integrated into a functional array for the purpose of monitoring, controlling and configuring the substation. The SA system components include voltage or potential, current and power transformers, protection, control and communication equipment and devices, etc. There are basically two types of equipment and devices in a substation: primary equipment and secondary equipment. Primary equipment includes transformers, switchgears, capacitor banks, etc. Secondary equipment includes protection, monitoring, control and communication equipment and deices. Further, secondary equipment are categorized into three levels in the IEC-61850 standards: station, bay and process level equipment. The integration of primary and secondary equipment and devices is realized through intelligently reforming the conventional primary substation equipment and devices with smart instrument transformers and condition-based monitoring technologies. Key equipment and devices such as transformers and switches with intelligence components are to achieve the integration of condition-based monitoring, measurement, protection, control and other functions. IEC 61850 (DL/T860) provides SAS unified information models and interface standard. Based on the unified standard, advanced applications for management and control units of substation can be realized by integrating and reusing substation information and data. The IEC-61850 standard main focus is to support the substation functions through the communication of sampled values for CTs and PTs, I/O data for protection and operation, control and trip signals, engineering and configuration data, monitoring and supervision signals, data and information provided to the control center, time-synchronization signals, etc. Other functions such as metering, condition monitoring and asset management are also supported in IEC-61850. An optimum automation system is the basis for a high level of smart substation functionality and flexibility. All substations also require proper physical and cyber protection and monitoring.

The SA components and elements present in smart substations include: microprocessor-based IEDs, specific function and control dedicated devices and equipment, substation display, communication connections and devices and cybersecurity protection. Microprocessor-based IEDs provide inputs and outputs to the system while performing some primary control or processing service. Common IEDs are protective relays, load survey and/or operator indicating meters, revenue smart meters, programmable logic controllers and power equipment controllers of various descriptions. Devices dedicated to specific functions for the SA system like transducers, position sensors and clusters of interposing relays are also used. Dedicated devices are often using a SA controller or interface equipment, such as conventional remote terminal unit (RTU) as a connection into the SA system. A substation display or user station, connected to a substation host computer, the local server is also included in the smart station equipment. Communication connections and equipment to the outer utility operations centers, maintenance offices and/or engineering centers. Most SA systems connect to a traditional SCADA system master station, serving the real-time needs for operating the utility network from one or more operations centers. SA systems may also incorporate a variation of SCADA remote terminal units for this purpose or RTU functions may appear in a SA controller or substation host computer. Other utility services are usually connected to the system through a firewall, also connected to the SCADA system. In electric substation automation, the operation center (master control center or SCADA master station) receives and processes data from several other substations, taking appropriate remote control actions and commands. The master station system may sometime use open and distributed architecture, but multiple master stations are used accordingly with different topologies to interconnect them for synchronizing grid operational data. Each master station is supported with a backup/emergency master station (automatic) and is synchronized with a primary master station database. Main SCADA master station elements are Human-Machine Interface (HMI) structures, application servers, firewall, communication frontend with RTU or data concentrators and external communication server gateway to communicate with other control centers. These elements are networked within the SCADA master via dedicated local area network (LAN). The application servers support all energy management system (EMS) or distribution management system (DMS) applications. Redundancy is provided for SCADA master hardware and software elements, e.g. redundant LAN and substations, e.g. redundant critical computer, as well as for master-to-master communication network.

Real-time data, the operational data, are instantaneous values of power system analog and status points such as voltages, currents, active power in MW, reactive power in MVAR, circuit breaker status and switch positions. Such information and data are critical in the power system field equipment protection, monitoring, operation, and control. Intelligent electronic devices (IEDs) are microprocessor-based devices, capable to exchange data and control signals with another device (IEDs, meters, controllers, SCADA, etc.) over a communication link. IEDs perform protection, monitoring, control and data acquisition functions in substations and along feeders, being critical to the operations of the grid. IEDs are widely used in substations for different purposes. In some cases, IEDs are separately used to achieve individual functions, e.g. differential or distance and overcurrent protection, metering, sensing and monitoring. There are also multifunctional IEDs that can perform several protection, monitoring, control and user interfacing functions on a hardware platform. IEDs are key components of substation integration and automation technology. Substation integration involves integrating protection, control and data acquisition functions into a minimal number of hardware platforms to reduce investment and operating costs, panel and control room spaces, eliminating redundant equipment and databases. Automation involves the substation deployment and feeder operating functions and applications ranging from SCADA and alarm processing to integrated the Volt-VAR control (IVVC), optimizing the management of assets and enhancing the operation and maintenance efficiencies with minimal human intervention. The main multifunctional IEDs advantages are fully compatibility with IEC 61850 standard, compact size and various functions are combined in a single design, allowing size reduction of overall systems, increasing efficiency, improvements in the robustness, providing extensible and flexible solutions based on mainstream communications technology. IED technology helps utilities improve reliability, operational efficiencies, enabling asset management programs, such as predictive maintenance, life extensions and improved planning. The sensor main functionality is to collect data from substation equipment, such as transformers, circuit breakers and power lines. With the advanced communication, new smart sensors are available to acquire several asset-related information and data. Important advantages of the smart sensors are higher accuracy, no saturation, reduced size and weight, safer, low-power requirements and environmentally friendly, avoiding oil or SF6, higher performances, wide dynamic ranges, lower maintenance requirements and costs. SG technologies generate tremendous information, real-time and operational data with the extended uses of sensors and the needs for more information on system operation. However, smart sensing, monitoring and communication are the topics of later book chapters.

6.4.3 Integrated and Intelligent Volt-VAR Control

The power distribution system is a core part of the electric grid, linking generation through transmission and distribution systems to the consumers. The current trends in meeting increased energy demands by adding DG units in the power distribution systems require new control technologies, methods, equipment and devices. The high DG penetration is one of the most attractive smart grid features aside from being automatic, modern, reliable, robust and efficient. However, higher DG penetration comes with challenges to deal with such as the voltage and reactive power control and increased power losses. The IEEE Standard 1547 specifies how much voltage violation is allowed with or without the DG presence. Voltage variations must lie within the permissible limits at the point of the distribution network where DG units are connected. The DG integration affects the power system voltage profile, system operation and stability. Modifying conventional Volt-VAR control methods, moving to smart techniques is a core requirement to mitigate such issues. Voltage control is difficult if it is only handled by the OLTC transformers and switched shunt capacitors (SC) due to the DG and RES unpredictable behaviors. The DG ability to inject the reactive power with the proper coordination of SC and OLTC can contribute to controlling the voltage and reactive power besides minimizing the system and load power losses. Performing the Volt-VAR control in an integrated and intelligent manner can provide a flat voltage profile over the feeder while minimizing the system power loss. In addition, a coordinated operation of VRs and CAP banks permits avoiding of an excessive and unnecessary tripping. A common Volt-VAR control objective is minimization of power losses or power demand (power loss plus power consumption) while satisfying the voltage and loading constraints, distribution substation power factor or reactive power limits.

Expanded objectives for smart grid Volt-VAR control, beside the basic requirement of maintaining acceptable voltage, are to support major SG objectives, i.e. improve the system efficiency (reduce system losses) through voltage optimization, reduce electrical demands, accomplishing the energy conservation through voltage reduction, promote a "self- healing" grid (VVC plays a significant role in maintaining voltage after "self-healing" action has occurred) and enable deployment of DG, renewable energy, energy storage and other distributed energy resources, through dynamic Volt-VAR control. As one of the main goals of the integrated Volt-VAR control (IVVC) is the demand reduction of demand and minimizing the energy losses, and the voltage profile improvements, the problem is formulated as an optimization problem, with the objective of maximal loss reduction, under a set of constraints, such as: $V_{min} \leq V(t) \leq V_{max}$, and $PF_{min} \leq PF_{Station} \leq PF_{max}$, while limiting capacitor operations during a day and that power factor at the substation to be higher or equal to 0.95. New VVC methods are using AI techniques, such as machine learning, fuzzy logic control, etc. VVC relates to switching of distribution substation and feeder voltage regulation equipment and capacitor banks with two main objectives: reducing VAR flow on the power distribution system and adjusting voltage at the customer delivery point within required limits. An effective VVC approach combines, coordinates and optimizes the control of both VAR flow and customer voltage level. Components of VVC are: VAR control, compensation and power factor corrections. Substation and distribution feeder capacitor banks are used to minimize VAR flow (improve power factor) on the distribution feeder during all load levels (peak or base). Reductions of the VAR flow are lowering the distribution system losses, which reduces the load on the substation and distribution feeders. Conservation voltage reduction represents the control of substation transformer LTCs (load tap changers) and distribution feeder voltage regulators to reduce customer delivery voltage within specified and safe margins at the customer service point during load peak periods, resulting in a customer load reduction, which in turn results in the overall load reduction at substation and distribution feeders. Voltage control is not only exercised for CVR purpose but also to comply with normal operation and regulatory compliance. In summary, IVVC has three basic objectives: reducing feeder losses by energizing or de-energizing the feeder capacitor banks, ensuring that an optimum voltage profile is maintained along the feeder during normal operating conditions and reducing peak load through feeder voltage reduction by controlling transformer tap positions in substations and voltage regulators on feeder. Advanced algorithms are employed to optimally coordinate the control of capacitor banks, voltage regulators and transformer tap positions. Volt-VAR optimization (VVO) techniques offer the capability to optimize the objectives of VAR flow minimization and losses, as well as the load reduction, within the voltage constraints, by using optimization algorithms and well-defined control objectives subject to various system constraints through centralized or decentralized decision makings and controls.

VVC and VVO techniques are dealing with the optimal scheduling of voltage and VAR compensation devices to be dispatched in power distribution systems subject to operating and environmental constraints. Performing the Volt-VAR control in integrated approaches provide flat voltage profiles (very beneficial for many load operations) over the entire feeder, while minimizing the system power losses. In addition, a coordinated operation of VRs and CAP banks permits avoiding excessive and unnecessary tripping. The centralized voltage and reactive power control is considered the most cost-effective function of real-time distribution automation method. Rule-based centralized capacitor control with an objective of a unity power factor has a relatively long history of implementation. With development of a more reliable real-time power flow, the power-flow-based optimal Volt-VAR control is attracting more and more attention. Optimal VVC allows a wider

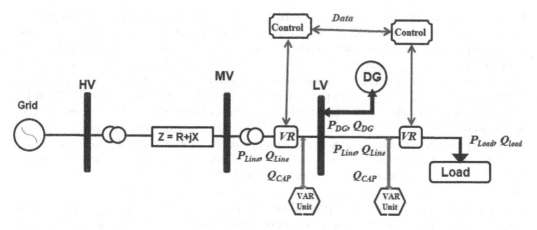

FIGURE 6.9 Radian power network with DG unit and distributed control.

choice of objectives which can be achieved with higher mathematical accuracy. Operation of the power distribution networks within voltage ad loading constraints serves as the primary objective, where other objectives are reduction of the power losses, demand, etc. In addition, distribution utilities are significantly investing in remotely controlled capacitors and step voltage regulators as part of their power distribution automation strategy. This offers the opportunity for periodic closed loop Volt-VAR control, which determines the optimal set of control actions and executes them immediately. Conventional Volt-VAR control methods have several downsides, such as low robustness to erroneous inputs, computational complexity, erroneous identification of states, etc. In addition, the classical optimization techniques, such as linear programming or decoupled Newton-based optimal power flow and mixed integer programming, usually provides limited results. Relatively new approaches to design are the applications of AI techniques, such as machine learning, multi-agent systems, fuzzy logic control, etc. These methods are classified as: analytical and numerical methods, heuristic approaches and AI methods and techniques. Automatic voltage control is commonly used in conventional power distribution feeders to maintain voltage level. For example, the secondary side voltage of the OLTC is monitored and compared with a reference voltage and action is taken based on the mismatch found. The control schemes used in smart power distribution are classified into two main categories: (i) autonomous (or local) and (ii) communication-based. Additional classification of the first category based on the way of exchanging the information between the participating entities is: (i) centralized, (ii) decentralized and (iii) distributed. Hybrid combinations among these schemes are also possible and used. However, the most used two types of control are the centralized control and decentralized control approaches. Decentralized control, as illustrated in Figure 6.9, is the one that facilitate the DG units to be employed into the voltage and VAR control, remembering that most of the renewable energy resources like wind energy, solar-thermal systems and PV plants are fluctuating energy sources and hard to control. Notice that most of research is focusing on decentralized control rather than the centralized control approaches.

The main VVO concept is to utilize various VVC and supporting devices in an optimal way for the purpose of minimal losses and maximum energy savings. From a mathematical point of view, the VVC optimization problem is a minimization problem with inequality constraints. By solving power flow, we can calculate the objective function with given settings of the control variables, satisfying the constraints and limits. There are three main types of objective function that may be formulated as follows: (i) distribution subsystem active power losses, the sum of the losses in lines, transformers and capacitors; (ii) distribution subsystem power demand, the sum of the power losses and customer demand; and (iii) the minimum number of control actions. The main constraints include voltage at each distribution transformer primary bus; voltage at each distribution transformer secondary bus; current flow in each line or transformer; power factor at the distribution substation bus and feeder; and reactive power at the distribution substation bus. In many cases, it is possible that all available control resources are not sufficient to satisfy all the above-mentioned constraints, so a compromise (decision) solution must be made. Since not all the constraints are necessarily rigid (hard) at all times, it is practical to relax some of the constraints in order to satisfy the more critical other constraints, leading to a *suboptimal* solution for objective function. A typical VVO problem is a mixed integer non-linear type optimization problem and may have a single or multiple objectives. Recently, new devices such as distributed generation, plug-in-hybrid electric vehicles, energy storage devices, demand response, etc. are introduced, affecting the performance of traditional devices (on-load tap changers, voltage regulators and capacitor banks) as well as could complicate optimization processes. From a mathematical standpoint, the Volt-VAR control optimization problem is a typical minimization problem with inequality constraints. The objective function is the sum of the all losses in a power distribution system for load variation including transformer and line losses. The value of the objective function is determined from power flow analysis. The objective function is given by the equation below, subject to a set of constrains. Objective of the optimization is to achieve maximal loss reduction, and mathematical process is expressed as:

$$F(x) = Min(Power Losses) \tag{6.25}$$

The main constraints refer to the voltage and power factor. The first function to be minimized is the total power drawn by the distribution system form the substation. It consists of the load power and the line losses which are represented as:

$$\Im(P) = P_{Line} + P_{Load} \tag{6.26}$$

$$V_{LB} \leq V_{Nominal}(t) \leq V_{UB} \tag{6.27a}$$

And

$$PF_{Min} \leq PF \leq PF_{Max} \tag{6.27b}$$

Here, V_{LB} and V_{UB} are the lower and upper bounds of the voltage range, PF_{min} and PF_{max} are the power factor lower and upper limits, as prescribed by the standards or specific requirements. It is known that each node/line has to be within the limits of the active and reactive powers. This fact creates operational limits for the nodes of the system. As such, the above equations are showing the minimum and maximum active (P_t) and reactive (Q_t) powers of nodes/line within their limits, as expressed by:

$$P_t^{min} \leq P_{Nominal}(t) \leq P_t^{max} \tag{6.28a}$$

And

$$Q_t^{min} \leq Q_{Nominal}(t) \leq Q_t^{max} \tag{6.28b}$$

OLTC is usually equipped with a line drop compensation (LDC) feature built in it. This feature keeps the remote bus voltage within limits. In practice, the LDC function is disabled to keep simpler operation and also to keep the OLTC free from the effect of the line X/R ratio. Shunt capacitors inject reactive power at the power distribution location where they are placed. Reactive power injected into the power distribution network boosts the voltage by mitigating reactive power demand. Shunt capacitors are added to power distribution networks to boost the voltage levels. These capacitors are also neutralizing the effect of degrading power factor created due to of inductive loads, a common presence. By adding capacitor banks in the network can boost system voltage and helping to keep it within desirable limits. Optimized voltage profile can be achieved by controlling the injected reactive power magnitude by capacitors. Capacitor banks, either fixed or switched, are set in a distribution network to correct the power factor of the load. When the feeder experiences leading power factor, a fixed capacitor can cause overvoltage. Fixed capacitors are used to fulfil the minimum reactive power requirement and are placed where minimum voltage boost is needed, while switched shunt capacitors are used to handle larger voltage fluctuations, being turned OFF or ON during heavy or light loading conditions. The switched capacitor can be controlled by either manual or automated mode. Conventionally, switched shunt capacitors are switched based on time, voltage-controlled, time-controlled, current-controlled and temperature-controlled and power factor control scheme. Power distribution systems are planned and designed considering transmission line parameters, loads and generation capability. When DG units are introduced in the power distribution without any action taken, it might have a significant impact on the system power flow and voltage profile. Based on the location, the DG size also depends on the load profile which can bring both positive and negative impacts. The connection of DG, if operated at leading power factor, may increase the voltage by adding reactive power to the network. The direction and magnitude of real and reactive power are altered based on DG location and size. Due to uncoordinated operations for regulation, more challenges may arise such as power loss increment and number of switching increment. Interaction between DG and OLTC and the OLTC influence on DG are important. The smart grid is bringing considerable opportunities for the advanced communication, automation of power distribution systems. It provides the electric utilities with more visibility and pervasive control over their assets and services. When the communication is not available, the VVC operates in dark, the states of the controlled devices are invisible to system operators and the solution performances are hard to verify. Some of these solutions are trying to overcome some of the shortcomings by coordinating the control settings. To overcome the VVC limitations, modern approaches employ the two-way communications.

Power distribution networks supply the electricity to the end-users of the feeders from the substations. A model of a radial power distribution network, Figure 6.9, has a stiff grid transmission lines, transformers and loads. Power losses occur due to the resistance of transmission lines and other factors. The main purposes of regulating voltage in a power distribution system and controlling reactive power are to reduce system losses and to provide acceptable service to the end-users. The conventional way of controlling voltage and reactive power in the power distribution network is done by using on-load tap changing transformers, switchable capacitors (SCs) and voltage regulators at various locations of the feeder. These devices are very reliable and efficient to control Volt-VAR until the DG units were integrated into power distribution. For example, OLTC units are used to regulate or maintain the voltage within the limits. Using the substation as a main source of power may help OLTCs provide electric power to the distribution network. Most distribution transformers have tap changers which regulate the voltage by changing the tap. Turns ratio is regulated on the primary and secondary side. A typical OLTC has 8–16 taps with a step size of 1.5% voltage variation. Overall, one OLTC has the capability to control the voltage to the range of ±12% of the nominal voltage. Loads located at the downstream of the network are more likely to experience voltage drop, and the OLTCs are likely to raise the voltage by tap changing to an appropriate level, keeping the secondary voltage V within the limit as prescribed.

6.4.4 Smart Distribution System Reliability and Service Restoration

Reliability refers to a system's ability to perform its required function under given conditions for a stated time interval. From power distribution system perspective, its function is to supply electrical energy to final customers without interruptions and within accepted tolerance margins (i.e. acceptable values and limits for voltage and frequency). Power distribution systems are responsible for approximately 90% of all service interruptions experienced by customers. Therefore, it is important to understand how power distribution system behaves and the effect that every system element has in terms of system reliability; an accurate evaluation of power distribution reliability

is essential to identify design weaknesses and areas within the system that require special attention. A power distribution system is composed of a great number of elements, while the failure in
one of these elements is affecting the continuity of service provided to the customers. The impact
of element failure depends on the element's statistical parameters, characteristics and the system
design. The most important statistical parameters are the failure rate and the repair time. A failure rate is defined as the number of expected failures per element in a given time interval, and
the repair time refers to the time required to restore the service, whether by repairing or replacing the failed element. The area affected by an element failure depends on system design, and a
reliability-based design includes protection and switching devices, whose main task is to reduce
the number of customers affected by a service interruption. The analysis of power distribution
reliability is performed by following either a predictive or a statistical approach. These approaches
are not mutually exclusive as they have different purposes and are carried out at different stages,
but are equally relevant to the reliability evaluation of a power distribution system. Predictive
analysis of power distribution system reliability can be used to validate system design as well as
ensure that the reliability level provided meets utility policies and customer requirements. A great
variety of methods have been developed for the predictive analysis of power distribution reliability,
which can be classified into several categories; these methods can be classified as analytical and
simulation-based methods.

In analytical methods all distribution system elements or devices are represented by means of
their mathematical models. Such approaches use analytical methods to estimate power system reliability indices. However, due to the complexity of the actual power distribution, analytical methods are relying on simplification assumptions. Analytical equations are rather straightforward and
results are found in a short time periods. However, developing the equations that model the system
behavior can be a very complex task, especially when considering system reconfiguration processes
and distributed generation. Simulation-based methods rely on power system simulation to analyze
the power system reliability. The Monte Carlo method has been extensively used in power system
reliability studies due to the random behavior presented by power system failure. The main disadvantage of the simulation-based methods is the large computational efforts and long simulation
times to obtain accurate results. Statistical analysis allows monitoring system performance from a
reliability point of view, providing also information that can help to validate predictive analysis and
to identify areas where reliability needs to be improved. Statistical approaches require historical
data of all service interruptions experienced by customers over a defined time period. System reliability is quantified by the distribution system reliability indexes or measures, and the most common
are the system average interruption frequency index (SAFI) and system average interruption duration index (SAIDI), defined by the following relationships:

$$SAFI = \frac{\sum_{j=1}^{n} N_j}{N_T} \qquad (6.29a)$$

And

$$SAIDI = \frac{\sum_{j=1}^{n} N_j \cdot H_j}{N_T} \qquad (6.29b)$$

However, these indices do not capture all information relating to the power system reliability,
notably omitting to capture of the load lost during outages. The above indices also suffer from the
fact that they are often calculated inconsistently and can corrupt the reliability analysis. Since these

indices are system averages, they may not give information on specific bus reliability. For specific bus reliability analysis, load point indices are required to be estimated. Additional indices are the customer average interruption duration index (CAIDI) and the average service availability index (ASAI), defined by these relationships:

$$CAIDI = \frac{\sum_{j=1}^{n} N_j \cdot H_j}{\sum_{j=1}^{n} N_j} = \frac{SAIDI}{SAFI} \qquad (6.30a)$$

And

$$ASAI = \frac{8760 - \dfrac{\sum_{j=1}^{n} N_j \cdot H_j}{N_T}}{8760} = \frac{8760 - SAIDI}{8760} = \frac{Customer\ hours\ service\ aialablity}{Custumer\ hours\ service\ demanded} \qquad (6.30b)$$

In all the above equations, j is the number of interruptions, N_j is the number of customer interrupted by a fault, N_T is the total number of system customers and H_j is the customer interruption duration by a fault. However, the needs of more detailed and extensive data for the calculation of the reliability indices may represent a drawback for certain utilities as they do not possess enough historical data and may not have the needed equipment to record of every interruption experienced in the systems or lack an application that allows a prompt and easy access to it, posing an obstacle for an accurate evaluation of the system reliability. Indices like SAIDI, SAIFI and others provide power system wide reliability measures. The reliability measures of distribution system components or individual buses (also known as load point indices in references) can be determined. These indices include the average failure rate, λ (expected number of failures/year), the expected outage time, Otg (h), the annual unavailability, AU (h/y), and the expectation of unserved energy, UE (kWh/outage or kWh/y), that is capturing the energy demanded by the system loads that cannot be delivered to those loads. These probabilistic data can be adjusted using historical data or factors unique to the area of study, such as weather effects, aging infrastructure or other data pertinent to assessing expected component lifetime. Load point indices can be used to evaluate the relative performance of alternative system designs, restoration plans or system topology changes. Notice that the standardized test beds are quite useful in calculating the reliability indices and in evaluating the reliability of power distribution systems. Moreover, as discussed in the literature, the power distribution is the most significant part of the integrated power system that negatively impacts system reliability indices.

Power distribution system restoration is intended to promptly restore as much load as possible in areas where electricity service is interrupted following an outage. It plays a critical role in SDSs. By operating normally open tie switches and normally closed sectionalizing switches, system topologies are altered in order to restore power supply to interrupted customers. A well-designed service restoration strategy can restore the maximum amount of load with a minimum number of switching operations. Thus, the outage duration is shortened and system reliability is enhanced. Faulted zones are isolated by opening adjacent switches. Then, the actuated breaker or recloser is reclosed to bring service back to the customers upstream the fault. The loads downstream to the fault are picked up by other feeders, distributed generators or microgrids (local power networks, a system that can operate in grid connected or in standalone mode, separate from the grid) through feeder or distribution reconfiguration. Constraints, such as limits on bus voltages and capacity of transformers, need to be evaluated. After the faulted component is repaired, switching actions will be taken to bring the distribution system back to its normal topology.

6.5 CHAPTER SUMMARY

The basic layout, structure and operation of the power distribution infrastructure remain quite the same over the last half of the 20th century. However, in the last three decades, the equipment and power distribution structure have undergone steady improvements, transformers are more efficient, cables are much less expensive and easier to use and metering infrastructure, control, monitoring and protection equipment and devices are more and more computerized. Utilities operate more distribution circuits at higher voltages and use more underground circuits. But the concepts are much the same: alternating current, three-phase systems, radial circuits, fused laterals, overcurrent relays and so on. Advances in computer technology have opened up possibilities for more automation and more effective protection, monitoring infrastructure, operation and control. In industrial environments, electricity is provided to loads from load centers, switchgear and motor control centers. Substations are critical components in the operation and control of power systems, connecting generation, transmission and power distribution sections. Operation of the generation and transmission systems is monitored and controlled by SCADA systems. Substation automation systems provide solid bedrock for the smart grid development. Implementation of high-quality SAS system enables one to experience less outage rate using the state-of-the-art computerized functions of monitoring, control and protection. This chapter is giving a comprehensive presentation of electrical protection systems, protection devices, and Volt-VAR control methods. Maintaining voltage within desired level throughout the electric distribution network is one of the major challenges that electric power utilities have always grappled with. For power distribution grids, the impact of high penetration of distributed energy resources is discussed, typical control strategies are reviewed, and the challenges for Volt-VVAR control is also discussed. Volt-VAR optimization and conservation voltage reduction provide an opportunity to improve the power system performance in terms of efficiency and usability. As the customer electricity demand varies throughout the day, each customer having a specific (characteristic) load profile. Hence, the power supplied by a power distribution network throughout a day has to vary to meet customer load profiles. These issues are further complicated by the presence of RES and DG units, often having fluctuating power outputs and the bi-directional power flow. These are the operating issues at the heart of modern power grids that is not desirable for the system operators due to its higher expenditure and hard operational condition. The electric power utilities prefer to avoid the change in the supply through controlling the demand and supply. The last chapter section briefly discusses the power system reliability and the indices and measures common used to characterize the system and power distribution (the most impacted power system component) reliability.

6.6 QUESTIONS AND PROBLEMS

1. What is the purpose of power distribution systems?
2. Discuss the three most common configurations of the power distribution systems.
3. List the advantages and disadvantages of overhead and underground distribution systems.
4. What is the purpose of power distribution substations?
5. What are the three main tapes of power distribution networks?
6. Discuss the advantages and disadvantages of the overhead and underground power systems.
7. List the advantages and disadvantages of the three-phase delta (Δ) transmission lines.
8. List the advantages and disadvantages of the three-phase star (Y) transmission lines.

9. List the voltage levels used in the various stages of transmission and distribution in electric power systems.
10. Define the load factor, the diversity factor and the utilization factors.
11. How is the utilization factor used in applications?
12. What is a load curve and a coincidence factor?
13. A sub-station has three outgoing feeders: feeder 1 has maximum demand 10 MW at 10:00 am, feeder 2 has maximum demand 12 MW at 7:00 pm and feeder 3 has maximum demand 15 MW at 9:00 pm, while the maximum demand of all three feeders is 33 MW at 8:00 pm. What is the sub-station diversity factor?
14. Four individual feeder-circuits with connected loads of 225 kVA, 210 kVA, 160 kVA and 360 kVA and demand factors of 90%, 85%, 82% and 88%, respectively. By assuming a diversity factor of 1.35, estimate the transformer size, supplying these feeders.
15. Briefly describe the importance and the use of diversity and demand factors.
16. What are the effects of high primary voltages on distribution systems?
17. Why is the protection critical in power distribution networks?
18. Explain what is meant by service drop and by service lateral.
19. How many voltages can be set from a three-phase wye-connected feeder transformer?
20. Briefly describe the conventional Volt-VAR control.
21. What are the applications of single-phase electrical power and of three-phase electrical power?
22. Why is the voltage drop important in electrical power distribution systems?
23. Briefly describe the smart grid Volt-VAR control approaches.
24. How does the capacitor improve not only the power factor but also the voltage regulation (i.e. by reducing the voltage drop)?
25. A 450 kVAR, 480 V, three-phase capacitor bank is installed at the main switchgear of 1.25 kA service. The service transformer is rated at 1500 kVA, 4.16 kV–480/277 V and has impedance referred to its low-voltage side of $0.003 + j0.0135$ Ω/phase. Compute the percentage voltage boost due the capacitor bank installation.
26. A three-phase 180 HP. 460 V, 60 Hz induction motor is started direct from 480 V supply. The combined source and cable impedance is $0.01 + j0.028$ Ω/phase. If the motor starting power factor is 0.28 lagging, determine the voltage drop of the line voltage.
27. A 480/277 V, wye-connected, four-wire service is supplied by a 500 kVA service transformer. The transformer impedance is $0.0065 + j0.0285$ Ω/phase. If a 250 kVAR, 480 V capacitor bank is installed at the distribution switchgear, what is the voltage rise?
28. An industrial facility has a power demand of 1500 kW at a power factor of 0.70 lagging. Determine the capacitor bank reactive power ratings required to improve the power factor to 0.85, 0.90, 0.95 and unit power factor, respectively. What is the voltage rise in each case, if the feeder inductive reactance is 0.015 Ω/phase and the line-to-line voltage is 460 V?
29. List and briefly describe the main types of substations.
30. Briefly describe the voltage issues due to the DER presence in power distribution networks.
31. List and briefly describe the main voltage control methods.
32. Briefly describe the effects on power distribution of the DG inclusions.
33. A power distribution section, equipped with a PV plant, has a resistance of 0.60 Ω and an inductive reactance of 0.15 Ω. If the line terminal voltage is 2.4 kV estimate and the distributed generation unit is injecting 600 kW active power and 90 kVAR reactive power, estimate the voltage rise.
34. List and briefly describe the VVC methods.
35. Briefly describe the smart grid VVC objectives.
36. List the benefits of the VVO techniques.
37. What is the purpose of conservation voltage reduction?
38. List and briefly describe the main substation types.

39. Briefly describe the main functions of a substation.
40. List the distribution substation main components and elements
41. List and briefly describe the substation primary and secondary equipment.
42. What are the objectives of the conservation voltage reduction methods?
43. List and briefly describe the analysis of power distribution reliability methods.
44. List the main reliability indexes used in power systems.

7 Smart Grid Communication, Sensing and Monitoring

7.1 INTRODUCTION, NEEDS FOR SMART GRID ADVANCED COMMUNICATION

Power networks were designed as centralized systems, with electric energy flowing unidirectionally through transmission and power distribution lines from power plants to the customer premises. The grid intelligence is concentrated in central locations and only partially in substations, while loads or end-users are almost passive entities. However, the smart grids (SGs) are providing higher and widely distributed intelligence embedded in local electricity production, two-way electricity and information flows, sensing, monitoring and control, thus achieving reliable, flexible, robust, efficient, economic and secure power delivery and uses. The new approach, the smart grid, requires a complex and extended two-way communication infrastructure, sustaining power flows between intelligent components and sophisticated monitoring, computing, management and information technologies as well as business applications. The SG applications are enabling new approaches in electric grid management by applying demand side management, providing energy storage needed for load balancing and to overcome energy fluctuations caused by the use of renewable energy generators, preventing widespread power grid cascading failures and enabling the integration of plug-in hybrid and electric vehicles (PHEV and PEV) and other devices, thus reducing pollutant emissions, and improved generation and load balance. Smart grid is envisioned to meet the 21st-century energy requirements and expectations in sophisticated ways with real-time approach by integrating the latest technologies in communications, monitoring, sensing and advanced, adaptive and smart control to the existing power grids. Rapid advancements into control, information and communications technologies have allowed the conversion of existing electricity grids into smart grids that ensure productive and complex interactions among all electric grid stakeholders. These multiple and enhanced interactions help to solve many of the existing power grid issues, e.g. congestion, stability, resilience, fast service restoration, adverse environmental impacts, efficiency, sustainability, electricity price control or self-healing. Control systems are needed across broad temporal, geographical and industry scales, from devices and components to power system wide, from fuel sources to consumers or from utility pricing to demand response. With deployment of advanced feedback and communication infrastructure, significant opportunities for reducing consumption, efficient use of renewable energy and for increasing reliability, supply security, transmission and power distribution performances arise.

Smart grid key components are smart meters, adaptive and intelligent control, smart sensors, intelligent monitoring and sensing, advanced data and energy management, controlling the power and information (data) flows among various stakeholders, making the electric grids bi-directional information and energy networks. Advanced metering infrastructure (AMI) is a major smart grid component, consisting of smart meters, energy usage monitoring, optimization and control, advanced data management, storage, processing and transfer. AMI helps to provide economic benefits, improved services and new approaches to the environmental issues. Smart meter has bi-directional consumer-energy provider communication, recording of energy usage, service connect and disconnect switches with capabilities to sense power disturbances. Other smart grid key applications include advanced energy and power distribution management, distributed generation and their reliable integration, component and equipment sensing, diagnostic, control and optimized asset management. Smart grids are exploiting various communication technologies, from fiber optics, cellular networks to wireline techniques, smart sensing, measurement, smart metering and monitoring. This chapter provides a comprehensive discussion of communication, smart sensing and metering technologies and procedures in the smart grid context with focus on the power transmission and distribution networks, considering how these equipment and devices are deployed and operated. This chapter provides a comprehensive examination of smart grid communication, sensing, metering and monitoring infrastructure challenges faced in supporting a diverse set of grid applications, each with varying network performance requirements, reliability characteristics and traffic characteristics, the challenges faced with supporting legacy applications and networks. While there are many legacy and evolving applications, the following classes of applications (not necessarily mutually exclusive) are discussed as examples in the communication network architecture and design for smart metering, control and management, e.g. the advanced metering infrastructure, automated demand response (ADR), protection, control and security, distribution automation or smart microgrid management and operation.

The smart grid technology becomes possible by applying smart sensors, sensing units, field automated devices, monitoring systems and smart meters to the electrical transmission and power distribution. These equipment and devices measure, monitor and control power grid conditions and communicate data and information to operators, utilities and customers providing opportunity to real-time and dynamically respond to the changes in the grid conditions. Many technologies are adopted by the smart grid have already been used in other industries, such as sensor networks in manufacturing or wireless networks in telecommunications or environmental monitoring, and are being employed for use in the new intelligent and interconnected paradigm. Smart grid communication technologies can be grouped in key domains as: advanced components, sensing and measurement, improved interfaces and decision support, standards, and integrated communications. An integrated high performance, highly reliable, scalable and secure communications network is critical for the successful deployment and operation of next-generation electricity generation, transmission and distribution systems, the smart grids. However, much of the last decade's research to define SG communications architecture focuses on high-level service requirements with less attention to the implementation challenges. Moreover, smart metering, sensing, monitoring demand-side techniques have become increasingly needed to control demand during peak and off-peak hours, requiring secure communications. A reliable, scalable, secure and pervasive communication infrastructure is essential and crucial in the structure, development, implementation, operation and management of a smart grid. For example, the electric power infrastructure is highly vulnerable against many forms of natural and malicious physical events, which can adversely affect the overall performance and stability of the grid. Additionally, there is an impending need to equip the age-old transmission line infrastructure with a high-performance data communication network that supports future operational requirements like real-time monitoring and control necessary for smart grid integration. Wireless sensor network based-monitoring of transmission lines can provide solution for many of these concerns like real-time structural awareness, faster fault localization, accurate fault diagnosis by identification and differentiation of electrical faults from the mechanical faults and cost reduction due to condition based-maintenance rather than periodic maintenance. The sensor network use has been proposed for several SG applications like mechanical state processing and dynamic transmission line rating applications. Given the vast geographical expanse of the transmission line infrastructure, wireless networking technologies are presenting a feasible, scalable and cost-effective solution for transmission of monitoring data.

One of the smart grid key features is the all-in-one integration and communication between power system elements as hardware and the information, sensing, monitoring, processing, analyzing and controlling electric grids as intelligent systems. In order to integrate such diverse systems, advanced information, computing, processing and data acquisition technologies are needed e.g. data sciences, Internet of Things (IoT), or smart sensing. Smart sensors, intelligent sensing and monitoring are key enablers for the smart grid in order to reach its full potential. A critical *smart grid* concept is that the grid is responding to real-time demands, and in order to do this it requires new sensors, intelligent sensing and advanced monitoring techniques to provide this *real-time information and data*. A smart sensor is one of the most important devices of the smart grids and also for many more energy industry applications. Smart sensors are devices that inform the control and management systems about certain parameters and what is actually occurring in the physical objects that are monitored. Smart sensor provides raw data for processing, analyzing, control and feedback. In transmission and power distribution monitoring system, smart sensors are used at multiple places and spread over the entire grids such as on the transmission tower, transmission and distribution lines, transformers or substations. There are two major smart sensor types currently used in power and energy sectors. The first type is used to measure weather condition around the assets, end-user locations or power system elements, e.g. wind speed and direction sensors, temperature sensors, humidity sensors, pollutant and rain sensors. These sensors are also responsible for natural disaster mitigation and prevention that is not available on the current grid. Other smart sensors are used for asset and power quality monitoring, e.g. GPS technology, strain sensors, accelerometers for vibration and transmission line monitoring, conductor temperature sensors and magnetic field sensors. Such sensors are monitoring the states of towers, transmission lines and substations. Power quality monitoring sensors include voltage and current sensors, temperature and fault sensors. By installing such sensors in the transmission and distribution lines, quality and conditions of transmission and distribution lines can be monitored on a real-time basis. In addition, the sensor can be used to monitor or identify particular information in substation transformers such as temperature, primary and secondary voltage or frequency.

Several communication and networking technologies can be used to support smart grid applications, including phone lines, cable lines, fiber optic cable, cellular, satellite, microwave, WiMAX, power line carrier, wireless networks and broadband over power line, as well as short-range in-home technologies such as Wi-Fi and ZigBee. The Smart Grid applications that might be built on such communications technologies include home area networks (HAN), neighborhood area network (NAN), facility area network (FAN) and networks for wide area situational awareness (WASA), supervisory control and data acquisition (SCADA) systems, distributed generation monitoring and control, demand response and pricing systems, and charging systems for plug-in electric vehicles. Moreover, the utilities employ various smart grid and demand response applications for several years, while these applications have traditionally used private communications networks. Commercial service providers are increasingly partnering with utilities to provide communications for various smart grid applications. Notice that often commercial carriers have encouraged technological changes, such as the trend toward integrated platforms and open standards for utility communications that have historically been proprietary. Commercial providers have thus facilitated opportunities for better communications systems, even where utilities ultimately opted for private networks. Utilities cited higher rates of survivability following extreme events, as the ability to maintain service throughout a service territory, the avoidance of prioritization of other services when recovering from outages, and the service costs, as major reasons why commercial services could not adequately replace the private networks. In addition, some utilities suggested that dedicated wireless spectrum may be advantageous to smart grid services. As smart grid technologies continue evolving, future and advanced applications may lead to both an increase and a qualitative change into sensing, metering, monitoring, control, energy management and communications requirements and applications.

7.2 SMART GRID APPLICATIONS

Smart grid technologies are defined as self-sufficient systems that can find solutions to power generation and delivery problems as quickly as possible in available systems, reducing the workforce requirements while targeting sustainable, reliable, secure, safe and quality electricity to all consumers. Smart grid infrastructure covers management, protection, smart sensing, monitoring, metering, control, information and communication systems, energy efficiency, pollutant emissions, power quality, supply security and enhanced service restoration. Essential smart grid features are extended uses of information, computing and communications technology to gather and act on information in an automated fashion to improve the efficiency, reliability, economics and sustainability of the electricity generation, transmission and distribution. Smart grid automated and distributed energy system relies on two-way electricity, data and information flows. Almost instantaneous supply and demand is balanced at equipment level due to the distributed computing, sensing and communication, enabling exchange of information and data in real time. Smart grid applications are categorized into six functional categories: demand response, advanced metering infrastructure, wide-area situational awareness, distributed energy resource and energy storage uses, electric transportation and distributed grid management. Several communications and networking technologies are used with SG applications. In addition to the advantages of smart grid applications, it can encounter problems and challenges. In a smart grid the energy infrastructure represents the physical infrastructure for energy generation, transmission and distribution; the communication infrastructure is responsible for transferring the critical information and data through the network, while the information technology provides modeling, analysis, web visualization and commercial transactions; potential applications are responsible for distinguishing the uses of the SG infrastructure. The communication infrastructure performs a critical role in the overall smart grid framework. Connecting all relevant components in the grid is very important to collect information about the conditions of the components for control, monitoring and maintenance purposes. Any problems related to the energy infrastructure can be avoided if proper action is taken immediately with the help of the communication infrastructure.

The smart grid, being a vast system, can utilize various communication, sensing and networking technologies within its applications, which include both wired (e.g. copper cables, fiber optic cables and power line carriers) and wireless communications (e.g., cellular, satellite, microwave, etc.). Short-range wireless communication technologies such as Wi-Fi and ZigBee can also be used in some smart grid applications, such as in home area networks or building networks. Wireless communication options are easy to implement and enjoy wide support, but present security concerns are that it could be compromised to disable and damage the infrastructure of the system. Installing wired systems would add excessive costs, and power line communication has obstacles to overcome before it can be used in a fully implemented smart grid capacity. One of the application areas in the smart grid communication is the so-called advanced metering Infrastructure. Unlike the traditional way, where technicians are sent to each consumer site monthly to record the data manually for the billing purpose, the smart meters in AMI provide real-time and remote monitoring capabilities of the electric loads. The information on power usage is collected periodically (e.g., every 15 minutes) by a data concentrator at the intermediate layer using wired or wireless communications and forwarded to a central location. The real-time data is efficient and precise. This allows utility companies to analyze consumer energy consumption data and to provide outage notification and billing information using two-way communications. Furthermore, through AMI systems, the consumers can be provided with historical data for energy consumption and dynamic pricing, as well as suggestion to reduce peak load, encouraging participation and response of the end-users in energy management. For example, a customer can adjust the power usage based on the detailed energy consumption information and the dynamic peak price, which can be displayed on some in-home display (IHD) units. Analysis of such data can also help the utility companies to better understand the pattern of consumer power consumption and to plan for reducing some of their financial burdens.

In summary, in order to fulfil the different SG requirements and to fully employ the applications, several enabling technologies, e.g. communication, information technologies, sensing, control or monitoring are or must be developed, designed, tested, adapted and implemented. *Information and communications technologies* must include, among others: two-way communication technologies and capabilities to provide connectivity between different components in the power system and loads; open network architectures for plug-and-play of home appliances, equipment, electric vehicles, distributed and micro-generation units; communications and the associated software and hardware to provide the customers with better and extended information, enabling the end-users to trade in energy markets and providing demand-side response; software capabilities and functions to ensure and maintain the security of information and data; and developing and setting standards to provide scalability and interoperability of the SG information and communication systems. In terms of sensing, measurement, monitoring, control and automation technologies, smart grids need to include: intelligent electronic devices (IEDs) to provide advanced protective relaying, measurements, sensing, power system fault and event records; phasor measurement units (PMU) and wide area monitoring, protection and control (WAMPAC) to ensure the security and reliability of the power systems; and integrated smart sensors, measurements, control and automation systems and information and communication technologies to provide rapid diagnosis and timely response to any event in different parts of the power system. These are supporting enhanced asset management and efficient operation of power system components to help relieve congestion in transmission and power distribution circuits and to prevent or minimizing the potential outages, enabling faster service restoration in event of outages and enable working autonomously when conditions require quick resolution. By using smart appliances, equipment, communication, controls and monitors to maximize the safety, comfort, convenience and energy savings of homes; smart meters, communication, displays and associated software allows the customers to have greater choice and control over electricity and gas use. All such technologies are providing the consumers with accurate energy bills, along with faster and easier supplier switching, to give consumers accurate real-time information on their electricity and gas use and other related information and to enable demand-side management and participation.

7.2.1 Smart Grid Communications Infrastructure

A smart grid communications infrastructure enables utilities to interact with devices on their electric grid as well as with users, distributed power generation and storage facilities. In order to satisfy the full SG concept, the communications structure has to be designed as a multilayer architecture that extends across the whole grid infrastructure. It also has to cover large geographical areas, consequently the SG communications infrastructure should connect a large set of nodes of the entire region and utilities need to use and accept several networks. Wide area network (WAN) for automation, distribution and for covering long-haul distances by providing communication links between the NANs and the utility systems to transfer information to the neighborhood area network (NAN) or facility area network (FAN) for connecting multiple HANs to local access points. Home area network (HAN) extends communication to end points within the end-user home or business. Each of the three networks is interconnected through a node or gateway: a concentrator between the WAN and NAN or FAN and a smart meter between the NAN and HAN or FAN. Each node communicates, exchanging information, data and eventually control signals through the network with adjacent nodes. The communication infrastructure is envisioned to be a multi-layer structure, extending across the whole smart grid, as depicted in Figure 7.1. SG communication is broadly divided into two major structures: the short range communication systems, such as home area network, facility area network, neighborhood area network, connecting the varied devices, installations and equipment of a home, building or an area to its smart meter, feeder or service entrance and long range transmission structures necessary spanning distances from the utility to the customer (end-users). In fact, smart grids require further communication between the interconnected utility generators, suppliers, transmission and substations. Ensuring security in each of these networks (which may each have different media) is required, as all of them must work together for the smart grid to function properly. Many of the communication and security components are common between these energy subsystems. SCADA is the core subsystem of smart grid. A second key component to smart grid is a number of secure, highly available wireless networks. Wireless technologies include Worldwide Interoperability for Microwave Access (WiMAX), wireless local area network (WLAN), WAN, all cellular technologies and wireless sensor networks. Comprehensive security solution is the third

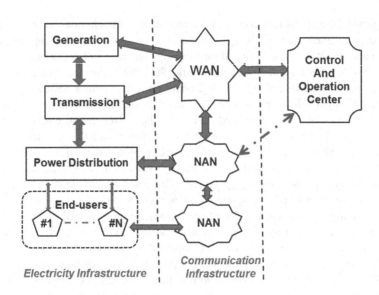

FIGURE 7.1 Smart gird multi-layer communication structure, showing the two parallel interdependent networks, the power system and the communication network, forming smart grid infrastructure.

key component as privacy remains a major issue for implementing smart grid technology. As any computer network-based structure, the smart grid must meet several requirements. High reliability and availability are essential for proper grid control and operation, so the system must be robust and include levels of redundancy for critical applications. The communication architecture must be able to span the large distances of the power grid and effectively handle the large number of nodes (all of the individual smart meters) and the data they produce without excessive delays. The network should also be maintainable, meaning that updates and improvements should be simple to perform and not require physical modifications at each node. Lastly and perhaps most importantly, the network should contain these foundational supports in regards to security management.

Various communication technologies are used for different purposes to cope with each specific application requirements. Information technology provides platforms for sharing the information coming from different business fields related to the smart grid by giving support to the collection of different information, analysis and advanced applications and providing the integration of information from different layers of the smart grid. Potential applications are better at management, automation, detection and generation techniques for the overall system; fundamental applications generally focus on reducing the electricity consumption in homes, offices and factories and changing the customer consumption behaviors with advances in-home displays and energy dashboards, etc. For communications, low-powered, short-distance technologies have been proposed for on-premises applications that include wireless communications (e.g. Wi-Fi, IEEE 802.11 standard or ZigBee, IEEE 802.15.4 standard) and power line communication networks such as HomePlug that uses building electrical wiring networks to carry data, as advanced infrastructures connecting electric devices, equipment and appliances. Overall, the smart grid communication is usually a hierarchical architecture with interconnected sub-networks, each taking the responsibility of separate areas. For example, the AMI communication systems can be organized into three classes, as wide area networks, field area networks and home area networks. Home area network is used to establish a communication link between the smart meter and the smart appliances, other building meters, in-home display and the micro generation unit. HAN provides centralized energy management, services, and facilities. The communication protocol can be a wired or wireless media. Neighborhood area network is used to transfer the data between neighboring smart meters, facilitating diagnostic messages, firmware upgrades and real-time messages. ZigBee communication protocol is widely

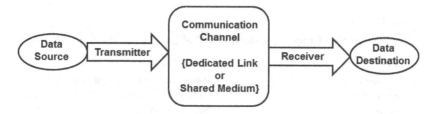

FIGURE 7.2 Point-to-point communication system configuration.

used in NAN due to high speed of data transferring and low cost. Wide area network is a sub-network, in which smart meters are connected to a remote server through a WAN. GSM, GPRS, 3G and WiMAX communication technologies can be used to connect the meter to the WAN.

Data and information communication systems are essential in any modern power system and even their importance is increased as the smart grid develops and expands. As a quite simple example, a data communication system can be used to send status information from an intelligent electronic device (IED) to a workstation (human-machine interface) for display in a control center. Any coordinated control of the power system relies on effective communication systems linking large number of devices, computers and other equipment. The smart grid is a large-scale energy system, extending from a power generation facilities, through transmission and power distribution to each and every power consuming entity and device, such as home appliances, computers, electric motors, installations, equipment, monitoring systems and phones. This large-scale nature of the smart grid has increased the possibilities of remote operation of power management and distribution system, requiring advanced, diverse and large-scale communication infrastructure. With energy being a premium resource, ensuring security against theft, abuse and many malicious activities in a smart grid is one of the prime security concerns, requiring sensors, smart meters, communication and control subsystems. Figure 7.2 shows a typical configuration of a simple point-to-point data communication in which the communication channel is the path along data travels as a physical signal. The communication channel could be a dedicated link between the source and destination or could be a shared medium, connecting them. Data communication channels are characterized by their maximum data transfer and error rates, delays and the technology. Certain applications require the transmission of data from one point to another and others uses may require the data transmission from one point to several points. When a secure communication channel is required, a dedicated link is used by the specific data source and destination. In contrast, when a shared communication channel is used, a message sent by a source is received by all the devices connected to the channel. An address field within the message specifies for whom the message is intended, other destinations ignoring this message. For example, dedicated communication channels can be used for differential protection of transmission lines. It is used to transmit a signal corresponding to the sum of the three line currents, added by using a summation transformer at one relaying point to another transformer for comparison with similar signal at that point.

7.2.2 SMART GRID COMMUNICATION CHANNEL PARAMETERS AND CHARACTERISTICS

Communication channels are characterized by their maximum data transfer speed, error rate, delay and communication technology used. Communication requirements for commonly used power systems applications are given in Table 7.1. Any coordinated control of the power system, and especially the smart grid, relies on effective and secure communications linking a large number of devices. Certain smart grid applications require the data transmission from one point to a single point and other uses may require the data transmission from one point to several points. When a secure communication channel is required from one point to another, a dedicated link is used exclusively by the *source* and *destination* only for their communication. In contrast, when a

TABLE 7.1

Power System Communication Physical Devices and Components

Component	Physical Device
Data Source	Potential/current/instrument transformer
Transmitter	RTU: Remote terminal unit
Receiver	Network interface
Communication Cannel	LAN or WAN
Destination	Operator workstation/and control IED

shared communication channel is used, a message sent by the source is received by all the devices connected to the shared channel. An address field within the message specifies for whom it is intended, while others are simply ignoring the message. For example, dedicated communication channels are used for differential protection of transmission lines. The communication channel is used to transmit a signal corresponding to the summation of the three line currents (they are added using a summation transformer) at one relaying point to another for comparison with similar signal at that point. Communication channels are supported by physical media between the source and the destination. In the case of dedicated channels, usually a single physical medium, as shown in Figure 7.3, is used. Shared communication channels may involve more than one medium, depending on the route the signal travels. A communication channel may be provided through guided media such as a copper cable or optical fibers or through an unguided medium such as a radio link or wireless network.

The performance and characteristics of a communication channel are described by a few important parameters described here. *Bandwidth (bit rate)* represents the difference between the upper and lower cut-off communication channel frequencies. In the case of an analogue system, it is usually measured in Hertz. In digital transmission, the term bit rate is used to express the channel capacity or digital bandwidth. The bit rate is measured in bits per second (bps) or often in kbs or Mbs. Transmission rate determines the data volume that can be sent to a destination through the communication network within a certain time, critical for smart grid operation and control. Smart grid communication system is very important for data acquisition, analysis, control and monitoring of the huge number of the smart grid components, equipment, and devices. As a result, there is a huge amount of data that has to be continuously and bi-directionally transferred by the communication system. Therefore, research efforts should concentrate on developing appropriate high data rate communication technology or improving existing technologies. *Attenuation* refers to the fact that as the signal propagates along a communication channel, its amplitude decreases. In long-distance transmission, amplifiers (for analog signals) and repeaters (for digital signals) are installed at regular intervals to boost the attenuated signals. For example, when transmitting digital signals in electric copper cables, repeaters are required every 10 km, whereas in the case of optical fibers a signal can propagate up to 100 km without significant attenuation. In any communication system, the electrical *noise* is an inherent problem. For

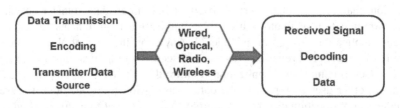

FIGURE 7.3 Typical data transmission media (e.g. wired network, optical or radio networks).

example, when digital communication channel, the noise can be sufficient to change the voltage level corresponding to logic "0" to logic "1" or vice versa. Noise level is described and characterized by the signal to noise ratio (SNR) and is measured in decibels (dB). The SNR is defined by this expression:

$$SNR = 10 \cdot \log_{10} \left[\frac{\text{Signal Power}}{\text{Noise Power}} \right] \tag{7.1}$$

Example 7.1 **For example, if the SNR = 20 dB, from Equation (7.1) the ratio of signal to noise power is 10(20/10) = 100.**

Latency refers to the time it takes a piece of data to reach a destination correctly. *Reliability* measures how likely a certain piece of data is received correctly. Since different smart grid functions have very distinct requirements on data transmission, being critical criteria in the technology selection. *Signal propagation delay* represents the finite time delay that it takes for a signal to propagate from data source to destination. In a communication channel, both the media and the repeaters that are used to amplify and reconstruct the incoming signals are introducing delays. Keeping in mind that the power grid goal is to maximize the power transfer from generation to end-users, without noise and information, the so-called communication objective is to transmit information (data), trying to minimize the power. The total power received (*P*) through a communication channel is the sum of signal (*S*) and noise (*N*), being expressed as:

$$P = S + N \tag{7.2}$$

Assuming the uncorrelated signal and noise, the previous relationship can be expressed, in terms of the RMS received signal (*V*), transmitted signal (*V_S*) and noise (*V_N*) voltages, as:

$$V^2 = V_S^2 + V_N^2 \tag{7.3}$$

Assuming the information is encoded in voltage levels, the total voltage range is divided into a number of voltage bands, 2^b, each of equal size. Larger number of bands allowing more information to be encoded, on the expenses of smaller band size, increasing the probability of noise interferences that can cause the voltage level to fall outside the intended band, leading to errors. The maximum number of bands, from the above equations, and with previous notations can be expressed as:

$$2^b = \log_2 \sqrt{1 + \frac{V_S^2}{V_N^2}} = \log_2 \sqrt{1 + \frac{S}{N}} \tag{7.4a}$$

Or

$$b = \log_2 \sqrt{1 + \frac{S}{N}} \tag{7.4b}$$

If we are assuming the *M* measurements of b bits can be made in a *T* period, the Nyquist theorem states that if *B* is the highest frequency component of signal than the can be reconstructed if the 2*B* is equal to *M* divided by *T*. According to Shannon's capacity formula, the maximum channel capacity in bps is given by $B \cdot \log_2[1 + (\text{signal power})/(\text{noise power})]$, where *B* is the bandwidth of a channel in Hz. So, the general bound of the communication channel *C* on the maximum rate, at which the information can flow through a cannel, relating the bandwidth (B), signal and noise levels is expressed as:

$$C = B \log_2 \sqrt{1 + \frac{S}{N}} \qquad (7.5)$$

The noise is often assumed to be Gaussian, with N_0 is the noise variance and E_s is the signal energy (the signal needs to be strong enough to overcome the noise, for the receiver to be capable of signal decoding), the maximum channel capacity, assuming infinite bandwidth (the bandwidth, B is not included in the following relationship), the maximum channel capacity for white Gaussian noise is:

$$C = \log_2 \left(1 + \frac{E_s}{N_0} \right) \qquad (7.6)$$

From above equation, the energy per bit, E_b is easy to obtain, as well as the minimum energy to transmit one bit, $E_b^{\min}$ which are expressed by the following two expressions:

$$E_b = \frac{E_s}{C} = N_0 \frac{2^C - 1}{C} \qquad (7.7a)$$

And

$$E_b^{\min} = N_0 \cdot \lim_{C \to 0} \frac{2^C - 1}{C} = N_0 \cdot \ln(2) \qquad (7.7b)$$

In terms of information theory and statistical physics, Equation (7.7b) shows that minimum energy to transmit one bit of information in a Gaussian channel is equivalent to the energy needed to store one information bit via an isothermal compression of an ideal gas. The information and energy, as well as the gas laws (thermodynamics) and communication are strictly and directly related. Notice that in a modern power system, the smart grid is employing quite extended scale concepts and relationships from the information and network theories. For example, information theory application can effectively increases communication network capacity, the AMI systems can benefit directly from decreased traffic by using compressed sensing. Concepts, like classical channel capacity, compression or comprehensive sensing are used in smart grid to decreases the load requirements, in synchrophasors, for data phasor compression or in AMI to decrease the meter loads. Other concepts, such as entropy, prediction or combined entropy and prediction are used for security, demand repose, distributed generation and energy storage management, to increase encryption strength, predict power outputs, smooth the peak energy demands and reduce the power variances. While concepts from quantum information theory, inference and spectral graph theory are used for security, state estimate, power grid and communication networks for quantum key distribution, inferring state by using less or minimum data, network structures or distribution network analysis. The relationships of the communication, energy and information can be the foundation for a much deeper relationship between communication theory and smart grid that has to be explored for the future smart grid functions. Even the communication infrastructure is critical for smart grid operation and management further SG advances are requiring a fundamental extension of the information theory to unite with power systems, what is viewed by some smart grid research as the power information theory. The smart grid communication architecture must evolve in a holistic way as a whole to be done in an efficient and cost-effective way. All smart grid components and elements in this approach can increase the information entropy and can determine the fundamental communication limits, given the underlying physical limits of the future power systems. However, these topics and applications are beyond the scope of this chapter and book.

Example 7.2   Estimate the channel capacity for the signal to noise power of Example 7.1, and typical bandwidth of 60 Mbs (Mega-bit per second).

SOLUTION

Form Equation (7.5), the channel capacity in this case is:

$$C = 60 \cdot \log_2 \sqrt{1+100} = 199.75\,\text{Mbs} \approx 200\,\text{Mbs}$$

Switching techniques and methods are used to establish a link between a source and a destination to facilitate and transmit the data across a shared medium communication channel, often a network consisting of nodes and link. A node, performing the tasks such as routing data or acting as gateway between two different network types can be a network adaptor, a switch or a router. Circuit switching, message switching and packet switching are commonly used switching techniques for data transfer between data source and destination. In circuit switching, a dedicated physical connection is set up for the exclusive source and destination uses during a communication session. Nodes and links allocated for a communication session cannot be used by any source or destination other than the two involved in that communication session, making circuit switching inefficient if the data transmission pattern is intermittent. In message switching, the source sends a message, e.g. a measurement data collected by a sensor or a control function to a node. The node stores the data in its buffer, and when the entire message has been delivered to the node, it then looks for a free link to another node and then sends the data (message) to this node. This process continues until the data is delivered to the destination. Packet switching is dominating in today's data communication networks for reasons of economy and reliability. A message is transmitted after breaking it into suitably sized blocks, the packets. When traversing network adapters, switches, routers and other network nodes, the packets are buffered and queued, resulting in variable delay and throughput depending on the network traffic load. There are two approaches to packet switching; virtual circuit packet switching and datagram packet switching.

In the power systems and specifically into the smart grids the communication infrastructure is used to support control systems and methods that are reducing power losses and improving the power quality with the grid. As is shown in Figure 7.4, the power grid losses are significantly increasing from generation sections to power distribution sections. At the same time, the communication channel capacity is decreasing in opposite direction to the power grid losses. A fundamental

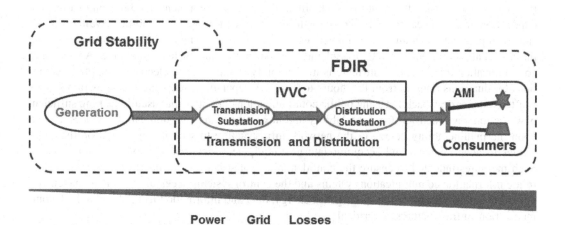

FIGURE 7.4   Simplified conceptual diagram of power system and communication infrastructure.

understanding of the relationship between the power grid efficiency, power losses and the capacity of the SG communication channels is critical and essential, while the communication links, extending smart grid communication infrastructure is keeping added in a somewhat haphazard, ad-hoc and integrative manner. In Figure 7.2, only applications such as stability control, integrate voltage-VAR control (IVVC), fault detection, isolation and recovery (FDIR) and advanced metering infrastructure were depicted as SG applications. Often, each application implements an independent control mechanism for SG optimization. It is proposed and extensively researched in terms of the power information paradigm the application of the networked control theory and approaches. However, this theory is still in research phase, not addressing the issue of the fundamental relationships between information, communication and power (energy). The previous concepts, as presented in the literature, are starting to play a significant and growing role into the smart grids. Moreover, the information carried by communication channels must be encoded and transferred in the most efficient format and secure way to reduce the bandwidth (source coding) and to have the capability of *self-healing*, e.g. error corrections or channel coding. The network science and engineering and information theory will play significant and extended roles in the future smart grid developments. Complete and overall smart grid control requires real-time information collection in order to implement dynamic control, ideally complete and continuous data sets of measurements, substations, transmission lines, power system devices and end-users (loads), however, today almost an impossible task. Future research on active networked control, combination of state estimated (used extensively in control) to fully understand the impacts on communication on the power grid intelligent control theory.

The smart grid communication infrastructure objectives are to provide bi-directional end-to-end, reliable and secure communications between the grid components and utilities and end-users; equipment, appliances and devices as required and expected. For example, some of the smart grid applications require real-time low-latency communication capabilities, it is important to consider and take into account the channel propagation delays. Notice, e.g. that interference and noise issues are inherently present in the power line communication (PLC) techniques. The communication channel variation (e.g. highly frequency selective and time invariant) is the key issue which still stands in the way of a complete PLC-based solution. This is because the channel characteristics vary drastically from location to location. The channel modeling techniques for such harsh and noisy environments are still under development, and some encouraging results have been put forward. On the other hand, the robustness and reliability are critical factors of a SG communication infrastructure. Robustness is a system capability to resist to the severe environment in which it must operate. Concerning the operational communications services, different aspects must be considered, such as reliable and stable hardware and software, energy supply autonomy, management of the cybersecurity risks, flexibility or scalability. The critical equipment, modules and subsystem duplication, in certain cases, of the whole platform increase the system availability. Availability is a statistical parameter that must be estimated across the whole chain and must be coordinated among the different system constituents, complementing but not replacing fault tolerance. An essential communication infrastructure attribute is its continuity in the event of electricity supply interruptions for durations ranging from few hours to few days depending on operational constraints. This is often a major drawback for using public communications facilities that usually lacking sufficient power autonomy. The robustness, scalability and flexibility of the communication services depend upon the invulnerability degree of the network infrastructure to security risks, and on the capabilities to adapt, expand and accommodate new equipment, devices, technologies and services. Such measures indented to ensure the confidentiality, integrity, and availability of the data, electronic information communication systems and the control systems necessary for the management, operation and protection of the smart grid's energy, data and information technology and telecommunications infrastructures are critical.

A smart grid uses two-way communication structures, digital technologies, advanced and intelligent sensing, latest computing infrastructure and software abilities in order to provide improved monitoring, protection and optimization of all grid components and elements including generation, transmission, power distribution and consumers. In order to do that the SG communication infrastructure must complain with specific requirements and must possess specific features and capabilities. The smart grid communication systems are critical and very important for data acquisition, data analysis and control of the smart grid's components, units and devices. Typical requirements for the smart grid communication networks include the following features and capabilities. *Reliability* is standing that the communication networks must be able to provide reliable communication links with similar or exceeding the power grids reliability. *Scalability* means that the communication networks are expected to last decades and serve ever increasing in equipment devices or equipment. Protection mechanisms and methods, redundancy and fault tolerance with self-healing capabilities must exist to guarantee a high degree of *availability*. *Real-time monitoring* means that the communication networks must be able to collect, transfer and pre-process in real-time significant amount of data collected form sensors located throughout the grid to the power grid control and operation centers. Smart grid communication networks must guarantee the end-to-end *security,* complete privacy from unauthorized accesses and communication confidentiality, across the networks. The low *latency* requirement of some of the smart grid applications are extremely demanding, beyond the conventional communication applications. *Hard quality of service* (QoS) must be provided for all SG applications with predictable latency and error rate. The smart grid communication networks must also be *cost effective*. Standards on communication networks must enable interoperability, the so-called *standard-based interoperability.* These requirements are general ones for the SG communication networks. Specific requirements vary for various applications in terms of the bandwidth, latency, security and priority. However, the smart grids are including communication networks from diverse technologies with a hierarchical architecture. Notice that the two-way communications allows energy consumers to receive accurate real-time prices, bills and grid status information, while the smart grid operator can receive consumers' real-time information about the amount of the consumed energy. The reliable real-time information flow between all grids' components is essential for smart grid's successful operation.

7.2.3 SMART GRID COMMUNICATION STANDARDS

In general, communication standards follow the seven-layer ISO/OSI (International Organization for Standardization/Open Systems Interconnection) reference model, which are mutually independent, allowing for various layer combinations. The seven layers are as follows: physical layer, data link, network, transport (data transfer control) layer, session (communication control) layer, data presentation layer and application (data models and service) layer. For example, it becomes possible to apply layers with long-term stability (e.g. the application layer covering the data models and the communication services) with layers which change according to the technology progress (e.g. link layer or physical layer). The layer functions are described with the analogy *Letter,* expressing the thoughts and necessary information for the receiver in sentences with a definite syntax and semantic of the chosen language within the application layer. The information is presented in written form, e.g. with black letters on white paper, while the transfer method of transfer needs to be defined at the session layer, e.g. air mail. The transportation layer requests the address of the receiver and the network layers defines which provider is used for the transfer. The link layer is mainly responsible for secure transfer of the information. In the case of a posted letter, this can be done by checking the transfer status online and by mandatory confirmation upon arrival at the addressee. The physical layer defines the physical media of the communication channel, in the case of the airmail letter this is the aircraft and the car. The exact definition of all layers builds the communication protocol. Simple communication protocols may use only the layers 1, 2 and 7. However, they are limited for definite point-to-point connections. The link or physical layers can be changed along the way through a communication network, usually a conversion is requested for the change over into the new protocol.

The addressing scheme of the previous layers (assigning each participant of the communication a definite identification) has to be kept stable to ensure that the message arrives at the right receiver. The application layer is important for the consistent, plausible and definite expression of the information. The standard needs to define the syntax and semantics of the data models because the computer intelligence is not capable of abstractions like the human intelligence. In summary, the transmission system operators in Europe and North America are following the method and transfer their data bases to the common information model (CIM) according to IEC 61970 "Energy management system application" standards.

If the control center is communicating with the entities representing the system components by using different application layers, all of these application layers need to be embedded into the control center computer network. The definition of uniform data models of the application layer is the mandatory pre-requisite of an efficient SG communication infrastructure. For historical reasons, various communication standards are currently used for the electric network operations, e.g. inside the substations for different types of assets (e.g. protection, switchgear, sensing, or meters) and between the substations and control centers. Such practice requires the conversion of data formats between the different system levels, which in turn requires higher technical efforts, being a source of inconsistencies. Furthermore, the detection of inconsistencies in the information exchange causes higher efforts for commissioning tests. Conventional remote control of the power system is structured in accordance with the importance of the system components for supply reliability. The remote control and supervision function based on communication facilities covers the transmission grid, the regional distribution (or sub-transmission) network and the MV busbars in the HV or MV substations. The smart grid challenges and requirements require a paradigm change in the area of supervisory control and data acquisition (SCADA) for the electric networks.

First, the communication has to penetrate the power distribution level down to the LV network users in order to perform the three smart power distribution pillars: distribution automation, intelligent energy management and smart metering. Second, the global standard protocols using uniform data models and services have to be applied to ensure system efficiency, interoperability and "plug and play" capabilities of the intelligent electronic devices (IEDs) from various providers, data consistency and information security at all levels of the electric power system. The transmission system operators often use internal communication networks for their SCADA systems. However, the enhancement of the distribution network operation may use the existing communication infrastructures or SG private infrastructures. The distribution system operators (DSOs) can either establish their own communication channels, e.g. by using the power line carrier (PLC) technologies or through a communication service provider who is able to offer of a separate communication domain with high information security and network performances related SCADA functions. The most efficient communication technologies depend on the local conditions and may be in different physical forms: copper cables, fiber optic cables or radio. The data exchange interoperability over all power system levels of the power system, using uniform data models and services is a pre-requisite and a critical requirement for the successful SG enhancement. The development of the appropriate communication standards began in 1980 and is still an ongoing effort. The development of the standard communication protocols had a deep impact on SCADA system architecture and performances. Both the SCADA technology and the communication standard were developed in a mutual context as described in a later book chapter. However, the development history caused the introduction of a number of proprietary, regional and international standard series which are still applied in the practice of the power system control and operation. In short, the electricity networks, as critical infrastructure systems are taking into account in their operation, management and standards that the remote control and supervision of electric networks are vulnerable to several security threats like: external attacks, internal attacks, natural disasters, equipment failures, carelessness, intentional data manipulation and loss of data. The reactions to these threats can physically damage the electric network assets and have tremendous legal, social and financial consequences. In order to be able to meet the requirements in terms of confidentiality, availability, integrity and non-repudiation standards for information security have to be developed and implemented. Advanced encryption methods and the objectives of the standard IEC 62351 "Power systems management and associated information exchange – Data and communications security" must be applied to ensure the security of the power system control via communication networks.

7.2.3.1   IEC 61850 Standards and Interoperability

IEC 61850 is an international standard defining communication protocols for intelligent electronic devices at electrical substations. It is a part of the International Electrotechnical Commission's (IEC) Technical Committee 57 reference architecture for electric power systems. It is a vendor-neutral, open systems standard for utility communications, set and designed to significantly improving the functionality, yielding to substantial customer savings. The standard specifies protocol-independent and standardized information models for various application domains in combination with abstract communications services, a standardized mapping to the communications protocols, a supporting engineering process, and testing definitions. IEC 61850 standard allows a standardized communication between IEDs located within electric utility facilities (e.g. power plants, substations and feeders), as well as outside of these utility facilities (e.g. wind farms, electric vehicles, PV plants, energy storage systems, monitoring and metering systems). The standard includes and specifies the requirements for database configuration, object definition, file processing and IED self-description methods. These requirements are facilitation the addition of devices to a utility automation system as simple as adding new devices to a computer system by using "plug and play" capabilities. With IEC 61850, utilities are benefiting from cost reductions in the system design, substation wiring, redundant equipment, IED integration, configuration, testing and commissioning. Additional cost savings are also coming from training, management information systems (MIS) operations and system maintenance. IEC 61850 has been identified by the NIST as a cornerstone technology for field device communications and general device object data modeling. The IEC 61850 standard was originally designed and developed for substation communications solutions and implementation. It was not specifically designed to be used over the slower communications links typically used in distributed automation. However, as wide area networks and wireless technologies (e.g. WiMAX) advance, the application of the IEC 61850 communications to devices in the distribution networks becomes possible. It is therefore possible that IEC 61850 eventually to be used in all aspects of the utility infrastructure. The IEC 61850 standard concepts and solutions are based on three cornerstones or capabilities. *Interoperability* represents the ability of IEDs from a single or several manufacturers to exchange information and data, and to use that information for their own functions and operation. *Free configuration*, meaning that standard supports different philosophies, allowing free function allocation, e.g. working well for centralized (RTU based) or decentralized (substation control system based) configurations. *Long-term stability* is stating that this standard is a future-proof entity, meaning that it is able to accommodate new communication technologies and evolving system requirements.

IEC 61850 standard is not prescribing the communication infrastructure, being the user's choice. In fact, the IEC 61850 standard aims, in principle to solve the interoperation between substation devices, which is accepted as an international unified communication standard. It takes advantage of a comprehensive object-oriented data model and the Ethernet technology, bringing in great reduction of the configuration and maintenance cost. A driving force behind the IEC 61850 is the data acquisition standardization and description methods to reduce the integration effort. Vendors can then support fewer standardized and proprietary communications methods and better focus on making the best IEDs possible to serve the electric power system. Network designers find mapping consistently named values into database locations and operator displays much easier than understanding a unique method for each IED. The process of IEC 61850 clients discovering available LANs within locally connected IEC 61850 servers are automated. Consequently, a key component of a communication system is the ability to describe itself from both a data and services (communication functions that IEDs perform) perspective. Other key requirements include high-speed

IED-to-IED communication, networkable throughout the utility enterprise, high-availability, guaranteed delivery times, standards based systems, multi-vendor interoperability, support for voltage and current samples data, support for file transfer protocols (FTP), auto-configurable/configuration support and overall support for the communication security. The major architectural construct that 61IEC 850 adopts is that of "abstracting" the definition of the data items and the services, creating data items, objects and services that are independent of any underlying protocols. The abstract definitions then allow "mapping" of the data objects and services to any other protocol that can meet the data and service requirements. The definition of the abstract services is found in part 7.2 of the IEC 61850 standard and the abstraction of the data objects (referred to as Logical Nodes) is found in part 7.4. In as much as many of the data objects are made up of common pieces (such as: status, control, measurement, and substitution), the concept of "common data classes (CDC)" was developed which defined common building blocks for creating the larger data objects. CDC elements are defined in IEC 61850 – part 7.3.

7.2.3.2   Smart Grid and Wireless Communication Standards

IEEE 802.11 standard refers to the collection of wireless communication technology known as WIFI used for WLANs networks. This technology has proved its success due to its simple access structure based on CASMA/CA and its operation in unlicensed frequency bands (2.4 GHz and 5 GHz). IEEE 802.11 is a standards family; the latest release is the IEEE 802.11n which supports the highest data rates up to 150 Mbps while IEEE 802.11a/g supports maximum 54 Mbps. Other standard like 802.11e appears to be important for the SG applications because its QOS features. The 802.11s standard is allowing multi-hop and mesh networks over physical layer and finally 802.11p standard for wireless networks for vehicle-to-grid (V2G) systems. The IEEE 802.16 standard known as WiMAX supports long distance up to 10 Km broadband with up to 100 Mbps of data rates. The WiMAX system was designed to handle thousands of synchronized users over large distances, being very suitable for smart grid applications. The 802.16j standard is the recent version of WiMAX supporting multicast and broadcast multi hop technique with seamless handover for mobile users it empowers flexible distribution and higher coverage which make it suitable choice for NAN and AMI applications. The WiMAX under development version named 802.16m will provide a greater mobility up to 350 km/h with 100 Mbps data rate, supporting handover with long-term evolution (LTE) and WIFI. The IEEE 802.15.4 is a standard for physical and MAC layers for low-rate wireless personal area networks (LRWPAN), and it offers up to 250 kbps over 10 m. Several network topologies are supported like star, tree or mesh multi-hop. IEEE 802.15.4 is the basis of radios for many other standards for monitoring and control applications; the most important are ISA 100.11a standard, ZigBee standard and wireless HART standard. These standards are replacing the 802.15.4 MAC protocol with TDMA based scheme. ZigBee is widely adopted for WPANs for both commercial and industrial environments, as well as in smart grids. IEEE P2030 is the first IEEE standard providing a roadmap and guidelines for better understanding and defining smart grid interoperability. Hence, IEEE P2030 has defined the SG interoperability reference model to expand the current knowledge base, characteristics and principles for SG interoperability and power grid architectural designs and operations to provide more reliable and flexible power system. IEEE P2030 is generally focusing on three integrated architectural perspectives, e.g., power systems, communication technology and information technology. There exist two additional complementary standards based on the IEEE P2030 standard. IEEE P2030.1 provides guidelines on the knowledge base addressing terminology, methods, equipment and planning requirements for transportation applications, which can be used by utilities, manufacturers, transportation providers, infrastructure developers and end-users of electric-sourced vehicles; IEEE Standard P2030.2 focuses on discrete and hybrid energy storage systems integrated with the electric power infrastructure.

The IEEE 1901 is a standard for high-speed (up to 500 Mbit/s at the physical layer) communication devices via electric power lines, often called broadband over power lines (BPL). The standard uses transmission frequencies below 100 MHz. This standard is usable by all classes of BPL devices, including BPL devices used for the connection to Internet access services as well as BPL devices used within buildings for local area networks, smart energy applications, transportation platforms (vehicle) and other data distribution applications. IEEE 1901 has been widely recognized as the standard that will enable universal communications in smart grid applications. IEC 62351 standard deals with cyber-security issues of the smart grid. The different security objectives include authentication of data transfer through digital signatures, ensuring only authenticated access, prevention of eavesdropping, prevention of playback and intrusion detection. IEC 62351 standard deals with cyber-security issues of the smart grid. The different security objectives include authentication of data transfer through digital signatures, ensuring only authenticated access, prevention of eavesdropping, prevention of playback and intrusion detection. IEC 62056 is a set of standards for electricity metering data exchange. It includes the following standards:

IEC 62056-42: physical layer services and procedures for connection-oriented asynchronous data exchange. IEC 62056-46: data link layer using HDLC protocol. IEC 62056-47: COSEM transport layers for IPv4 networks. IEC 62056-53: COSEM Application layer. IEC 62056-61: object identification system. IEC 62056-62: interface classes. PLC G3 supports high-speed, highly reliable IP-based communications across existing power lines, allowing data and control messages to flow across the generation, transmission and distribution systems that comprise a regional smart grid. It was developed to provide robust connections between smart grid elements to allow the application of advanced billing and demand management techniques to customer loads and to efficiently integrate conventional and renewable-based distributed energy resources, including solar power plants or wind farms. IEC 61970 and IEC 61968 are the reference model standards to carry out the information, diverse legacy devices, and applications integration in the distribution grid and to accomplish successful SG deployment. Another important standard is the IEC 62351 standard covering the security issues related to power system management and associated information and data exchanges.

IEEE 802.11, a family of standards, is governing WLAN and Wi-Fi communication technologies, specifying the physical layer and the MAC (medium access control) layer of the WiFi technologies, with IEEE 802.11b and IEEE 802.11n most popular among these versions. IEEE 802.11b allows working in the ISM band at 2.4 GHz, employing Direct Sequence Spread Spectrum (DSSS) modulation with data rate up to 11 Mbs for indoor environments and up to 1 Mbs for outdoor environments. The range for indoor is 30–40 m, while the outdoor range is 90–100 m. IEEE 9802.11g version allows working at the same 2.4 GHz ISM band frequency by using orthogonal division multiplexing technique with data rates up to 54 Mbs. The IEEE 802.15.4 standard defines MAC and physical layers in low-rate personal area networks (PANs). In 2008, Smart Utility Networks (SUN) Task Group 4g was created to define new physical layers to provide a global standard to facilitate very large-scale process control applications, such as smart grids. WiFi provides robust performances in shared spectrum and noise RF channel environments, supporting all IP-based protocols and a broad range of applications, including Smart Energy Profile 2.0. A wide range of data rates is supported along with point-to-point and point-to-multipoint communications, while security features are also implemented making them a strong candidate for SG communication, being one of the prime choices for HAN, FAN and AMI of smart grids. WiMAX technologies were developed under IEEE-802.16 standards for Wireless Broadband, with IEEE 802.16e the most used. IEEE 802.16 specifies the Physical and MAC layers for WiMAX technologies. WiMAX uses two frequency bands, one for line-of-sight (11–66 GHz) and other for non-line-of-sight operation (2–11 GHz). IEEE 802.15.5 standard for WPAN defines mesh architecture in PAN networks based on the IEEE 802.15.4 standard. Notice also that ZigBee specifies several high-level communication protocols used in low consumption electronics, which are also based on the IEEE 802.15.4 standard.

IEEE 1815-2012 standard, Distributed Network Protocol (DNP3) is a popular communication protocol for SCADA, aiming for minimal bandwidth use, making remote devices intelligent, sharing the best features of the existing protocols, ensuring the compatibility with existing standards and providing a high degree of reliability. NISTR 7628, a three-volume report provides the guidelines for smart grid cyber-security, by helping organization to make effective cyber-security strategies was developed in 2021 by National Institute of Standards. IEEE 1711-2010, known as Trial USE Standard for Cryptographic Protocol for Cyber-security of Substation Serial Links is regulating the serial communication among SCADA substations. The standardized protocol provides integrity and operational confidentiality of serial links connecting substations. Notice that applications or systems are required to tolerate message losses, since serial SCADA protection protocols is designed to discard susceptible messages. IEEE 2010-2011 standards, sponsored by the IEEE Standards Coordinating Committee 21 on Fuel Cells, Photovoltaics, Dispersed Generation and Energy Storage, IEEE STD 2030 Guide for Smart Grid Interoperability of Energy Technology and Information Technology Operation with Electric Power Systems, End-use Applications and Loads has been developed as a guideline for supporting smart grid interoperability. The guideline introduces smart grid as "a complex electric system of systems", listing 12 architectural principles, e.g. openness, extensibility, scalability, interoperability, security and privacy. It is worth to notice that the communication interoperability represents a view of the control processes and data management, emphasizing the interconnectivity and communication among systems, applications and devices.

7.2.4 ADVANCED METERING INFRASTRUCTURE

Advanced metering infrastructure (AMI) is not a single technology; it is rather a configured and complex infrastructure, integrating a number of technologies to achieve its goals. AMI is not limited to electricity distribution, covering also gas and water networks too. However, electric meters are typically fed from the same electric feed that they are monitoring. Smart electric meters also have embedded controllers to manage the metering sensor, a display unit and a communication module which is generally a wireless transceiver. AMI enables a two-way communication, so the communication, issuance of command or price signal from the utility to the meter or load-controlling devices are also possible. This infrastructure includes smart meters, communication networks in different levels of the infrastructure hierarchy, meter data management systems (MDMS) and means to integrate the collected data into software application platforms and interfaces. End-user devices are comprised of state-of-the-art electronic hardware and software capable of data collection or measurement in desired time intervals and time stamping. These devices and equipment have an established communication with remote data center and are capable of transmission of such information to various parties in required time slots set by system administrator. Unlike automatic meter reading (AMR) units, AMI communication is bi-directional and smart devices or load-controlling devices can accept command signals and act accordingly. At the consumer level, a smart meter communicates consumption data to the user and to the service provider. Utility (electricity, gas, water) pricing information supplied by the service provider enables load-controlling devices (e.g. smart thermostats) to regulate consumption based on pre-set user criteria and directives. Where distributed energy resources and/or energy storage units are available, power systems can come up with optimized solutions that each source can share part of demand. From the measurement point of view, smart meters are of three distinct broad categories: electrical, fluid and thermal. There are also a number of sensors including smart sensors that measure atmospheric parameters such as humidity, temperature and light, which are directly influencing to the energy consumption. The sensors could be expanded based on the user needs, desire or system designer, considering cost and functionality. For example, home automation systems deal with the proper selection, placement and utilization of various sensors within the home premises or facilities. Smart meters have two main functions: measurement and communication, and therefore each meter have two subsystems: metrology and communication. The metrology part varies depending on a number of factors including region, measured phenomenon, required accuracy, level of data security and applications. There are also multiple factors, including security and encryption, which define the suitable communication method. There are a number of essential functionalities of smart meters, regardless of the type or quantity of their measurement. The meter sensors must be able to accurately measure the medium quantity, using physical principles, topologies and methods. Control and calibration, although varies with the type, the meter should be able to compensate the small system variations. Meter must communicate, by sending stored data and receiving operational commands, having also the ability to receive firmware upgrades. Versatile power management, in the event of a primary energy source is going down, the system should be able to maintain its functionality. Customers are able to see the meter information since this information is the base for billing by proper display. A display is also needed as energy demand management at customer premise is not possible without the knowledge of the real time consumption. Timing synchronization is critical for reliable transmission of data to central hub or other collector systems for data analysis and billing. Timing synchronization is even more critical in case of wireless communication. Based on previous reasons, key features of smart electricity meters can be summarized as follows: time-based pricing, providing consumption data for consumer and utility, net metering, failure and outage notification, remote command operations, load limiting for demand response purposes and power quality monitoring including: phase, voltage and current, active and reactive power, power factor, energy theft detection, communication with other intelligent devices, improving environmental conditions by reducing emissions through efficient power consumption.

Smart metering is one of the first smart grid applications deployed by utilities, encompassing much more than energy measurement. Smart meters are able to send collected information and data to the analyzing computer and to receive operational commands from operation center. Therefore, the communication is an AMI important part. Considering the number of users and smart meters at each center, a highly reliable communication network is required for transferring such high data volume. Design and selection of an appropriate communication network is a complex process, requiring consideration of such key factors: massive data amount transfer, data access restriction, sensitive data confidentiality, complete information of customer's consumption, showing the grid status, authenticity of data and precision in communication with target device, cost effectiveness, ability to host modern features beyond AMI requirements and supporting future expansion. Various topologies and architectures are used for AMI communication. The most practiced architecture is to collect the data from groups of meters in local data concentrators, and then transmit the data using a backhaul channel to central command where the servers, data storing and processing facilities as well as energy management and billing applications reside. As different types of architectures and networks are available for realization of AMI, there are various mediums and communication technologies for this purpose as well. Examples include: power line carrier (PLC), broadband over power lines (BPL), copper cables or optical fibers, cellular, WiMax, Bluetooth, general packet radio service (GPRS), Internet, satellite communications, peer-to-peer technologies or ZigBee.

Many of the SG applications require frequent power (active and reactive) and power quality (e.g., voltage, frequency) measurements. Such measurements, provided by smart meters, can be required as often as once every 15 minutes or so to support energy management applications, real-time pricing (RTP), time-of-use (TOU) pricing and critical peak pricing (CPP) features for billing and demand response applications. Advanced metering infrastructure (AMI) is an integrated system of smart meters, communications networks and data management systems that enables two-way communication between utilities and customers. It is one of the most important components of smart grid which aggregates data from smart meters (SMs) and sends the collected data to the utility center (UC) to be analyzed and stored. In traditional centralized AMI architecture, there is one-meter data management system to process all gathered information in the UC; therefore, by increasing the number of SMs and their data rates, this architecture is not scalable and able to satisfy SG requirements, e.g., delay and reliability. Since scalability is one of most important characteristics of AMI architecture in SG, the scalability of different AMI architectures have been proposed and studied in last two decades. The system provides a number of important functions that were not previously possible or hard to be performed manually, such as the ability to automatically and remotely measure electricity uses, connect and disconnect service, detect tampering, identify and isolate outages and monitor voltage. Combined with customer technologies, such as in-home displays and programmable communicating thermostats, AMI enables the utilities to offer new time-based rate programs and incentives that encourage customers to reduce peak demand and manage energy consumption and costs. In the context of home and office applications of AMI, the utilities should have four tiers: (i) the backbone, which is the path to utility data center; (ii) the backhaul, which is the aggregation point for neighborhood data; (iii) the access point, which is most likely the smart meter; and (iv) the home area network, building or facility area network. Important components, supporting AMI applications, as well as other major smart grid applications may include the followings. Control centers are responsible for supervising the overall smart grid operations, automating the data collection process from smart meters, evaluating the data quality, generating edits where errors and gaps exist and transmitting the price information or demand response commands. Base station communicates wirelessly with smart meters and other field devices, connected by using fiber optic to connect directly with the control center. Data concentrators are a combination of software and hardware unit that collects information and data from smart meters and forwards the data and information to the utility. Data concentrators are very often used in densely populated areas. Field devices are devices that allow remote control from a central location to accomplish selected smart grid applications, such as distribution automation. Example field devices include remotely controllable voltage regulators, capacitor banks, switches, protection equipment, etc. A smart meter is an electronic meter, with some processing power capabilities that can be used to record consumption of electric power/energy and transfer (transmit) the consumption data and information to a utility (control and management centers). It can also be used to receive control commands, price signals and other information from the utility. In many cases, these devices incorporated short-range radio transmitters which could be read by a utility employee driving nearby with proper equipment. Some more advanced applications used low bandwidth data transmission over the power lines to send this data back to an aggregation point. As with most projects, cost is normally a major factor and smart meter installations are no exception. To minimize costs, they usually do not incorporate anti-tampering devices such as pressure sensors to alert the system that a problem exists or a system breach may have occurred. In addition, due to the priority of maintaining low cost devices, the processing power is comparatively small. This restricts the complexity of the security algorithms that can be used by such devices. With smart meters sending data to the utilities automatically, there are no needs to have the meter mounted outside the customer premises. Placing the meters inside the building provide a much more protected location and aid in the security of the smart grid. This would require moving or extended the power line terminus from their normal location to the interior which may add considerable expense and most likely be prohibitive for many extensive smart grid projects.

The HAN, BAN or FAN networks are envisioned to connect the smart meters, smart appliances, smart equipment or generic smart loads and electric vehicles and electricity generators and storage units. The idea is to incorporate data communication for IHDs and load controls for automated energy management during peak hours. Normally, each device in the HAN, BAN or FAN transfers the data indicating its instantaneous electricity use; therefore, communications needs can be considered as modest. However, any communication technology selected for this application should be scalable and flexible to meet the requirements of large home, office buildings or a facility. Reliable and standard communication systems are critical features of the smart grid. Smart meters and advanced metering infrastructure must have the capability to communicate to control and distribution operators the power information, data for reporting and analysis. AMI can be leveraged to provide consumers with historical energy consumption data, comparisons of energy use in similar households, dynamic pricing information and suggested approaches to reducing peak load via in-home displays. Notice that AMI networks, however, require a significant investment to build out fully, and are not required to enable most consumer-facing applications. The communication networks which transfer this massive amount of information and data to the concentrator and database management require a high degree of operational reliability. The communication network also serves as the control or information link to the operational command center to different end devices for control or customer services display and alarm. These end devices can be control switches for capacitor banks or load control devices. Hence, the communication system, by implication, is a two-way system and practically serves as a system neural network. The measured data are used for conventional customer billing or new customer billing strategies such as time of use rates, pre-pay metering, etc. Customer access to information about how one uses the electricity imposes new requirements on how the metering data is stored, its format and accessibility. To handle demand response, energy consumption profiles must be maintained for energy demand reduction and conservation. At AMI level, devices within the premises of the house communicate with each other as well as the utility network through smart meters and could be called in-home network. At upper layer, HAN communicates with the utility provider, forming another network that could be called utility network. HANs connect smart meters, smart devices within the home premises, energy storage and generation (solar, wind, etc.), electric vehicles as well as in-home display and controllers together. Since their data flow is instantaneous rather than continuous, HANs require that bandwidth varies from 10 to 100 Kbps for each device, depending on the task. The network however, should be expandable as the number of devices or data rate may increase to cover office buildings or large houses. The calculated reliability and accepted delay are based on the consideration that the loads and usage are not critical. Given these requirements and considering the short distances among nodes that enable low power transmission, wireless technologies are the dominant solutions for HANs. These technologies include 2.4 GHz WiFi, 802.11 wireless networking protocol, ZigBee and HomePlug. ZigBee is based on the wireless IEEE 802.15.4 standard and is technologically similar to Bluetooth.

Reliable and standard communication is an important feature of smart grids or smart metering or advanced metering technologies. Among others, the smart grid communication infrastructure for smart grid has to meet requirements for time synchronization, reliability, latency, criticality of data delivery and support for multicast. Furthermore, a major issue in networking communications in smart grid is interoperability, as discussed previously. Standardization of smart grid communication has received significant attention. Several organizations that are working on this include IEEE, International Electrotechnical Commission and the National Institute of Standards and Technology. Several relevant standards from these organizations focusing on the smart grid communication are issued. The IEEE defined standards include: IEEE C37.1 standard, describing SG specific requirements of SCADA and automation systems; IEEE 1379 standard concentrates on the communications, interoperations of IEDs and remote terminal units in substations operating into the smart grids; the IEEE 1547 standard specifies the requirements of electric interconnection of distributed energy resources with the power grid; while the IEEE 1646 standard on communication delivery time for substations. The IEC defined standards include: IEC 60870 standard is regulating the communication systems for power system control and specifies requirements for power system interoperability and performance; IEC 61850 standard, discussed previously defines automated control related to substation management; IEC 61968 and IEC 61970 standards are dealing on model for data exchange between devices and networks); and IEC 62351 refers on the cyber-security issues of the IEC protocols. NIST published standards include NIST 1108 standard describes, among others, smart grid interoperability and requirement of communication networks; and the NIST 7628 standard is describing the smart grid information security issues.

The AMI system must help the utilities to manage their resource and business process in the most efficient and secure way. AMI system need to support the following minimal set of functionalities and features: (a) remote meter data reading at configurable intervals; (b) time of day metering; (c) pre-paid functionality; (d) net metering or billing; (e) alarm and event detection, notification and reporting; (f) remote load limiter and connection/disconnection at defined/on demand conditions; (g) remote firmware upgrade; (h) integration with other existing systems like billing and collection software, GIS mapping, consumer indexing, new connections and disconnection, analysis software, outage management system, etc. (i) import of legacy data from existing modules where possible; and (j) security features to prevent unauthorized access to the AMI including smart meter and meter data etc. and to ensure authentication of all AMI elements by third party. This is only an indicative and not exhaustive AMI list of functionalities. The system should be capable to support the other functionalities, upgrading, changes and additional features as per the requirements of the utilities in seamless way. AMI and smart metering system should accurately maintain system time synchronization across all devices to ensure accuracy of data. The system must support the interfacing with the current and future smart grid functionalities and devices like outage management system, distribution automation, self-healing system, distribution transformer monitoring units, electric vehicle, distributed energy resources, smart microgrids and DC nanogrids, smart monitoring and sensing, smart homes, appliances, installations and equipment, etc. The communication network must preferably be able to support multiple and concurrent applications. AMI is based on smart meters, which have two functions – measurement and communication. For quantitative measurements, the smart meter must be able to accurately measure the used energy or other quantity of the medium using physical principles, topologies and methods. Smart meters consist of control and calibration facilities and features. Customers must be able to see the meter information and data since this information is the base for utility billings. A display is needed as demand management at customer side. Timing synchronization is critical for reliable data transmission to central hub or other collector systems for data analysis, reporting and billing. All needed data and variables can be measured and then can be collected by storing in various storage units. After these data and information which stored is analyzed, the variables and instant conditions can be used via display screens. Automatic smart reading (AMR) systems are separated from AMI, operating with less developed technologies such as turn off appliances and equipment manually, collecting data without real time, etc. the advanced metering infrastructures have real time operations, customer/pricing or utility options, remote control and management units, etc.

Moreover, AMI became an important contributor to outage management, service restoration and voltage monitoring for many SGIG projects, particularly those that implemented AMI systems alongside investments in distribution automation technologies. AMI enables utilities to isolate outages faster and dispatch repair crews more precisely, reducing outage duration, limiting the inconveniences and reducing labor hours and truck rolls for outage diagnosis and service restoration. Utilities facing regular, severe weather events and storm-induced outages have greater incentives for using AMI for outage management than those that do not. AMI data integration with other information and management systems, including outage management systems (OMS) and geographic information systems (GIS), enabled utilities to create detailed outage maps, and in some cases posted these maps on utility websites to keep the public informed on service restoration progress. Voltage monitoring provides another promising benefit stream to include in business case analysis of AMI investments. Utilities can use AMI voltage monitoring capabilities to enhance the effectiveness of automated controls for voltage and reactive power management, particularly for conservation voltage reduction (CVR) programs. AMI and customer systems provided utilities with new capabilities to offer time-based rate, incentives and direct load control (DLC) programs. This enabled utilities to reduce peak demand, lower wholesale power purchase costs, sell excess electricity to regional markets and defer investments in new generation and delivery capacity. Notice that AMI system implementation costs and benefits varied widely across the projects, for a variety of reasons, such deployment and technology costs, investment priorities, adequate communication infrastructure, etc.

7.2.5 AUTOMATED DEMAND RESPONSE

One major smart grid component is the possibility of customer participation in the overall energy grid management, performed via the notion of demand response or demand-side management, in which the utilities and energy companies are providing incentives for customers to efficiently use the energy or shift their load over time. In the past, utilities were vertically integrated, in full control of the energy supply, while maintain a careful equilibrium between energy (power) demand and electricity supplied through the grid. Operating philosophy was and is still to use large and centralized generation to reduce the cost and supply all power demanded, regardless on the demand level and time. However, the large generation cost, centralized power stations made the operation at less than the full capacity uneconomical, and the idea to control the energy demand rather the supply, since shifting the demand can be a cheaper alternative can be a better solution. Moreover, with the electricity market deregulation, different organizations are managing different components of power generation and transport. Each of them can offer their product at the market price, adding a strong motivation to explore ways of influencing the power demand through price motivation. A common exemplification is to reduce the peak power demands by increasing prices when demand is high in attempt to level the power demand. In the smart grid context, the customers are provided with partial autonomy to participate in managing, buying and selling energy from and to the grid. Thus, in any smart grid mechanism, it is imperative to factor in demand response models and their associated challenges. Demand response activity or application is an action taken to reduce electricity demand in response to price, monetary incentives, or utility directives so as to maintain reliable electric service or avoid high electricity prices. It can be viewed as an additional variable for smart grid control and management. Demand responsive loads can be adjusted, interrupted or shifted out to provide and overall load reduction during utility grid peak energy demand periods. Introduction of smart sensors, smart metering in distribution networks and the two-way communication connecting control centers and end-users are one of the drivers for incorporating demand response into smart grid paradigm. Demand-side management techniques are expected to be a major driver into the development, realization, and operation of the smart grids. Enabling interconnection and interactions of the consumers, electric vehicles, distributed generation and energy storage, microgrids and utilities can only be and it is made possible with efficient demand-side management, intelligent sensing, monitoring and two-way communications. Due to the complex interactions between customers and the power companies, as well as the need for pricing schemes, demand-side management has often been researched and studied using several tools from game theory, optimization, decision-making and microeconomics.

Demand response (DR) activity represents temporary changes in the electricity consumption in response to supply conditions, electricity price structure, changes and/or other events in the grid and energy infrastructure to get the maximum benefits for the end-users. DR includes quite broadly any intentionally designed attempt to change the consumer power uses in terms of time, instantaneous demand or total power consumption and uses. There are three common used approaches to doing this. First, the consumers can be motivated to use less energy only during critical periods of power grid stress. Second, the consumers can respond continuously to market price fluctuations and changes. Third, consumers that have generation and energy storage capabilities are encouraged to use them to reduce the grid power demands. Generally speaking, demand response refers to a set of actions initiated by the utility, end-users or both to lower energy consumption level of individual consumers, thereby lowering the total power system demand. These actions often are taking place during the grid peak energy consumption periods, when the utility gets closer to its generation capacity, trying to avoid generation shortages and/or load shedding. Like many smart grid aspects, DR has received considerable research as a solution to reduce costs and environmental negative impacts, while improving grid reliability operation. Moreover, the inclusion of new energy sources and energy storage elements combined with the need to reduce peak loads and conserve energy has driven the introduction of distributed automated demand response (ADR) applications. ADR applications, e.g. can be used to reduce the amount of energy consumed by appliances, installations and equipment during peak power periods. While demand response has been used by utilities over the years through scheduled load shedding and manually managed consumption reduction with a few large consumers, ADR is much wider in scope, bringing dynamic load management directly to residential and low-power consumers and users. ADR often works in concert with distributed energy resources closer to the point of consumption or other energy sources and energy storage connected into the grid. Thus, in some cases, ADR may not necessarily reduce overall energy consumption but only reallocate and transfer the source of some of the consumed energy to DER. Such load shifting is resulting in reducing emissions if DER units are renewable energy resources. Notice that the use of the energy storage units as a key and cortical element in demand-side management. By employing energy storage, there are possibilities of selling energy stored at the customer premises to the grid and/or to other customers, when making economic sense. Moreover, there are several challenges related to the demand-side management. One of the key challenges of designing demand-side management models refer to the need for modeling customer behavior. Other challenges that must be overcome before fully deploying and employing demand response models include modeling customer participation, developing decision-theoretic tools, optimizing pricing, incorporating time-varying dynamics (e.g., fluctuating demand) and accounting for power grid constraints. All of these issues motivate the need for advanced decision-making tools, such as game theory, optimization or stochastic control to properly model and analyze arising demand response situations. It can be expected that demand response is an important stepping-stone towards practical deployments and the operation of the smart grid.

Historically, the demand response fundaments originate from supply and demand economic analysis of the demand and supply profiles or curves. These profiles (curves) are assumed to be independent and are the price per any unit quantity, assuming the quantity is available for sale on the supply curve and the given quantity is purchased by consumers on the demand curve (profile). With the perfect competition assumption, the supply is determined by the marginal cost, e.g. a supplier will continue to produce additional units as long as the unit cost is less than the selling price (in order to make profit). The demand curve represents the product amount that a consumer will purchase taking into account the product utility and the alternatives the consumer could make with regard to satisfying the utility by other means. This is the so-called opportunity costs, and as long as the marginal utility to the consumer to purchasing more of the product is greater than opportunity cost, the consumer will keep purchasing. The point where the curves intersect is the equilibrium point, where the cost per unit satisfies both demand and supply. The DR algorithms are trying to find in optimum ways such points or areas. For economics and power operation points of views, the demand response is one of the most crucial part of the smart grid, According to the US Department of Energy, DR is "a tariff or a program established to motivate changes in electric use by end-user customers, in response to changes in price of electricity over time of high market prices or when grid operation is jeopardized". From the DR definition and its origin, it is clear the DR is beneficial not only to the end-users but also to the utilities. Traditional DR methods, limited by the lack of proper communication, do not have the ability to make real-time decisions, making it difficult to improve grid energy efficiency. However, smart grids have the ability to interact in real-time with customers, the advanced communication and management, helping the users and utilities to use energy efficiently, keeping the demand response of electricity in check, reducing power system stresses, as well as the harmful emissions produced during electricity production. A typical DR action consist of a specific desired schedule (start and end times), target power demand reduction level, list of involved end-users and other attributes, depending on the DR programs selected (utility and/or consumer initiated programs).

In the case of utility-initiated demand response programs, the utility initiates the action, informing the end-users of the DR event attributes, e.g. demand reduction (duration, amount, etc.) signals, updated electricity rates, to which consumers react adjusting consumption level. The former leads to involuntary demand reduction or incentive-based DR, while the latter encourage users to voluntarily reduce their demands to maximize their savings. The most common utility-initiated DR programs are *direct load control* (DLC), *interruptible/curtailable load* (I/CL) and *emergency demand reduction*. In the first program, the utility sends a DR command to the controllable loads to adjust the energy usage. Such type of DR is more suitable to residential users, and one of the main driver is the home automation (HA) systems. In an enhanced version of this program, the utilities are able to shift the operation of certain appliances and home equipment to non-peak demand hours, program referred as DR shifting. In I/CL program the signal sent to the consumers is a request not a command to reduce the demand at a certain level. Compliance is not compulsory, but depending on the contract provisions may incur penalties if the request is rejected. Such programs are more suitable for large-scale consumers (industrial and large commercial facilities). The third one is similar to the previous class of programs, with the exception that it is mainly used during emergency conditions, i.e. very short advance notice as a step before load shedding. Under such programs, large-scale consumers provide ancillary services to the utilities, behaving as virtual spinning reserves that can reduce the energy demand upon request. The voluntarily demand reduction (price responsive DR) is initiated by the utility, by issuing updated electricity prices, but is executed by costumers as part of a local demand-side management (DMS), being more in the form of demand shifting from peak to non-peak demand hours. Consumers may subscribe to different time-based rate structures, e.g. time of use, critical peak pricing or real-time pricing and/or by adjusting their load patterns or increasing price responsiveness, while extended implementation of time-based rates can reduce the frequency and severity of price spikes and reserve short-age and the needs for incentive-based DR programs. Costumer-initiated programs are proactive strategies, in which the customers bid to reduce their energy demands in exchange for financial incentives. The bids along attributes, e.g. demand reduction available, duration, starting time, are sent to the utilities, where they are analyzed and shortlisted in agreement with the utility technical and financial criteria. Technically, the demand response engines (subsystems) may be implemented in DMS environment. The DR engine would receive load forecasts, weather forecasts, load flow and smart meter data from external subsystems and grid modules and generate the DR commands/signals to be transmitted to different DR resources and subscribers.

For demand response the use of advanced communication technologies is essential and critical. DR engines are transmitting signals/commands to customers via communication media, which can be fiber optics, wireless communication networks, power lines or even telephone lines. The DR signals consist of updated electricity rates, demand reduction requests, or turn ON/OFF control commands. The old style traditional demand response managers, at the customer sites perform their action manually or by semi-automated processes, while the modern energy management subsystems are fully automated within the home, building or facility automation systems able to regulate the energy uses for equipment, appliances and installations. In customer-initiated and in voluntary DR programs, in the framework of the smart grids the energy use is reported back to the utility via advanced metering infrastructure, the customer being billed accordingly with the actions taken and the agreement provisions. Notice that the incentive-based reduction programs require faster two-way communication networks since the utility needs to validate the response to the individual customers to the DR event (action) in order to determine whether the supply shortage is met or additional actions are required, e.g. emergency DR or load shedding. The estimation can be done by measuring pre- and post-event energy consumption levels. Regardless of the communication technology used in any DR program, it must be dependable, reliable, secure and robust. Proper selection of the communication media and protocols adopted are able to solve these potential issues. Higher success rates of DR programs are beneficial the utilities by avoiding the use of alternative options, such as emergency demand reduction and/or load shedding. To achieve this objective, the utilities must perform comprehensive statistical analysis on the available consumers' historical data, patterns and load profiles, inferring information and prediction of the consumer behavior and the associated attributes.

7.3 SMART GRID COMMUNICATIONS TECHNOLOGIES

This subsection briefly highlights the key smart grid communication infrastructure motivations, as well as the SG features, requiring advanced communication and information SG infrastructure. The reasons and motivations are related to power system's operation and environment aspects in the emerging smart grid paradigm through communication infrastructures. The smart grid communication networks use a broad range of communication technologies from wired, wireless and hybrid networks. Conventional electric grid has already embedded communication networks, supporting its operation between generation units, substations and control centers. However, this communication networks is expensive, uncompromising and insufficient, covering mostly generation and transmission segments. In the smart grid approach, all segments and entities are envisioned to be covered by the communication infrastructure, with a special focus on power distribution and end-users. Therefore, a new adequate and extended communication infrastructure is needed to support and backup the smart grid application, subsystems and components. The smart grid communication is likely to be hybrid, composed form various networking technologies, topologies and protocols. Moreover, in order to monitor, control and have bi-directional data flows between subsystems, components, devices and smart grid utilities, integrated communication networks must cover all smart grid entities and domains. It is also very important in smart grid communication to select the most effective communication topology and structure in agreement with specific requirements and local conditions. Another important aspect is that the smart grids will have huge number of sensors, actuators and relays than the legacy of conventional power grid, deployed at all levels of grid components (power plants, substation equipment, generators, transformers or end-users) collecting and transmitting huge amounts of data and information with control centers and other smart grid entities. Furthermore, smart grid communication infrastructure must have wide bandwidth in order to ensure high rates of information flow and has to be self-healing and automatically adaptive to changes.

To develop and implement a smart grid communication system, keys are to understand its requirements, classified into four categories: *data transmission, cyber-security, data privacy* and *interoperability*. Data transmission requirements of the communication systems primarily include transmission rate, latency and reliability. Transmission rate determines the volume of data that can be sent to a destination through the communication network within a certain time. Latency refers to the time needed for a data piece to reach correctly a specific destination. Reliability measures how likely a certain data piece is received correctly. The smart grid functions have very distinct requirements on data transmission and the requirements for each of these smart grid functions must be discussed separately. Cyber-security requirement specifies that each domain is composed of a certain number of actors or agents which may be subsystems, applications, devices or other participants in smart grid. Each actor may exchange data with other actors within the same domain or in other domains. In such a large and highly interconnected complex system, cyber-security is a critical issue for the reliability of the whole power system, especially when the public communication network (e.g. Internet) is adopted. Compared with physical attacks, cyber-attacks, which are not constrained by distance, are generally less risky, cheaper, and much easier to coordinate and replicate. With a little basic knowledge about the structure and operation of the network, adversaries are able to launch various attacks wherever they are through a set of interconnected computers or even just smart phones. Notice that beside such security issues, virus, worms and other malwares raise other cyber-security issues in smart grid. In order to mitigate these concerns, cyber-security in smart grid aims at maintaining availability, integrity and confidentiality of the entire system. It is worth noting that, different from an information and communication system, a power system is concerned more about availability than integrity and confidentiality. However, with the increasingly complex interactions among different components in smart grid, massive sensitive data are produced and propagated. Confidentiality is becoming an increasingly important issue in the development of smart grids.

To achieve higher intelligence and automation in a smart grid system, smart meters are required to provide some fine-grained details in the user power consumption data, within a much shorter time interval (tens of minutes) than before. This transformation from aggregate data to granular data brings in the privacy issues. Using some of the newest monitoring technologies like non-intrusive load monitoring, daily activities of end-users are recorded. Due to the interoperability requirement of smart grid, many parties and actors besides the utilities have access to such records, raising data privacy questions. More seriously, it is not difficult for data analysts to derive further information that may invade user privacy. An electricity consumption of a residence may show whether it is occupied or not. The daily usage of electricity may reflect when the homeowner wakes up, leaves for work and so on. Also, the charging data of a PEV may be used to track the traveling schedule of an individual and his/her location. Many third-party companies may be interested in these sensitive private data to exploit their commercial benefits. This is undoubtedly giving rise to the privacy concerns during various data-processing phases, like monitoring, collection, aggregation and analysis. For data privacy, some relevant laws and regulations have to be enacted first, showing very clear who owns and controls the data, who has the right to access the user data, how do utilities or other parties share the user data without compromising privacy and how to manage the huge amount of sensitive data. Secondly, information and communication technologies are potential alternatives to cope with the data privacy issues. As a very complex and diverse system of systems, the smart grid has to ensure interoperability among the power systems, information systems and communication systems. A good design of a particular system without proper consideration of other subsystems may not achieve the best result for the whole power system. Smart grid interoperability allows utilities, consumers and other stakeholders to communicate securely and effectively, through the variety of different information systems over geographically dispersed regions.

Moreover, a smart grid communication infrastructure needs the scalability of accommodating more and more devices and services into it and more end-user interaction real-time monitoring of energy meters. According to the Electric Power Research Institute (EPRI), one of the emergent requirements facing the smart grid development is related to cyber-security of the involved systems and structures. As indicated in the EPRI reports, cyber-security is a critical issue due to the increasing potential of cyber-attacks and incidents against this critical sector as it becomes more and more interconnected. Cyber-security must address not only deliberate attacks, such as from disgruntled employees, industrial espionage and terrorists, but also inadvertent compromises of the information infrastructure due to user errors, equipment failures and natural disasters. Vulnerabilities might allow an attacker to penetrate a network, gain access to control software and alter load conditions to destabilize the grid in unpredictable ways. To achieve the characteristics of the desired SG features, there are several key technology areas that must be developed and implemented. Some of these smart grid features are discussed here:

- *Integrated communications:* High-speed, fully integrated, two-way communication technologies will make the modern grid a dynamic, interactive "mega infrastructure" for real-time information and power exchange. Open architecture will create a plug-and-play environment that allows the networks' grid components to talk, listen and interact securely.
- *Sensing, measurement and monitoring*: These attributes are enhancing power system measurements and monitoring capabilities, enabling the transformation of data into information. They evaluate the health of equipment and the grid integrity, while supporting advanced protective relaying, are eliminating meter estimates and preventing energy theft. They enable demand response, helping to relieve congestion.
- *Smart components*: Advanced and smart components are playing active roles into the smart grid behavior. The present and next generations are using the latest research in materials, solid-sate devices, control, computing, energy storage and power electronics, with higher power densities, greater reliability and power quality, enhanced efficiency bring environmental benefits and improved real-time diagnostics.
- *Advanced control methods*: Adaptive and intelligent methods are used to monitor essential components, enabling a rapid diagnosis, timely and appropriate response to any event, while they are also supporting market pricing, enhancing asset management and operation efficiency.
- *Improved interfaces and decision support*: In many situations, the available time for grid operators to make decisions is only seconds. The smart grid requires wide, seamless, real-time use of applications and tools, enabling grid operators and managers to make decisions quickly. Decision support methods and approaches with improved interfaces will enhance human decision-making ability at all levels of the grid.

The major attributes and capabilities of a smart grid, as summarized in the literature are:

1. Use of digital information and controls;
2. Dynamic optimization of grid operations and research;
3. Development and integration of distributed energy resources, especially renewable sources;
4. Development and use of demand response;
5. Deployment of "smart" technologies for metering, communications and automation;
6. Integration of "smart" appliances and consumer devices;
7. Use of peak-shaving technologies, including advanced storage technologies;
8. Providing consumers with timely information on pricing for control of their energy intake;
9. Development of standards for communication and interconnection of "smart" devices; and
10. Identifying and lowering potential barriers to adoption of the smart grid.

Secondly, the attributes proposed by the Electric Power Research Institute (EPRI) IntelliGrid vision focuses on communications and computer control. This vision aims for an automated, reliable, intelligent, system, integrated with the existing power grid. The major attributes addressed by EPRI are as follows:

1. Reliability: Smart grid is self-healing, adaptive, automatically re-routing power flows around faults.
2. Control: Digital switching enables precise control of power flows, eliminating congestion, loop flows and bottlenecks that impede long-distance wholesale transactions.
3. Advanced Customer Services: Two-way customer communications enables real-time pricing, net metering, demand response, active participation, premium service and more.

Smart grid or intelligent electricity grid visions are focusing on the advanced two-way communications, intelligence or smart sensing, measurement and monitoring, data manipulation and processing aspects, while leaving out a few details which must be included in the overall ideas and its definitions. The smart gird sensing, measurements, monitoring and control functions and interactions are depicted in Figure 7.5. A smart grid implementation is employing a variety of communication, monitoring and sensing methods and technologies, many of which have multiple diverse applications. On the other hand, the information and the data of the grid needs and availability, and extended sensing, monitoring and measurement data are assisting in power balancing, peak shaving and better services by avoiding undesired power injections and performing load shaving during peak hours. For such reasons, the advanced SG communication is being created, in order to facilitate information and data exchange. However, because the smart grid is a very complex network with nonlinearity, randomness, bi-directional power flow and bi-directional communication, despite of the technologies of smart devices, sensors and communication protocols, supervising the

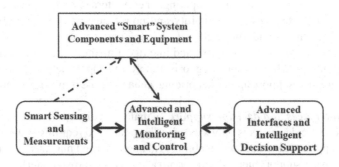

FIGURE 7.5   Smart grid key sensing, monitoring and communication architecture.

status of the whole system, and dealing with the large-scale real time data, remain an open problem. Major smart grid attributes are self-healing, high reliability, power quality, resistance to cyber-attacks, accommodating to a wide variety of DG and storage options, optimizing asset utilization while minimizing operations and maintenance expenses. These functions are substantiated by the addition of resilience to cyber-attacks. The following SG characteristics, found in the literature summarized are:

1. Enables active consumer participation: Consumer choices and increased grid interactions bring tangible benefits to both the grid and the environment, while reducing the cost of delivered electricity.
2. Accommodates all generation and energy storage options: Diverse energy resources with plug-and-play connections will extend generation and energy storage options, including opportunities for more efficient, resilient and cleaner power production.
3. Enables new products, services and markets: Grid open-access market reveals waste and inefficiency, helps drive them out, offering new consumer choices such as green energy products or new generation of electric vehicles, while reduced transmission congestion leads to more efficient electricity markets.
4. Provides power quality for the digital economy: Digital-grade power quality for those who need it avoids production and productivity losses, especially in digital-device environments.
5. Optimizes asset utilization and operates efficiently: Desired functionality at minimum cost guides operations and allows full uses of assets. More targeted and efficient grid-maintenance programs result in fewer equipment failures and safer operations.
6. Anticipates and responds to system disturbances (self-heals): Smart grid performs continuous self-assessments to detect, analyze, respond to and as needed, restore grid components or network sections.
7. Operates resiliently against attack and natural disaster: Smart grid deters or withstands physical or cyber' attack and improves public safety. After investigating these definitions, characteristics and visions of the smart grid there is obvious overlap and missing elements.

In order to create a solid, comprehensive, detailed SG definition, it was proposed to combine them into one list encompassing all as follows:

1. Use of digital information and controls (intelligence);
2. Dynamic optimization of grid operations and research (efficiency in controls);
3. Development and integration of distributed energy resources, especially renewable sources (Efficiency in generation and use of renewable resources);
4. Development and use of demand response (optimization in costs);
5. Deployment of "smart" technologies for metering, communications and automation;
6. Integration of "smart" appliances and consumer devices (intelligence, control and cost);
7. Use of peak-shaving technologies, including advanced storage technologies (efficiency and cost);
8. Providing consumers with timely information (Intelligence to provide customer options);
9. Development of standards for communication, and interconnection of "smart" devices;
10. Identifying and lowering potential barriers to the SG adoption (marketing, research and feasibility);
11. Creating a secure network for information and control (Intelligence, cyber-security and control);
12. Develop a self-healing framework which can withstand, both the system disturbances or major disruptions (physical security and adaptive);
13. Self-restorative after major blackouts (intelligence and adaptive); and
14. Integration of electric vehicles (energy storage).

Therefore, smart grid refers to use of intelligence and secure communications to efficiently control, manage and deliver reliable, cost-effective energy from a variety of sources, including renewable energy, to customers, giving all parties more choice to control and optimize utilization. However, the proliferation of wireless/wired sensors and communication devices and the emergence of embedded computing represent an opportunity to develop applications for connected environments in general, and especially management systems, addressing SG communication infrastructure challenges. These include the deployment of large-scale embedded computing, power grid legacy, smart appliances and next-generation communications that will provide the foundation for a post-carbon society. In this section, the context that gives these challenges urgency as well as the technical challenges that need to be addressed by smart grid communication infrastructure are discussed. Communication is the key component of the smart grid infrastructure. With the integration of advanced technologies and applications for achieving a smarter electricity grid infrastructure, a huge amount of data from different applications will be generated for further analysis, control and real-time pricing methods. Hence, it is very critical for electric utilities to define the communications requirements and find the best communications infrastructure to handle the output data and deliver a reliable, secure and cost-effective service throughout the total system. Electric utilities attempt to get customer's attention to participate in the smart grid system in order to improve services and efficiency. Demand-side management and customer participation for efficient electricity usage are well understood. Furthermore, the outages after disasters in existing power structure also focus the attention on the importance of the relationship between electric grids and communication systems.

In a complex smart grid system, through wide deployment of new smart grid components and the convergence of existing information and control technologies applied in the legacy power grid, it can offer sustainable operations to both utilities and customers. It can also enhance the efficiency of legacy power generation, transmission and distribution systems and penetrate the usage of clean renewable energy by introducing modern communication systems into smart grids. The smart grid cornerstone is the ability for multiple entities (e.g. intelligent devices, dedicated software, processes, control center, etc.) to interact via a complex communication infrastructure. The development of a reliable and pervasive communication infrastructure represents crucial issues in both smart grid structure and operation. A strategic requirement of this process is the development of a high quality and reliability communication for establishing robust real-time data and information transfer through WANs to the distribution feeder and customer level. Various communications technologies supported by two main physical media, i.e., wired and wireless networks are used for data transmission between smart meters, sensors and electric utilities. In some instances, wireless communications have some advantages over wired technologies, such as low-cost infrastructure and ease of connection to difficult or unreachable areas. However, the nature of the transmission path may cause the signal to attenuate and interference. On the other hand, wired solutions do not have significant interference issues and their functions are not dependent on batteries, as wireless solutions often do. Basically, two types of information infrastructure are needed for information flow in a smart grid. First information flow stream is from sensors and electrical appliances to smart meters, the second is between smart meters and the utility data centers. Usually, first data flow can be accomplished through power-line communication or wireless communications, such as ZigBee, 6LowPAN, Z-wave, and others. For the second information flow, cellular technologies or the Internet can be used. However, there are key limiting factors that should be taken into account in the smart metering deployment process, such as time of deployment, operational costs, the availability of the technology and rural/urban or indoor/outdoor environment, etc. The technology choice that fits one environment may not be suitable for the other. In the following, some of the smart grid communications technologies along with their advantages and disadvantages are briefly explained, in the subsequent subsections of the entree.

Different communications technologies supported by two main communications media, i.e., wired and wireless, can be used for data transmission between smart meters and electric utilities.

In some instances, wireless communications have some advantages over wired technologies, such as low-cost infrastructure and ease of connection to difficult or unreachable areas. However, the nature of the transmission path may cause the signal to attenuate. On the other hand, wired solutions do not have interference problems and their functions are not dependent on batteries, as wireless solutions often do. Basically, two types of information infrastructure are needed for information flow in a smart grid system. The first information flow stream is from sensor and electrical appliances to smart meters; the second is between smart meters and the utility's data centers. Usually, he first data flow can be accomplished through power-line communication or wireless communications, such as ZigBee, 6LowPAN, Z-wave and others. For the second information flow, cellular technologies or the Internet can be used. However, there are key limiting factors that should be taken into account in the smart metering deployment process, such as time of deployment, operational costs, the availability of the technology and rural/urban or indoor/outdoor environment, etc. The technology choice that fits one environment may not be suitable for the other. In the following, some of the smart grid communications technologies along with their advantages and disadvantages are briefly explained, in the subsequent chapter subsections. Potential communication media for power distribution system networking include power line carrier (PLC), wireless and dedicated wired. When the substation is considered due to the confined physical space, a dedicated wired medium such as Ethernet is the best choice. Substation communication networks use the well-established IEC 61850 and thus this paper will not discuss the requirements for substation level/distributed source level communication and medium. When feeders are considered, PLC is well-suited because it is a medium that is available throughout the distribution system. PLC has potential to transmit data at a maximum rate of 11 kbs; when the PLC has sufficient robustness and reliability, this maximum data rate can be achieved only in a narrow frequency range of 9-95 kHz. Wireless communication is another promising alternative for distribution level communication. One of the important characteristics of wireless communication is the feasibility of communication without a physical connection between two nodes, ensuring the continued communication even with a few distribution poles down. In other words, redundant paths for communication are possible without additional cost. Another wireless communication advantage is that the utility has to own only the terminal units, which are relatively cheap and could be integrated with cost-effective local processors. When multi-hopping is used in wireless communication, especially in WiFi and ZigBee, the range of communication can be extended and the nodes located in the feeder could be able to communicate with the control center. Disadvantages of wireless communication would be interference in the presence of buildings and trees which could result in multi-path; this can be avoided with improved receivers and directional antennas, which will increase the cost. Another major concern with wireless medium is easy accessibility, which could result in security issues. This can be avoided by using secure protocols. Rural feeder sections would be long and range of communication could become a concern; however, directional antennas could mitigate this issue. However, both PLC and wireless communication are promising in the power distribution level communication

Many applications such as smart grid energy metering have emerged from a decade of research in wireless sensor networks. However, the lack of IP-based network architecture precluded sensor networks from interoperating with the Internet, limiting their real-world impact. The IETF chartered the 6LoWPAN and RoLL working groups to specify standards at various layers of the protocol stack with the goal of connecting low-power devices to the Internet. Several standards are proposed by the working groups, emphasizing the research community contributions, actively participating in this process impacting design and implementations. The new communication infrastructures evolve toward ubiquitous data transport networks able to handle power delivery applications along with vast data amounts coming from the smart grid applications. These networks are scalable in order to support the present and the future set of functions characterizing the emerging smart grid communication technologies, and highly pervasive in order to support the deployment of last-mile communications (i.e. from a backbone to the terminal customers locations). In the rest of this section, we discuss several key factors for smart grid systems including power line communications, distributed energy resources, smart metering and monitoring and controlling. Conventional power system communication infrastructure typically consists of SCADA systems with dedicated communication channels to and from the System Control Centers and a wide area networks. Some long-established power utilities may have private telephone networks and other legacy communication systems. The SCADA connect all the major power system operational facilities and main elements, such as central generating stations, transmission grid substations and the primary power distribution substations to the System Control Centre. The WAN is used for corporate business and market operations. These form the core communication networks of the conventional power systems. However, in the Smart Grid, these two communication infrastructure elements are merged into a utility WAN subsystem. An essential SG development is to extend communication throughout the power distribution systems by establishing two-way communications with customers through neighborhood area networks covering the areas served by distribution substations. Moreover, in order to monitor, control and have bi-directional data flows between end devices and smart grid utilities, a highly reliable, integrated communication network should cover all the SG domains. NAN networks are installed at customer premises, offering access to in-home smart devices and appliances. IED send data readings over HAN to AMI applications through the home smart meter or threw the residential gateway. The HAN give also to home automation networks different services like home motoring and control, demand response applications allowing efficient power management and user comfort. The interfaces of the home and neighborhood area networks are usually set through smart meters and/or smart interfacing devices. The various SG communication sub-networks are employing different technologies and a key challenge is to integrate them effectively. The most suitable communication HAN network technology is the wireless technology since its ease of implementation for a large number of nodes, simple configuration and cost effectiveness. Every single in-home appliance generates particular data flow and may have specific communication requirement, but in general in-home wireless solution should be realized with multi-path environment due to surface reflection and interference with other intelligent devices at home.

A NAN network is a distribution domain network; it can be considered as a mesh of smart meters, connecting the AMI applications access point to smart meters in customer domain and various gateways in the distribution domain. The main purpose of this network is data collection from smart meter for monitoring and control. It covers long distances, and areas up to 10 square miles and the data rate is around 1000 kbps. Both wired and wireless communication technologies could be appropriate for NAN networks. However wireless communication technologies such as networks WiMAX, LTE, 3G and 4G are today candidates, while wired technologies such as PLC and Ethernet could be right solutions for NAN networks too. Often, NANs are deployed and operated within area of hundreds of meters which is actually few urban buildings. Several local HANs can be connected to one central NAN and they transmit data of energy consumed by each house to the NAN network. The NAN network delivers this data to Local Data Centers for storage and pre-processing. This data storage is very important for charging the consumers and data analysis for energy generation-demand pattern recognition. Field area network (FAN) is the communication network for power distribution domain in the smart grid, the electrical power control centers and application use FAN networks to collect data, monitor and control different applications in distribution domain such as IED devices, PHEV charging stations, AMI applications in NAN networks and WSNs networks in feeders and transformers. A WAN is deployed and operated within a vast area of tens of kilometers, and it consists of several NANs and local data collectors (LDCs). WAN enables communications systems between smart grid and core utility system. It is composed of two types of networks: backhaul and core network. The core network offers the connectivity between substations and utility systems, while backhaul network connect the NAN networks to the core network, this network is extended over thousands of square miles and data rates reaches hundreds of Mbps or even higher. A variety of technologies such as WiMAX, 4G and PLC could be used in WAN networks.

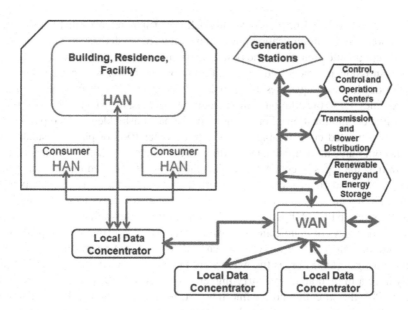

FIGURE 7.6 Typical smart grid communications infrastructure.

Also, virtual technologies like IP/MPLS could be used for the core network. Moreover, the two-way communication of all smart grid components including operator control and operation centers, generation station and renewable energy generation units, energy storage systems, transmission and power distribution is based on WAN. The WAN has very high transmission data rate up to a few Gbps. The WAN can be implemented by Ethernet networks, WiMAX, 3/4/5G/LTE and microwave transmission.

In summary, the smart grid communications infrastructure is based on three main types of networks as previously mentioned: HAN, NAN and WAN structure, interconnected, as shown in Figure 7.6, by the schematic diagram of the smart grid communication infrastructure based on these three communication networks. In a typical implementation, a HAN consists of a broadband Internet connection, shared between several users through a wired or often through wireless modems, enabling the communication and sharing of resources between computers, smart meter and sensing, mobile devices over a network connection. In SG implementation, all smart home devices, equipment and meters are HAN connected. The acquired data is transmitted through HAN to the smart meters, allowing efficient home energy management. A HAN network can be implemented by using ZigBee, Ethernet or other technologies.

7.3.1 Wireline (Wired) Communication Technologies

7.3.1.1 Power Line Communications

One of the earliest initiatives for the automation of the electricity grids was by using the power line communication (PLC) technology. PLC technology involves the introduction of a modulated carrier signal over the existing power transmission line (cable) infrastructure for two-way communication. PLC is classified into two major categories: narrow-band PLC and broad-band PLC. Key PLC applications involve automation of the medium-voltage grids, transmission and distribution substations. Power line communications or power line carriers are using the power feeder line as physical communication media, offering the possibility of sending data simultaneously with electricity over the same medium and can be considered an open wire communication system method. PLCs use different frequency bands, and a line matching unit to inject signals into a high voltage transmission or distribution lines, and the injected communication signal is prevented from spreading to other parts of the power network by line traps (LC filters). It is worth mentioning that in later 20th-century decades, common wired communication physical media were twisted cables and coaxial cables for low-range data communication. Since the power distribution networks were originally intended for AC power transmission at typical frequencies of 50 or 60 Hz, power wire circuits have only a limited ability to carry higher frequencies. It is problematic to establish high-frequency communication through power lines due to the dilution of high-frequency signals. Propagation issue is a limiting factor for power line communications. For example, in the past unshielded twisted pair (UTP) cables, consisting of two twisted copper cables, each with an outer PVC or plastic insulator, were used extensively in communication circuits. Depending on the data rate (which is influenced by the cable material as well as the types of connectors), UTP cables are categorized into several categories. In coaxial cables a shielded copper wire is the communication medium. The outer coaxial conductor provides effective shielding from external interference and also reduces losses due to radiation and skin effects. Bit rates up to 10 Mbps are possible over several meters. First-generation ripple control systems provide one-way communications, in which centralized load control and peak shaving have been performed for many years. PLC systems generally operate by transmitting a modulated carrier signal on the wiring system. But, as the wires were intended to deliver AC power, the power wire circuits only have a limited ability to carry higher frequencies. As a special case, broadband over power lines (BPL), the so-called power line Internet, is an application of PLC technology which provides broadband Internet access through ordinary power lines. A computer needs to plug a BPL modem into an outlet in an equipped building to have high-speed Internet access. However, there are frequency ranges and bands restrictions for PLC, as ones set by the EU standards, restricting the use of frequencies between 3 kHz and 95 kHz for two-way communications for electricity distributor use. Several second-generation PLC systems with low data rates have been proposed since the later 1990s. The automatic metering systems have been deployed based on this technology. Third-generation systems based on OFDM with higher data rates are currently deployed for smart grids, distribution automation and advanced metering management.

Example 7.3   If a PLC system uses a carrier frequency of 120 kHz and if value of the capacitance in the line trap is 10 nF, what is the required inductance?

SOLUTION

A line trap is a LC filter with the resonance frequency is given by:

$$\omega = 2\pi f = \frac{1}{\sqrt{LC}} \tag{7.8}$$

Solving for inductance yields to:

$$L = \frac{1}{(2\pi f)^2 C} = \frac{1}{(2\pi 120 \times 10^3)^2 10 \times 10^{-9}} = 0.000176 \, H \, or \, 0.176 \, mH$$

With the smart grid development, the PLC systems on transmission and distribution networks have become one of the main technologies to exchange the information between the end-users and the utilities. In order to provide communication services with different priorities under the smart grid environment, it is a must design that a PLC system with variable data rates supported, which means understanding of the PLC physical channel characteristics become vital. Notice that it is almost impossible to use the frequency range from 10 to 20 MHz for the reliable communications from distribution transformer to end-user, without the aids such as the repeater and the modulation schemes. Beside the fact that feeder cables are not designed for data transmission, they are also prone to be interfered by the inverters. Therefore, PLC modems for domestic applications may not be suitable. Limitations and difficulties that obstruct transmission are not common. There is a possibility of communicating in such an environment and discusses the possible solutions such as the use of a pulse width modulation filter to overcome those limitations. The recent PLC research focused on high-data-rate applications, such as Internet access or multimedia communication serving a relatively small number of users. However, it lacks the PLC considerations for sensing, control and automation in large systems comprising hundreds of components spread over wide areas. PLC is still a promising technology for smart grid applications because of the existing infrastructure decreases the installation cost of the communications infrastructure. The standardization efforts on PLC networks, the cost-effective, ubiquitous nature and widely available infrastructure of PLC can be the reasons for its strength and popularity. Data transmissions are broadcast in nature for PLC; hence, the security aspects are critical. Confidentiality, authentication, integrity and user intervention are some of the critical issues in smart grid communications. HAN application is one of the biggest applications for PLC technology. Moreover, PLC technology can be well suited to urban areas for smart grid applications, such as smart metering, monitoring and control applications, since the PLC infrastructure is already covering the areas that are in the range of the service territory of utility companies. However, there are some technical challenges due to the nature of the power line networks. The power line transmission medium is a harsh and noisy environment that makes the channel difficult to be modeled. The low-bandwidth characteristic (20 kbs for neighborhood area networks) restricts the PLC for applications that need higher bandwidth. Furthermore, the network topology, the number and type of the devices connected to the power lines, wiring distance between transmitter and receiver, all adversely affect the quality of signal that is transmitted over the power lines. The sensitivity of PLC to disturbances and dependency on the quality of signal are the disadvantages that make PLC technology not suited for data transmission. However, there have been some hybrid solutions in which PLC is combined with other technologies, i.e. GPRS or GSM, to provide full-connectivity not possible by PLC technology alone.

7.3.1.2   Digital Subscriber Lines

Digital Subscriber Lines (DSLs) is a high-speed, high-band digital data transmission technology that uses the telephone network for data and video transfer. Its main advantages consist of its simple utilization in the smart grid context since electric utilities can make immediate advantage of them without any extra cost for additional deployment, the infrastructure being there. Coexistence is possible as telephony signals are using the frequency range 300–3500 Hz, and DSL operates around 25–30 kHz, the exact operating frequency depending on the type of DSL system. Splitter, low-pass filters are used on cable to avoid audible DSL interference. There are number of DSL alternatives like ADSL for Asymmetric DSL that supports up to 8 Mbps for downstream and 640 Kbps for upstream, the ADSL 2+ with up to 24 Mbps and 1 Mbps for downstream and upstream, respectively. And VDSL (for Very high bit DSL) providing up to 52 Mbps for downstream and 16 Mbps for upstream but only for short distances. It is common to see frequencies greater than 1 MHz through an ADSL enabled telephone line. Hence, many companies chose DSL technology for their smart grid projects. Several smart metering projects have been carried out using DSL technologies. Usually, a communication box is installed at the customer premises and consumption data is transmitted over DSL architecture. There are several DSL architectures between the service provider and end-user premises, consisting of a mixture communication media, e.g. optical fibers, phone lines, etc. However, the throughput of the DSL connection depends on how far away the subscriber is from the serving telephone exchange and this makes it difficult to characterize the performance of DSL technology. The widespread availability, low-cost and high bandwidth data transmissions are the most important reasons for making the DSL technology the first communications candidate for electricity suppliers in implementing the smart grid concept with smart metering and data transmission smart grid applications. *However, there are some disadvantages and issues of the DSL technologies.* The reliability and potential down-time of DSL communication networks may not be acceptable for smart grid mission critical applications. Distance dependence and lack of standardization may cause additional problems in smart grid applications. The wired DSL-based communications systems require communications cables to be installed and regularly maintained, and thus cannot be implemented in rural areas due to the high cost of installing fixed infrastructure for low-density areas. Notice that the data rates up to 200 Mbs of the modern DSL are more than sufficient to cover many of the smart grid needs and applications.

7.3.1.3 Optical Fiber Communication

Optical fiber communication has been widely used to connect substations to operation and control centers in the backbone networks, thanks to its multiple advantages and characteristics, such as robustness against electromagnetic interferences, broadband or data rates, making it a suitable choice for high voltage environs, and its capacity to transmit over large distances with very high bandwidth. Optical fiber transmission is used both inside substations and for long-distance data transmission in wide grid areas. Optical fibers are often embedded in the stranded conductors of the shield (ground) wires of the overhead power lines. These cables are known as optical ground wires (OPGW). An OPGW cable contains a tubular structure with several optical fibers in it, surrounded by layers of steel and aluminum wire. Optical fibers may be wrapped around the phase conductors or sometimes a standalone cable, when an all-dielectric self-supporting (ADSS) cable, is used. An optical fiber consists of three layers: core, cladding and buffer layer. The thin glass fiber center, where the light travels is the core; the outer optical layer surrounding the core, reflecting the light back into the core is called the cladding. In order to protect the optical surface from moisture and damage, it is coated with a buffer coating layer. Compared to other communication media, optical fiber cables have a large bandwidth, less susceptible to signal degradation than copper wire and weight less than a copper cable. Unlike electrical signals in cables, light signals from one fiber do not interfere with those of others in the same cable, being also immune to external electromagnetic interference (EMI). This is important in many power system applications since data transmission through the electrically hostile environments is usually required. The main disadvantages of optical fiber communication include the cost, the special termination requirements and its mechanical vulnerability, being more fragile than cables. A light signal from the optical source is first incident on surface and then is refracted inside the core by surface between core and cladding, and the subsequent light beam path, according to laws of optics depends on refractive indexes and the incident angle, θ_1, as:

$$\sin(\theta_a) = n_1 \cos(\theta_1) \tag{7.9a}$$

And

$$n_1 \sin(\theta_1) > n_2 \tag{7.9b}$$

After some algebraic manipulation, the following relationship is obtained:

$$\theta_a < \arcsin\left(\sqrt{n_1^2 - n_2^2}\right) \tag{7.10}$$

Here, θ_a is the acceptance angle, n_1 and n_2 are the refractive indices of core and cladding, while for air the refractive angle is 1. The right-side term in previous equation is referred to as the numerical aperture of the optical fiber. Depending on the core diameter, there may be multiple transmission paths or a single transmission path within the core of a fiber. Optical fiber cables having core diameters of 50–400 μm reflect light entering the core from different angles, establishing multiple propagation paths. On the other hand, fiber with a smaller core diameter, 5–10 μm, supports a single transmission path, being single mode fibers, having advantages such as low dispersion, low noise and can carry signals at much higher speeds than multimode fibers, being preferred for long-distance applications. Optical fiber communication methods are playing a crucial role in smart grid infrastructure. For example, it is very suitable the use of technologies like optical power ground wire (OPGW) in the power distribution and transmission lines, in smart grid context since the combination of grounding and optical communications allow long distance transmissions with high data rates. Another application of optical fiber technology is to provide services to customer domain with the use of passive optical networks (PON) since they use only splitters to collect optical signals and do not require switching equipment. Ethernet PON is also interesting grid operators and seems to be suitable technology for smart grid access segment meanwhile its enable using interoperable IP-based Ethernet protocols over optical networks technology.

To conclude, wired technologies, such as DSL, PLC, optical fiber, are costly for wide area deployments, but they have the ability to increase the communications capacity, reliability and security. On the other hand, wireless technologies can reduce the installation costs, but provide constrained bandwidth and security options.

7.3.2 SMART GRID WIRELESS COMMUNICATION TECHNOLOGIES

7.3.2.1 Radio Communication, Microwave Radio and Wireless Technologies

Power plants, substations and other major components of the power systems are often widely distributed and far from the control center. For such long distances, the use of cables or fiber optics is quite expensive. Radio links provide an alternative for communication between the Control Centre and substations. Even though radio communication is not provide the bandwidth offered by wired technologies, the reliability, performance and costs of radio networks have improved considerably over the past decades, making it an attractive option for smart grids. Radio communication may be multipoint or point-to-point, operating typically either at ultra-high (UHF) frequencies, having range 300 MHz to 3 GHz, or microwave frequencies ranging from 3 to 30 GHz. UHF radio represents an attractive choice for smart grid applications where the required bandwidth is relatively low and where the communication end-points are widespread over harsh terrain. Unlike microwave radio, UHF does not require a line of sight between the Source and Destination. The maximum distance between source and destination depends on the size of the antennae and is likely to be about 10–30 km with a bandwidth up to 192 kbps.

Microwave radio operates at frequencies above 3 GHz, offering high channel capacities and transmission data rates. Microwave radio is commonly used in long-distance communication systems, with parabolic antennas are mounted on masts and towers at the source transmitters are sending a beam to another antenna situated at the destination, tens of kilometers or miles away. Microwave radio offers capacity ranging from a few Mbps to hundreds of Mbps. However, the capacity of transmission over a microwave radio is proportional to the frequency used; thus, the higher the frequency, the bigger the transmission capacity but the shorter the transmission distance. Microwave radio requires a line of sight between the source and destination antennas; hence, high masts are required. In case of long-distance communications, the installation of tall radio masts will be the major cost of microwave radio installation and operation.

The 4G standard or LTE (Long Term Evolution) is a wireless communication standard providing an enhancements of the deployed today introducing capabilities, e.g. bandwidth, easing handover between different networks and advanced networking proficiencies. LTE has multiple advantages that make it a good candidate for NAN or LAN networks such as end-to-end quality of services, peak upload rates near 75 Mbs, and download rates reaching 300 Mbs. The LTE technology is implemented into smart grids in two approaches. The first one is the most simple for immediate implementation, efficient and cost effective solution, consisting of carrying the data over the actual mobile network architecture of MNOs (mobile network operators) with piggy-backing technique from smart grid end devices in the HANs over the NAN network to the WAN and eventually to the utility. The second approach consists of the use of a special network architecture for data transfer, in a similar way to the MVNO (mobile virtual network operator) method. Specifically, through the rental of a part of the MNO core network by the utility, or by the implementation of the core network architecture by the utility using the LTE technologies like the MNOs but totally disconnected from the MNO core network. The cost effectiveness security and the simple implementation make LTE a good choice as a NAN communication technique.

Cognitive radio (CR) technology is a standalone radio based on IEEE 802.22 standard. It is a key technology for optimizing the spectrum underutilization due to the spectrum increasing demands by advancement of wireless communication technologies. CR networks enable to the secondary users the spectrum access, when the primary licensed user does not use it efficiently, without causing any interference with primary users. This spectrum sensing technique could be widely deployed in SG WAN, backhaul and distributions networks over large geographic areas. The CR technique consists of opportunistic access to unused spectrum, and it is believed that this technique will have a great future for SG applications, since it delivers a high-performance, high-speed data transmission, scalability, robustness, reliability, sustainability and fault tolerant broadband access.

7.3.2.2 Cellular Networks

Cellular networks are largely deployed in most countries and have well-established infrastructure. Moreover, they allow high data rate communications of up to 100 Mbps. There are several existing technologies for cellular communication such as GSM, GPRS, 2G, 3G, 4G and WiMAX. The WiMAX technology is the most interesting for smart grid implementation. It is working on 2.5 and 3.5 GHz frequencies, with data exchange rate of 70 Mbs and coverage up to 50 km. The WiMAX chips are integrated inside the smart meters that are deployed through the smart grid. Major advantages of the cellular networks are already existing infrastructure with wide area of deployment, high rates of data transfer, available security algorithms that are already implemented in the cellular communication. The cellular technologies main advantages over wireless technologies is the larger coverage area, that why the utilities have used them especially in AMR systems and SCAD; however, the high cost of this technology with problems such latency if a large number of users are served by the same base station has to be solved. The major disadvantage is that cellular networks are shared with other users and are not fully dedicated to the smart grid communications. This can be a serious problem in case of emergency state of the grid. Therefore, the cellular networks can be used for communication between different smart grid components and devices. Existing cellular networks can be a good option for communicating between smart meters and the utility and between far nodes. The existing communications infrastructure avoids utilities from spending operational costs and additional time for building a dedicated communications infrastructure. The advantages of the cellular networks, as mentioned, are already existing infrastructure with wide area of deployment, high rates of data transfer, available security algorithms that are already implemented in the cellular communication. The major disadvantage is that cellular networks are shared with other users and are not fully dedicated to the smart grid communications, a serious problem in case of grid emergency. Cellular technologies are sustained significant advancements in the recent decades with the development of 3G or 4G standards such high packet access standard (HSPA+) providing data rates up to 168 Mbps in the downlink and 22 Mbps in the uplink.

7.3.2.3 WiMAX Technology

Worldwide interoperability for microwave access (WiMax) is the trade name for a broadband wireless access medium defined under IEEE 802.16. Initially, the standards were limited to fixed, line-of-sight transmission but have since also incorporated standards for mobile, non-line-of-sight access. While the range for WiMAX transmission can be up to 30 miles or 50 km, three to five miles is more typical in practice. This does provide significantly greater range than WiFi, making it a more useful medium for this application. In addition WiMAX uses the 2.5 GHz spectrum making it less prone to interference from other sources and wireless devices. The latest standards allow for data rates up to 1 Gbps although this bandwidth must be divided between the users. WiMAX service is provided by a base station which then allows connections by subscriber stations which would be thing smart meters in our application. This is essentially the same as cellular phone service and WiMAX is one of the technologies being used for 4G implementations by the cellular providers. For security, WiMAX utilizes the strong advanced encryption standards (AES) encryption method and also includes key management and authentication. The first version of its key management protocol was subject to attacks but this has since been strengthened. It is assumed that any new devices for the smart grid would include the latest protection. Like any wireless medium, WiMAX is susceptible to denial-of-service attacks by jamming and scrambling attacks where the jammer only transmits during certain periods to disrupt control information. Other attacks have been proposed involving a weakness in the initial network entry and key exchange. By eavesdropping at the proper time, one could gain enough information to implement a men-in-the-middle attack, thereby compromising the security of all traffic routed through the node. An intruder may also be able to gain enough access to the system to transmit various control messages which could effectively overload the system. However, since the devices tied to the smart grid are not mobile and constantly changing WiMAX base stations (and therefore exchanging keys and performing authentication routinely), there should be little opportunity to exploit these vulnerabilities in this application.

Cellular networks are largely deployed in most countries and have well-established infrastructure, a major advantage, while allowing high data rate communications of up to 100 Mbps. Therefore, the cellular networks can be used for communication between different smart grid components and devices. Cellular network solutions also enable smart metering deployments over a wide area. All the 2G, 2.5G, 3G, 4G, WiMAX and LTE are the cellular communication technologies available to utilities for smart metering deployments. The WiMAX technology is the most interesting for smart grid implementation. It is working on 2.5 and 3.5 GHz frequencies, with data exchange rate of 70 Mbs and coverage up to 50 km. The WiMAX chips are integrated inside the smart meters that are deployed through the smart grid. When a data transfer interval between the meter and the utility of typically 15 min is used, a huge amount of data is generated and a high data rate connection is required to transfer the data to the utility. For example, if T-Mobile's Global System for Mobile Communications (GSM) network is chosen for the deployment of Echelon's Networked Energy Services (NES) system, an embedded T-Mobile SIM within a cellular radio module is integrated into Echelon's smart meters to enable the communication between the smart meters and the backhaul utility. Since T-Mobile's GSM network can handle all communication requirements of the smart metering network, there is no need for an investment of a new dedicated communications network. Code-division multiple-access (CDMA), wideband code-division multiple-access (WCDMA) and universal mobile telecommunications system (UMTS) wireless technologies are also used in smart grid projects. Widespread and cost-effective benefits make cellular communication one of the leading communications technologies in the market. Due to data gathering at smaller intervals, a huge amount of data will be generated and the cellular networks will provide sufficient bandwidth for such applications. When security comes into discussion, cellular networks are ready to secure the data transmissions with strong security controls. To manage healthy communications with smart meters in rural or urban areas, the wide area deployment capability of smart grid becomes a key component and since the cellular networks coverage has reached almost 100%. In addition, GSM technology performs up to 14.4 kb/s, GPRS performs up to 170 kb/s and they both support AMI, Demand Response, and HAN applications. Anonymity, authentication, signaling protection and user data protection security services are the security strengths of GSM technology. Lower cost, better coverage, lower maintenance costs and fast installation features highlight why cellular networks can be the best candidate as a smart grid communications technology for the applications, such as demand response management, advanced metering infrastructures, HAN, outage management, etc. Some of the power grid mission-critical applications need continuous availability of communications. However, the services of cellular networks are shared by customer market and this may result in network congestion or decrease in network performance in emergency situations. Hence, these considerations can drive utilities to build their own private communications network. In abnormal situations, such as a windstorm, cellular network providers may not provide guarantee service. Compared to public networks, private networks may handle these kinds of situations better due to the usage of a variety of technologies and spectrum bands.

7.3.2.4 ZigBee

ZigBee is based on an IEEE 802.15 standard. ZigBee is used in applications that require a low data rate, long battery life, low cost and secure networking. Applications include wireless light switches, electrical meters with in-home-displays, traffic management systems and other consumer and industrial equipment that require short-range wireless transfer of data at relatively low rates. ZigBee allows connection of up to 60,000 devices to its network. ZigBee has a defined rate between 20 to 250 kbs, best suited for periodic, intermittent data or single signal transmission from sensors or input devices. The technology defined by the ZigBee specifications is intended wireless communication technology that is relatively low power, data rate, complexity and deployment cost. It is an ideal technology for smart lightning, intelligent energy monitoring, home automation and automatic meter reading, etc. The technology defined by the ZigBee specifications is intended to be simpler and less expensive than other wireless personal area networks (WPANs), such as Bluetooth or WiFi. ZigBee networks are secured by 128-bit symmetric encryption keys. There is a "ZigBee Smart Energy" application that allows integration of smart meters into the ZigBee network together with other devices. By using this application, smart meters can collect information from the integrated devices and control them. Moreover, the consumers can view their energy consumption in real-time. It also allows better energy consumption and real-time dynamic pricing. The advantages of ZigBee application in smart grid are low price, small size and it uses relatively small bandwidth. The disadvantages of the ZigBee are small battery that limits its lifetime, small memory, limited data rate and low processing capability. Moreover, its operation in unlicensed frequency of 868 MHz and 2.4 GHz may have interference with other WiFi, Bluetooth and microwave signals. ZigBee and ZigBee Smart Energy Profile (SEP) have been realized as the most suitable communication standards for smart grid residential network domain by the National Institute for Standards and Technology (NIST). The communication between smart meters, intelligent home appliances and in home displays is very important. Many AMI vendors, such as Itron, Elster and Landis Gyr, prefer smart meters that the ZigBee protocol can be integrated into AMI system. ZigBee integrated smart meters can communicate with the ZigBee integrated devices and control them. ZigBee SEP provides utilities to send messages to the homeowners, and home owners can reach the information of their real-time energy consumption.

ZigBee has 16 channels in the 2.4 GHz band, each with 5 MHz of bandwidth; 0 dBm (1 mW) is the maximum output power of the radios with a transmission range between up to 100 m with a 250 kb/s data rate and specific modulation. ZigBee is a good option for metering and energy management, ideal for smart grid implementations along with its simplicity, mobility, robustness, low bandwidth requirements, low cost, its operation within an unlicensed spectrum, easy implementation, being a standardized protocol based on the IEEE 802.15.4 standard. ZigBee SEP has some advantages for gas, water and electricity utilities, such as load control and reduction, demand response, real-time pricing, real-time system monitoring and advanced metering support. There are some ZigBee technology constraints, e.g. low processing capabilities, small memory size, small delay requirements and being subject to interference with other appliances, which share the same transmission medium, license-free industrial, scientific and medical (ISM) frequency band ranging from IEEE 802.11 wireless local area networks (WLANs), WiFi, Bluetooth and Microwave. Hence, the concerns about ZigBee robustness under noise conditions increase the possibility of corrupting the entire communication channel due to the interference in the ZigBee vicinity. Interference detection and avoidance schemes and energy-efficient routing protocols should be implemented to extend the lifetime and provide a reliable and energy-efficient network performance. Similar to ZigBee, *Bluetooth* is the trade name for a wireless personal area network (WPAN). Although this technology holds a commanding position in the cell phone accessory industry, it has not made a large impact in the HAN/smart home market. However, the industry consortium behind the technology is promoting itself for smart grid applications. Unlike ZigBee, Bluetooth is not a mesh network, but it is able to transmit at greater speeds (~1 Mbps) while using only slightly more power with a range up to 100 m. For one aspect of security, Bluetooth utilizes frequency hopping during communication but there are devices available on the market, which can match the changes, therefore eavesdrop on any transmissions. One of Bluetooth features is the ease of "pairing" two devices to communicate and exchange data. This can be accomplished by placing the device in a discoverable mode. However, if let in such a state, the device may be open to attacks. Another possible avenue of attack is that Bluetooth addresses are not encrypted during transmission, even when in a secure mode that encrypts the rest of the message. Like other wireless technologies, Bluetooth uses the 2.4 GHz spectrum so interference is possible with Wi-Fi and ZigBee. Without a mesh network, Bluetooth is more limited than ZigBee in the total coverage area. It is also at a disadvantage since many smart meters are already designed with embedded ZigBee chips. However, it does have the advantage that most computers and smart phones can communicate via Bluetooth, so it can be integrated directly with the controls of the smart home devices.

7.3.2.5 Wireless Mesh

A mesh network is a flexible network consisting of a group of nodes, where new nodes can join the group and each node can act as an independent router. The self-healing characteristic of the network enables the communication signals to find another route via the active nodes, if any node should drop out of the network. Especially, in North America, RF mesh-based systems are very popular. In PG&E's SmartMeter system, every smart device is equipped with a radio module and each of them routes the metering data through nearby meters. Each meter acts as a signal repeater until the collected data reaches the electric network access point. Then, collected data is transferred to the utility via a communication network. A private company, SkyPilot Networks, uses mesh networking for smart grid applications due to the redundancy and high availability features of mesh technology. Mesh networking is a cost effective solution with dynamic self-organization, self-healing, self-configuration, high scalability services, which provide many advantages, such as improving the network performance, balancing the load on the network, extending the network coverage range. Good coverage can be provided in urban and suburban areas with the ability of multi-hop routing. Also, the nature of a mesh network allows meters to act as signal repeaters and adding more repeaters to the network can extend the coverage and capacity of the network. Advanced metering infrastructures and home energy management are some of the applications that wireless mesh technology can be used for. Network capacity, fading and interference can be counted as the major challenges of wireless mesh networking systems. In urban areas, mesh networks have been faced with a coverage challenge since the meter density cannot provide complete coverage of the communications network. Providing the balance between reliable and flexible routing, a sufficient number of smart nodes, taking into account node cost, are very critical for mesh networks. Furthermore, a third-party company is required to manage the network, and since the metering information passes through every access point, some encryption techniques are applied to the data for security purposes. In addition, while data packets travel around many neighbors, there can be loop problems causing additional overheads in the communications channel that would result in a reduction of the available bandwidth.

7.3.2.6 Satellite Communication

Satellites have been used for many years for general telecommunication industry, being also adopted for the SCADA systems. Satellite communication is still the only technology to deliver global coverage and is playing significant roles in smart grids. For example, the information and data from weather satellites are used to predict wind farms and solar parks electricity production, while satellite remote sensing data, GPS or satellite image services are used to provide information on the transmission line status, fault localization, grid management and maintenance. A satellite communication network is considered as part of microwave networks with a satellite acting as a repeater. The nature and characteristics of the satellite communications are determined by the satellite orbit. Many of the communication satellites that are in operation are placed in geostationary orbit (GEO) technology. A GEO satellite or GEOS is usually at 35,786 km above the equator and its revolution around the Earth is synchronized with the Earth's rotation. The high altitude of a GEO satellite allows communication coverage of approximately one-third of the Earth's surface. Even though GEO satellite-based communication offers technical advantages in long-distance communication, it still is presenting a few drawbacks, such as the challenge of transmitting and detecting the signal over the long distance between the satellite and the user, and the large distance traveled by the signal from the source to destination, resulting in an end-to-end delay or propagation delay of about 250 ms. Low earth orbiting (LEO) satellites are positioned from 200 to 3000 km above the Earth, which reduces the propagation delay considerably. In addition to the low delay, the proximity of the satellite to the Earth makes the signal easily detectable even in bad weather. LEO satellite-based communication technology offers a set of intrinsic advantages such as: rapid connection for packet data, asynchronous dial-up data availability, reliable network services and relatively reduced overall infrastructure support requirements when compared to GEO. In addition, LEO satellite-based communication channels can support protocols such as TCP/IP since they support packet-oriented communication with relatively low latency. However, closer proximity to the Earth a LEO satellite can be seen only within a radius of 1000–2000 km around the sub-satellite point, and a high-density ground stations (inter-satellite links) are required to establish a bi-directional communication path.

7.3.2.7   Wireless Local Area Networks

A wireless local area network (WLAN) connects several devices in a geographical area, such as homes, office buildings or research facilities, with distances in the order of hundreds of meters, by using spread-spectrum or orthogonal frequency division multiplexing (OFDM), usually providing a connection through an access point to the wider Internet. This gives users the ability to move around within a local coverage area and still be connected to the network. Most modern WLANs are based on IEEE 802.11 standards, marketed under the WiFi brand name. WLANs have become popular in the home due to ease of installation, and in commercial complexes offering wireless access to their customers. This standard supports tow network configurations. In first configuration, one device act as *access point* (AP), and the rest of devices are connected to AP, the so-called *infrastructure*. The AP is connecting the WLAN to other networks via wired or wireless connections. The other alternative is the so-called *ad-hoc*, in which there is no AP and the devices are connecting directly among them. WLAN could be easily integrated into smart grid due to its vast deployment around the world. WLAN works in 2.4–3.5 GHz frequencies. The advantages of WLAN are low cost, technology maturity, vast deployment around the world and across each region, capabilities of plug and play devices. The major disadvantage of WLAN is high potential for interference with other devices that communicate on the same frequencies.

Dash 7 is a technology for wireless sensors networks based on ISO/IEC 18000-7 standard and promoted by Dash 7 Alliance. It is made for active radio frequency identification devices (RFIDs). Dash 7 operates at a 28 kbps rate up to 200 kbps and it has coverage of about 250 m extendable to 5 km. It is a low power technology, with tiny sensors stacks and long live battery (up to several years), which make it cost effective solution. The Dash 7 uses very small amount of energy for wake up signal up to 30–60 mW and it is low latency with around 2.5–5 s. Its several advantages like interoperability, robustness and cost effectiveness made it widely deployed for military and commercial applications, such as building automation, smart energy systems, smart home, PHEVs, logistics control and monitoring. In the context of smart grid, Dash 7 seems to be a good alternative to ZigBee, having several advantages like its wide range avoiding the multi-hop technique for HAN solution, less number of nodes and communication time.

7.3.2.8 Wireless Sensor Networks

Wireless sensor networks (WSNs) are a cost-effective solution for monitoring, control, measurement and fault diagnosis in various domains of smart grids. A sensor node mainly contains (smart) sensors, memory, processor, power supply, transceiver, actuator and power management unit. Sensors are used to measure various quantities like humidity, temperature, current, etc. Generally, WSN nodes are battery powered. Figure 7.6 shows the basic structure or topology of a WSN node. WSN can facilitate many of the smart grid sensing, monitoring and communication requirements. Several small sensor nodes collectively form a large sensor network which can be used for remote wireless communication in HAN, NAN and WAN. Large-scale deployment of wireless sensor nodes can be used to communicate the conditions of various generation, transmission and distribution elements and units. Wireless sensor nodes can provide economical solution for smart microgrid monitoring which facilitates high penetration of renewable energy sources. WSN is a significant part of advanced metering infrastructure. Sensing and communication are also crucial for the plug in hybrid electric vehicle (PHEV) systems which are one of the most ingenious components of the smart grid technology. PHEV contains gasoline or diesel engine with an electric motor as well as a large rechargeable battery. PHEV can be recharged from an electrical power outlet. For example, WSN can be used to communicate PHEV statistics to upstream network layers for the operation and control of smart grid components. This information can be made available online to various stakeholders through a web of sensor nodes. An effective remote monitoring, diagnosis and control can prevent cascaded disastrous events and breakdowns. Wireless sensor networks can be used for accurate monitoring of generation, transmission, distribution and consumption of electricity. WSN is the most suitable solution for HAN, NAN, WAN and smart microgrid applications for integration and operation of renewable energy sources. Wireless sensor networks can be used at the generation side for monitoring and management of produced energy. WSNs are a prominent solution for smart microgrid applications using renewable energy resources, e.g. solar farm, wind farm, biogas plant, etc. to monitor and control intermittent energy. One of the main smart grid objectives is to expedite the renewable energy use. Renewable energy resources are often situated in remote locations and harsh environments. Moreover, their unpredictable behavior creates operation and management challenges. WSN nodes are economical solutions for monitoring the behavior of renewable energy resources. Various parameters of generating equipment can be effectively measured, communicated and controlled using WSN.

Transmission and distribution of power contains various components like overhead transmission lines, underground cable network, transmission or power distribution substations and distribution transformers. WSN is an essential element of SCADA system. Real-time remote monitoring of these components is inevitable to prevent power failures due to equipment breakdown or malicious attacks. Wireless sensor networks can be used for power monitoring, fault detection and isolation, location discovery and outage detection. Wireless sensor network is an effective and prominent solution for home automation systems. It can be used for complete energy management of customer premises. Consumer plays an active role in smart grid technology. Consumers have the power to decide the time of use and rates of energy usage in smart homes. For these applications, wireless sensor networks are inevitable for communication and processing of information. WSN is the backbone of smart home applications and HAN. WSNs are a vital part of self-healing smart grid network as sensor nodes communicate parameters pertaining to conditions of various equipment and energy sources. However, there are several challenges in deployment, uses and operation of WSN due to limitations of the sensor nodes, energy supply and complexity of heterogeneous smart grid network. These challenges include: severe environmental conditions, various network topologies, errors, limited capabilities, security of the sensor networks or nodes and quality of service necessities and requirements for the smart grid environment. For example, heterogeneous network topologies in

energy distribution network due to various features and failure of sensor nodes may cause technical challenges in design of sensor nodes. The restricted processing and memory capabilities may
cause various challenges in the design and deployment of wireless sensor networks, while detection
and correction of errors require greater memory and processing facilities which make the design of
sensor network challenging. The parameters like high data rates, latency, reliability and authenticity are vital for quality of service necessities of smart grid applications. Wireless sensor networks
must fulfill these smart grid criteria and requirements for successful implementation of various SG
applications.

7.3.3 SMART GRID COMMUNICATION REQUIREMENTS

The communication infrastructure between energy generation, transmission and distribution and
consumption requires two-way communications, interoperability between advanced applications
and end-to-end reliable and secure communications with low-latencies and sufficient bandwidth.
Moreover, the system security should be robust enough to prevent cyber-attacks and provide system stability and reliability with advanced controls. In the following, major smart grid communication requirements are presented. Secure information storage and transportation are extremely
vital for power utilities, especially for billing purposes and grid control. To avoid cyber-attacks,
efficient security mechanisms should be developed and standardization efforts regarding the
security of the power grid should be made. Providing the system reliability has become one
of the most prioritized requirements for power utilities. Aging power infrastructure, increasing
energy consumption and peak demand are reasons creating unreliability issues for the power
grid. Harnessing modern and secure communication protocols, the communication and information technologies, faster and more robust control devices, embedded intelligent devices (IEDs)
for the entire grid from substation and feeder to customer resources, will significantly strengthen
the system reliability and robustness. The choice and availability of the communication structure is based on preferred communication technology. Wireless technologies with constrained
bandwidth and security and reduced installation costs can be a good choice for large-scale smart
grid deployments. On the other hand, wired technologies with increased capacity, reliability and
security can be costly. To provide system reliability, robustness and availability at the same time
with appropriate installation costs, a hybrid communication technology mixed with wired and
wireless solutions can be used.

Scalability is another important feature for the SG itself as well as the SG communication systems. A smart grid needs to be scalable enough to facilitate the operation of the power grid. If
smart meters, smart sensor nodes, smart data collectors and renewable energy sources are joining
the communications network. Hence, smart grid should handle the scalability with the integration
of advanced web services, reliable protocols with advanced functionalities, e.g. self-configuration,
security aspects. Secure and robust communication between the energy supplier and power customers is a key issue and of the smart grid. Performance degradation like delay or outage may compromise stability, therefore, a *quality-of-service* (QoS) mechanism must be provided to satisfy the
communications requirements (e.g. high-speed routing) and a QoS routing protocol must be applied
in the communications network. This incurs two important questions unique to smart grid: (1) how
to define the QoS requirement in the context of smart grid; and (2) how to ensure the QoS requirement from the home appliance in the communications network. To answer the first question, the
detailed mechanism of power price, based on the dynamics of the load, must be investigated. Then,
a reward system is built for the home appliance based on the power price and the utility function of
the appliance, thus obtaining the impact of delay and outage on the reward of the home appliance.
Finally, the QoS requirement is derived by optimizing the reward. To answer the second question,
routing methodologies meeting the derived QoS requirement are focused on. Due to the requirements of high computing and storage capabilities imposed by the heterogeneity of the smart grid,
multiple QoS-aware routing within multiple (more than two) constraints must be considered (e.g.

a greedy algorithm with K-approximation, where K is the number of constraints). A QoS requirement usually includes specifications, like average delay, jitter and connection outage probability. To derive the QoS requirement, it is important to describe the probabilistic dynamics of the power system, to evaluate the impact of different QoS specifications on the smart grid system and to derive the QoS requirement from the corresponding impact. The power price is typically determined by locational margin price (LMP) driven by the load that varies with time. A constrained optimization problem can be used to derive the LMP from the load and other parameters, where the Lagrange factors of the constraints are considered as prices. To efficiently link together the large number of smart grid components, a powerful data communications infrastructure will be provided. It is expected that part of this infrastructure will make use of the power distribution lines themselves as communications carriers using PLC technology. It is also expected to have a combination of wireless technologies to establish a reliable communications infrastructure. Also, recent standardization efforts under the umbrella of IEEE (P1901.2), ITU (ITU-G.hnem) and others are dedicated to PLC technology for smart grid applications. One of the challenges of employing PLC in power distribution grids is multi-hop transmission message routing. The basic idea is that network nodes, i.e., PLC enabled devices act as repeaters of messages in order to achieve sufficient coverage. The focus in these two previous studies is on reliable delivery of messages taking into account unpredictable and possible sudden changes of communications links and network topology. In this regard, for flooding of messages, the concept of single-frequency network (SFN) transmission is an option. However, the problem of routing in PLC networks was revised taking into account that network nodes are static and thus, their location is known *a priori*. In other words, the nodes know in which direction a message is intended to flow. Specifically, if a node receives a packet, it can decide whether to forward it or not. Such routing algorithms are known as geographic routing in the wireless communications literature, where they have been applied mainly in the context of wireless sensor networks. These algorithms present high performance for the application at hand: they close the gap between flooding on the one hand and improved shortest path routing on the other.

7.4 SMART SENSING, MONITORING AND MEASUREMENT

The smart grid environment requires the upgrade of tools for sensing, metering, monitoring and measurements at all levels, sections and components of the grid. To some degree, the smart grid kernel combines advanced and smart sensing and measurement technology, information and data communication technology, analysis and decision-making technology, automatic control technology and electrical energy technology, to realize the objectives of building a highly reliable, secure, economic, efficient and environmentally friendly energy network. Smart sensing, monitoring and advanced metering infrastructure are key factors in the smart grid which is the architecture for automated two-way communications between end-users and utilities. One of the most obstructive barriers to the smart grid development is that there are not enough deployed sensors to provide an information interface for the implementation of a smart grid. In order to transfer the electricity in more optimal ways to respond to a very wide range of conditions in a smart manner, it is necessary to deploy a huge number of various sensors in the power system to obtain more information on its status. With the objective in mind of being smart, it is necessary to add more new types of sensors to the existing infrastructure of power system, not just the traditional sensors (such as current and voltage transformer). The added sensors in a smart grid must possess such characteristics as: electrical or non-electrical sensing (e.g., mechanical, chemical, or video or image); contact or noncontact sensing; compatible and integrable with current and future technology; low power consumption and preferable self-powered; and advance communication capabilities. The sensors to be beneficial to the development of smart grids must include: electrical measurements (voltage, current, electric field and magnetic field), weather and environmental sensing (temperature, wind, humidity, pressure and solar radiation), mechanical measurements (pressure, tension, displacement, acceleration) and finally other environmental sensors, such as chemical sensors.

Sensing and measurement technology plays a fundamental role in smart grid monitoring, analysis, operation and control. Another area is the development of test technology for smart grid. It is always important to prove that the individual items of equipment, forming the systems, circuits and substations are suitable for their intended purpose and acceptable for services, this being achieved by testing.

Sensors, smart sensing and monitoring are the key enablers for the smart grid to reach its potential. Sensing provides outage detection and response evaluates the health of equipment and the integrity of the grid, eliminates meter estimations, reduces energy thefts and enables consumer choice, demand-side management and various grid monitoring functions. Advanced ad smart sensing, monitoring, and measurement in smart grid may extend to the whole electricity production and consumption chain, including power generation, transmission, distribution and end usage. The fundamental functionalities of an advanced sensing and measurement system in a smart grid can be categorized into the following four applications: (1) enhancing power system measurements and enable the transformation of data into information; (2) evaluating the health of equipment, the integrity of the grid and support advanced protective relaying; (3) enabling consumer choice and demand response and help relieve congestion; and (4) advanced metering infrastructure, which provides the interface between the utility and its customers (loads) for bi-direction control, real-time electricity pricing, accurate load characterization and outage detection and restoration. The current measurement technology in smart grid can be categorized into four types: energy measurement methods, power quality measurements, grid monitoring and measurements for optimization and control. For example, current measurement is one of the fundamental tasks in power system instrumentation, serving functionalities such as metering, control, protection and monitoring. The energy measurement technologies, including smart meter and other metering systems, may facilitate the interaction between end-users and utilities, providing metering for utility and essential support to ensure electricity supply security, grid stability, power quality and fair trade between commercial parties employing the grid. Other applications, e.g. renewable energy integration, optimizing power consumption, are energy balance or load balance, being highly dependent on energy measurement technology. The power quality measurement are essential to anticipate, detect and respond to system and power quality problems, and service disruptions and the key to a continuous improvement of power supply and its use, by knowledge of the relevant local system circumstances in detail, using the information of all available sources connected to the system. In a modern power system, utility companies and end-users are not only concerned with the steady power quality issues such as voltage fluctuation, flicker, frequency fluctuation and harmonics, but also with transient power quality issues such as voltage swells, sags and interruptions. The monitoring covers a wide range of applications such as measurement of the system state, monitoring of the operation parameters such as stability or efficiency, monitoring of status of the devices including primary HV devices and secondary electronic or communication devices, and other weather and environmental parameters such as temperature, humidity, wind, cloudiness, precipitation type and intensity and solar radiation. Measurement for control and optimization includes the measurement for system operation, such as control, protection, regulation, dispatching and planning. In this category, the measurement is usually in closed-loop feedback control. The SCADA (supervisory control and data acquisition) and PMU (Phasor Measurement Unit) in WAMS (Wide Area Measurement System) are key elements of the emerging smart grid, handling a wide range of data collection, sharing, and coordinated control actions that make the system more efficient and reliable.

Current measurement is one of the fundamental tasks in power system instrumentation and sensing, supporting functionalities such as metering, control, protection and monitoring. Traditionally, current measurement in power system has been performed with current transformers (CTs). However, because of the fast development of the modern power system, CTs, due to their intrinsic disadvantages (e.g., nonlinearity of the CT magnetic core characteristic, not work for DC measurement, narrow bandwidth, etc.), are inadequate to be used for either quantitative or qualitative evaluation of current transformer. For instance, in modern power transmission, high-voltage direct-current (HVDC), known as the most energy-efficient and economical, is popularly used for long-distance power transmission. The disadvantages and deficiencies of CTs are further amplified when used in modern power quality measurements, monitoring, which is becoming more crucial as suitable theoretical approaches and indices in power quality have been properly defined. The smart grids are bring additional requirements to the current measurement subject. For example, because the 1547 IEEE standard on interconnecting distributed resources requires that no more than 0.5% of the rated current be injected as DC, it seems likely that at least some current measurement systems are require a frequency response extending down to DC. Magnetic sensors have long been used for applications such as current sensing, encoders, gear tooth sensing, linear and rotary position sensing, rotational speed sensing and motion sensing, etc. Solid-state magnetic field sensors, which are generally employed for this purpose, have an inherent advantage in size and power consumption compared with search coil, fluxgate and more complicated low-field sensing techniques, such as superconducting quantum interference detectors and spin resonance magnetometers. This holds true even for high-frequency applications. These sensors work on the principle of converting the magnetic field into a voltage or resistance. The sensing can be easily done in an extremely small area, further reducing size and power requirements. One of the advanced types of solid-state magnetic field sensors are the magneto-resistive (MR) sensors, based on the magneto-resistive effect (the change of the resistivity of a material due to a magnetic field). The first generation of such sensor is the anisotropic magneto-resistance, in which the resistance of a piece of iron increases when the current is in the same direction as the magnetic force and decreases when the current is at 90°. This effect is used in a wide array of sensors for measurement of Earth's magnetic field (electronic compass), for electric current measuring (by measuring the magnetic field created around the conductor), for traffic detection and for linear position and angle sensing. In recent years, novel types of magneto-resistive materials with much higher sensitivity to small changes in the magnetic fields have been found, manufactured and used with much higher sensitivity. This measurement technique is directly compared to the CT measurement, which relies on the measurement of the secondary current, induced by the magnetic field generated by the primary current. MR sensors can measure primary current directly from its radiated magnetic field without the need of secondary induced current. Because of the factors related to CTs measurement accuracy, such as burden, rating factor, magnetic core saturation, external electromagnetic field, temperature and physical configuration, CTs will be replaced by MR devices for the 21st-century smart grid. In terms of safety, MR sensor measurement is a noncontact method which means convenient installation and safe operation.

A smart meter is an advanced meter, which identifies power consumption in much more detail than a conventional meter and communicates the collected information back to the utility for load monitoring and billing purposes. Consumers can be informed of how much power they are using so that they could control their power consumption and the consequent carbon dioxide emission. By managing the peak load through consumer participation, the utility will likely provide electricity at lower and even rates for all. AMI has already gained great attraction within the industry, with the advantages in accuracy and process improvement of on-line meter reading and control. Additional benefits are suggested to be gained in managing power quality and asset management with AMI. There also are vivid discussions in the smart grid and power engineering communities how reliability, operational efficiency and customer satisfaction can be addressed with an AMI deployment. However, the AMI benefits are countered by the cyber-security issues. Modern technologies require that communication infrastructure provides interconnectivity. Hence, the vulnerabilities that expose other internetworking systems will ultimately lead to security threats to AMI systems. To monitor the status of the power system in real time, sensors are placed into various grid components in the power networks. These sensors are capable of taking fine-grained measurements of a variety of physical or electrical parameters and generate a lot of information. Delivering this information to the control center in a cost efficient and timely manner is a critical challenge to be addressed in order to build an intelligent smart grid. Network design is a critical aspect of sensor based transmission line monitoring due to the large scale, vast and complex terrains, uncommon and diverse topology, and critical timing requirements. The design goals are to deploy multiple different sensors in critical locations of the transmission lines to sense the status and mechanical properties of its various components and transmit the sensed data through a suitable communication network to the control center. At the control center, the received data and information can be combined with existing electrical data in the system for optimum operation, and to achieve ideal preventive or corrective control decision. Often a hybrid hierarchical network that spans wired, wireless and cellular technologies to provide cost optimized delay and bandwidth constrained data transmission is employed.

7.4.1 SMART GRID SENSING AND MEASUREMENTS

A sensor is a device that generates an electrical signal that is proportional or correlated to a physical quantity. The output signal could be in analog (continuous) or digital (discrete) form. Digital sensors typically have more capabilities than analog sensors, such as signal conditioning, analog-to-digital conversion and data processing capability and digital signal output. To some degree, digital sensors have some "smart" capabilities. Smart sensors are digital sensors having integrated microprocessors, include some logic functions, and/or can make some types of decisions, being defined as sensors that can manipulate and compute sensor-derived data and communicate the data through a bi-directional digital bus to users via a communication interface. Generally speaking, sensors are devices that responds to a physical stimulus, e.g. heat, light, sound, pressure, magnetism, motion, fluid flow, etc., and convert the physical input that into an electrical signal, performing an input functions into a device or equipment. Devices which perform an output function are generally called actuators, and are used to control some external device, e.g. movement. Both sensors and actuators are collectively known as transducers. Transducers are devices used to convert energy of one kind into energy of another kind, being the most important element of measurement system. The sensors, used in power grids, are enabling the remote monitoring of equipment such as transformers, capacitor banks, feeders and power lines. They are improving the performance and extending the life of grid components to ensure a safe and reliable operation of the electricity networks. Integrated electricity distribution infrastructure monitoring systems require various types of sensors and transducers to build smart or intelligent sensing capabilities. For example, the IEEE C37.118 standard defines a transmission format for reporting synchronized phasor measurements in power systems. Accurate and precise time-stamping of sensor captured data is critical. As specified for the IEEE C37.118-based PMUs, the timing reference and signal must be traceable and aligned to Coordinated Universal Time (UTC) with an uncertainty of less than 1 μs. The other important requirement for the sensor and smart sensing is the need for a network communication interface to transport the data to a data repository and management system. Monitoring and control applications require different information types for optimizing power distribution network operations. Therefore, sensor data output requirements should meet the anticipated future SG monitoring and control applications. Some requirements of the sensors for used in smart grids are:

1. High-accuracy timing and time synchronization to the UTC;
2. High-speed data processing and intelligent algorithms, e.g. producing synchronized phasor, frequency, and rate of frequency change estimates from measured voltage and current signals along with time synchronizing signals;
3. High measurement accuracy and sensitivity for current and voltage magnitudes and phase angle;
4. High-speed, secure, and reliable network communications and standards-based data transmission;
5. Standardized interfaces and test methods to help achieve smart sensor interoperability and plug-and-play capability;
6. A wide range of sensors with high bandwidth/dynamic range, such as measuring voltage from 600 V to 69 kV of medium voltage, current from amperes to kilo-amperes, frequencies from 50 Hz to 5 MHz;

7. Multiple sensing capabilities of electrical and physical parameters including voltage, current, power flow, temperature, humidity, wind velocity, cloudiness and solar radiation; and
8. Intelligent capabilities for sensors, such as self-identification, self-localization, self-awareness, self-diagnostics and self-calibration.

7.4.2   SYNCHROPHASOR MEASUREMENTS

The introduction of phasor measurement units (PMUs) based on the GPS technology has made it possible to obtain synchronized measurements of important power systems quantities with significant benefits for power system operation and control by providing better and more reliable information about power system operating status in real time. The ability to monitor grid conditions and receive automated alerts in real-time is essential for ensuring power system reliability. Synchronized PMUs take sub-second readings system-wide and through visualization and advanced applications provide an accurate picture of grid conditions. With synchrophasor technology, it is possible to monitor reliability metrics, grid dynamics, identify and diagnose system problems, system stresses, oscillations and other abnormal situations that may occur in the power systems to enable them to take proactive actions to prevent blackouts, reduce the footprint of blackouts and enable faster service restoration and recovery after such events. In addition, this technology has the potential to improve efficiency by allowing operation closer to inherent thermal physical limits with equal or greater safety margins and by increasing the capacity of the transmission lines use based on dynamic ratings and margin assessment. PMUs provide voltages, currents and frequency change rates. Such data and information can be used to provide early detection to prevent grid disturbance events, assess and maintain system stability following a destabilizing event, as well as alerting system operators to view precise real-time data, reducing the probability of an event causing widespread grid instability. In addition, the phasor measurements are very valuable for postmortem event analysis to understand the cause and impact of system disturbances. In addition, PMU data is improving power system fault location detection and analysis. Phasor data is also useful in validating the dynamic models of generation resources, energy storage resources and system loads for use in transmission planning programs and operations analysis, such as dynamic stability and voltage stability assessment. Synchrophasor technology has also an important role in determining dynamic system ratings, allowing for more reliable deliveries of energy, especially from renewable energy generation location to load centers.

PMU technologies were developed from symmetrical component distance relay (SCDR), a recursive algorithm for calculating symmetrical components of voltage and current. The PMU unit records the sequence currents, voltages and/or their phase values and time-stamps the reading with GPS time, achieving accuracy of synchronization of 1 μs or 0.021°, for a 60 Hz signal. PMUs are operating under the IEEE C37.118-2005 standard, attempting to address all factors that PMUs can detect in power system dynamic activity. PMUs typically sample grid conditions at a rate of several hundred times per second and use this sampled data to calculate phasor values for electric voltage and current, at a rate of 30 or more per second. PMUs are deployed across an area as wide as an entire interconnection, phasor data concentrators (PDCs) are collecting data and display it for operators to understand grid conditions indicating possible levels of stress in the grid, such as areas of low voltage, frequency oscillations or rapidly changing phase angles between two locations on the grid. PMUs are measuring the power system frequency, a key indicator of the balance between generation and load in the power system. Abrupt changes in frequency due to major generation and/or load losses can compromise power system stability and may lead to a blackout. Synchrophasor units can be used to monitor, predict and manage the voltage on the transmission subsystems. Some transmission systems are voltage stability-limited, the voltage cannot exceed a certain level without causing system stability problems, voltage collapse can happen if the voltage stability limits are reached or exceeded. The high resolution of synchrophasor measurements provides the ability to map changes in voltages at a bus to power flow changes on connected lines, measuring the point voltage sensitivity, an indicator of possible voltage stability problems. The phasor measurement based voltage stability analysis application assesses the power (or current)-to-voltage system operating point and sensitivities at a sub-second resolution. The calculated adaptive voltage stability margin, expressed as active and reactive power, provides not only the power transfer limit to a load center, in terms of active power, but also the reactive power support needed at this load center.

The phase angle difference (an angle difference within a predetermined range is acceptable but needs to be monitored closely for early warnings) between buses measured by PMUs on the transmission network indicates of system stability and system stress. An increasing phase angle difference margin can be a serious problem when the deviation is large enough to cause voltage and system instability. System oscillations are expected in a large network of loads and generators. Moreover, during normal operations, generators have the ability to retain synchronization with other interconnected generators, maintaining system stability. If the system is properly designed and operated and a disturbance occurs, the oscillations are damped and the system naturally returns to equilibrium. However, if the disturbance is large or the interconnected system is weak, the oscillations caused by disturbance can grow, causing the system to become unstable, breaking apart and leaving large areas without power. PMU data can be used to compute the damping ratio and determine the magnitude and energy associated with oscillations with certain frequency, so-called oscillation modes in real-time operations. Intermittent, variable generation units can strain grid stability and may become a much more serious treat. Monitoring, detecting and identifying low damping oscillations in systems with renewable energy systems is critical, allowing starting control measures to damp out the oscillations by re-dispatching or forcing reduction in power generation. When oscillations are detected, a control signal is generated and sent to generator's excitation system to regulate its voltage and bring it into synchronization or reduce its output to a level that is no longer a system threat. Data obtained from PMUs can help identify the causes of some of the system stability issues, such as some of the unstable modes are local where a small group of generators is out of synchronization with the rest of the generators in the system or inter-area where many generators in one part of the system are out of synchronization with the rest of the system. There are also some higher frequency oscillation modes that are caused by poorly tuned exciters, power system stabilizers, governors or static VAR compensators. PMUs are expected to be particularly useful for improved monitoring, managing, and integrating of distributed generation and renewable energy resources. One of the challenges in integrating such energy resources is how to identify and respond to their power generation variability, and the PMUs can be a way to identify, control and manage such issues.

7.5   CHAPTER SUMMARY

Smart grids (SGs) are electrical power grids that apply information, advanced networking and real-time monitoring and control technologies to lower costs, save energy and improve security, interoperability and reliability. A smart grid is characterized by the bi-directional connection of electricity and information flows to create a *smart* or intelligent, automated and a widely distributed electricity delivery network. Its infrastructure covers the following areas: management, protection, information and communication systems, energy efficiency, emissions, power quality and supply security. The smart grid uses two-way communications, digital technologies advanced sensing and computing infrastructure and software abilities in order to provide improved monitoring, protection and optimization of all power grid elements, including generation, transmission, power distribution and end-users. The smart grid is based on combination of legacy power grid with advanced smart metering, remote sensing, remote control of all key components and equipment. The success of the smart grid depends directly on reliable, robust, versatile, scalable and secure communication systems with high data rate capabilities. Smart grids incorporate the legacy of the conventional electricity grid and the benefits of modern information and communications technologies to deliver real-time data and information, enabling the near-instantaneous balance of energy supply and demand management. Key features of the smart or intelligent grid involve monitoring, smart control and protection, automation, optimization, integration and security of the power flows from utility generators to the end-user equipment, devices and appliances. This eventually results in conservation of energy and its efficient utilization for both power and infrastructure applications. A sophisticated, reliable and fast communication infrastructure is necessary for the connection among the huge amount of distributed elements, devices and equipment such as generators, transmission lines, substations, energy storage systems and end-users, enabling a real time exchange of data and information necessary for the management, control and operation of power systems and for ensuring improvements in terms of efficiency, reliability, flexibility and investment return for all those involved in a smart grid: producers, operators, stakeholders and customers. One of the most important structure of smart metering, the advanced measurement infrastructures are equipped with information and communication technologies. Key elements of the AMI which are the smart meters allowing real-time dynamic pricing, incorporate wireless communication protocols, unlike conventional automatic meters. AMI enables bi-directional data communication, includes the stages of measuring, collecting, storing, analyzing, using and managing. Another major smart grid component refers to the customer participation in the grid energy management. This participation is done via demand response or demand-side management, in which (a) power companies provide incentives to customers to shift their load over time, and (b) customers are provided with autonomy to participate in energy market. Thus, in any smart grid mechanism, it is imperative to factor in demand response models and their associated challenge. Smart sensors, intelligent sensing and monitoring can provide real-time data, information and the status of the power grids for real-time monitoring, management, protection and control of grid operations. The uses of synchrophasor technology for smart grid wide-area measurements, monitoring, analysis and control is enhancing electric power system reliability, efficiency and resilience. PMUs enable finer and faster control of the network, identifying and developing ways to mitigate or prevent evolving grid events and collapse, increasing grid asset utilization and system efficiency by expanding grid throughput and operating closer to the margin reliably and securely, while enabling reliable and secure dynamic grid operation with changing resources mix, including DG and RES integration, demand response and variable load characteristics.

7.6   QUESTIONS AND PROBLEMS

1. Briefly discuss the needs for advanced and diverse smart grid communications infrastructure.
2. List the main smart grid applications
3. List the most common smart grid communications technologies.
4. Briefly discuss the most important standards dealing with the smart gird communications.
5. List the attributes of the smart grid communications.
6. In your own words state the main reasons why the smart grid communication is more likely to be an integrated hybrid communication network.
7. Briefly describe in your own words the demand-side management.
8. List the main requirements of the sensors used in smart grids.
9. What are the main characteristics of smart meters?
10. List the minimal set of smart metering infrastructure functionalities.
11. If the SNR is 30 dB, what is the ratio of signal to noise power?
12. Estimate the channel capacity for the signal to noise power of the previous problem, and typical bandwidth of 60 Mbs (Mega-bit per second).
13. Briefly describe the minimal smart grid sensing, monitoring and control capabilities.
14. Compare and estimate the maximum channel capacity of twisted copper and coaxial cables. For copper cable, the bandwidth is 250 kHz and the SNR is 20 dB. For coaxial cable, the bandwidth is 150 MHz and the SNR is 22 dB.
15. List and briefly discuss the smart grid communication requirements.

16. Briefly describe the major attributes of the communication channels in the smart grid context.
17. Estimate the channel capacity for the signal to noise power of 25 dB, and typical channel bandwidth of 60 Mbs.
18. What are the three common approaches of demand response paradigm?
19. Briefly describe the utility-initiated demand response.
20. Briefly describe the customer-initiated demand response.
21. Briefly describe the main reasons for the development of a power information theory.
22. List the main requirements of the IEC 61850 standard.
23. List the key features of the smart meters.
24. Briefly discuss the major functionalities of the smart meters.
25. List the most common wired communication technologies used in smart grids.
26. Briefly describe the power line communication, list the major advantages and disadvantages.
27. If a PLC system uses a carrier frequency of 90 kHz. If the value of the inductance in the line trap is 0.30 mH, what is the value of the capacitance required in this case?
28. Briefly discuss the main advantages and disadvantages of the DSL technology.
29. Calculate the maximum allowable angle for a multimode fiber that has a core of refractive index 1.5 and cladding of refractive index 1.475.
30. List the ZigBee power distribution and home energy management applications.
31. List the main Bluetooth advantages and disadvantages.
32. What are the main smart grid communication requirements?
33. Briefly describe the major wireless sensor networks' smart grid applications.
34. What are the reasons that magneto-resistive sensors may replace CTs for current measurements?
35. List the fundamental functionalities of an advanced sensing and measurement system in a smart grid.
36. What are the major benefits of the power system applications of synchrophasor measurements?
37. Briefly describe the PMU operation.
38. How PMUs can help the grid integration of renewable energy sources?

8 Smart Grid Transmission, Operation and Power Flow Analysis

8.1 INTRODUCTION

Smart grid, the 21st-century electrical energy infrastructure integrates the latest digital communications, monitoring and advanced, adaptive and smart control technologies to the existing power grids. Most of the initial smart grid focus has been limited to making the distribution system smarter and to installing the advanced metering that can make the consumer an active participant in the grid's operation. With 21st-century technological advances paving the way, many regulators have stressed this initial focus on advanced metering infrastructure (AMI) and demand response (DR). Most of the smart grid previous work has placed great emphasis on the power distribution systems and demand side as evidenced by the wide range of applied emerging technologies. The vision of the whole transmission grid, in the context of smart grids, is still somewhat unclear and well defined. Moreover, the fact is, while the entire power distribution systems has seen fewer technological changes over most of its existence, there is now huge needs and opportunities for improvements, made possible by digital communications, computing and control tools. However, transmission subsystem situation is somewhat different. Technical advances have occurred throughout transmission history, with advances in monitoring, protection, analysis and control, accompanied by periodic breakthroughs in transmission capacity. Power electronics, communication, smart sensing, monitoring and computing have played important and critical roles, by enabling DC transmission, operation at capacity limits, smart protection, optimized management and operation, and a variety of Flexible AC Transmission Systems (FACTS) enhancements and applications. It is this continuous technical progress that has perhaps placed transmission on the smart grid backburner; it is already pretty smart. While recognizing that transmission challenges are large, the first requirement for any transmission system remains the same, it must be extremely reliable and robust. The importance of this tenet is perhaps best illustrated by the events of the 2003 Northeast blackout. That blackout was not so much the result of an inadequate or failing transmission infrastructure; it was faulty computer and monitoring systems, human errors, and lack of proper maintenance that were mostly to blame.

Rapid advancements in controlling Information and Communications Technologies (ICTs) have allowed the conversion of traditional electricity grids into smart grids that ensure productive and complex interactions among energy providers (utilities), consumers and other stakeholders. These multiple and enhanced interactions are helping to solve the issues raised in the existing power grids. Control systems are needed across broad temporal, geographical and industry scales, from devices, components to power system wide, from fuel sources to consumers, from utility pricing to demand response, etc. With increased deployment of feedback and communication infrastructure, opportunities arise for reducing consumption, for better and more efficient exploiting renewable energy sources and for increasing the reliability and performances of the transmission and power distribution networks. Key components of the smart grids are smart meters, adaptive and intelligent, sensors, monitoring systems and data management systems that control the flow of power and information among various stakeholders, making the grids two-way communication and power networks. Other smart grid applications include energy management, demand-side management and distributed generation and its reliable integration into power systems, equipment diagnostics,

DOI: 10.1201/9780429174803-8

control, overall optimized asset management etc. The fact is, while the entire distribution system has seen very little technological changes, there is now a huge need and opportunity for improvement, made possible by today's digital communications, smart sensing and monitoring and control tools. The transmission situation is somewhat different because technical advances have occurred throughout transmission's history, with advances in monitoring, protection, analysis and control, accompanied by periodic breakthroughs in transmission capacity. Power electronics has also played an important role, by enabling DC transmission and FACTS enhancements. It is this continuous technical progress that has perhaps placed transmission on the smart grid backburner; it is already quite smart. However, an efficient, reliable and secure transmission system has had, and will continue to have, an essential role in satisfying the electricity demands. Advanced digital technology and power electronics can raise transmission to a new performance level, even as the emergence of remote renewable energy and increasing electricity market applications create new challenges. Transmission in modern power systems are facing several challenges and more demanding operating conditions due to increasing penetration of distributed generation, larger power flows and trans-border electricity exchanges. To place this in context, the historic role of transmission was to connect generation to loads and to interconnect power systems. Interconnection provides several benefits by exploiting the diversity between systems:

1. Diversity between peak load occurrences allows the same generation equipment to supply more loads.
2. Diversity of outages, allowing spare equipment to support more than one power system.
3. Diversity of fuel sources, allowing the most economical fuel choice and use for any given situation.

Load or power flow is the backbone of any power system analysis, operation and management, including the smart grids, and any evaluation in power systems. A variety of methods, tools and approaches are employed in power system operation, management and planning, and are also available in the literature to perform power flow, transient and voltage stability analyses considering uncertainties associated with electrical parameters. In order to design and plan an efficient power system for future modification, the power flow studies are critical and inevitable processes. These studies provide information regarding the magnitude of voltage and its phase angle at each bus and also the active and reactive power flow in each part of the system. The power flow has a significant importance in estimating the system contingencies and the measures to be taken to modify the system in order to serve an added load in the system. In transmission planning, the power flow studies are utilized for evaluating the effects of peak load on transmission system and also to determine the limits violated during peak and off peak condition in order to find the solutions to overcome that situation. The process of analyzing the effect of loads and generators during transmission is necessary for evaluating the deliverability of generation and load. Power transfer between system areas is a major function of an efficiently running an electric power system, being set by the fact that the transmission networks have limited capability to transfer power, and power flow studies are critical here. However, new and emerging requirements find transmission in roles it was not initially designed to perform. One example is the role of market environment, connecting buyers and sellers across very large geographic regions. Excessive transmission-use variability and far less predictability are the results. Further complicating factors come with the penetration of renewable generation such as wind and solar. These intermittent and non-dispatchable sources of power affect the operation and control of the power system because of the uncertainties associated with their output power. Depending on the penetration level, the electric grid may experience considerable changes in power flows and synchronizing torques associated with system stability, because of the variability of the power injections, among several other factors. In these new visions of the smart grids, each smart transmission network is regarded as an integrated power transmission system that functionally consists of three interactive smart components, i.e. smart control centers, smart electricity transmission networks and smart substations. The features and functions of each of the three functional components as well as the enabling technologies to achieve these features and functions are discussed in detail in this chapter.

8.1.1 POWER SYSTEM CONSTRAINTS AND REQUIREMENTS

Growth in electricity demand and new generation capacities, lack of investment in new transmission infrastructure and the incomplete transition to fully efficient and competitive wholesale markets have allowed transmission bottlenecks to emerge. However, power transmission subsystems are facing ever increased demanding operating conditions and requirements with increasing penetrations of renewable energy generation, larger power flows and greater cross-boundary trading of electricity. The variability of the power output of renewable energy sources and unplanned power flows through transmission grids are causing difficulties and stresses on the power system operations, requiring additional measures to maintain the system stability. Excessive power flows in transmission networks and large variations in bus voltages may arise during steady-state operation, so that when faults and network outages occur they can lead to system collapse. Moreover, the market deregulation, enabling the power delivery within and between regions, facilitating the access to interconnected competitive generation may put additional stresses on the power system operation and management. However, the existing power systems are not designed for open-access power delivery, distributed and dispersed generation, creating inefficiencies and instabilities in power delivery. Additionally, there are few or no market-based incentives for large transmission investments, which have contributed to significant power system capacity deficiencies, bring significant stress on the power system operation and management. New transmission line permitting, siting and construction are difficult, expensive, time-consuming and typically politically charged, and the likelihoods that such new transmission lines alone are resolve these problems are smaller. The demands being placed on the transmission system can result in several operating limits being reached, thus creating serious reliability concerns. These characteristics include among others: steady-state power transfer limit, contingency limit, voltage stability limit, dynamic voltage limit, transient stability limit, power system oscillation damping limit, thermal limit and short-circuit current limit.

8.2 SMART GRID TRANSMISSION

Conventional grid transmission is under significant pressure for several environmental, customers and market challenges and needs, as well as the existing infrastructure issues and problems. Transmission systems in many regions or countries are facing quite a few demanding operating conditions with increasing energy demand or penetrations of renewable energy generation, larger flows and greater cross-border electricity trading. The variability of the power output of renewable energy sources and unplanned flows through transmission grids are causing difficulties for the system operators, who are responsible for maintaining the system stability. These issues, needs and challenges are important and urgent than before, driving the grid to expand and enhance its functions and applications towards smarter features with leverage in the newest technologies. Moreover, if excessive power flows in transmission circuits and large variations in busbar voltages may arise during steady-state operation so that when faults and network outages occur, they can lead to system collapse. In order to aid the transmission system operators to monitor, control and optimize the performance of generation and transmission systems, a set of applications collected into an Energy Management System (EMS) is employed. As the monitoring and control functions for EMS are often provided by SCADA system, these systems are also usually referred to as EMS/SCADA. The EMS is usually located in the System Control Centre and effective real-time monitoring and remote control are operating between the Control Centre and the generating stations and transmission substations. With the increased availability of measurements from Phasor Measurement Units (PMUs), it is expected that, PMU measurements will be integrated with EMS. However, at present,

PMUs are mainly incorporated into separate Wide Area Applications. It is expected that EMS and Wide Area Applications will still coexist separately for a while, but the integration is in full steam.

The envisioned smart features and functions of the transmission subsystems include: digitalization, intelligence, flexibility, resilience, sustainability, scalability and customization. Owing such smart features and functions, the smart grid transmission is able to deal with challenges, issues and needs discussed previously, e.g. environmental challenges, market or customer needs and expectations, infrastructure issues and new and innovative technologies. Smart transmission networks are employing and will make use of digital platforms with friendly user interface, visualization error-tolerant capabilities for fast, timely, reliable and accurate measurements, commutation, sensing, communication, control, protection, visualization, operation and maintenance of the transmission subsystems. Intelligent technologies, smart devices and human expertise will continue to be incorporated and embedded in smart transmission networks. System operation state self-awareness will be fully available with the aid of online time-domain analysis, such as security analysis, voltage and angular stability. Fully implemented self-healing capabilities will enhance the smart transmission grid security through coordinated protection and control equipment. Smart transmission flexibility involves four major aspects: (1) scalability and expandability for future developments, including seamless integration of diverse and innovative generation technologies; (2) adaptability to any geography and climate; (3) multiple control strategies; and (4) compatibility with various market operations and plug-in-play hardware and software capabilities. The smart grid is and will be even more capable of delivering electricity to customers securely and reliably in case of any events, with fast service restoration. A fast and advanced self-healing capability enables the system to dynamically reconfigure in order to recover from natural disasters, attacks, component failure or blackouts. Smart sensing, online computation and analysis are enabling fast and flexible electric network operation and control, load prioritization and islanding and sectionalization in the emergency with fast service restauration. Smart transmission sustainability features efficiency, sufficiency and environment-friendly operations. Increasing energy demands should be satisfy by using alternative energy sources, energy storage, smart energy usage and mitigation of network congestion. Smart transmission grids are customer-tailored structured, further liberating the electricity market.

The transmission voltages are to a large extent determined by the system loss considerations. High voltage transmission requires conductors of larger cross-section, resulting in lower resistance values, hence less energy losses on the transmission lines. The load centers are usually located away from generating stations. Therefore, the power is transmitted to the load centers where it is stepped down to 66 kV, 33 kV or other values appropriate for the power distribution level. The load demand determines the voltage at which power is to be supplied. The loads may be residential, industrial or commercial, which can bring issues about the peak load and off peak load hours. Power is transmitted from low demand areas to high load demand in the grid network. The control of generation, transmission, distribution and area exchange is performed from a centralized location. In order to perform the control functions satisfactorily, the steady state power flow must be known. The entire system is modeled as an electric network and a solution is simulated using a digital program. Such a problem is called power flow analysis. The power flow analysis (the load flow problem) is a very important and fundamental tool involving numerical analysis applied to a power system. The results play a major role in the day-to-day operation of any system for its control and economic schedule. The analysis is also employed during power system design procedures, planning expansion and development of control strategies. Notice that even though the electric power networks are composed of components or elements, which are or can be approximated to by linear models, the electric power flows, real and reactive powers, are nonlinear quantities. Load or power flow studies are likely the most common of all the power system analysis calculations, being critical the system planning and operation. Load flow studies identify out of range line loads, bus voltages, inappropriately large bus phase angles, potential associated stability problems, component loads (usually transformers), proximity to the reactive power limits at generation buses and other parameters having potential to create operation issues. Power flow studies are assisting system operators in computing power levels at each generation unit for economic dispatch, analyzing outages and other forced operating conditions, e.g. contingency analysis and coordinating power pools. Besides giving real and reactive power, the load flow provides information about line and transformer loading and losses throughout the system and voltages at different system points for the performance evaluation and regulation of the power systems. Further study and analysis of future expansion, stability and reliability of the power system networks can be easily analyzed through this study. Growing demand of the power and complexity of the power system networks, power system study is a significant and critical tool for power system operators in order to take corrective actions in timely manner. However, usually power flow analysis is employed to assess system performances and operations under a given (specific) condition.

8.2.1 SMART TRANSMISSION FRAMEWORK AND TECHNOLOGIES

This section introduces some of the main components and technologies of the power transmission subsection, focusing on conventional power grids and smart grids with a great deal of attention how transition to smart grid is taken place. Technologies of the transmission and sub-transmission systems are classified according to the mechanical construction principle as: overhead lines and underground or submarine lines which can be categorized and perform as cable lines or gas insulated transmission lines (GILs). If they are related to the voltage, levels are classified as: high-voltage (HV) transmission lines, operating from 60 kV up to 220 kV; extra-high voltage (EHV) lines with voltage levels from 220 kV up to\800 kV); and ultra-high voltage (UHV) lines with voltage range from 800 kV up to 1200 kV. According to their signal type or physical transmission principle, they are classified as AC lines and DC lines. The AC line transfer capacity is limited by parameters, such as line electrical phase shift, voltage drop, and thermal effects in the line. Figure 8.1 shows the diagram of the power transmitted between two sections of an electric network sections. Neglecting the line resistance the power transfer between the network sections is given by:

$$P = \frac{V_1 \times V_2}{X} \sin(\delta_1 - \delta_2) \tag{8.1}$$

The angle difference between the section voltages of the network is important for the grid static and dynamic stability. If the angle difference $(\delta_1 - \delta_2)$ is close to or higher than $90°$, the static stability is lost. Furthermore, in the case of faults the voltages decline and the transfer capacity is reduced. The subsequent power swing is better dampened if the transmitted power is significantly lower than the power curve peak. Consequently, a higher voltage level V allows a larger power transfer capability $(P_3 > P_2)$. On the other hand, the longer the transmission line, the larger total line impedance, Z, resistance, R, and reactance X, so the transmission line length is also limiting the energy transfer capacity $(P_1 > P_2)$. On the other hand, the DC transmission lines do not have the stability-caused capacity limitation. However, the second power transfer limitation is valid for both AC and DC transmission lines, namely in this case the line resistance causes power losses proportionally to the current squared, I^2. Higher transmitted power means higher energy losses and a higher temperature. For example, the power losses of a transmission line that is 140 miles (or about 230 km) long with 400 kV are about 6.3 MW or 1.05% for a 600 MVA power transfer, and about 25.2 MW or 2.1%

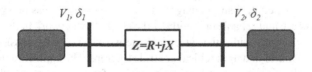

FIGURE 8.1 Power transfer and the impact of parameters on the AC line power transfer capacity.

for a 1200 MVA power transfer. The line conductor temperature is also limiting the power transfer capacity. The power transfer capacity and transmission distance are selection criteria for transmission lines, regarding the voltage levels and the physical principles, by setting the capacity of the line's power transfer.

AC aerial transmission lines are usually constructed with steel towers, porcelain, silicone or glass insulators, ground and phase conductors consisting of a steel core providing the mechanical stability and aluminum housing providing a lower electric resistance. Double lines with two-phase systems on one tower are often used up to a level of 400 kV. Furthermore, a 400 kV double line tower may also be extended with two systems of 110 kV each. Such a combination is efficient option for countries or states where the track territory is limited and the permission for line erections requires longer permitting procedures. A transmission line is protected against atmospheric lightning by a ground conductor. Each line can be represented by an equivalent scheme of the resistance R, the series inductive reactance X_L and the shunt capacitive reactance X_C, respectively. The line resistance causes energy losses and voltage drops along the line. The line reactance X_L causes reactive power losses and voltage drops proportional to the current flow, while the X_C generates reactive power proportional to the voltage and increases the voltage at the line end. For each line construction a natural load or surge impedance load (SIL) exists, where the voltage at both ends is equal (reactive power losses are equal to the capacitive reactive power generation). The rated transmission line capacity is selected above the SIL, so the voltage decreases along the line in the direction of the active power flow.

8.2.2 CONTROL CENTERS

The electricity flows, stability and security of the power system to prevent any service interruptions to the end-users or the power system collapse, while the system must be protected from being damaged due to the failure of any component or equipment. In addition, power systems must operate efficiently and economically to ensure the best electricity rates for the costumers. The power system control is structured in the hierarchical levels transmission, subtransmission (or regional power distribution) and local power distribution. To effectively operate, manage and control a power system requires and extensive monitoring and sensing. All major components, equipment and functions in power plants, substations, transmission lines, protection equipment, etc. must be monitored, evaluated and controlled mostly in real time. Besides the extensive monitoring, the power system operation is continuously evaluated, future demands are forecasted and power system studies and planning are conducted. These functions require the use of sophisticated methods, algorithms, hardware and software applications. All the above tasks and functions are managed and executed in the control centers, which are the power system brain. The power system control is structured in the hierarchical levels transmission, subtransmission (or regional distribution) and local distribution. The vision of the smart control centers is built on the conventional grid control centers developed over the last six decades. It is the highest-level monitoring, control and management location for large grid sections. Control centers are the sites where is managed the *power system health*, and where not only the management and control operation take place, but also the strategic decisions are also made. The transmission system operation at the highest level performs two basic and essential functions: Energy Management System (EMS) and Network Supervision, Control and Data Acquisition (SCADA). The control center of transmission systems is often called "dispatching center" in the context with the combination of both functions. As supervisors of the transmission subsystem operation, control and management, the TSOs (transmission system operators) are responsible for the secure and reliable operation of the electricity network in their respective control area and for interconnections with other electricity networks. The transmission system operation may be managed in a central or a hierarchical control structure. The control centers at utilities today mainly perform the following main tasks: grid monitoring, fault management, planning, optimizing the grid topology, in order to achieve optimal grid configuration, the management of the customer requests and providing customer information, generation management to some or lesser extent, management of switching operations, operation statistics including fault and interruption statistics, database management and change power system variables and set-points (e.g. set-points of regulators, voltage control, etc.). Notice that the management of customer requests and providing customer information is typically shared with the customer call centers. The information received by the call center that could have value for the control center (interruption information, outage observations, etc.) is routed to the control center.

The control or dispatching center is a key element in coordinating operation of a power system, such as monitoring, controlling and switching. During outages especially when many network customers are affected, the most important decision making process at the control center is to prioritize the reconnection of the customers. For reconnection of the customers, two methods are available: rerouting, finding alternative paths from the energy source to the customer; and repair, the replacement and/or restoration of the faulty component(s) and sections. The technology of the control centers (or dispatching centers) is mainly based on components which are commercially available on the markets for computer and communication technologies. The specific of the vendor solutions consists in the selection and configuration of such hardware components and the software solutions and applications. Any control center has a large display "mimic board" representing main power system components, such as transmission lines, transformer or protection equipment, displaying the power system current conditions, e.g. power flows in all lines and status of each circuit breakers. System operators are monitoring all the time the power system and implement corrective measures, if needed, shifting power flows on transmission networks and lines to ensure that all system components operate within their specified voltage and current ranges. The high security and reliability requirements are also supported by the redundancy of components in the dispatching or control center. For example, the telecommunication interfaces (TCI) and the SCADA system are normally operated redundantly in a hot stand-by mode.

8.2.3 FLEXIBLE AC TRANSMISSION IN ACTIVE AND REACTIVE POWER CONTROL

Smart grids are expected to be flexible, secure, scalable, self-healing, cost-effective and environmentally compatible compared to the conventional power grids. Some or combinations of these tasks and attributes can be handled with the help of intelligent solutions and methods. FACTS systems and devices are playing increasingly important and critical roles in the smart grids and future developments of the power systems. This is achieved by continuously changing underlying impedance by using solid-state devices, rather than simple switching power through transformers and capacitors. FACTS use advances in solid-state devices and power converters, underlying power system physical proprieties to control power flows. FACTS devices consist essentially of the power reactors, capacitors for reactive power management, control and the power electronic equipment, i.e. converter switches and power electronic controllers together with their dedicated control and protection system. FACTS' major application includes reactive power compensation, phase shifting and power flow control. The three function principles of FACTS may be explained employing the equation for AC active power transfer, as depicted in Figure 8.2. The impact of FACTS on the power transfer is achieved through: (1) parallel compensation, the control of the voltages at one or both sides of a transmission line, higher voltages are leading to a higher power transfer; (2) series compensation, the reduction of the transmission line reactance through serially connected capacitors; and 3) load flow control, by the effect of the voltage angle difference between the transmission line ends. A FACTS device can shift the voltage phase angle to smaller differences, allowing larger amount of electrical energy to flow without loss of stability. In parallel compensation, parallel connected capacitors and reactors at one or both transmission line ends are used to generate reactive power (+Q), increasing voltage while the reactor sinks reactive power (-Q), reducing voltage. The electronic control is changing the reactive power balance from +Q to –Q, within about 40 ms. Depending on converter technology two types of compensators are employed:

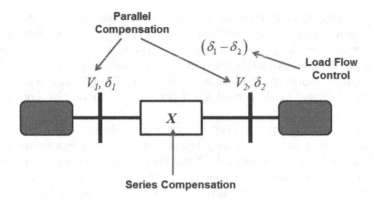

FIGURE 8.2 FACTS' functions and actions.

static VAR compensator (SVC), based on light triggered thyristors, Static synchronous compensator (SATCOM), based on IGBT transistors.

Parallel compensation performs the voltage control and reactive power control. Furthermore, the fast control opportunities are creating a positive impact on the system stability, facilitating a fast damping of post-fault power and voltage oscillations after faults, which are one of the blackout's root cases. The series compensation connects a capacitor serially with the transmission line (inductive element), decreasing in this way the line reactance (X), increasing the line current. For control purposes the capacitor may be connected to a parallel circuit containing a reactor and converter equipment. The effects of such installation are the power transfer capacity is significantly increased, the stability limits are extended and the post-fault power oscillations are damped much faster. However, the connection of capacitors or inductors in shunt changes the flow of reactive power in a circuit and so changes the network voltage. In general, a limited shunt reactive compensation level is benign, posing fewer risks to the power system. In a similar mode, power electronic shunt compensators change the reactive power flows. If the capacitors are connected to a circuit in series to reduce the line inductive reactance, a more hazardous approach as very high voltages can occur across the capacitor during faults and electro-mechanical resonances can be stimulated by the rotating machines. Notice that for a given value of capacitance, series compensation altering the reactance of the circuit is more effective in controlling voltages than shunt compensation that changes the reactive power flows. The FACTS technology is fully established, being now a mature technology and is operating in thousands of projects in the United States and worldwide.

8.3 POWER SYSTEM ANALYSIS

Successful operation of electrical energy systems requires that: generation must supply the load demand plus losses, and bus node voltage magnitudes must remain close to rated values, generators must operate within specified real and reactive power limits and transmission lines and transformers should not be overloaded for long periods. To transmit the electricity from power plants to those points where required transmission power systems are needed. These transmission subsystems must transmit the power in a safe way to the consumers maintaining some parameters within of a range of values. Examples of these parameters are the frequency, the voltage level and the voltage and current phase inter alia. Transmission systems operators (TSO) are in charge of transmitting the power into the consumers safely controlling the system elements values. However, the most critical challenge of power systems revolves around the most efficient and secure transfer of large electrical energy amounts over long distances. Power flow analysis is crucial during the power-system planning phases as well as during periods of expansion and change for meeting present and future load demands. It also is useful (and efficient) at determining power flows and voltage levels under normal operating conditions, while providing insight into system operation and optimization of control settings, which leads to maximum capacity at lower the operating costs. Therefore, it is important that voltages and power flows in an electrical system can be determined for a given set of loading and operating conditions. Power system states and the methods of the state calculations are extremely important in the system evaluation, control and planning of future expansion or restructuring. Optimum power flow studies are conducted to minimize power distribution losses and electrical energy costs, without affecting the on the voltage regulation, being also very important for power system planning, and for grind expansion or restructuring decisions. Frequency dynamics is one of the most important parameters of electrical power systems to understand power system dynamics, for which accurately measured wide-area frequency and frequency analysis are needed. Sudden changes in load demands or generation leads to significant frequency changes. These changes propagate throughout the grids, requiring monitoring and analysis for optimum operation and performances. One of the major smart grid capabilities is the extensive uses of distributed generation and renewable energy systems. However, the integration of renewable energy resources introduces a challenge in frequency control of smart grids.

8.3.1 POWER FLOW AND LOAD STUDIES

Under normal conditions, the electrical transmission systems operate in their steady-state operation mode and the basic calculation required to determine the characteristics of this state is called power flow. The objective of power flow calculations is to determine the steady-state operating characteristics of the power generation system for a given set of loads. Power flow calculations provide active and reactive power flows and bus voltage magnitude and their phase angle at all the buses for a specified power system configuration. A bus is a node at which one or several transmission lines, loads, capacitor banks and generators are connected. Each node or bus is associated with four quantities (parameters), such as voltage magnitude, voltage phage angle, active or real power and reactive power. In any load or power flow problem, two out of these four quantities are specified, while the other two are required to be determined through the solution of power flow equations. Loads are usually specified by their active and reactive power requirements, assuming that are unaffected by the small variations of voltage and frequency expected during normal steady-state operation. The power flow solution is consisted of two steps: first step involves finding the complex voltage at all buses, while the second step of computing active and reactive power flow. The solution has to be obtained for ill-conditioned problems, in outage studies and for real time applications. The mathematical formulation of power flow problem results in a system of algebraic nonlinear equations, which must be solved by iterative techniques. The purpose of power flow studies is to plan and account for various hypothetical situations. The basic steps involved in power flow studies are: determine element values for passive network components, locations and values of all complex power loads and the generation specifications and constraints. Next phases consist of the development of a mathematical model describing power flow in the network, solving for the voltage profile of the network, the power flows and losses in the network and finally checking for the constraint violations. The power flow model of a power system is built using the relevant network, load and generation data. Outputs of the power flow model include voltages at different buses, line flows in the network and system losses. These outputs are obtained by solving nodal power balance equations. Since these equations are nonlinear, iterative numerical techniques to be used to solve this problem. The problem is simplified as a linear problem in the DC power flow techniques.

Power flow analysis objective is estimated the power amount and characteristics transmitted through each path within a network of transmission lines. Power flow (or load flow) analysis provides the steady-state solution of a power network for specific network conditions which include both network topology and load levels. The power flow solution gives the nodal (bus) voltages and phase angles and hence the power injections at all buses and power flows through lines, cables and transformers. It is the basic tool and is critical for analysis, operation and planning of distribution networks. For example, data on peak load conditions assists the power system planners to determine the size of the system components, e.g. conductors, transformers, reactors, capacitor banks, protection devices or in setting new generation and transmission, and planning inter-ties with other systems to meet predicted demand in agreement with codes and regulations. Each node or bus with the power network is a common conductive point within the power network is characterized by single common current and voltage value. The term busbar, a wider strip of conductive material (metal), from which additional power lines can be connected to branch out other locations is originated from the term bus. A bus (node) can have several connections with power entering and leaving the node. In a power system, each busbar is associated with four quantities: the magnitude of voltage, |V| and its phase angle, θ, real power injection, P and reactive power injections, Q. For power flow analysis, only two of these quantities are specified, and the remaining two are obtained by the power flow solution. The starting point of solving power flow problems is to identify known and unknown variables in the power system. Depending upon the specified and unspecified quantities, the buses are classified into three types, slack, generation, and load buses as shown in Table 8.1. Power flow studies are one of the basic tools in power system analysis, forming the basis for stability analysis. Power flow problem is usually formulated as a vector equation, as described here:

$$f_{PF}(x_{PF}, p_{PF}) = 0 \qquad (8.2)$$

This vector equation represents a system of nonlinear equations referred to as the power flow equations, where x_{PF} represents the vector of bus voltage magnitudes, angles and other relevant unknown variables such as generator reactive powers, and p_{PF} stands for the vector of specified active and reactive powers injected at each node, as well as terminal generator voltage set points.

Simply stated, the load-flow problem is as: at any bus or node, there are four quantities of interest: |V|, θ, P and Q. If any two of these quantities are specified, the other two must not be specified, otherwise we end up with more unknowns than equations. Because records or measurements enable the real and reactive power to be accurately estimated at loads, P and Q are specified quantities at loads, which are termed as PQ buses. Most of the buses in practical power systems are load buses. Likewise, the real power output of a generator is controlled by the prime mover and the magnitude of the voltage is controlled by the exciter, so and P and |V| are specified at generators, which are called PV buses, meaning that |V| and θ are unknown at each load bus and θ and Q are unknown at each generator bus. Since the system losses are unknown until a solution to the load-flow problem has been found, it is necessary to specify one bus that will supply these losses. This is called the slack (or swing, or reference) bus and since P and Q are unknown, |V| and θ must be specified. Usually, an angle of $\theta = 0$ is used at the slack bus and all other bus angles are expressed with respect to slack. Load buses both net real and reactive powers of the loads are specified. The slack (swing) bus is required to provide the mismatch between scheduled generation and the total system

---

TABLE 8.1

Bus Types for Power Flow Analysis

Bus Type	Specified Parameter	Parameters to be Estimated	Remarks
Slack/Swing Bus	|V|, θ	P, Q	|V|, θ: are assumed, if not are Specified, they are set as 1.0 and 0°.
Generation PV (Real Power; Voltage) Bus	P, |V|	Q, θ	A generator is present at the bus.
Load PQ (Real Power; Reactive Power) Bus	P, Q	|V|, θ	About 80% of buses are of PQ type
Voltage Controlled Bus	P, Q, |V|	a, θ	"a" is the % tap change in tap-changing transformer

load including losses and total generation. The slack bus is usually considered as the reference bus because both voltage magnitude and angles are specified; therefore, it is also called the swing bus. Notice that the slack bus is usually a PV bus with the largest capacity generator of the given system connected to it. The swing bus generator supplies the *difference between the specified power in the system at other buses and the total system output plus losses*. It needs to supply additional real and reactive power to meet the losses. The generator buses are called regulated or PV buses because the net real (active) power is specified and voltage magnitude is regulated. For PQ buses, voltage magnitudes and angles are unknown, whereas for PV buses, only the voltage angle is unknown. As both voltage magnitudes and angles are specified for the slack buses, there are no variables that must be solved for. A system with n buses and g generators has 2n-g-1 unknowns. To find these unknowns, real and reactive power balance equations are used. To write these equations, the transmission network is modeled in terms of admittance matrix (Y-bus). The power flow is so formulated as a set of nonlinear equations and then appropriate mathematical methods, e.g. the Gauss-Seidel, Newton-Raphson or fast-decoupled method are used for the solution of the equations. The purpose of any load flow analysis is to compute precise steady-state voltages and voltage angles of all buses in the network, for specified load, generator real power and voltage conditions. Once this information is known, the real and reactive power flows into every line and transformer, as well as generator reactive power output is analytically determined. The power analysis is performed in terms of node voltage, current, and apparent power as:

$$S_i = V_i \cdot I_i^* = P_i + Q_i \tag{8.3}$$

Here, the index i is keeping track the of which bus is analyzed and described, P_i and Q_i are the bus real and reactive powers, respectively, and the star indicates the complex conjugate of the bus current. The admittance matrix of a power system is a mathematical model of the system, consisting of admittance values of both lines and buses. The Y-bus is a square matrix, symmetrical along main diagonal with dimensions equal to the number of buses in that system, being expressed as:

$$[Y] = |Y_{ij}|, \; i = 1, \cdots, n \; \text{ and } \; j = 1, \cdots, n \tag{8.4}$$

The values of diagonal elements (Y_{ii}) in the admittance matrix are equal to the sum of the admittances connected to bus i. The off-diagonal elements (Y_{ij}) are equal to the negative of the admittance connecting the two buses i, and j. It is worth noting that with large systems, the admittance matrix, Y-bus is a sparse matrix. Relationship between per-unit real power and reactive power supplied to the power system at bus i and the per-unit current injected into the system at that bus is given by the Equation (8.3). Then the current node (bus) relationship at bus I, applying Ohm's Law, can be written as:

$$I_i = \sum_{j=1}^{n} Y_{ij} V_j \tag{8.5}$$

The load flow problem is restricted, here to a balanced three-phase power system, so that the analysis can be carried out on a single-phase basis. Notice that quantities are expressed in per-unit. The first step in the analysis is the formulation of suitable equations for the power flows in the system. The power system is a large interconnected system, where various buses are connected by transmission lines. At any bus, complex power is injected into the bus by the generators and is drawn by the loads. Either generators or loads may not be present at any bus. The power is transferred from one bus to other via the transmission lines. At any bus i, the complex (apparent) power S_i (injected), as shown in Figure 8.3, is defined as:

$$S_i = S_{Gi} - S_{Ldi} \tag{8.6}$$

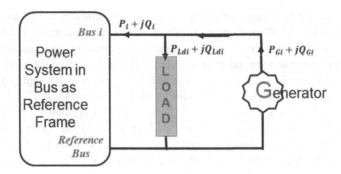

FIGURE 8.3   Power flow at bus i.

Here, S_i is the net complex power injected into bus i, S_{Gi} is the complex power injected by the generator at bus i and S_{Ldi} is the complex power drawn by the load at bus i. According to conservation of complex power, at any bus i, the complex power injected into the bus must be equal to the sum of complex power flows out of the bus via the transmission lines, as expressed by:

$$S_i = \sum_j S_{ij}, \quad \forall i =, 1,\ldots, n \tag{8.7}$$

where S_{ij} is the complex power of line j, and the sum is over all lines connected to the bus and n is the number of buses in the system (excluding the ground). The bus current injected at the bus-i is defined as:

$$I_i = I_{G,i} - I_{Ld,i}, \quad i = 1,\ldots,n \tag{8.8}$$

where $I_{G,i}$ is the current injected by the generator at the bus and $I_{Ld,i}$ is the current drawn by the load (energy demand) at that bus. In the bus frame of reference, power flow in matrix form is given by:

$$I_{Bus} = \begin{bmatrix} I_1 \\ I_2 \\ \vdots \\ I_n \end{bmatrix} = \begin{bmatrix} Y_{11} \ Y_{12}\ldots Y_{1n} \\ Y_{21} \ Y_{22}\ldots Y_{2n} \\ \ldots\ldots\ldots\ldots \\ Y_{n1} \ Y_{n2}\ldots Y_{nn} \end{bmatrix} \cdot \begin{bmatrix} V_1 \\ V_2 \\ \vdots \\ V_n \end{bmatrix} = Y_{Bus} \times V_{Bus} \tag{8.9}$$

The admittance matrix of a power system, as the one in Equation (8.9) is an abstract mathematical model of the system. It consists of admittance values of both lines and buses. The Y-bus is a square matrix with dimensions equal to the number of buses. A power system admittance matrix is an abstract system mathematical model, consisting of admittance values of both lines and buses. The Y-bus is a square matrix with dimensions equal to the number of buses, expressed by:

$$Y_{ii} = \sum_{\substack{j=0 \\ j \neq i}}^{n} y_{ij}, \quad \text{and } Y_{ij} = Y_{ji} = -y_{ij} \tag{8.10}$$

The net injected power at any bus can be calculated using the bus voltage (V_i), neighboring bus voltages (V_j), and admittances between the bus and its neighboring buses (y_{ij}). Equation (8.9) can be considered for the sake of computation simplification, as expressed per elements, as:

$$I_i = \sum_{j=1}^{n} Y_{ij} \cdot V_j, \quad \forall i = 1, \ldots, n \tag{8.11}$$

Then the complex (apparent) power S_i as in Equation (8.3) is expressed as:

$$S_i = V_i \cdot I_i^* = P_i + jQ_j = V_i \sum_{j=1}^{n} Y_j^* \cdot V_j^* \tag{8.12}$$

Or by using the complex conjugate of the bus apparent power, Equation (8.11) is expressed as:

$$S_i^* = P_i - jQ_i = V_i^* \left(V_i Y_{ii} + \sum_{\substack{j=1 \\ j \neq i}}^{n} Y_{ij} V_j \right) \tag{8.13}$$

With the notation for bus voltages, expressed in magnitude and voltage angle, and by using bus conductances, G and susceptances, the apparent power can be expressed in a form that is more suitable for power flow calculations.

$$V_i \triangleq |V_i| < \delta_i = |V_i|(\cos\delta_i + j\sin\delta_i)$$
$$\delta_{ij} = \delta_i - \delta_j$$

And

$$Y_{ii} = G_{ij} + jB_{ij}$$

Hence from Equations (8.11) and (8.12), we get after some algebraic manipulation:

$$S_i^* = \sum_{j=1}^{n} |V_i||V_j|(\cos\delta_{ij} + j\sin\delta_{ij})(G_{ij} - jB_{ij}) \tag{8.14}$$

Separating real and imaginary parts in Equation (8.14), the real and reactive powers are obtained:

$$P_i = \sum_{j=1}^{n} |V_i||V_j|(G_{ij}\cos\delta_{ij} + jB_{ij}\cos\delta_{ij}) \tag{8.15a}$$

$$Q_i = \sum_{j=1}^{n} |V_i||V_j|(G_{ij}\sin\delta_{ij} - jB_{ij}\sin\delta_{ij}) \tag{8.15b}$$

Alternate forms of the active and reactive powers, Pi and Qi can be obtained by representing voltages and the bus admittances Yij in phasor (polar) forms.

$$P_i = \sum_{j=1}^{n} |V_i||V_j||Y_{ij}| \cos(\delta_i - \delta_j - \theta_{ij}), \quad \forall i = 1,2,\ldots,n \tag{8.16a}$$

$$Q_i = \sum_{j=1}^{n} |V_i||V_j||Y_{ij}| \sin(\delta_i - \delta_j - \theta_{ij}), \quad \forall i = 1,2,\ldots,n \tag{8.16b}$$

And the current (I_i) can be written as a function of the active and reactive powers, the bus voltages and admittances as expressed by this relationship:

$$I_i = \frac{P_i - Q_i}{V_i^*} = V_i Y_{ii} - \sum_{\substack{i=1 \\ i \neq j}}^{n} Y_{ij} V_j \tag{8.17}$$

Here, θ_{ij} is the bus admittance angle. Equations (8.11) trough (8.17) are the *power flow equations* or the *load flow equations*, in two alternative forms, corresponding to the n-bus system, where each bus-i is characterized by four variables, P_i, Q_i, $|V_i|$ and δ_i. Thus, a total of 4n variables are involved in the power flow equations. The load flow equations can be solved for any 2n unknowns, if the other 2n variables are specified. These equations are establishing the need and provide the way for the classification of buses of the power system for load flow analysis into, PV bus, PQ bus, slack bus, etc. Irrespective of the method used for the solution, the data required is common for any load flow. All data is normally in per units. The bus admittance matrix is formulated from these data. The various data required are system data, generator bus data, load data, transformer data, transmission line data and shunt element data. System data includes the total number of buses, n, the number of PV buses, loads, transmission lines, number of transformers, shunt elements, the slack bus number, voltage magnitude of slack bus (angle is usually taken as $0°$), tolerance limit, base MVA and maximum permissible number of iterations. Generator bus data for every PV bus, i, includes the bus number, active power generation P_{Gi}, the specified voltage magnitude $|V_{sp,i}|$, minimum reactive power limit $Q_{i,min}$ and maximum reactive power limit $Q_{i,max}$. For all loads the data required includes the bus number, active power demand P_{Ldi} and the reactive power demand Q_{Ldi}. For every transmission line connected between buses i and k the data includes the starting bus number i, ending bus number k, resistance of the line, reactance of the line and the half line charging admittance. For every transformer connected between buses i and k the data to be given includes the starting bus number i, ending bus number k, resistance of the transformer, reactance of the transformer and the off nominal turns-ratio a. The data needed for the shunt element includes the bus number where element is connected, and the shunt admittance ($G_{sh} + jB_{sh}$). In summary, due to growing demand of the power and complexity of the power system networks, the power system studies are significant tools for power system operators in order to take corrective actions in time. The power flow models are built using the network, load and generation data. Outputs of the power flow model include voltages at different buses, line flows in the network and system losses. These outputs are obtained by solving nodal power balance equations, a set of nonlinear equations. Since these sets of equations are nonlinear, iterative techniques must be employed to solve them. The most common techniques for iterative solution of non-linear power (load) flow analysis equations are Gauss-Seidel, Newton-Raphson and Fast Decoupled methods.

The Gauss-Seidel (GS) method, also known as the method of successive displacement, is the simplest iterative technique used to solve power flow problems, being used in a broad range of the engineering problems. In general, GS method follows the following iterative steps to reach the solution for the function $f(x) = 0$. In the context of a power flow problem, the unknown variables are voltages at all buses, with the exception of the slack bus. Both voltage magnitudes and angles are unknown for load buses, whereas voltage angles are unknown for regulated/generation buses. In the simplest method variant (known sometimes as Gauss-iterative method), the voltage, V_i at a bus i can be calculated, by using the iterative relationship, in terms of system parameters, voltage, admittance and powers, expressed as:

$$V_i = \frac{1}{Y_{ii}} \left(\frac{P_i^{sch} - Q_i^{sch}}{V_i^*} - \sum_{\substack{j=1 \\ i \neq j}}^{n} Y_{ij} \cdot V_j \right)$$

(8.18)

where Y_{ij}, is the admittance between buses i and j, an element of the Y_{bus}, P_i^{inj} is the net scheduled injected real power, Q_i^{inj} is the net scheduled injected reactive power, and Q_i^* is the conjugate of V_i. The net injected quantities are the sum of the generation minus load. In this method, the initial estimates of V_i is assigned to $1\angle 0°$. The iterative voltage equation is as follows:

$$V_i^{(k+1)} = \frac{1}{Y_{ii}} \left(\frac{P_i^{sch} - Q_i^{sch}}{V_i^{*(k)}} - \sum_{\substack{j=1 \\ i \neq j}}^{n} Y_{ij} \cdot V_j^{(k \text{ or } k+1)} \right)$$

(8.18)

The Gauss iterative method uses the same set of the voltage values throughout a complete iteration, instead of immediately substituting each new value obtained to compute the voltage at the next bus. Both real and reactive powers are scheduled for the load buses, and Equation (8.17) is used to determine the voltage magnitudes and the angles ($|V_i| < \delta$) for every of the iterations of the bus voltage ($V_i^{(k+1)}$). Notice that for regulated buses, only the real power is scheduled. Therefore, net injected reactive power is calculated based on the iterative voltages ($V_i^{(k+1)}$), by using this equation:

$$Q_i^{(k+1)} = -\sum_{j=1}^{n} |V_i|^{(k)} |V_j|^{(k \text{ or } k+1)} |Y_{ij}| \sin\left(\theta_{ij} - \delta_i^{(k)} + \delta_j^{(k)}\right)$$

(8.19)

where $|V_i|$ and $|V_j|$ are the voltage magnitudes at buses i and j, respectively, while δ_i and δ_j are the associated voltage angles, $|Y_{ij}|$ is the magnitude of the Y-bus element between the two buses, and θ_{ij} is the corresponding angle. Since the voltage magnitude ($|V_i|$) is specified at regulated PV buses, Equations (8.16a and b) can be used to determine the voltage angles only. The iterative process ends when the voltage reaches acceptable limit, the differences between successive voltages is less than a threshold.

$$\left| V_i^{(k+1)} - V_i^{(k)} \right| \leq \varepsilon$$

(8.20)

In Gauss-Seidel, the upgraded iterates are used as soon as they are available, thus the number of iterations is reduced. For a system with n buses, the calculated voltage at any bus i is given by:

$$V_i^{(k+1)} = \frac{1}{Y_{ii}}\left(\frac{P_i^{sch} - Q_i^{sch}}{V_i^{*(k)}} - \sum_{\substack{j=1 \\ i\neq j}}^{i-1} Y_{ij} \cdot V_j^{(k)} - \sum_{\substack{j=i+1 \\ i\neq j}}^{n} Y_{ij} \cdot V_j^{(k+1)} \right)$$

(8.21)

Example 8.1   Calculate the voltage at bus 2 for the simple system shown in figure below, by using the Gauss iterative method, if the voltage at bus 1 (PS 1) is $V_1 = 1.0{<}0°$ p.u.

SOLUTION

The injected apparent power, the voltage, and the admittance bus matrix are:

$$S_{Bus2} = j1.0 - 0.75 - j1.0 = 0.75$$

$$V_2 = 1.0 < 0°$$

$$Y_{Bus} = \begin{bmatrix} -j2 & j2 \\ j2 & -j2 \end{bmatrix}$$

The voltage, V_1 is specified, a is a constant through all the iterations, the initial voltage at bus 2 is:

$$V_2 = 1.0 + j0.0 = 1.0 < 0° \text{ p.u.}$$

The estimated voltage at bus 2 is computed by using the iteration of Equation (8.17):

$$V_2^{(1)} = \frac{1}{Y_{22}}\left(\frac{P_2^{sch} - Q_2^{sch}}{V_2^{*(k)}} - Y_{12} \cdot V_1 \right) = \frac{1}{-2j}\left(\frac{-0.75}{1 < 0°} - j2 \times 1 < 0° \right) = 1.0 - j0.375 = 1.068 < -20.6°$$

$$V_2^{(2)} = \frac{1}{-2j}\left(\frac{-0.75}{1.068 < 20.6°} - j2 \cdot 1 < 0° \right) = 1.1235 - j0.3287 = 1.1706 < -16.3°$$

After four more iterations, the difference in the voltage magnitudes is less than 10^{-6} p.u., the process is stopped and the voltage is 1.1324<-14.6° p.u.

The Newton-Raphson method is a procedure to solve nonlinear algebraic equations system of the form F(x) = 0. Using Taylor series and neglecting terms with order over first order, next expression used to solve nonlinear equations systems is found. Solving this system and by actualizing the variables at each iteration allows finding a system solution. This method is often used to perform transmission networks studies. This method is very effective for large systems; however, it does not take the advantage of the radial structure of the power distribution, being quite ineffective. The method fails when the Jacobean matrix is singular or the system is in ill-condition as in the case of a low distribution reactance-resistance, X/R ratio. In power flow studies, the Newton-Raphson method assumes an initial starting voltage used to compute the apparent power mismatch, the mismatch power, ΔS. Solving this system and actualizing the variables at each of the iteration allows finding the system solutions. As variable vector this method use voltage arguments and modules of PV and PQ buses. The equations to solve are Equations (8.16a) and (8.16b). From these equations, it is possible to obtain the so-called power mismatch equations in order to calculate the Jacobean matrix and functions vector:

$$\Delta P_i = P_i - |V_i| \sum_{j=1}^{n} |V_j||Y_{ij}| \cos(\delta_{ij} - \theta_{ij}), \quad \forall i = 1,2,\ldots,n \tag{8.22a}$$

$$\Delta Q_i = Q_i - |V_i| \sum_{j=1}^{n} |V_j||Y_{ij}| \sin(\delta_{ij} - \theta_{ij}), \quad \forall i = 1,2,\ldots,n \tag{8.22b}$$

Then, this system of equations is solved, and the solution actualized:

$$[J] = \begin{pmatrix} J_1 & J_2 \\ J_3 & J_4 \end{pmatrix} = \begin{pmatrix} \partial \Delta P / \partial \delta & \partial \Delta P / \partial V \\ \partial \Delta Q / \partial \delta & \partial \Delta Q / \partial V \end{pmatrix} = \begin{pmatrix} \Delta P(x) \\ \Delta Q(x) \end{pmatrix} \tag{8.23}$$

To start the iterative process, a solution estimation is needed, with typical value for voltage magnitude is $V_i = 1.0$ and the voltage argument are $\delta_i = 0$. The function values and Jacobean matrix, J are estimated for these initial values (guesses), the equations are solved and the variables actualized. The process is repeated until the errors were smaller than a certain error value, the threshold values, ε, as expressed by:

$$\max|\Delta P_i(x)| \leq \varepsilon \tag{8.24a}$$

$$\max|\Delta Q_i(x)| \leq \varepsilon \tag{8.24b}$$

Fast-decoupled methods were developed in order to obtain a fast power flow solution. This method is based on Newton-Raphson equations with some simplifications, reason why it is less accurate. The fast-decoupled methods simplify the Jacobean matrix by using small angle approximations to eliminate, relatively small Jacobean elements, being developed in order to obtain a fast power flow solution. These simplifications are based basically on the relative small phase shift between two adjacent buses, the bus admittances are nearly pure imaginary. With these simplifications, the Jacobean matrices are:

$$J_2 = J_3 = 0$$

$$J_{ij}^1 = J_{ij}^4 = -V_i V_j B_{ij} \tag{8.25}$$

$$J_{ii}^1 = J_{ii}^4 = -V_i^2 B_{ii}$$

where B_{ii} and B_{ij} are the admittance imaginary terms, the susceptances. Then, the system of equations is reduced into two decoupled systems of equations, expressed in matrix form as:

$$-V \cdot B \cdot V \cdot \Delta \delta = \Delta P \tag{8.26a}$$

$$-V \cdot B \cdot V \cdot \frac{V}{\Delta V} = \Delta Q \tag{8.26b}$$

With the additional simplification of the voltage at bus i can be approximate to 1, the second V matrix of Equations (8.26a) and (8.26b) is replaced by an identity matrix. Then, the two decoupled systems are:

$$-B' \cdot \Delta\delta = \Delta P \tag{8.27a}$$

$$-B'' \cdot \Delta V = \Delta Q \tag{8.27b}$$

Here, the two matrixes B' and B" don't depend on the iteration and they are calculated and inverted only once. This is the reason why the decoupled method is very fast. As in Newton-Raphson method, the iterations should continue until the errors were smaller than the desired threshold error.Although power flow computation for transmission systems is well understood, the distribution's power flow analysis can encounter difficulties in an ill-conditioned distribution network due to that the network structure of a distribution system is often radial or weakly meshed, the low X/R ratio in distribution feeders, often the power distribution loads are unbalanced, the distribution networks consists usually of a mix of short line segments with low impedance and long feeders with high impedance and the connection of distributed and renewable energy generators. Therefore, several of power flow methods have been developed, or adapted from transmission power flow methods, for the analysis of distribution networks.For example, the adapted Newton-Raphson method uses power mismatches at the ends of feeders and laterals to deal with the nodal voltages and can accelerate the convergence of the algorithm. The adapted Gauss-Seidel method uses the bus-impedance matrix to deal with the branch currents. In the *forward-backward sweep method*, the distribution system is modeled as a tree network and the slack bus is the tree root, the branch networks as layers and weakly meshed networks are converted in radial structures. The algorithm starts with the slack bus selection and assuming of the initial voltage and phase angle at the root, node and other buses, followed by the current injection calculation at certain iteration, m, from the scheduled active and reactive powers (P_i^{sch}, Q_i^{sch}), and the latest bus voltage, $V_i^{(m-1)}$ as expressed by:

$$I_i^{(m)} = \left[\frac{P_i^{sch} + jQ_i^{sch}}{V_i^{(m-1)}} \right]^* = \left(\frac{S_i^{sch}}{V_i^{(m-1)}} \right)^* \tag{8.28}$$

Next, by starting from the feeder and lateral ends towards the root end, the voltages are calculated by:

$$V_i^{m+1} = V_j^m + \frac{1}{Y_{ij}} I_{ij}^m \tag{8.29}$$

where j is the adjacent down-stream bus to bus i, and two busbars are connected by a branch having admittance of Y_{ij} or impedance of Z_{ij}. The iteration ends, when the termination criterion, the calculating the power mismatch (or the voltage mismatch) is less than the threshold, ε, as expressed by:

$$\Delta S_i^m = S_i - V_i^m \left[I_i^m \right]^* \le \varepsilon \tag{8.30}$$

Example 8.2   If the line impedance connecting bus 2 and bus 3 is $Z_{23} = 0.05 + j0.08$, perform first iterations of the forward-backward method for the electric network of Figure 8.5.

SOLUTION

In first step of the iteration, the iteration all bus voltages are assumed equal to 1<0° p.u.

$$V_1^{(1)} = V_2^{(1)} = V_3^{(1)} = 1.0 \, pu$$

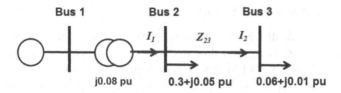

FIGURE 8.4 Diagram of the network for Example 8.2.

In the forward seep the currents from bus 2 to bus 3 and bus 3 to bus 4 are calculate:

$$I_1^{(1)} = \left[\frac{0.3 + j0.06 + 0.05 + j0.01}{1.0}\right]^* = 0.35 - j0.07 \, pu$$

And

$$I_2^{(1)} = \left[\frac{0.05 + j0.01}{1.0}\right]^* = 0.05 - j0.01 pu$$

In the backward sweep the voltages at bus 2 and bus 1 are computed as:

$$V_2^{(2)} = V_3^{(1)} + Z_{23} \times I_2^{(1)} = 1.0 + (0.05 + j0.08) \cdot (0.05 - j0.01) = 1.0033 + j0.0035$$

And

$$V_1^{(2)} = V_2^{(1)} + Z_{12} \times I_1^{(1)} = 1.0 + j0.08 \cdot (0.35 - j0.07) = 1.0056 + j0.028$$

In second step, the forward sweep, the currents are updated with the new bus voltages, followed by calculation in the backward sweep of the new voltages at bus 2 and bus 1. The process can continue until the desire accuracy is reached, as specified by Equation (8.29).

Notice in the second step of the previous power flow algorithm, the lateral voltages are either computed or assumed, which involves the bus voltage, $V_i^{(m-1)}$ of previous iteration and the injected current of the present iteration, $I_i^{(m)}$. Computation starts from the last branch of the lateral feeder, moving back through the tree root (node), by using as preceded before the following expression:

$$I_i^{(m)} = \left[\frac{P_i^{sch} + jQ_i^{sch}}{V_i^{(m)}}\right]^* = \left(\frac{S_i^{sch}}{V_i^{(m)}}\right)^* \tag{8.31}$$

Other commonly used alternative methods are the *load flow based on the sensitivity matrix for mismatch calculation* and the *bus impedance method*. The first method is an improved version forward-backward sweep method, employing a sensitivity matrix scheme to compensate the mismatch between the slack bus power injection and the power (load) flow at the feeder and lateral ends. This is an adapted Newton-Raphson method for power distribution load flow. It assumes the slack bus as root bus, the active and reactive power injected in the slack bus is equal to the sum of all system loads and the load flows in each branch is equal to all downstream connected loads. Then, the latest powers and voltages, together with the power losses are computed. From losses powers and voltages are updated, and when the mismatch power is approximately zero the load flow converged. The second approach uses the bus impedance matrix to solve the network equations in a distribution system. It employs a simple superposition to compute the bus voltages through the system. The bus voltages are computed after specifying the slack bus voltage, followed by the calculation of the incremental voltage change, ΔV due to the current injected into network. The method assumption include: initial no load in the system, the bus voltages are initialized using the slack bus voltage and then the nodal voltages are updated function of connected loads.

8.3.2 Smart Grid Power Flow Studies

Power (load) flow methods discussed previously for transmission and distribution grid sections are not sufficient for smart grid power flow studies for several reasons. Adequate analysis tools are thus required to properly analyze power system stability when subjected to penetration of intermittent power sources and distributed generation. A variety of methods and approaches are available in the literature to perform power flow, transient and voltage stability analyses considering uncertainties associated with electrical parameters. The smart grid, consisting in the integration of various technologies such as dispersed generation, dispatchable loads, communication systems and storage devices which operates in grid-connected and islanded modes are require new tools and methods for analysis and operation. As a result, traditional optimization techniques in these new power systems have been seriously changed during the last decades. One of the most important technical and economical tools in this regard is the optimal power flow (OPF). As a fundamental optimization tool in the operation and planning fields, OPF has an undeniable role in the conventional power systems and smart grids. Optimal power flow for minimizing the cost of power generation subject to operating constraints and meeting the load or energy demands provides one of the most important smart grid applications. The main OPF objective is to meet the load demand while keeping the minimum generation cost. It also includes the economic load dispatch between the generating units by assigning the load to each unit so the fuel cost and the losses are minimized. OPF also maintain system security by maintaining the system in desired operating range at steady state. Maximum and minimum operating range is decided by the control centers, so that at the time of overload, necessary action can be taken easily. However, the OPF only deals with steady state operating of power system, not with transient stability, contingency analysis of power system. The OPF applications are used to calculate optimum generation pattern and to achieve the minimum generation cost of generation, by using current state of short- and long-term load forecasting, OPF can provide preventive dispatch, while at the time of overload, when voltage limits get violated, a corrective dispatch action is provided by OPF solution, being also used to provide optimum generation voltage setting for switched capacitor and for static VAR compensators. Optimal power flow solution is also used for calculation of bus incremental cost. Bus incremental cost tool is generally used to determine the marginal cost of power at any bus. Load or power flow studies, incorporating the stochastic and random studies of the smart grid are modeled in the way that the conditioning load flow topology is including the feeders and the time-dependent power flows. The RES and DG interoperability with smart grid specifications can account for the extension of the current methodologies to perform the analysis in both usual and emergency situations. Such approaches are proven useful in terms of characteristics and uses in power system planning and operation. There are two types of modeling distribution networks in smart grid: balanced equivalent single-phase modeling, which aims at naively approximating the network by a balanced system of three decoupled single-phase subsystems, and unbalanced three-phase modeling, which preserves the unbalanced structure of the electric network for constructive power flow analysis.

In power system, interconnections were primarily used for pooling of power between generation stations and load centers along with the added advantages of reduction of overall generation capacity, minimum generating cost with increased reliability, supply security and better utilization of energy reserves. Power transfer between areas is a main function or application of a running and operating an electric power system. However, transmission networks have limited capability to transfer power. Available transfer capability (ATC) is the capability of transmission present in the power system network for any more transfer of energy above the contracted usage. ATC gives the overview of the method used for finding the capability of transmission line for transmitting the electrical power which can be further used for finding the appropriate method for calculating each component's contribution in the transmission network. The ATC importance in deregulated market is critical in maintaining the system reliability, security and restoration. It gives the information about the available capacity in the transmission network, which it is helpful in making future decisions regarding the next bunch of transaction to be undertaken.

Total transfer capability (TTC) is the capacity of the transmission network in which electrical power transfer can be carried out reliably taking care of all the contingency conditions simultaneously.

Transmission reliability margin (TRM) refers the capacity left in the transmission line to ensure secure transmission under any uncertainty in the system. Existing transmission commitments (ETC) denotes the current power transferred between the concerned power system areas. Capacity benefit margin (CBM) is the reserve transfer capacity by the network to ensure the reliability requirement of the generation. Mathematically, these parameters are related to a simple relationship, expressed by:

$$ATC = TTC - TRM - ETC - CBM \tag{8.32}$$

According to North American Electric Reliability Corporation (NERC 1996), transfer capability is the function of total generation, demand of customers and the transmission network condition for the particular time period. In summary, the transfer capability refers to the ability of the power system to reliably transfer power between two power system areas. These areas may be those formed by individual electric systems, power pools, control areas or sub-regions. The transfer capacity is related to the rating of equipment used for transmission, i.e. the capacity of network cannot change but capability changes with time and condition of the system. ATC evaluation methods are of two broad categories: deterministic methods and probabilistic methods. The determination of transfer capability is mostly based on computer simulations of various scenarios of operations. These simulations are performed by power-flow solutions.

8.3.3 SYNCHROPHASOR MEASUREMENTS, TIME SYNCHRONIZATION AND FREQUENCY ANALYSIS

The electric power system is a complex interconnected system, where the loads and the network topology of the power system change randomly in real time, and the generation must change instantaneously in order to satisfy the demand. Any slight mismatch gives rise to instantaneous frequency fluctuation in different areas of the network. Electromechanical oscillations occur when the system responds to large disturbances such as those triggered by faults and component failures. System frequency is normally the same at all points of the power network at steady state and in normal operation conditions. Frequency in a power system is a real-time changing variable that indicates the balance between generation and energy (load) demand. The transmission system operators are usually responsible for maintaining the frequency response of the power system within acceptable limits. For example, in UE, two main levels define these limits: the operational limit, which is equal to ± 0.2 Hz (i.e. 49.8 Hz to 50.2 Hz), and the statutory limit, which is equal to ± 0.5 Hz (i.e. 49.5 Hz and 50.5 Hz). Under a significant drop in the frequency (i.e. below 49.2 Hz), a disconnection by low-frequency relays is provided for frequency control of both the generators and load demands. The introduction of phasor measurement units (PMUs) in power systems significantly improves the monitoring and analyzing power system dynamics. Synchronized measurements make possible to directly measure phase angles between corresponding phasors in different locations within the power system. Improved monitoring and remedial action capabilities allow more efficient uses of the power system. Improved information facilitates fast and reliable emergency actions, reducing the needs for relatively high transmission margins required by potential system disturbances.

Synchrophasors are precise measurements of the power systems and are obtained from PMUs, measuring the voltage, current and frequency in terms of magnitude and phasor angle at a very high speed (about 30 measurements per second). The main PMU components are the current and voltage sensors, and units to provide accurate and absolute timing, and data processing. It may be a standalone unit or can be integrated in other power system devices. Its purpose is to accurately measure a waveform with a reference to an absolute time, enabling it to be comparted in meaningful ways with other waveforms. A PMU measures 50 Hz or 60 Hz sinusoidal waveforms of voltages and currents at a high sampling rate, up to 1200 samples per second and with high accuracy. From these voltage and current samples, the magnitudes and phase angles of the voltage and current signals are calculated by the PMU phasor microprocessor. PMU measurements can be used locally or transmitted to central locations. Each phasor measurement recorded by PMU devices is time-stamped based on universal standard time, so the phasors measured by the PMUs installed in different locations can be synchronized by aligning time-stamps. The PMUs use the clock signal of the Global Positioning System (GPS) to provide synchronized phase angle measurements at all measurement points, the measured phasors are often referred to as *synchrophasors*. The phasor measurements are transmitted either via dedicated links between specified sites or over a switched link that is established for the communication purpose. Synchrophasors measured at different network points are transmitted to an area phasor data concentrator (PDC) at a rate of 30–60 samples per second, which is sending the collected data to a grid PDC system where there is application software for data visualization, storing the data in a central database and for integration with Energy Management Systems, SCADA and Wide Area Application Systems. Notice that a synchrophasor is meaningful as long as the frequency of the measured parameter is constant. The main PMU sources of errors are related to the timing and frequency variations. However, the system frequency varies in connection with the load and demand relationship. If the generation is not able to keep with the demand the frequency is decreasing, while when the generation is exceeding the demand, the frequency is increasing. So, it is important that the synchrophasor measurement system to be able to provide the frequency and frequency time change rate (FCR) along with the phasor value as given by these relationships:

$$f(t) = \frac{1}{2p}\frac{df(t)}{dt} \tag{8.33a}$$

And

$$FCR = \frac{df(t)}{dt} \tag{8.33b}$$

Smart grids fundamentally change the grid and substation operations to automated systems, depending on precise and accurate timing, the automation system a mission-critical task and must be synchronized across large-scale distributed power grid switches, enabling smooth power transfer and maintain power supply integrity. Precise time synchronization guarantees that grid and substation devices have accurate clocks for system control and data acquisition, making possible to perform precise global analysis of network response and identification of when, where and why any faults and large disturbances have occurred. Synchronized timestamping is especially important for sampled values (IEC61850-9-2 standard) of current, and voltage values requiring accurate clocks inside the merging units. IEDs and merging units' (MUs) internal time clocks, Ethernet switches and wherever processes need to be synchronized should be precisely synchronized, with synchronization requirements for various power system applications, ranging from 25 ms to 1 μs. Time-synchronized networks are vital for the proper operation of network applications with optimal performance. The time synchronization ultimate goal is to bring the local clocks of servers and other instrumentation, devices and equipment in a network into phase so that their time differences is zero or very close to zero. Network time protocol (NTP) is a networking protocol for clock synchronization between computer systems over packet-switched, variable-latency data networks, intended to synchronize all participating computers to within a few milliseconds of Coordinated Universal Time (UTC). Nowadays NTP provides a well-tested method to enable reasonably accurate time synchronization across TCP/IP networking infrastructure.

8.4 STABILITY AND CONTINGENCY ANALYSIS IN SMART GRIDS

The smart grid operation, security assessment and control application includes, among others: security monitoring, security analysis, preventive control, emergency control, fault diagnosis and restorative control. The tools and methods required include: network topology analysis, external system equivalent modeling, state estimation, on-line power flow, security monitoring (on-line identification of the actual operating condition, such as secure or insecure), fault and contingency analysis. State estimation and contingency analysis are the two most fundamental tools for monitoring the power system. State estimation is the process of fitting data coming in from sensors in the field to a system model and determining an estimate of the power system state. By its nature, state estimation depends on the communication infrastructure, sensing, monitoring, information and computing technology. Contingency analysis is a critical activity in the context of the power infrastructure because it provides a guide for resiliency and enables the grid to continue operating even in the case of failure. It is directly connected to the fault analysis, localization, corrective actions and finally service restoration.

8.4.1 FAULT ANALYSIS

Any power system can be analyzed by calculating the system voltages and currents under normal and abnormal scenarios, e.g. in the case of faults. A power distribution network and a transmission subsystem are designed such that it is safe and remains undamaged under both normal and abnormal conditions. However, the dimensioning, cost effectiveness and safety of these systems depend to a great extent on being able to control short-circuit currents. Notice that different types of faults may occur in power systems. Fault analysis involves understanding and characterization of various fault types, designing the power system sections and subsystems to efficiently protect and clear the most likely or severe faults, localizing where the faults occur, restoring the service and/or repairing the fault affected section or component(s). With the increasing power of loads and the ratings of distributed generators, the importance of short-circuit current calculation increases. Accurate fault calculations are a prerequisite for the correct dimensioning of electrical equipment, setting of protective devices and ensuring stability. Two types of short-circuit faults are usually considered: symmetrical (balanced) faults and asymmetric (unbalanced) faults. The fault currents caused by short-circuits may be several orders of magnitude larger than the normal operating currents and are determined by the system impedance between the generator voltages and the fault, and under the worst scenario, if the fault persists, it may lead to long-term power loss, blackouts and permanently damage to the equipment. To prevent such an undesirable situation, the temporary isolation of the fault from the whole system it is necessary as soon as possible. The process of evaluating the system voltages and currents under various types of short-circuits is called fault analysis which can determine the necessary safety measures and the required protection system to guarantee the public safety. The analysis of faults leads to appropriate protection settings that are computed in order to select suitable fuse, circuit breaker size and type of relay. The severity of the fault depends on the short-circuit location, the path taken by fault current, the system impedance and its voltage level. In order to maintain the continuation of power supply to all customers which is the core purpose of the power system existence, all faulted parts must be isolated from the system temporary by the protection schemes.

There are two types of faults which can occur on any transmission lines; balanced faults and unbalanced faults also known as symmetrical and asymmetrical faults respectively. Most of the faults that occur on power systems are unbalanced faults. In addition, faults can be categorized as the shunt faults, series faults and simultaneous faults. The most common types of faults are single line-ground (SLG) and line-line (LL). Other types are double line-ground (DLG), open conductor and balanced three-phase fault. The electrical power system operates in a balanced three-phase sinusoidal operation. When a tree contacts a line, a lightning strikes a conductor or two conductors swing into each other and a fault occurs. When a fault occurs the system goes from a balanced condition to an unbalanced one. In order to properly set the protective relays, it is needed to calculate currents and voltages in the system under such unbalanced operating conditions. In order to analyze any unbalanced power system, C.L. Fortescue introduced a method called symmetrical components in 1918 to solve such system using a balanced representation. This method is considered the base of all traditional fault analysis approaches of solving unbalanced power systems. Fortescue theory suggests that any unbalanced system can be represented by a set of balanced systems equal to the number of its phasors, the so-called symmetrical components. In three-phase system, there are three sets of balanced symmetrical components: the positive, negative and zero sequence components. The key idea of symmetrical component analysis is to decompose the unbalanced system into three sequences of balanced networks. The networks are then coupled only at the point of the unbalance fault. In this way, the currents flowing in each line are determined by superposing the currents of these symmetrical components: positive sequence current I_+, negative sequence current I_- and zero sequence current I_0. Therefore, the phase current I_a, I_b, and I_c can be represented by these relationships:

$$I_a = I_+ + I_- + I_0 \tag{8.34a}$$

$$I_b = a^2 I_+ + a I_- + I_0 \tag{8.34b}$$

$$I_c = a I_+ + a^2 I_- + I_0 \tag{8.34c}$$

where the operator a and its power are expressed by:

$$a = \frac{1}{2} + j\frac{\sqrt{3}}{2} = 1 < 120^2, \quad a^2 = -\frac{1}{2} - j\frac{\sqrt{3}}{2} = 1 < 240^2, \text{and } a^3 = 1 \tag{8.35}$$

Similar expressions are obtained for the phase voltages. A symmetrical fault affects all three phases in the same way, and the symmetry of the system is retained. The single-phase representation of the three-phase system is used and models of lines, cables, transformers are the same as for normal operating conditions and load flow calculations. All parameters are expressed on a common per unit base. For simple fault calculations, as voltage of all the generators are taken as $V_n = 1\angle0°$ p.u., and all parallel-connected sources are replaced by the Thevenin equivalents, the fault current is calculated by:

$$I_F = \frac{cV_n}{Z_{Th} + Z_F} \tag{8.36}$$

where Z_F is the fault impedance; Z_{Th} is the Thevenin equivalent impedance; $V_n = 1\angle0°$ is the nominal voltage of the fault location; and c is a voltage factor, as defined in IEC 60909 standard to adjust the value of the equivalent voltage source for the maximum and minimum fault current conditions. An alternative approach is to determine the pre-fault voltages of the network using a load flow.

Example 8.3 **Given the phase voltages, $V_a = 5\angle 53°$, $V_b = 7\angle -164°$, $V_c = 7\angle 105°$, find the symmetrical components.**

SOLUTION

Using Equations (8.34a), (8.34b), and (8.34c), the zero, positive and negative sequence components are computed as:

$$V_0 = \frac{V_a + V_b + V_c}{3} = 3.5 < 122°$$

$$V_+ = \frac{V_a + a V_b + a^2 V_c}{3} = 5.10 < -10°$$

And

$$V_- = \frac{V_a + a^2 V_b + a V_c}{3} = 1.90 < 92°$$

8.4.2 STATE ESTIMATE METHODS

Provision of the accurate system state information to the network operators is critical for the operation of the power system in a safe, prompt and cost-effective manner, while making the best decisions and use of the assets. State estimation represents methods and techniques designed to estimate the system internal states from measurements performed on several locations throughout the power system. It is used to clean up measurement errors and to estimate the system state. The understanding, management, operation and control of power systems requires knowledge of its actual state, e.g. active and reactive power, voltages, currents, phase differences and several other system properties and parameters. The reasons of the state estimations arise from the fact that the actual states of the entire power grid cannot be directly measured. However, this is changing in the smart grids due to the smart sensing, huge deployment of sensors throughout the system, at more locations, equipment and components, increased communication and data processing capabilities. State estimation techniques are widely used in transmission subsystems where redundant measurements are available. Power system state estimation is traditionally formulated and computed in a centralized way at regional control centers. However, with the smart grid advent and deregulation of electricity market, requiring monitoring of the power system over large geographical areas, decentralized state estimation schemes are the choice, and they can enhance the computational performance and the reliability of the estimation algorithms. But at the same time they require more efficient and reliable communication techniques and need to solve time skewness issue. Both distributed and multi-area state estimations focus on interconnected systems, the difference between them lies on the structure of the state vector. In the distributed state estimations several nodes or areas estimate a common state or parameter vector through local collaborations, while in the multi-area state estimations the measurements of each area only relate to a small part of the whole state or parameter vector. Multi-area state estimations can be formulated as either a hierarchical process or a fully distributed manner.

Power system state estimation aims to find the best match between the real-time measurements and the power system states, i.e. voltage phasors at buses. Thereby, most state estimators firstly formulate the mathematical model that describes the relationship between the system states and the measurements as:

$$z = h(x) + e \tag{8.37}$$

Here $z \in \mathbb{R}^m$ is the measurement vector, $z \in \mathbb{R}^n$ is the unknown true state vector, m and n are the numbers of measurements and states ($m \geq n$), respectively. Here, $h(s) : \mathbb{R}^n \rightarrow \mathbb{R}^m$ is a function relating the measured quantities to the state variables, which is called the network model. The parameter $e \in \mathbb{R}^m$ is the unknown measurement error vector. As Equation (8.29) contains measurement vector, it is usually called the measurement or observation model. To solve state estimation problems requires selecting an x that makes z most likely to be observed, in other words, to find x that maximizes the likelihood of the observed measurements z, i.e. maximum likelihood estimate (MLE). As different measurement sources can possess different likelihood functions, which are often defined by probability density function, we need to take measurement probability distributions into account when selecting suitable state estimation algorithms. Usually, it is assumed that the measurement noise has a Gaussian distribution. State estimation is then solved by minimization of least square of the objective error function, expressed as:

$$\underset{x}{\text{miminzin}} \ J(x) = \frac{1}{2} \sum_{i=1}^{m} \frac{[z_i - h(x)_i]^2}{\sigma_i^2} \tag{8.38}$$

Here, $h(\mathbf{x})i$ is the ith element of $h(\mathbf{x})$, σ_i^2 is the variance of the corresponding ith measurement z_i. Besides the field measurements, there are two other kinds of measurements, known as the pseudo-measurement and the virtual measurement. Pseudo-measurements are manufactured data, such as generator output or substation load demand, that are based on the historical data or the dispatcher objective guesses. Virtual measurements are values that do not require metering, e.g. zero injection at a switching station. These three types of measurements own different variances, particularly the variance of the zero injection is zero given the correct topology information, so the covariance matrix could become ill-conditioned. Especially when numerically solving WLS, the ill-condition problem becomes more stringent. The most commonly used state estimation methods include: weighted least square (WLS) method, weighted least absolute value (WLAV) method and robust state estimation method. In WLS, the state estimation, determining the estimated value $\hat{x}$ of a true measured value, x is an optimization problem, which minimizes the weighted measurement residuals, defined as in Equation (8.37), being achieved by minimizing an objective function of the measurement residual, as given by:

$$\min(\Psi(x)) = \min \sum_{i=1}^{m} |z_i - h(x_i)|^2 \tag{8.39}$$

where m is the number of measurements, z_i is the measured value and w_i is the weight. In WLAV, e.g. the residuals are defined as:

$$|z_i - h(x_i)| \tag{8.40}$$

The solution of the optimization problem, as expressed by Equation (8.39) provides the estimated state $\hat{x}_i$ that must satisfy the necessary optimality conditions. Because a large number of pseudo measurements must be used to make the power distribution system observable, the distribution state estimator needs to be robust to the presence of errors and bad data in the pseudo measurements. For conventional WLS estimators, bad data are detected, identified and removed from the system state estimation through editing the measurement residuals, such as the largest normalized residual test, developed to detect and identify bad data, the normalized residuals higher than a given detection threshold are removed. In addition, hypothesis testing identification techniques have been used for bad data identification to overcome the deficiency of the largest normalized residual test method.

Notice that the measurement configuration, type, location and accuracy has large impacts on the quality of the estimates.

8.4.3 CONTINGENCY ANALYSIS AND CLASSIFICATION

Contingency analysis (CA) is one of the well-known methods to paint the future scenarios for any contingencies in the power system. It is widely used in power transmission system control, operation and management. CA is a powerful tool for transmission power system (TPS) operations. A CA program is usually performed to simulate the power flows of the TPS in case of contingencies and activate alarms, if the operational limits of some components are violated (e.g. bus voltage limits or line power flow limits), to help operators to maintain the security of the TPSs. Steady-state contingency analyses are used to predict power flows and bus voltage status and conditions, after events, such as: generator, transmission line and/or transformer outages. The reasons for such outages are planned maintenance, switching operations to control power flows and/or to overcome the voltage control issues. The usually standard approach in contingency analysis is the *normal-minus-one* (N–1), requiring that the power system remains operational after one contingency. Remember that line flow limitations are due to thermal limits (power capacity) and stability limits. A line power flow computes and provides the information on the line as the line goes down and the loads are shifted to other transmission lines. In almost all current power transmission CA (TCA) programs, power distribution networks (DNs) are simply treated as load injections into the transmission buses (these connection buses are referred as *boundary buses*), and the inner power flows of the DNs are no longer considered. In future smart grids, since distribution networks more frequently have loops in their operation, so the traditional TCA that neglects the interaction of transmission and distribution may give inaccurate alarms. Such issues need to be addresses by the smart grid contingency analysis. However, the CA approach, described before, was extended to the power distribution system with the development and implementation of the smart distribution networks. This tool performs N–K contingency analysis in a distribution network (N–1 and N–2 are normally used), which informs the network operators of the vulnerability of the distribution system in real time. It is mainly used for operation planning of a distribution network. For any set of contingencies (possible faults), contingency analysis tool returns the ranked severity of all the contingencies considered and for each contingency, it also returns the optimum remedial actions and procedures. The tool can be run periodically, triggered by events (topology changes, load condition changes, control availability, etc.), or operated in study mode for operator training and planning. The factors needed to be considered in power distribution contingency analysis for the smart grids include the impact of distributed energy resources and microgrids.

North American Reliability Corporation requires the power system operators to maintain the N–1 contingency criterion. However, multiple outage contingencies are becoming increasingly relevant because of the way the power system is being operated and the growing threats from cyberspace that attackers are gaining more useful information to knock down multiple grid components. Usually, a power system is guaranteed to be N–1 secure due to computational complexity in evaluating multiple contingencies for a large power system. In particular, for system having N components, the list size or possible number of events, C for k contingency is given by this combinatorial expression:

$$C = \binom{N}{k} = \frac{N!}{k!(N-k)!}$$

Here, the symbol "!" signifies the factorial of that number, N or k, for example. Notice even for power systems of modest size the number of possible events, C is huge. So, it is required to model the power systems optimally, efficiently and effectively for simulating their behavior during multiple and various components failure and scenarios. Although the smart grid networks introduce significant enhancements and improved operation capabilities compared to the conventional power grid, they are becoming more complex and vulnerable to different kinds of physical, and especially cyber-attacks. A smart grid is also considered as a typical cyber-physical system due to tight coupling between ICT and physical power system. Vulnerabilities with the information and computing technologies allow the attackers to access the network and break data and information confidentiality and integrity for service interruption. However, a large number of possible $(N-k)$ combinations make their assessment computationally prohibitive.

8.4.4 STABILITY ANALYSIS

The power system stability is the ability to reestablish the initial state after a disturbance or interruption causing deviations from the initial parameter values. Notice that there is distinction between static and dynamic stability, the ability to reestablish after smaller and larger disturbances. The stability is critical for the reliable and secure operation of a power system. A system disturbance occurrence produces a deviation of the system parameters, and if the power that is expanded by a load $W_{Load} + \Delta W$ after a disturbance is greater than the maximum power that the outside supply can replace, it is needed to restore the system to the initial state. A system that can naturally restore to its initial operating conditions after a disturbance is categorized as stable. Power system stability is essentially a single operation problem. However, in order to properly understand and effectively deal with the various forms of power system instability, it is convenient to make simplifying assumptions which allow analyzing them using the right degree of detail of system representation and appropriate analytical techniques. The main stability types are related to marinating generator synchronization, a constant nominal AC current frequency and the keeping the voltages within a constant and correct range. Each of them can be separated as short-term and long-term category. Usually, the short-term instabilities are damped and die out rapidly. The instability forms that a power system may undergo are classified as: rotor angle, frequency and voltage instability.

Voltage stability deals with the ability of the power system to maintain acceptable voltage levels under normal operating conditions and after the system is perturbed by small or large disturbances, being associated with short-term or long-term phenomena. Short-term voltage stability is studied considering system dynamics, whereas long-term voltage stability is commonly studied by means of steady-state solution techniques. For instance, voltage collapse, usually associated with long-term phenomena, is commonly analyzed by power flow techniques and linearization of the system equations. Even though voltage stability is a dynamic process, steady-state analysis techniques are used to identify the absence of a long-term equilibrium in the post-contingency state, which leads to voltage instability. The maximum load that the system can withstand without experiencing a voltage collapse is commonly used to compute voltage-stability indices. This maximum loadability is associated with a saddle-node bifurcation or a limit-induced bifurcation point. Saddle-node shape bifurcations refer to the operating point where the system state matrix becomes singular. Such system state matrix singularity usually coincides with the singularity of the power flow Jacobean and thus, no power flow solution can be found using conventional power flow techniques. On the other hand, limit-induced bifurcations arise when generators reach their reactive power generation limits, thus losing voltage control capability.

One term used in conjunction with voltage instability is voltage collapse. It refers to the process by which the sequence of events accompanying voltage instability leads to a blackout or abnormally low voltages in a significant part of the power system. Loads are the driving force of voltage instability, and for this reason this phenomenon has also been called *load instability*. Note, however, that loads are not the only responsible for instability. A transmission system has a limited transfer capability, as is well known from power engineering theory. This limit, also being affected by the generation system, marks the onset of voltage instability. The cause of voltage instability is the attempt of load dynamics to restore power consumption beyond the capability of the combined transmission and generation systems. The voltage stability is described in terms of the following two main categories. *Small-disturbance voltage stability* refers to the system ability to maintain steady voltages when subjected to small perturbations such as incremental changes in system load. This form of stability is influenced by the characteristics of loads, continuous controls and discrete controls at a given instant of time. It can be studied with steady-state approaches that use linearization of system dynamic equations at a given operating point. *Large-disturbance voltage stability* refers to the system ability to maintain steady voltages following large disturbances such as system faults, loss of generation or transmission line outages. This ability is determined by the system and load characteristics, and the interactions of both continuous and discrete controls and protections. It can be studied by using nonlinear time-domain simulations. The timeframe of interest for voltage stability problems may vary from a few seconds to tens of minutes, according to the speed of load restoration. Therefore, the analysis of voltage stability can be decomposed in two time scales: *short-term voltage stability*, corresponding to a timeframe of several seconds, and *long-term voltage stability* corresponds to a timeframe of several minutes or even more. Notice, however, that small-disturbance and large-disturbance voltage instability manifests in the same way that is as a progressive and uncontrollable fall of voltages. The timeframe for short-term voltage stability is also the time scale (frame) of the angle stability (described in the next paragraph), so there is not always a clear separation, distinction between grid voltage stability and angle stability problems.

Angle stability refers to the ability of synchronous machines of an interconnected power system to remain in synchronism after being subjected to a disturbance. It depends on the ability to maintain or restore equilibrium between electromagnetic torque and mechanical torque of each synchronous machine in the system. A power system is considered stable if its generators are operating synchronously, so their voltages are in phase and have the same frequency. Angle instability occurs in the form of increasing angular swings of some generators leading to the synchronism loss with other generators. Small-disturbance angle stability is concerned with the ability of the power system to maintain synchronism under small disturbances. In practice, this form of instability is usually associated with insufficient oscillation damping due to the lack of damping torque. Large-disturbance rotor angle stability or transient stability is concerned with the power system ability to maintain synchronism when subjected to a severe disturbance, such as a short circuit. Such instability usually occurs in the form of aperiodic angular separation due to insufficient synchronizing rotor torque. Both forms of angle stability can be captured by simulating the system behavior during 10–20 seconds following a disturbance. Transient stability refers to the ability of a power system to remain in synchronism when subjected to large, unexpected sudden disturbances. The disturbance severity makes it impossible to linearize the system dynamic equations; thus, in practical systems numerical techniques are used to solve such nonlinear equations.

Power angles also indicate, as shown in Equation (8.1), how much real power is injected or extracted from a power system. Stability is involved with power angle changes, which is a dynamic variable, while in a stable power system the generators operate at the same frequency and phase. If the generators fall out of synchronization, the resulting large circulating currents can trip the protection devices and may damage the transmission lines. Considering the system of Figure 8.1, the power angle is the relative difference between the angles of the line ends. Assuming the voltage magnitudes and the line impedance fixed, to transmit more power is to have larger power angles, with the largest value 90°. However, as the power angle keeps increasing, the feedback in terms of circulating current decreases. If by accident the power angle exceeds 90°, the transferred real power starts decreasing leading to instability, faster generator keeping running faster. If the power angle keeps increasing, assuming the voltages has roughly the same magnitude, but more out of phase with one another, the circulating current phase tends to shift between the voltage phases of the generators. This results in power oscillating between the generators, rather than flowing in one direction. So, the only practical way to transmit more real power is to lower the line impedance, while avoiding also instability. Power angles are in fact describing the voltage phase shifting, and beyond a maximum threshold, the phase difference (the rotor speed difference) became too large that the leading rotor (generator) appears behind the other rotors, resulting in loss of power and even in loss of stability if the system is not able to stabilize after perturbation. If the transmission line outages are of relatively short duration, the power angle tends to oscillate around the equilibrium position, eventually damping out, returns to equilibrium and the system remains stable. However, if the line outages last too long, the power angle could become too large, the restoring power weaker and weaker at large values of δ and likely that the stability is lost, requiring additional generation to restore the stability.

Frequency stability concerns the ability to maintain/restore equilibrium between the total generation and the total load powers, with minimum unintentional loss of load. Frequency instability typically occurs in the form of frequency decay or rise leading to tripping of generating units and/or loads. Changes in demand or power station disturbances impair this balance and cause a deviation of the system frequency. To meet the frequency control requirements, transmission system operators need access to the control power so that they are able to comply with their responsibility for the system stability in general. In large interconnected power systems, this type of situation is most commonly associated with extreme conditions following splitting of systems into islands. Historically, improvements in protections and voltage regulators have reinforced considerably the system against angle instability. With enhancement of the angle stability limit, in many power systems, voltage instability has become more limiting. The frequency protection detects abnormally high and low frequencies in the electric networks. If the frequency lies outside of the permissible range, for that power system section then appropriate actions are initiated such as load shedding (if frequency f is lower than the prescribed) or separating a generator from the electric network (if frequency is higher than the range upper limit).

8.5   CHAPTER SUMMARY

Electric power flows through several transmission lines, substations at different voltage levels on its way between the large power generation stations and the end-users. Optimal power flow is important for operation and planning of smart grids. The objective of power flow or load flow analysis is to compute the amount and the characteristics of the power transmitted (power flow) through each path in a network of power transmission lines. The networks at the transmission and sub-transmission levels are always mesh operated and structured. On the transmission level advanced technologies are requested to enhance the transfer capability of the network and to ensure a flexible and smart operation management in the case of congestions. Flexible AC Transmission Systems are used to increase the power transfer capability of existing AC lines, to control steady state and dynamic power flow through an AC circuit, to control reactive power and voltage and to enhance voltage and angle stability. Unbalanced faults in power systems require a phase-by-phase solution method or other techniques. One of the most useful techniques to deal with unbalanced networks is the "symmetrical component" method, developed in 1918 by C.L. Fortescue. Any three unbalanced sets of voltages or currents can be resolved into three balanced systems of voltages or currents, referred to as the system symmetrical components. The results of the state estimation is the basis for a great number of power system applications, including automatic generation control, load forecasting, optimal power flow, corrective real and reactive power dispatch, stability analysis, security assessment, contingency analysis, etc. Fast and accurate determination of the system state is a critically important for the secure and safe operation of power systems. State estimation and power system stability are critical components of the power system management, control and operation. Operational control and management require comprehensive knowledge of grid states, while the control is required to maintain the system stability. Stability appears in various forms within the power systems. Power system state estimation plays key role in the energy management systems of providing the best estimates of the electrical variables in the grid that are further used in functions such as contingency analysis, power flow analysis, automatic generation control, dispatch, control, etc. In smart grids, the state estimation involves the collection and processing of huge amount of information and data from the large number of sensors, deployed all over the grid. How this data and information are utilized, processed and transferred are significantly impacting the smart grid communication. Contingency analysis is a critical function widely used in energy management systems to assess the impact of power system component failures. Its outputs are important for power system operation for improved situational awareness, power system planning studies and power market operations. It is a traditional approach to testing all contingencies sequentially to evaluate system performance and reliability. CA simulates the outage of particular grid components and evaluates the consequences following the outages. The power system stability is the ability to reestablish the initial state after a disturbance causing deviations from the initial parameter values. The stability issue is acritical one for the reliable and secure operation of a power system. The stability is critical for the reliable and secure operation of a power system.

8.6 QUESTIONS AND PROBLEMS

1. List and briefly describe the main bus types.
2. What are the power flow analysis applications?
3. List and briefly discuss the main features and functions of the smart transmission.
4. Briefly describe the major control center functions.
5. List the main components of the transmission subsystems.
6. What re the factors affecting the transmission line capacity?
7. Calculate the voltage at bus 2 for the simple system shown in Figure P8.1, by using Gauss-Seidel method. If the voltage $V_1 = 1.0 < 0°$, the apparent power of the generator #1 is SGen1 $= 1.0 + j0.5$, and load 1 draws $0.5 + j0.5$. All quantities are expressed in per-unit system.
8. Repeat Problem 7, but using Gauss-Seidel method. Compare the number of iterations.
9. For the three-bus topology of Figure P8.1, write the impedance matrix
10. If the base is 100 MVA, apply Gauss-Seidel method to determent the voltage magnitude and angle at bus 2 and bus 3.
11. List and briefly describe the most common methods to solve power flow problems.
12. What are the main advantages of Gauss methods for power flow studies?
13. List the power flow analysis methods for power distribution.
14. Briefly describe the power flow fast-decoupled method.
15. List the main FACTS devices.
16. Briefly describe the main FACTS functions.
17. What is main issue of the power distribution load studies?
18. What are the critical roles of smart grid power flow studies?
19. Perform two additional iterations for the network of Example 8.2.

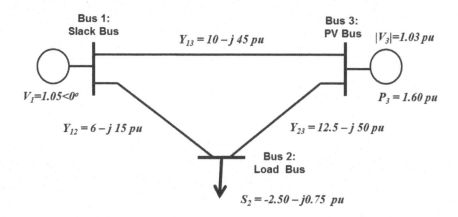

FIGURE P8.1 A three-bus topology.

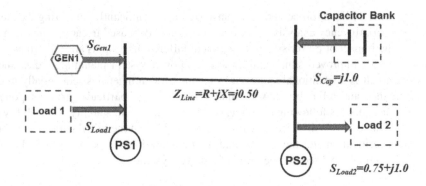

FIGURE P8.2   Modified network of Example 8.1

20. One more bus is added to the network of Figure 8.4. If the impedance of the line connecting bus 3 and bus for is $Z_{34} = 0.05 + j0.08\,p.u.$, the load connected to the bus 4 is 0.06 + j0.01 MVA perform first four iterations for this network.
21. What is the main objective of active transfer capability?
22. List the most common unbalanced faults in power systems.
23. If the values of the fault currents in a three phase system are: $I_a = 150{<}45°$, $I_b = 250{<}150°$, and $I_c = 100{<}300°$. Find the symmetrical components?
24. Briefly describe the Fortescue symmetrical components methods.
25. Given $V_0 = 3.5{<}122°$, $V_+ = 5.0{<}{-}10°$, and $V_- = 1.9{<}92°$, find the phase sequence components.
26. Briefly describe the main reasons (causes) of contingencies.
27. What the main PMU components?
28. What are the power system parameters measured by PMUs?
29. Why the time synchronization is critical for smart grid operation?
30. Briefly describe the role and importance of contingency analysis in power systems.
31. Why there is higher contingency likelihood in smart grids?
32. Why the state estimations are needed in power system operation and management?
33. List the commonly used state estimation methods.
34. Define the power system stability in the most general sense of the definition.
35. How the synchrophasors are related to the state estimation in smart grids?
36. List the main types of power system instabilities.
37. Briefly describe the voltage instability. What is the voltage collapse?
38. What is the angle instability?

9 Power Electronics for Smart Grid

9.1 INTRODUCTION

Smart grid integrates the latest communication, information, smart sensing and monitoring technologies, computing advanced, adaptive and smart control. However, by adding advanced communication and information technology to the existing electric grid only yield to a limited increase into its efficiency, stability, supply security and improved power quality, because the grid components are inflexible and of limited efficiency. The smart grids are increasingly including more controllable power electronic devices, circuits and equipment to make the best use of existing circuits, maintain flexibility and optimum operation of the power system and to facilitate the connection of renewable energy resources at all voltage levels. Moreover, the advances in solid-state devices and power electronics, combined with smart sensing and monitoring enable the electric grids to operate in a more efficient, stable and flexible way. For example, power electronics can reduce the large transformer inductive losses or allow renewable energy generators to be seamlessly grid integrated. The rapid development of solid-state devices and associated control techniques has allowed a number of applications of power electronic converters in all sections of the electric power system, from generation to the end-users. The smart grids increasingly include more controllable power electronic devices and circuits to make the best use of existing equipment and power system components, maintain flexibility and optimum operation of the power system and facilitating the connection of renewable energy units at all voltage levels. Moreover, the advances in power electronics are changing the ways in which communication, energy management and monitoring are used in smart grids. Power electronic devices are discrete, with several control dimensions than the old analog ones, being more flexible and able to accomplish more tasks. There is also a higher resolution along the dimensions in which the electrical energy can be controlled and transformed, implying that communication bandwidth to be greater and the reaction times to be faster.

The main task of power electronics is the conversion of one form of electrical energy to another, which involves the conversion of voltage and current in terms of magnitude or RMS values, frequency changes and number of phases. Electronic circuits doing such conversions are referred to as power converters. Power converters are used in different power and voltage ranges, their spectrum of converting power ranges from a few mW to several hundred MW, the voltage range extends from a few volts to several 10 kV or even hundreds of kV and current ratings ranges from mA to kA. Power converters are found wherever there is a need to modify a voltage, current or frequency. For example, variable-speed wind generator systems or PV arrays need power electronic interfaces to the electrical grid. Even for fixed-speed wind generator systems, when the energy storage system is connected to the grid power electronics devices are essential. Power semiconductor devices that are used as switches in power electronic circuits. Most power semiconductor devices are used only in commutation mode and are therefore optimized for this operation. Common power devices are the power diode, thyristor, power metal-oxide-semiconductor field-effect transistors (MOSFET) and insulated gate bipolar transistor (IGBT). A power diode or a power MOSFET transistor are operating similarly to their low-power counterparts but can carry a larger current and typically are able to support a larger reverse-bias voltage in OFF state. Structural changes are made in power devices to accommodate the higher current density, higher power dissipation or higher reverse breakdown voltages. Newer trends in modern power electronics for the integration of wind energy conversion systems, small hydropower, solar thermal energy systems and photovoltaic systems. There are

DOI: 10.1201/9780429174803-9

several reasons for these developments, among others are: increasing number of renewable energy sources and distributed generators, into transmission and power distribution sections of the grid, new strategies for the operation and management of the electricity grids, improved the power-supply reliability and quality, and liberalization of the grids leads to these newer management structures.

Power electronics technologies, together with suitable protection configurations, methods and the modern and adaptive control schemes and procedures are envisioned to lay an important and critical role into the integration and extended use of the distributed generation and renewable energy sources into the future electrical grids. Power electronics technology as part of the smart grid infrastructure enables a fuller exploitation of existing distribution energy resources, maintaining and even improving the hitherto state of the power supply security, power system resilience and power quality. The use of power electronics (PE) circuits in the smarter electricity systems can be divided into two main categories: (a) electrical energy transmission systems and (b) electrical energy distribution system. The transmission system is composed of two complementary technologies for energy transfer: (a) a system using conversion to DC current, high-voltage direct current (HVDC) devices and installations and (b) direct AC transfer, such as flexible alternating current transmission systems (FACTS) devices. A HVDC advantage is the capability to transmit energy between systems of various frequencies. However, in the case of conventional HVDC, i.e. with the use of SCR devices, it is necessary to use large filters and there is no possibility of supplying power to end-users on the side from which the source is disconnected. This drawback does not occur when modern solid-state devices, such as GTO thyristors or IGBT transistors are employed. Notice that with HVDC devices the energy from one system flows to the other through power converters. Because of this, the cost is high, even in single-station installation. FACTS devices are static equipment which helps in not only for compensating reactive power but also control AC transmission parameters. There are basically three types of FACTS devices. Devices connected in shunt with the power system, devices connected in series with the power system and a combination of shunt and series connected devices. While in FACTS devices, e.g. SVC (static VAR compensator), STATCOM (static synchronous compensator), TCSC (thyristor-controlled series compensator), TSSC (thyristor switched series compensator), SSSC (static synchronous series compensator), SPS (static phase shifter) and UPFC (unified power flow controller), only part of the power flows through converter. These devise have allowed compensate reactive power and mitigate problems occurring in transmission lines. However, such devices can be applied to the control of energy flow only in AC single frequency systems. Notice that using FACTS devices and high-voltage direct-current transmission, coupled with advanced monitoring and control create smart power transmission systems. This enhanced the features of electric transmission, such as reliability, supply and data security, resilience, capacity and efficiency. The ever-increasing progress of HV high-power fully controlled semiconductor technology continues to have a significant impact on the development of advanced power electronic apparatus used to support optimized operations and efficient management of electrical grids, which in many cases, are fully or partially deregulated power networks. Developments advance both the HVDC power transmission and the FACTS technologies. They allow for a change in the system voltage and line impedance. Other power electronics application in the smart grid's power transmission systems include line monitoring and energy harvesting.

A significantly greater variety of PE arrangements and settings occurs in power distribution systems. In these systems PE converters and controllers are applied usually for: matching parameters and coupling of distributed energy sources and energy storage units with power lines or local end-users, controlling electrical energy consumption, and the exchange of energy between energy storage systems and power lines (see block diagram of Figure 9.1). Power electronic is also contributing to the improvements in power supply quality, such as compensation of voltage sags and swells, asymmetry and distortions of supply voltages, or compensation for the distortion, asymmetry and phase shift in load current. Moreover, the power output of some of the fuel cells, renewable energy and energy storage systems is DC (for example, photovoltaic systems or batteries) so inverters are needed to interface them to the AC grid. Even though renewable energy sources using

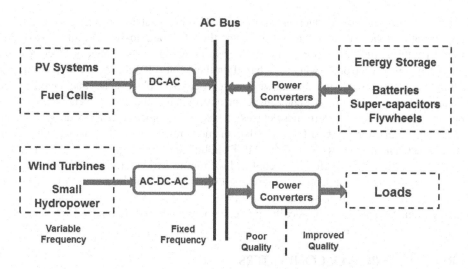

FIGURE 9.1   Power electronics for renewable energy, energy storage units and loads.

an AC generator (e.g. wind turbines) cannot be connected directly to the grid, they are often being connected to a conversion AC-to-DC and DC-to-AC in order to improve the power quality and prevent instabilities. Some power system operating conditions demand rapid independent control of the active and reactive power output of the renewable energy generators. The power electronic interface between a renewable energy source and the grid can be used to control reactive power output and hence the network voltage as well as curtailing real power output, and so enable the generator to respond to the requirements of the grid. These control actions and methods can only be achieved conveniently using power electronic interfaces. This chapter provides an overview and basic understanding of various power electronics devices, converters: rectifiers, inverters, DC-DC choppers, cyclo-converters, pulse width modulation (PWM)-based voltage source converters (VSCs) and current source inverters (CSIs). Details one the major power electronics applications in smart grid transmission, power distribution, power quality, voltage regulations, etc. are also included. For detailed smart grid power electronics studies and applications, interested readers are directed to the power electronics textbooks, listed in the references book section or elsewhere in the literature.

Power electronic devices are the electronic (solid-state) devices that can be directly used in the power (better said, energy) processing circuits to convert and/or control electric power (energy). Notice that the electric power that power electronic devices are dealing with usually much larger energy than that the information (digital) electronic device does. The progress of the power electronics industry has been contingent on the level of progress in the semiconductor or solid-state power device industry, otherwise known as the power electronic device industry, and particularly on the advancements in power electronic switches. Power electronic devices are commonly used as switches, making it possible to control power flows, by operating very close to an ideal switch, being able to handle large power amounts at low power dissipation, through their internal structure, construction and switching control. Development of power semiconductors with very high voltage and current ratings has enabled the use of power electronic converters in many industrial and utility applications. The range of applications continues to expand in areas such as power supplies, motor speed control, factory automation, transportation, energy storage, industrial drives and electric power transmission and distribution. Applications in power transmission include high-voltage DC (HVDC) converter units, FACTS devices and equipment and static-VAR compensators, etc. Modern power electronic devices can be classified in a number of ways based on the number of terminals, the number of PN junctions, level of controllability, bi-directional capability, device losses and the gate signal requirements, to name just a few. Notice that there are three types of power losses in solid-state switched devices: conduction (during the ON state), switching (during the transition for ON to OFF states or vice versa) and turn-off (during the OFF state) losses. Thermal characteristics are critical for the design, operation and efficiency of power electronic devices. A critical classification of the power electronic switches is based on their controllability that includes three main controllability groups: the uncontrollable switches, semi-controllable switches and the fully controllable switches. Power electronic switches are acting as a switch without any mechanical movement, and the most common power electronics devices include power diodes, metal-oxide-semiconductor and field-effect transistors (MOSFETs), bipolar junction transistor (BJT), insulated-gate bipolar transistor (IGBT) and thyristors (SCR, GTO, MCT). Solid-state devices are completely made from a solid material and their flow of charges is confined within this solid material. The following are the main advantages of power semiconductor devices: faster in operation by a factor of at least a hundred times; less routine maintenance; do not stick, bounce or wear; no sparking on break; longer life expectancy when used properly; physically smaller devices; and they are relatively cheaper devices.

9.2 REVIEW OF POWER CONVERTERS

The power electronics applications to perform power conversion with minimum losses usually eliminates the usage and needs of resistive elements or other similar elements with high dissipative losses (at least not in the critical paths). Power electronic converters can be classified into four different types on the basis of input and output: DC-to-DC, DC-to-AC, AC-to-DC and AC-to-AC, named with the first part referring to their input and the second to their output. Notice that the electronic components such as transistors, thyristros are not operated in their linear regions because this would also incur higher power losses, but as electronic switches. Development of power semiconductors with very high voltage and current ratings has enabled the use of power electronic converters in many grid applications. Only the components with minimal power losses, which are essentially reactive elements such as: capacitors, inductors and transformers, together with the joints to these elements, the electronic switches are employed in grid applications. Switches are key components in power electronics because they are the only elements that can control the flow of currents and voltages selectively, either actively (by gate pulses, externally driven) or passively (as results of external electrical behavior of load or network, i.e. load or source commutated). The simplest (idealized) switch is the two-pole switch, it can assume two states – ON and OFF. The power loss of the ideal switch is always zero, because either its voltage or its current is zero, so their product, the switch power is zero. The circuits connected to input and output of the switch must be capable of such step-like switch parameter changes. The most important candidates of two-pole switches are diodes, MOSFET and bipolar transistors. Many devices can only conduct the current in one direction and withstand blocking voltage only in reverse direction. Exceptions are the MOSFETs which can conduct the current in both forward and reverse directions, and the thyristor which is capable both of forward and reverse blocking voltage. For many applications, these shortcomings of the real devices are no serious handicaps, because handling of both voltage and current polarities is often not required. If that is necessary in some cases, several devices are combined in order to cover all needed specifications. The main goal of electronic power convertors is to convert electrical energy from one form to another, from the source to the load using power semiconductor devices with, higher efficiency, higher availability, higher reliability, smaller size, lighter weight and lower cost possible. The role of power electronic converter and equipment are summarized in Figure 9.2.

9.2.1 DC-DC POWER CONVERTERS

The conversion of direct current into direct current is intended to improve the power from a DC source and to match the voltage of the source and the consumers. DC-DC converters are power electronic circuits that convert a DC voltage to a different voltage level. There are different types of

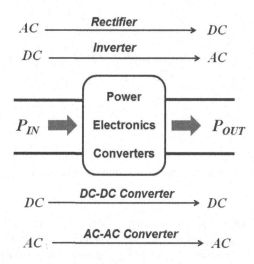

FIGURE 9.2   The role and conversion types of power electronic converters.

conversion method such as electronic, linear, switched mode, magnetic, capacitive. The purpose of DC voltage power converters is not only the conversion of direct current but also the regulation or stabilization of the voltage (or current) in the load. Converters used only for stabilization are known as stabilizers. We may distinguish between two types of DC converters: continuous converters and pulsed converters. Continuous converters stabilize the voltage in a DC circuit with variation in the source or load voltage. Continuous converters employ transistors operating in the active region of the output voltage-current characteristics. These circuits, classified as switched mode DC-DC converters, are electronic devices that are used whenever change of DC electrical power from one voltage level to another is needed. In most cases the grid voltage is first rectified by a rectifier and then adjusted by a power-electronic converter, which together are called a switch mode power supply (SMPS), because the use of a switch or switches for the purpose of power conversion can be regarded as an SMPS. Even though many of the renewable energy sources and energy storage units have a DC output they need to be connected to the microgrids through DC-DC converters to regulate and improve the output DC signals. For example, PV panels are connected through a DC-DC Boost converter to raise the output voltage level at the nominal microgrid voltage level and to track the maximum power point of the solar PV panel output which fluctuates depending on the solar irradiation. A power-electronic converter adjusts the grid voltage to a voltage with amplitude and frequency which is required by the load. There are different kinds of DC-DC converters. A variety of the converter names are included here: the Buck converter, the Boost converter, the Buck-Boost converter, the Čuk converter, the Flyback converter, the Forward converter, the push-pull converter, the Full Bridge converter, the Half Bridge converter, the current Fed converter and the multiple output converters.

The use of DC-DC power converters is extensive in renewable energy, energy storage devices and industrial electronics applications. Such DC-DC power converters are also used in cases where a DC voltage produced by rectification or supplied by a power sources is used to supply secondary loads, and the DC voltage is needed to be changed as required by the loads. The conversion is often associated with stabilizing, i.e. the input voltage is variable but the desired output voltage stays the same. The converse is also required, to produce a variable DC from a fixed or variable source. Since the conversion function is achieved by switching the DC power *ON* and *OFF* (chopping the power) at high frequencies, these power converters are also known as *DC choppers* or *switch-mode power supplies*. There are three basic configurations of the DC-DC power converters – Buck, Boost and Buck-Boost converters and a few variants of them. This chapter subsection describes the circuit topologies and the operation characteristics of the main types of DC-DC power converters. The

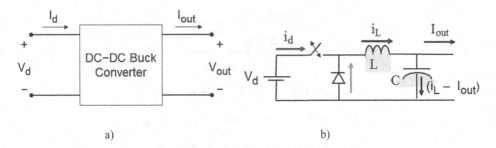

a)   b)

FIGURE 9.3   (a) Block diagram of a Buck converter and (b) Boost converter circuit.

basic circuit of a Buck DC-DC converter is shown in Figure 9.3 connected to a purely resistive load. By using the fact that the average voltage across the inductor is zero, assuming perfect filter, the voltage across the inductor is V_d during t_{ON} and $-V_{out}$ the remaining of the cycle. If the diode output voltage $v_{out}(t)$ is equal to the input voltage V_d when the switch is closed and to zero when the switch is open, an average output voltage V_{out} is computed as:

$$V_{out} = \frac{1}{T_{SW}}\left[\int_0^{t_{ON}} V_d \cdot dt + \int_{t_{ON}}^{T_{SW}} 0 \cdot dt\right] = \frac{t_{ON}}{T_{SW}} V_d = D \cdot V_d \tag{9.1}$$

Here, $T_{SW} = 1/f_{SW}$ is the converter switching period, f_{SW} is the switching frequency (Hz) and $D = t_{ON}/T_{SW}$ is the duty ratio. The output voltage of the Buck converter is always lower than the input voltage, so it is a step-down converter. A low pass filter is used to attenuate the high frequencies (multiples of the switching frequency) and leaves almost only the DC component. Assuming ideal components and that the input and output powers are the same, power conservation then:

$$V_d \cdot I_d = V_{out} \cdot I_{out} \Rightarrow I_{out} = \frac{I_d}{D} \tag{9.2}$$

The output current is the same as the inductor current, the inductor L being in series with load. Finally, the consideration on the output voltage ripples, which are assumed that the ripple current is absorbed by the converter capacitor, C, i.e. the voltage ripples are small, being given by:

$$\frac{\Delta V_{Out}}{V_{Out}} = \frac{1-D}{8 f_{SW}^2 LC} \tag{9.3}$$

Example 9.1   A Buck DC-DC converter has the following parameters: $V_d = 50$ V, $D = 0.4$, $L = 400$ μH, $C = 100$ μF, $f_{SW} = 20$ kHz, and the load, $R = 20$ Ω. Assuming ideal components, calculate (a) the output voltage and (b) the output voltage ripple.

SOLUTION

a. The inductor current is assumed to be continuous, and the output voltage is computed from Equation (9.41), as:

$$V_{out} = D \cdot V_d = 0.5 \times 50 = 25\,\text{V}$$

b. The output voltage ripple is computed from Equation (9.3):

$$\frac{\Delta V_{Out}}{V_{Out}} = \frac{1-D}{8f_{SW}^2 LC} = \frac{1-0.5}{8\times\left(20\cdot10^3\right)^2\times400\cdot10^{-6}\times100\cdot10^{-6}} = 0.003906 \text{ or } 0.39\%$$

For the set-up Boost converter, the average inductor current is the average output current, and the output voltage is always higher than the input voltage. Its topology is shown in Figure 9.4a, and the diagram of Figure 9.4b is showing the converter connected to a resistive load. Similarly, there are two different topologies based on the switch conditions of the switch, and again the way to calculate the relationship between input and output voltage we have to take the average current of the inductor to be zero, and the output power equal to the input power hence the output voltage and current are:

$$V_d \cdot t_{ON} + \left(V_d - V_{out}\right)\cdot\left(T_{SW} - t_{ON}\right) = 0$$

And

$$V_{out} = \frac{V_d}{1-D}, \quad \text{and} \quad I_{out} = I_d\left(1-D\right) \tag{9.4}$$

Following an analysis similar to that of a Buck converter, the output voltage ripples are given by:

$$\frac{\Delta V_{Out}}{V_{Out}} = \frac{D}{f_{SW}\cdot RC} \tag{9.5}$$

It is important to note that the operation of a Boost converter depends on parasitic components, especially for duty cycle approaching unity. These components limit the output voltage to levels. The Buck-Boost converter, having the topology as shown in Figure 9.4b, can provide output voltage that can be lower or higher than that of the input, depending on the duty ration range. Again, the operation of the converter can be analyzed using the two topologies resulting from operation of the switch. By equating the integral of the inductor voltage to zero, power conservation, the output voltage and current are estimated as:

$$V_d \cdot DT_{SW} + \left(-V_{out}\right)\cdot\left(1-D\right)T_{SW} = 0$$

And

$$V_{out} = \frac{D\cdot V_d}{1-D}, \quad \text{and} \quad I_{out} = I_d\frac{1-D}{D} \tag{9.6}$$

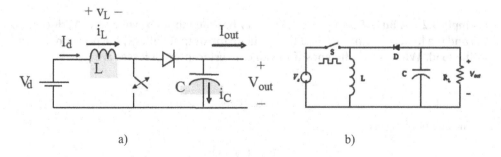

a)   b)

FIGURE 9.4 (a) Boost converter circuit diagram and (b) Buck-Boost converter circuit diagram.

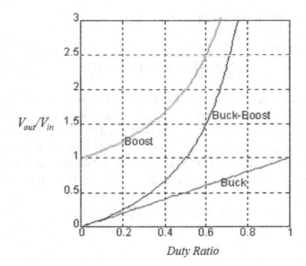

FIGURE 9.5 Voltage ratio vs. duty ratio for Buck, Boost and Buck-Boost converters.

The output voltage ripples are calculated by using Equation 9.5, the one used for the Boost converters. The voltage ratios achievable by the DC-DC converters are summarized in Figure 9.5. Notice that only the Buck converter shows a linear relationship between the control (duty ratio) and output voltage. The Buck-Boost can reduce or increase the voltage ratio with unit gain for a duty ratio of 50%. All the DC-DC converters are transferring energy between input and output by the inductor, so the analysis is based of voltage balance across the inductor. In many DC-DC applications, multiple outputs may be required and output isolation may be needed. In addition, input-to-output isolation may be required to meet safety standards and/or provide impedance matching. The DC-DC converter topologies discussed before can be easily adapted to provide isolation between input and output. The Flyback converter can be developed as an extension of the Buck-Boost converter provides insulation to the inclusion of a transformer into its circuit. The Buck-Boost converter works by storing energy in the inductor during the ON phase and releasing it to the output during the OFF phase. With the transformer the energy storage takes place into the magnetization of the transformer core, while providing also insulation between converter input and output stages. To increase the stored energy a gapped core is often employed.

Neglecting the energy consumption in the operation of the converter control subsystem, considering only the losses into the inductor and MOSFET transistor the converter efficiency can be expressed as:

$$\eta = \frac{P_{Out}}{P_{Out} + Losses} \qquad (9.7)$$

Example 9.2   A Buck-Boost power converter has the input DC voltage 42 V, duty ratio 0.6 and the load resistance R_L is 10 Ω. Calculate the output voltage, the power delivered to the load. What is its efficiency if the all converter internal losses are 31.1 W?

SOLUTION

The output voltage is:

$$V_{Out} = \frac{D \cdot V_D}{1-D} = \frac{0.6 \times 42}{0.4} = 63 \, V$$

The power delivered to the load is calculated as:

$$P_{Load} = \frac{V_{Out}^2}{R} = \frac{63^2}{10} = 396.9\,W$$

Applying Equation (6.11), the converter efficiency is:

$$\eta = \frac{396.9}{396.9 + 31.1} = 0.927 \text{ or } 92.7\%$$

Energy storage units are based usually on batteries or other devices require a bi-directional DC-DC power converter to provide power when the energy sources are not present or able to operate and to accumulate energy when the energy storage systems are not needed. The characteristics and block diagram of the DC-DC bi-directional converter looks like the ones in Figure 9.6.

Example 9.3   Design a Buck-Boost power converter, supplied from a PV array whose voltage varies from 60 V to 120 V, depending on the available solar radiation. This power converter supplies a load of 9 kW at fixed 90 V DC. Compute the range of duty ratio needed to supply the load at the rated voltage.

SOLUTION

The converter output rated current is computed from:

$$I_o = \frac{P_{Rated}}{V_{rated}} = \frac{9000\ W}{90\ V} = 100\,A$$

The maximum average input current rating is computed form output power and the lower voltage limit as:

$$I_o = \frac{P_{Rated}}{V_{Min}} = \frac{9000\ W}{60\ V} = 150\,A$$

The duty ratio for an input voltage of 60 V (lower voltage limit) can be computed from Equation (9.6) by:

$$D_{max} = \frac{V_{Out}}{V_{In} + V_{Out}} = \frac{100}{60 + 100} = 0.625$$

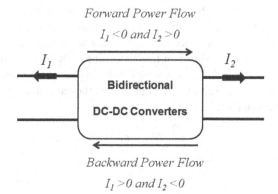

Forward Power Flow

$I_1 < 0 \text{ and } I_2 > 0$

I_1 I_2

Bidirectional

DC-DC Converters

Backward Power Flow

$I_1 > 0 \text{ and } I_2 < 0$

FIGURE 9.6   Bi-directional DC-DC power converter characteristics and schematics.

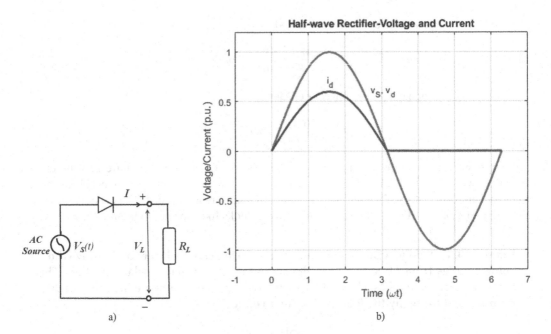

FIGURE 9.7   (a) Half-wave rectifier circuit diagram and (b) current and voltage waveform.

Similarly, for upper voltage limit 120 V can be calculated by:

$$D_{min} = \frac{V_{Out}}{V_{In}+V_{Out}} = \frac{100}{120+100} \approx 0.454545 \approx 0.46$$

Therefore, the duty ratio range is:

$$0.46 \leq D \leq 0.625$$

9.2.2   RECTIFIERS AND OTHER POWER ELECTRONIC DEVICES

A rectifier is a circuit that converts AC signals to DC, a process known as rectification. The simplest rectifier is the half-wave configuration, consisting of a diode, a single-phase AC source and a purely resistive load (Figure 9.7a). If the load includes an inductance and a source (e.g. charging a battery), the diode continues to conduct even when the load voltage is negative as long as the current is maintained, changing the rectifier waveforms. When the source voltage is positive, the current flows through the diode and the load voltage equal the sources voltage (Figure 9.7b). However, usually half-wave rectifiers are not common but rather a single-phase diode bridge rectifier, such as shown in Figure 9.8. The load can be modeled with one of two extremes: either as a constant current source, representing the case of a large inductance that keeps the current through it almost constant, or as a resistor, representing the case of minimum line inductance. The rectifier analysis, assuming that the instantaneous input voltage is $v(t) = V_m \cdot sin(\omega t)$ including its waveforms are determine the following expressions for the output voltages and currents in the case of diode full-wave rectifier and the thyristor full-wave rectifier, as:

$$V_d \approx \frac{2 \cdot V_m}{\pi}, \quad \text{and} \quad I_d = I_S \tag{9.8}$$

And respectively:

$$V_d \approx \frac{V_m}{\pi}(1+\cos(\alpha)), \quad \text{and} \quad I_S = 0.816 I_d \tag{9.9}$$

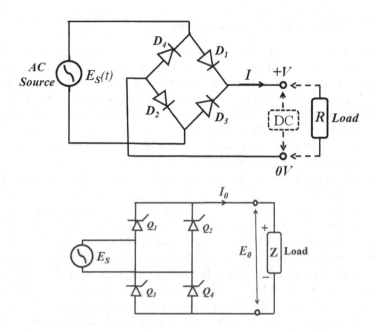

FIGURE 9.8 Single-phase full-wave rectifier circuit, right panel shows diode-based and left shows thyristor-based.

Where V_m is the amplitude of the sinusoidal voltage, and I_s are the RMS values of the input AC voltage and current, while α is the delay angle, corresponding to the time we delay triggering the thyristors after they became forward biased. On the DC side, only the DC voltage component carries power, since there is no harmonic content, while on the AC side the power is carried by the fundamental, since there are no harmonics in the voltage. If the current on the DC side is sustained even if the voltage reverses polarity, then power is transferred from the DC to the AC side. The DC side voltage can reverse polarity when the delay angle exceeds 90°, and the current is maintained (e.g. a battery). Another important measure is the RMS voltage, more useful in estimating the effect on heating or incandescent lighting equipment. The two mean and RMS values normalized to their base when $\alpha = 0$. The RMS voltage is given by:

$$V_{RMS} = V_m \sqrt{1 - \frac{\alpha}{\pi} + \frac{\sin(\alpha)}{2\pi}} \tag{9.10}$$

Example 9.4 Find the peak (amplitude) source voltage that a full-wave diode rectifier is producing 100 V (DC) across a 100 Ω. What is the power absorbed by this load?

SOLUTION

The output rectifier DC voltage is computed by using Equation (9.8), so the peak AC voltage is calculated as:

$$V_m \approx \frac{\pi \cdot V_d}{2} = \frac{3.14 \times 100}{2} = 157\,V$$

The power absorbed by the 100 Ω resistive load, is then:

$$P_{100-\Omega} = \frac{V_d^2}{R} = \frac{157^2}{100} = 246.5\,W$$

9.2.3 DC-AC POWER CONVERTERS OR INVERTERS

A power inverter, or inverter, is an electronic device or circuitry that changes a direct current (DC) into an alternating current (AC), more precisely, inverters transfer power from a DC source to an AC load. A power inverter can be entirely electronic or may be a combination of mechanical effects and electronic circuitry, while the so-called static inverters are not using moving parts in the conversion process. Inverters can generate single or poly-phase AC voltages from a DC supply, while into the class of poly-phase inverters, three-phase inverters are by far the largest group. These power converters are used in applications such as electric motor drives, uninterruptible power supplies and utility applications such as grid connection of DC-type renewable energy sources or energy storage systems. Inverters are classified into two main categories: voltage source inverter (VSI), which has stiff DC source voltage, DC voltage has limited or zero impedance at the inverter input terminals; and current source inverter (CSI), which is supplied with a variable current from a DC source that has high impedance. The resulting current waves are not influenced by the load. Single-phase inverters are of two main types: full-bridge inverter and half-bridge inverter. Half-bridge inverter is the basic building block of a full-bridge inverter. It contains two switches and each of its capacitors has a voltage output equal to $V_{dc}/2$. The switches complement each other, that is, if one is switched ON the other one goes OFF. Full-bridge inverter achieves the DC-to-AC conversion by switching in the right sequence. It has four different operating states which are based on which switches are closed. A three-phase inverter converts a DC input into a three-phase AC output. Its three arms are normally delayed by an angle of 120° so as to generate a three-phase AC supply.

The input voltage, output voltage and frequency, and the power handling depend on the specific device or circuitry design. Notice that the inverter power is provided by the DC source. Figure 9.9a is showing the inverter block diagram. An inverter is usually able to set the converted AC signal characteristics at any required voltage and frequency levels with the use of appropriate transformers, switching and control circuits. Solid-state inverters have no moving parts and are used in a wide range of applications, from small switching power supplies in computers to large electric utility high-voltage direct current applications that transport bulk power. Inverters are commonly used to supply AC power from DC sources such as solar panels or batteries. A typical power inverter device or circuit requires a relatively stable DC power source capable of supplying enough current for the intended power demands of the system. The input voltage depends on the design and purpose and application of each inverter. The output voltage $v_{out}(t)$ can be theoretically any value into the range – V_{DC} to +V_{DC} (the DC source voltage), or zero, depending on which the state of the inverter switches. The current waveform in the load depends on the load components. For the resistive load, the current waveform matches the shape of the output voltage, while inductive loads have a current that has more of a sinusoidal quality than the voltage because of the inductance filtering property. A separate class of inverters is the line-commutated inverters for multi-MW power ratings that are using thyristors (silicon-controlled rectifiers, or SCRs). SCRs can only be turned *ON* or *OFF* on command. After being turned on, the current in the device must approach zero in order to turn the device off. Many other inverters are self-commutated. Such systems, converting DC into AC through the use of switching devices, such as GTOs, BJTs, IGBTs, and MOSFETs, are allowing the transfer of power from the DC source to any AC load, and gives considerable control over the resulting AC signal. Line commutated inverters need the presence of a stable utility voltage to function.

Figure 9.9b shows the operation of the one leg of an inverter regardless of the number of phases. To illustrate its operation, the input DC voltage is divided into two equal parts. When the upper switch, S_1, is closed, S_2 is open, the output voltage is +$V_d/2$, and when the lower switch, S_2 is closed, S_1 is open, the output voltage is –$V_d/2$. To control the output voltage waveform the inverter switches are controlled by pulsed width modulation (PWM), where the time each switch is closed is determined by the difference between a control waveform, and a carrier, usually a triangular waveform. When the control waveform (signal) is greater than the triangular waveform, the carrier, S_1 is closed, and S_2 is open. When the control wave is less than the triangular wave, S_1 is open and S_2 is closed.

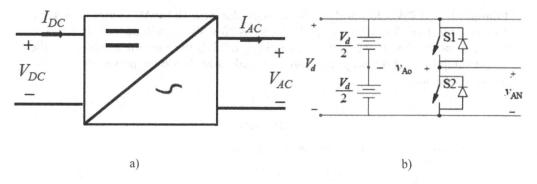

a)   b)

FIGURE 9.9   (a) Block diagram of an inverter and (b) one leg of an inverter.

In this way, the width of the output is modulated (hence the name, PWM). If the frequency of the triangular wave is f_{tri}, and the reference (sine) frequency is f_{ref}, then their ratio defines the frequency modulation index, m_f, while the ratio of the control voltage, V_{ref} to the triangular waveform voltage, V_{tri} the amplitude modulation index, m_a, is defined as:

$$m_f = \frac{f_{ref}}{f_{tri}} \tag{9.11a}$$

And

$$m_a = \frac{V_{m,ref}}{V_{m,tri}} \tag{9.11b}$$

In full inverter, the diagonal switches are operating in tandem, such that S_1 and S_3 open and close together, and S_2 and S_4 open and close together, while the inverter output oscillates between $+V_d/2$ and $-V_d/2$. Applying the Fourier transformation of a PWM square wave shown, the amplitude of the fundamental is a linear function of the amplitude index $V_o = m_a \cdot V_d/2$ as long as $m_a \le 1$, and the RMS output voltage is:

$$V_{o1} = \frac{m_a}{\sqrt{2}} \cdot \frac{V_d}{2} = 0.353 m_a \cdot V_d \tag{9.12}$$

When voltage modulation index becomes larger than 1, the output voltage increases also but not linearly with m_a. The output voltage amplitude reaches a peak value of $(4/\pi)\cdot V_d$, when the reference signal becomes infinite and the output voltage is a square wave waveform. Under this condition, the RMS value of the output voltage is:

$$V_{o1} = \frac{\sqrt{2}}{\pi} \cdot V_d = 0.45 \cdot V_d \tag{9.13}$$

Assuming the ideal components and no losses, the DC side power is equal with the AC side power:

$$P = V_d \cdot I_{do} = V_{o1} \cdot I_{o1} \cdot PF \tag{9.14}$$

Here, PF is the power factor, thus for normal operation, the RMS output current is:

$$I_{do} = 0.353 m_a \cdot I_{o1} \tag{9.15a}$$

And in the limit for a square-wave waveform is:

$$I_{do} = 0.45 \cdot I_{o1} \cdot PF \tag{9.15b}$$

Example 9.5 A PWM full-bridge inverter is producing a 50 Hz AC voltage across of a 10 Ω resistive load. The DC input to the bridge is 90 V, the amplitude modulation ratio m_a is 0.9 and the frequency modulation ratio m_f is 25. Determine the amplitude of the 50 Hz component of the output voltage and load current, and the power absorbed by the load (resistor).

SOLUTION

The amplitude of the 50 Hz is computed from Equation (9.12):

$$V_1 = \sqrt{2} \cdot V_{o1} = \sqrt{2} \cdot 0.353 m_a \cdot V_d = \sqrt{2} \cdot 0.353 \cdot 0.9 \times 90 \approx 40.5\,V$$

For the fundamental frequency, being a resistive load the current is:

$$I_1 = \frac{V_1}{R} = \frac{40.5}{10} = 4.05\,A$$

The power absorbed by the load (assuming ideal conditions), and for a resistive load PF = 1, is:

$$P = V_1 \cdot I_1 = 40.5 \times 4.05 = 182\,W$$

A single-phase half-bridge DC-AC inverter is shown in Figure 9.10a. The analysis of the DC-AC inverters is taking into accounts the following assumptions and conventions. The switches S_1 and S_2 are unidirectional, i.e. they conduct current in one direction. The RMS output voltage is given by:

$$V_{out} = \frac{V_{in}(DC\text{ source voltage})}{2} \qquad (9.16)$$

The instantaneous output voltage $v_{out}(t)$ is rectangular, not a purely sinusoidal waveform and its amplitude is set by the DC source voltage, V_{DC}. The instantaneous output voltage, $v_{out}(t)$ and output current, having angular frequency, ω, for the case of a resistive load, R are expressed in Fourier series format as:

$$v_{out}(t) = \sum_{n=1,2,3,\ldots}^{\infty} \frac{4 \cdot V_{DC}}{n\pi} \sin(n\omega t) \qquad (9.17a)$$

And

$$i_L(t) = \sum_{n=1,2,3,\ldots}^{\infty} \frac{4 \cdot V_{DC}}{R \cdot n\pi} \sin(n\omega t) \qquad (9.17b)$$

A single-phase full bridge DC-AC inverter is shown in Figure 9.10b. The analysis of the single-phase DC-AC inverters is taking in account following assumptions and conventions. The switches S_1, S_2, S_3 and S_4 are unidirectional, i.e. they conduct current in one direction. A single-phase square wave-type voltage source inverter produces square-wave output voltage for a single-phase load. Such inverters have very simple control logic and the power switches need to operate at lower frequencies compared to switches in other inverter types. The first generation inverters, using thyristor switches, were almost invariably square-wave inverters because thyristor switches could be switched on and off only at lower frequencies. However, the present day switches are much faster and used at switching frequencies of several kilohertz. Single-phase inverters mostly use half-bridge

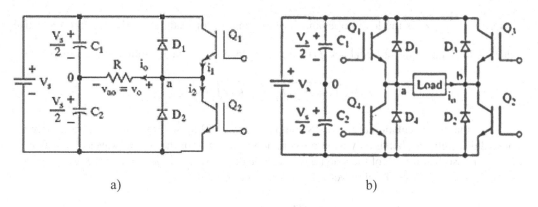

FIGURE 9.10 (a) Half-bridge and (b) single-phase full-bridge inverter schematics.

or full-bridge topologies. All inverter topologies, discussed hare, are analyzed under the assumption of ideal circuit conditions, assuming that the input DC voltage is constant, while the switches and inverter components are lossless. With the ideal component assumption and constant DC voltage, the power supplied by the source must be the same as absorbed by the load. Power from a DC source (source current, I_S) is calculated by the well-known relationship:

$$P_{DC} = V_{DC} \cdot I_S \tag{9.18}$$

Far a single-phase full bridge inverter, the output voltages are computed by using Equation (9.17a), while the load currents for the case of a purely inductive load, having and inductance, L and for the case of a resistive-inductive (R-L) load are given by these relationships:

$$i_L(t) = \sum_{n=1,2,3,...}^{\infty} \frac{1}{n \cdot \omega L} \frac{4 \cdot V_{in}}{n\pi} \sin\left(n\omega t - \frac{\pi}{2}\right) \tag{9.19a}$$

And

$$i_L(t) = \sum_{n=1,2,3,...}^{\infty} \frac{4 \cdot V_{in}}{n\pi \sqrt{R^2 + (n\omega L)^2}} \sin(n\omega t - \theta_n) \tag{9.19b}$$

Where, the phase angle, θ_n, is given by:

$$\theta_n = \tan^{-1}\left(\frac{n\omega L}{R}\right)$$

Power absorbed by a load with a series resistance is determined from well-known relationship:

$$P_{Load} = R \cdot I_{RMS}^2 \tag{9.20}$$

where the RMS current can be determined from the RMS currents at each of the components in the Fourier series by:

$$I_{RMS} = \sqrt{\sum_{n=1}^{\infty} I_{n,RMS}^2} = \sqrt{\sum_{n=1}^{\infty} \left(\frac{I_n}{\sqrt{2}}\right)} \tag{9.21}$$

Here $I_n = \dfrac{V_n}{Z_n}$, Z_n is the load impedance at harmonic n, for example in the case of a resistive inductive load, operating at frequency, f (Hz), and the load impedance amplitudes are expressed as:

$$Z_n = \sqrt{R^2 + (n\omega \cdot L)^2} = \sqrt{R^2 + (2\pi nf \cdot L)^2}$$

Notice that the calculation of the RMS component by dividing square root of 2 is valid only for sinusoidal waveforms, being only an approximation in the case of the square waves. Equivalently, the power absorbed in case of the load resistor can be determined for each frequency in the Fourier series, through Ohm law. Total power is then computed through:

$$P = \sum_{n\geq 1} P_n = \sum_{n\geq 1} R \cdot I_{n,RMS}^2 \tag{9.22}$$

Example 9.6 **A full-bridge inverter has a switching sequence that generates a square-wave voltage waveform across a series *R-L* load. The switching frequency is 50 Hz, the DC source voltage is 280 V, and the load resistance is 20 Ω and load inductance is 20 mH. Determine the load voltage and current amplitudes, and the power absorbed by the load.**

SOLUTION

The load voltage amplitudes for a resistive-inductive load, Equation (9.17a) is:

$$V_n = \frac{4 \cdot V_{DC}}{n\pi} = \frac{4 \times 280}{n \times 3.14} = \frac{356.7}{n} \; V, n = 1,3,5,...$$

For example: $V_1 = 356.7, V_3 = 118.9 \, V, V_5 = 71.3 \, V$

The load current amplitudes, Equation (9.19b) are:

$$I_n = \frac{V_n}{Z_n} = \frac{V_n}{\sqrt{R^2 + (n\omega L)^2}} = \frac{356.7}{n\sqrt{100^2 + (n \cdot 100\pi \times 20 \cdot 10^{-3})^2}}, \; n = 1,3,5,....$$

And for example $I_1 = 0.812 \, A, I_3 = 0.157 \, A, I_5 = 0.0515 \, A$

Power at each frequency is determined from Equation (9.22) as:

$$P = \sum_{n=1,3,5,..} R \cdot I_{n,RMS}^2 = \sum_{n=1,3,5,..} R \cdot \left(\frac{I_n}{\sqrt{2}}\right)^2$$

And first three power terms are:

$$P_1 = 20 \times \left(\frac{0.812}{\sqrt{2}}\right)^2 = 6.59 \, W, P_3 = 0.248 \, W, and \, P_5 = 0.0265 \, W$$

Basically, only first term has a significant contribution to the load power, and an approximate value of the power delivered to the load is:

$$P \approx P_1 + P_2 + P_3 = 6.59 + 0.248 + 0.0515 = 6.863 \, W$$

The voltage waveforms of practical inverters are, however, non-sinusoidal and contain certain harmonics. Square-wave or quasi-square-wave voltages are acceptable for low- and medium-power applications, and for high-power applications, low, distorted, sinusoidal waveforms are required. The output frequency of the inverter is determined by the rate at which the semiconductor devices are switched on and off by the inverter control circuitry and consequently, an adjustable frequency ac output is readily provided. By sequentially switching ON and OFF, the voltage across the load changes polarity, producing an alternating the voltage and current. Pulse width modulated (PWM) inverters are among the most used power-electronic circuits in practical applications. PWM is a technique which is characterized by the generation of constant amplitude pulse by modulating the pulse duration by modulating the duty cycle. PWM control requires the generation of both reference and carrier signals, fed into the comparator, which are based on some logical output, the final output is generated. The reference signal is the desired signal output maybe sinusoidal or square wave, while the carrier signal is either a saw-tooth or triangular wave at a frequency significantly greater than the reference. These inverters are capable of producing ac voltages of variable magnitude as well as variable frequency. The quality of output voltage can also be greatly enhanced, when compared with those of square wave inverters. The PWM inverters are very commonly used in adjustable speed ac motor drive loads where one needs to feed the motor with variable voltage, variable frequency supply. For example, the fundamental of the output voltage of single-phase, as function of the DC source voltage, V_{DC} is expressed by this simple relationship:

$$V_{o1(rms-fundamental)} = \frac{4V_{DC}}{\pi\sqrt{2}} = 0.90 \cdot V_{DC} \tag{9.23}$$

The three-phase bridge-type VSI with square-wave pole voltages has the output fed to a three-phase balanced load. This circuit may be identified as three single-phase half-bridge inverter circuits put across the same DC bus. The individual pole voltages of the three-phase bridge circuit are identical to the square pole voltages output by single-phase half-bridge or full-bridge circuits. The three-pole voltages of the three-phase square-wave inverter are shifted in time by one third of the output time period. The three-phase square-wave inverter can be used to generate balanced three-phase AC voltages of desired frequency. However, harmonic voltages of 5th, 7th and other non-odd multiples of fundamental frequency can severely distort the output voltage. In many cases such distortions in output voltages may not be tolerable, and it may also not be practical to use filter circuits to filter out the harmonic voltages in a satisfactory manner. In such situations the inverter discussed here is not a suitable choice. Fortunately, there are some other kinds of inverters, namely pulse width modulated (PWM) inverters, providing higher quality of output voltage. The square-wave inverter discussed here may still be used for many loads, notably AC motor type loads, the inductive ones in their nature with the inherent quality to suppress the harmonic currents in the motor.

9.2.4 CURRENT SOURCE AND VOLTAGE SOURCE CONVERTERS

Two common types of power converter – the current source converter (CSC) and the voltage source converter (VSC) – are in uses in power systems. In a CSC circuit, the converter DC side current is kept constant with a very small ripple using a large inductor, thus forming a current source on the DC side, power flow direction, through a CSC is determined by the polarity of the DC voltage while the current flow direction remains the same. Today, the CSC is mainly used in high-power applications, especially for HVDC transmission. In a CSC, the power electronic switches (thyristor valves) are turned on by control circuits but switch off through natural commutation when the current through them drops to zero. In a current source converter (CSC), the current on the DC side is kept constant with small ripples, only by using a large inductor, forming a current source on the DC side. The power flow direction of through the CSC is set by the polarity of the DC voltage while the direction of current flow remains the same. The CSC systems are mainly used in high-power applications, especially for HVDC transmission with a capacity of up to 7000 MW for a single link. In a

CSC the power electronic switches are turned on by control circuits but switch off through natural commutation when the current through them drops to zero. In a VSC unit, the DC side voltage is maintained constant by using a large capacitor. The VSC systems are widely employed for low and medium power applications, e.g. grid connection of micro-generators, renewable energy conversion systems and energy storage units. However, the VSC systems are also used at power levels up to 1000 MW for HVDC transmission. In a VSC controllable switches are controlling the current in the forward direction and anti-parallel diodes for current flow in the reverse direction. It is believed that with the rapidly increasing in sizes and reducing in costs of the semiconductor devices used in VSC circuits, the VSCs will dominate the near-future high-power DC applications. They are offering the advantages such as freedom to operate with any combination of active and reactive powers, the ability to operate in a weak grid and even black-start, fast acting control, the possibility of using voltage-polarized cables, generating improved sinusoidal wave-shapes.

A CSC consists of six thyristors, two on each branch, connected in series with each phase of the three-phase supply connected to the thyristor branch mid-point. During the positive period of the voltage, upper thyristors in each CSC branch can be turned ON, while the lower ones are turned ON during the negative part of the voltage. The DC voltage, V_d, of the CSC depends on the upper and lower thyristors that are conducting. The average value of the DC voltage can be obtained by integration. Since V_d is determined by the line-to-line voltage, V_{LL}, its peak value is $\sqrt{2}V_{LL}$ and the average DC voltage is given by:

$$V_d = \frac{3}{\pi}\int_{-\pi/6}^{\pi/6}\sqrt{2}V_{LL}\cdot\cos(\theta)\,d\theta = \frac{3\sqrt{2}}{\pi}V_{LL} \approx 1.35V_{LL} \tag{9.24}$$

If there is a firing angle, α of the CSC thyristors, the average DC voltage is given by:

$$V_d = \frac{3\sqrt{2}}{\pi}V_{LL}\cdot\cos(\alpha) \approx 1.35V_{LL}\cdot\cos(\alpha) \tag{9.25}$$

Since α can vary from $0°$ to $180°$, V_d can have values in the range from $+1.35V_{LL}$ until $-1.35V_{LL}$. When V_d is positive, the power flows from the AC side to the DC side, thus the CSC acts as a rectifier, while when V_d is negative, power flows from DC side to AC side, and the CSC is acting as an inverter. If the inductor in the DC side is large enough, the DC current, I_d is constant with very small ripples.

Example 9.7   If the angle, α, of CSC, connected to a supply source with line-to-line voltage, 460 V is limited to a range of $20°$–$110°$, estimate the range of V_d.

SOLUTION

Applying Equation (9.26), the V_d range is given by:

$$V_{d1} \approx 1.35\times460\cdot\cos(20°) = 583.6\,\text{V}$$

And

$$V_{d2} \approx 1.35\times460\cdot\cos(110°) = -211.8\,\text{V}$$

A VSC employs controllable switches, controlling the current in the forward direction and an anti-parallel diode is provided for current flow in the reverse direction. The current in the forward direction can be switched ON and OFF. The semiconductor switches, used in CVS circuits, include

Metal Oxide Semiconductor Field Effect Transistor (MOSFET), Insulated Gate Bipolar Transistor (IGBT), Gate Turn-off Thyristor (GTO) and Insulated Gate Commutated Thyristor (IGCT). The most common law and medium power VSC applications include inverters for PV and energy storage systems, back-to-back VSCs for wind power generators or active filters and DVRs. Some of these applications use a single-phase VSC whereas others use a three-phase VSC configuration. The common single-phase VSC topology is the H-bridge converter, as shown in Figure 9.11. In this topology the switch pairs (S1 and S3) and (S2 and S4) are turned on and off in a complementary sequence. When the two switches S1 and S3 are ON, the voltage across the load is V_{dc}, whereas when the two switches S2 and S4 are ON, the voltage across the load is $-V_{dc}$. Even though the output is a square wave, its fundamental is a sinusoidal waveform. Due to higher harmonics content, produced by the square-wave voltage output, this simple switching strategy is only employed with off-grid low-power generators. In many applications a sine-triangular pulse width modulation (PWM) technique is used to control the four VSC switches. The output voltage has a fundamental component and a series of harmonics, whose frequencies are depending on the frequency modulation index, m_f, defined as the ratio of the carrier signal, f_c and f_m. the frequency of the modulating signal). As each switch is turned ON and OFF several times in a cycle, a PWM-switched VSC produces larger switching losses than a square wave VSC. However, square-wave VSC operation is less attractive due to the low order harmonics it generates and the filtering needed to control these harmonics. The two-level three-phase VSC, so-called a six-pulse VSC, is basically a three-limb configuration of two complementary switches, employing a number of different modulating techniques can be used to generate the AC output, however the sine-triangular PMW technique is a commonly used approach.

An attractive topology for smart grid high-power and medium-power applications is multi-level converters due to the reduced switching operation frequency and so lowers switching losses. A multi-level converter output is a step-like waveform, and by using multi levels, a waveform closely resembling a sinusoid can be generated. The main topologies in use are the diode-clamped topology and the capacitor clamped topology. The simplest configuration is a three-level VSC, in which the auxiliary devices are used to clamp the output terminal to the potential of the DC-link mid-point. Additional levels can be achieved by adding extra switches and diodes; however, the number of components increases with the level number. An m-level DCC would require (m − 1) capacitors and (m − 1) (m − 2) clamping diodes. Each switch is only required to block a voltage level of V_{DC}/(m − 1). The clamping diodes need to have unequal voltage ratings. When the VSC is used for real power transfer, balancing of the capacitor voltages is one of the major challenges of such configuration. Notice that under unity power factor, at each half cycle the capacitors will be charged for an uneven period, resulting in an imbalance of the capacitor voltages. The voltage imbalance issue can be solved by connecting separately controlled DC sources across each capacitor with the trade-off is the increase of the circuit complexity and cost. The capacitor-clamped multi-level converter employs a large number of capacitors to form the step-like output. Another topology is the multi-modular

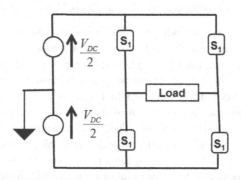

FIGURE 9.11 H-bridge VSC configuration.

converter (MMC), in which several H-bridge converters are connected to form the configuration. Due to its modular nature, redundancy can be incorporated and standard modules are used. Sometimes the MMC configuration is called a VSC chain circuit. The arrangement has two identical H-bridge converter cells with separate DC sources, allowing independent control of the converter cells, and this configuration requires only two DC capacitors. In such configurations, the number of levels can be defined as the number of DC voltages it can produce across the single-phase terminals, each H-bridge generating up to three voltage levels. Some MMC system controls vary the DC-link voltages to control the fundamental voltage component, beneficial for voltage regulation.

9.3   SMART GRID POWER ELECTRONICS APPLICATIONS

Traditional power systems employ large power generation plants situated at often large distances from the major load centers, supplying most of the grid power that is then transmitted to large consumption centers and then distributed between different customers, via substations, feeder and power distribution networks. This grid structure has started to be changed towards new scenarios at which DG units are spread over distribution networks. These DG systems utilize renewable energy resources such as wind turbines, photovoltaics, fuel cells, biomass, small hydro-plants, energy storage units and devices, etc. Beside their environmental benefits, DGs present a cheap way into market since they do not suffer vast transmission losses and the excess heat may be handled in useful purposes such as water and space heating. Loads that are more flexible are expected to support the grid by accepting varying supplies of energy from renewable energy sources and by controlling peaks in demand. For sensitive loads such as computers and high value manufacturing plants, the quality of supply will be important. Therefore, visibility, controllability and flexibility will be essential features throughout the future power system with power electronics playing a key role. The power output of some of the renewable energy sources is always DC (for example, photovoltaic systems or fuel cells) and an inverter is needed to interface them to the AC grid. Even though renewable energy sources using an AC generator (for example, wind turbines) can be connected directly to the grid, often some form of AC to DC and then DC to AC conversion is used. Some power system operating conditions demand rapid independent control of the active and reactive power output of the renewable energy generators. These control actions and methods can only be achieved conveniently using a power electronic interface. With the connection of a large number of distributed generators, including micro-generators, energy storage units and electric vehicles traditional methods of active power/frequency control and reactive power/voltage control will no longer be effective. The traditional voltage control method in a distribution circuit is an on-load tap changer and automatic voltage control relay, sometimes with line drop compensation. This control system may not operate satisfactorily when power flow in the power distribution circuit is flowing in reverse the power flow direction.

9.3.1   Power Electronics for Renewable Energy and Grid Integration

Fuel cells, small hydropower systems, wind energy conversion systems, biomass (solid biomass, biofuels and biogas), tidal stream, solar-thermal energy and photovoltaic (PV) are the usually choices. Variable speed turbines are often used for wind, mall hydropower and tidal generation units. These are usually using an AC-DC-AC power conversion system because the turbines are rotating at optimum speed to extract the maximum power from the fluid flow or to minimize the turbine mechanical loads. The variable frequency power output of such turbine-generator unit is converted to DC and then DC is converted to 50Hz or 60 Hz AC. The PV system output is DC and therefore a DC-AC power converter is required for grid connection. Biomass technologies use steam or gas turbines and conventional synchronous generator, with no specific power converter needs. Reciprocating engines may be fueled by biogas, connected to a synchronous generator which may be connected to the grid directly. Power electronic interfaces between renewable energy sources and the grid can be used to

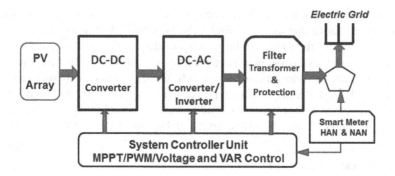

FIGURE 9.12 Block diagram of a residential grid-connected PV system.

control reactive power and hence the network voltage, to curtail the real power output, enabling the generator to respond to the grid requirements. Figure 9.12 shows the main components of a residential grid-connected PV system, often consisting of: a DC-DC converter for maximum power point tracking (MPPT) to improve the DC signal quality and to increase the voltage at desired level, a single-phase DC-AC inverter, an output filter and sometimes a transformer and controller and protection devices. The system is connected to a smart meter and HAN and/or NAN networks. Various DC-DC converter types are used, e.g. Boost, Buck-Boost, push–pull or flyback converters. The DC voltage on the inverter side of the DC-DC converter is usually maintained constant by the system control. The MPPT algorithm is used to find and set the maximum power point of the PV array, while the PV cell temperatures, insolation and operating conditions of the module change. The DC voltage obtained from the DC–DC converter is inverted to 50 or 60 Hz AC signal, also using a pulse width modulation (PWM) switching technique to minimize harmonic distortions. Several MPPT algorithms are employed, e.g. perturb and observe, incremental conductance, constant voltage or temperature method. A voltage source inverter is the most common choice. Finally, a filtering subsystem is included to eliminate the harmonics to be fed into the power system, and sometime a transformer is also used at the output of the inverter to ensure no DC component is injected into the electric grid.

Wind, hydropower, tidal and wave energy generation systems all involve converting the potential and/or kinetic energy in a fluid, water or air into electrical energy. In the last three decades, there has been a dramatic increase in electricity generation from the wind energy with the capacity of wind energy conversion systems, installed worldwide in hundreds of GW. Hydropower is a mature technology with units varying in size from a few kW and MW, in the case of small hydropower systems to hundreds of MW of the large settings. Wave energy and tidal stream generation are more recent developments and the subject of considerable research efforts. Wind farms are now being developed both onshore and offshore, with by far more challenging and expensive of offshore settings. However, the offshore wind farms enjoy a stronger and more consistent wind resource and reduced environmental impacts. The majority of generators used in offshore wind turbines are variable speed, while in onshore wind farms various generator types are used. Small and micro hydropower plants are one of the few power generation options in many remote or developing counties. The efficiency of these smaller units across a wide range of water flows can be improved by using a variable speed generator with a power electronic interface and controller. Different turbine designs and generation units are available for wave energy and tidal stream technologies, all requiring some sort of power electronics. The design of a hydropower turbine is optimized for a specific rotational speed, hydraulic head and discharge. As the hydraulic conditions are changing continuously, the conversion efficiency of the turbine also changes, making the variable speed operation changes the turbine speed so as to maximize its efficiency over a range of different hydraulic conditions. In tidal stream units, in order to extract maximum power, the torque presented by the generator to the prime mover varies with the tidal flow conditions. There are two options, a maximum power tracking and control concepts similar to that used for wind turbines, where the power extracted is determined by off-line calculations of the rotor angular velocity relative to the tidal stream flow. For variable speed operation of wind, hydro and tidal stream turbines, doubly fed induction generators (DFIG) or full power converter (FPC) based generators can be used. For example, in such cases the generator speed is controlled to track the maximum power and the generator output frequency varies with wind, hydraulic (river/stream) or tidal flow conditions. The variable frequency power is then converted to DC using the power electronic converter and then inverted back to 50 or 60 Hz AC signals.

9.3.2 FAULT CURRENT LIMITING DEVICES

A fault current limiter (FCL), also known as fault current controller (FCC), is a device which limits the fault current when a fault occurs (e.g. in a power transmission network) without complete disconnection. FCLs are considered serious candidates to be inserted into electrical grids in order to prevent short-circuit damage and lowering the needs for upgrading of the system equipment. The term includes solid-state, superconducting and inductive devices or topologies. Both superconducting and non-superconducting FCLs have been extensively applied in transmission and distribution networks and renewable energy systems for purposes such as stability enhancement, protection improvement, fault current reduction and fault ride through capability enhancement. Electric power distribution systems include circuit breakers to disconnect power in case of a fault, but to maximize the reliability, in order to disconnect the smallest possible portion of the network. This means that any type of circuit breakers, as well as all wiring topologies to them, must be able to disconnect large fault currents. Problems arise if the electricity supply is upgraded by adding new generation capacity or by adding cross-connections. Because these increase the power amount that can be supplied, all branch circuits must have their busbars and circuit breakers upgraded to handle the new higher fault current limit. This poses a particular problem when distributed generation and renewable energy, such as wind farms or solar power plants are added to an existing grid. It is desirable to be able to add additional power sources without large power system-wide upgrades. However, more and more distributed generation, energy storage and renewable energy units are connected to the smart grids, mostly in power distribution. Many of these generators are of asynchronous types that may inject sustained currents into a short circuit fault, rising significant operation problems. The asynchronous (induction) generators that are used, in such generation options, may have fast symmetrical fault current decays within less than a second, affecting the protection units operation. When large number of distributed generators are connected through power electronic converters whose fault current contributions is limited by the ratings of the electronic switches that they are employing and by their control systems. Moreover, the connection of different types of distribution generation and renewable energy units introduce issues such as: the synchronous generators can increase the fault current through switchgear, which has a limited interruption capability, requiring replacement of the switchgear or other measures to limit the short circuit current, while in circuits, rich in power electronic connected distributed generators, the fault current contribution may not be adequate to ensure detection of the fault by existing over-current protection, demanding new and different protection methods. FCL equipment and devices in distribution circuits limit the fault current. Different designs such as current limiting fuses, superconducting, magnetic, static and hybrid configurations of fault current limiters exist and are employed in smart grids.

A fault current limit circuit is a nonlinear element which has low impedance at normal (rated) current levels, but presents higher impedance at fault current levels. Further, this change is extremely rapid, before a circuit breaker can trip, in a few milliseconds later, while high-power circuit breakers are synchronized to the alternating current zero crossing to minimize arcing. Superconducting fault current limiters exploit the extremely rapid loss of superconductivity (quenching) above a critical threshold combination of temperature, current density and magnetic field. In normal operation conditions, the current flows through the superconductor element without resistance and negligible impedance. If a fault develops, the superconductor quenches, its resistance increases sharply and the current is diverted to a parallel circuit with the desired or required higher impedance. There are several FLC designs proposed or under research. However, a common design that has been implemented in a real system is the scheme consisting of inductor in series with circuit of capacitor in parallel with a pair of thyristors, in parallel with a switch and in series with second inductor. In this circuit, under normal operation, both thyristors are off and current flows through first inductor and capacitor, and their impedance is selected such that at 50 Hz or 60 Hz is zero. This introduces zero impedance in series with the transmission line. When there is a fault, both thyristors go into full conduction mode and second inductor is connected in parallel with the capacitor, thus increasing the total inductive impedance in series with the line. A varistor is connected to limit the transient voltage across the thyristors, and the rate of rise of the transient voltage is limited by the snubber circuit. Notice that the optimal location for FCLs in a power network has several potential benefits. These include enhancing the system reliability and security, reducing fault current and voltage sag, improving fault ride through capability and increasing the interconnection of renewable energy. Several optimal placement techniques have been reported and discussed in the literature.

9.3.3 SHUNT AND SERIES COMPENSATION

A device that that is connected in parallel with a transmission line used to compensate for the reactive power in the AC system is called a shunt compensator. It can improve the voltage profile, power-angle characteristics, system stability margin and can provide damping to power oscillations. The ideal shunt compensator is represented by an ideal current source that supplies only reactive power and no real (active) power. For decades shunt capacitor banks, reactors and synchronous condensers have been employed in power systems, often with electronic control, in order to manage and control reactive power. However, shunt compensation devices based on voltage source inverters such as STATCOMs, active filters and voltage source converters with energy storage (VSC-ES) have begun to be used in the power systems and smart grids. STATCOM devices are now extensively used to provide reactive power compensation in both transmission and distribution circuits in order to manage network voltages, reduce losses and overcome possible instabilities. The voltage changes across a distribution network, characterized by the resistance, R and the inductive reactance, X is estimated by this relationship:

$$\Delta V = \frac{PR+QX}{V} \tag{9.26}$$

Here, P and Q is the network active and reactive power flows, and V is the nominal (rated) voltage. As the power output of a distributed generator driven by a renewable energy source varies, the voltage change across the circuit to which it is connected also varies. Mitigating these voltage fluctuations can be effected with shunt compensation devices such as a STATCOM or a VSC-ES that can vary Q or P and Q with the change of voltage. The two most common such devices are the static VAR compensator (SVC) and the static compensator (STATCOM). Basically, the aim of these devices is to provide voltage support to the system by exchanging reactive power with the system and maintaining the bus voltage close or equal to a certain reference level. Notice that many of the electronic loads draw non-sinusoidal currents that can lead to unacceptable voltage distortion levels. Standards such as IEEE 519-1992 and ER G5/4 specify limits to either the harmonic current which may be injected, or to the resulting harmonic voltages on the network. If these limits are exceeded, then connection to the network is refused. If the load equipment cannot be modified economically to approach the sinusoidal current, then power electronic active filters may be used to correct the current drawn from the network. A STATCOM device is the power electronic counterpart of the conventional rotating synchronous condenser. Static synchronous compensator is one of the static component devices, coming under the family of FACTS devices. It can sink or supply reactive power in the single- or three-phase AC electric systems. When a STATCOM is connected to the distribution circuits, it is normally called a D-STATCOM. The basic operating principle of a D-STATCOM for reactive power control consists of two phases. First, when the amplitude of the VSC_{OUT} voltage is less than the terminal voltage, the D-STATCOM draws a current lagging the terminal voltage, absorbing reactive power. When the VSC generates a voltage higher than the terminal voltage, the D-STATCOM generates reactive power. The VSC of the D-STATCOM produces a controllable three-phase voltage in phase with the terminal voltage. Two main control approaches for STATCOM control are employed, one based on the phase shifting control, and the other one on using PWM. In the first approach, if the STATCOM voltage slightly leads the terminal voltage, net real power flows from the D-STATCOM to the AC system, decreasing the DC capacitor voltage and thus STATCOM voltage, and then reactive power is absorbed by the STATCOM. The converse occurs when the STATCOM output voltage slightly lags the system voltage, the capacitor voltage rises and the reactive power is injected by STATCOM. Note that a D-STATCOM can be used for power factor correction and to balance the current drawn by an unbalanced load. A control system that is used with thyristor-based static VAR compensators may be used with a D-STATCOM for voltage control in power grids. With recent advances in energy storage technology, the application of a voltage source converter with energy storage has now become a feasible option for steady state voltage control and elimination of power system disturbances. The VSC-ES can be controlled to exchange both real and reactive power with the AC system. The real and reactive power can be controlled independently of each other and any combination of real power and/or reactive power generation or absorption is possible and achievable (within the equipment ratings). As with a D-STATCOM device, the reactive power generation or absorption capability of the VSC-ES can be used for load compensation or steady state and transient voltage control. The smart grid applications of VSC-ES include: load compensation, steady state voltage control and sag mitigation.

Sensitive loads such as manufacturing plants of high-value products, e.g. semiconductor manufacturing facilities, paper mills or chemical plants, require the voltage to be strictly maintained within specified limits. A simplest and common topology of series reactive power compensation device is a series-connected capacitor, relying on its control system to avoid resonances and ensure the power system stability. A series-connected converter is used as a dynamic voltage restorer (DVR) to maintain the voltage within specified limits. If the incoming feeder voltage fluctuates beyond the voltage limits that a sensitive load could operate, then the DVR adds a voltage in series to compensate for the voltage fluctuations. The two commonly used compensation techniques are in-phase compensation and freeze-phase locked loop (PLL compensation) approaches. In the in-phase compensation technique, the load voltage phasor always kept in-phase with the supply voltage by introducing a sudden phase shift to the customer supply when the voltage sag causes a phase jump to the supply voltage vector. In the freeze PLL compensation approach, the DVR maintains the load voltage level as the same as the pre-sag condition by injecting the difference between the pre-sag supply voltage and the sagged supply voltage. Notice that the connection of inductive or capacitive reactances in shunt changes the flow of reactive power in a circuit and the network voltage. However, a modest level of shunt reactive compensation is posing little risks to the power system. Similarly, power electronic shunt compensators change the reactive power flows. In contrast, if the capacitors are connected to a network in series to reduce its inductive reactance can be a more hazardous approach as very high voltages can occur across the capacitor during faults and electro-mechanical resonances can be induced in the rotating machines. However, for a given capacitance, series compensation changing the circuit reactance is more effective in controlling voltages than shunt compensation that changes the reactive power flows and controlling the network voltages.

9.4 POWER ELECTRONICS FOR BULK POWER FLOW CONTROL

An electric power grid is a network of interconnected transmission and distribution lines carrying electrical energy from the generating locations to the loads (end-users). The smart grids, among other new technologies, involve connection of a larger number of renewable energy, distributed generation and energy storage units, many of medium and large power size, needed to reduce pollut-ant emissions, diversify energy sources, increase transmission capacity and maintain the continued supply security. These new connections and subsequent bulk power flows require electric network reinforcement, while the conventional methods of increasing bulk power transfer capacity are re-conducting existing transmission networks, upgrading to a higher AC voltage, capacity and/or con-structing new lines. However, these options, particularly the ones which involve new overhead lines, are difficult to implement due to planning constraints and environmental concerns. On the other hand, FACTS devices can increase the capacity of AC circuits, HVDC transmission may be used for the addition of new capacity, while additional interconnections can solve some of the issues. Notice that, for example, balancing of the energy supply and demand over larger areas would benefit from the diversity of the renewable energy resources and of the load through the spread of time zones. Unfortunately, the conventional approach to expanding the transmission system consists of building new high-voltage transmission lines, which may take years to complete. Therefore, the immediate solution is to utilize the existing transmission system more efficiently. One way is to identify the underutilized transmission lines and increase their power flows to the lines rating limits. This can be achieved using a full power electronics-based solution, an electromechanical solution or a hybrid of the two. It is also anticipated that with large hydropower in some regions, wind energy and solar energy in other areas the proposed interconnected network could ensure regional smoothing of electrical energy generation, increased supply security and reduced dependency on fossil fuels. Balancing of supply and demand would benefit from the diversity of the renewable energy resources and of the load through the spread of time zones.

The unified power flow controller (UPFC) is an AC-to-AC converter which can change the net-work parameters, e.g. end-line voltage, the phase angle between the two busbars or the apparent line reactance. It is implemented by two AC-to-DC converters (rectifiers) operating from a common DC link capacitor. One converter is connected in series and the other one is connected in shunt with the transmission line. Second VSC of this configuration generates a voltage $V_X(t) = V\sin(\omega t - \alpha)$ at the fundamental frequency (ω) with variable amplitude ($0 \le V \le V_{max}$) and phase angle ($0 \le \alpha \le 2\pi$) which is added to the AC system voltage by the series connected coupling transformer. The receiv-ing end apparent power, consisting of the apparent power associated with the uncompensated line and the second term, is the apparent power associated with the series injection by the second VSC of the UPFC is given by:

$$S_R = V_R \cdot I^* = V_R \left[\frac{V_G - V_R}{jX} \right]^* + V_R \left[\frac{V_X}{jX} \right]^* \tag{9.27}$$

Here, V_G and V_R are the generating (source) end and receiving end, V_X the VSC voltage, and X is the line reactance. The real power demanded by second VSC is supplied by the first VSC of the UPFC topology through the DC link capacitor. First VSC can also act as a shunt reactive power compensator, thus injecting or absorbing reactive power independent of the operation of the series connected converter, second VSC. Second configuration VSC can be controlled to obtain terminal voltage regulation, line impedance control or phase angle control. Terminal voltage regulation is achieved by injecting a voltage in phase with the sending (source) end voltage, while the line imped-ance can effectively be varied by injecting a voltage perpendicular to the line current. Phase angle control can be achieved by injecting a voltage to shift the sending end voltage by a desired angle. An interline power flow controller (IPFC) connects several VSCs in series to different lines. The DC side is connected in parallel, enabling power to be transferred from one transmission line to another through the series links.

In normal operation, the electricity supply is matched with its energy demands of the loads and power losses in various components such as transmission lines, power transformers, generators, etc. throughout the process, while the highest reliability is maintained. The electricity flows in a transmission line depends largely on the line impedance. If the impedance of a line is larger as compared to that of the transmission lines connected in parallel, the current and the resulting power flow through the high-impedance line is lower compared to that in the neighboring lines and vice versa. Sometimes it is desirable and required to decrease the impedance of a specific transmission line so that more current can flow through the line up to the allowable limits, resulting in higher line uses, to meet greater customer needs, to integrate new energy sources and to avoid building of new transmission lines, at least for the time being. In other instances, the opposite takes place, it is desirable to increase the line impedance, so that less current can flow through the power line. This is particularly important when a line becomes overloaded with a level of current that can trip it or a fault current that must be limited. If an overloaded line trips, its current will be redirected in the available lines proportionately, depending on the line impedances. This may cause a previously under-loaded line to become overloaded and tripped, which may create a possible cascaded failure of the grid, resulting in a blackout. In summary, line power flow is inversely proportional to its reactance. Power flow control is the ability to change the power flows through the grid by actuating line switching hardware or by controlling high-voltage devices connected in series or in shunt with transmission lines. It includes the ability to control the impedance on a major transmission line, inject a controlled voltage in series with a line, provide reactive voltage support for long lines so that they can be loaded to their thermal limits and to switch line circuit breakers to redirect power to other lines. FACTS devices are extensively used in modern power systems and smart grids to enhance power flows on the existing transmission lines. For a transmission line, as one shown in Figure 9.13, if source end voltage, $V_S\angle\delta_S$, the receiving end voltage, $V_R\angle\delta_R$ and the equivalent impedance of the parallel connected lines, X, the power transfer through the two transmission lines is given by the well-known relationship:

$$P_{Line} = \frac{V_S V_R}{X} \sin(\delta_S - \delta_R) \tag{9.28}$$

In the topology of the line configuration of Figure 9.13, the TCSC unit can change the line impedance, the STATCOM controls the voltage magnitude at the receiving line end to which it is connected by injecting or absorbing reactive power and the UPFC changes the phase angle of the sending end voltage, thus power flow through a line can be controlled in several ways, in agreement with Equation (9.28).

In transmission networks, the loads at the receiving end are varying continuously, and the receiving end (load terminals) voltages are also changing continuously. To maintain the voltage within the prescribed and required limits, reactive power compensation devices need to be employed. They are controlling the voltage at a node by injecting or absorbing reactive power. In addition to steady state voltage control, reactive power compensation devices are used to mitigate dynamic voltage variations caused by line switching, load rejection, faults and other system disturbances. In smart grids and modern power systems, reactive power compensation is becoming essential to meet the

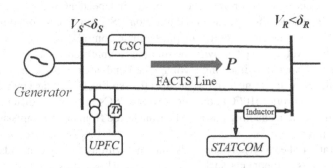

FIGURE 9.13 Enhanced power flow of an existing transmission line with a FACTS complementary line.

requirements set by utilities and specified by standards for the connection of renewable energy generators, sensitive load and distributed generation units. In most of the countries the national codes, defining the minimum operational and technical requirements for grid connection of generation stations, stipulates steady state and dynamic reactive power requirements. These requirements need to be satisfied at the point of connection of the generating plant to the utility network. Reactive power compensation devices are also needed to meet fault ride through network requirements. Shunt reactive power compensation devices commonly found in the modern power system include, capacitors, reactors and power electronic compensation devices. Fixed capacitors provide a constant capacitive shunt reactance to the network, and in some applications a shunt inductor can be used to obtain an inductive reactance. A reactor can be connected in series with the fixed capacitor to form a harmonic filter to improve network power quality. Power electronics controlled capacitors and reactors provide a variable reactive impedance. The reactive power output of these devices is proportional to the square of the line voltage divided by the reactive impedance. STATCOM devices are providing a variable reactive current that is injected into the transmission networks, thus their reactive power output are the product of the transmission line voltage and reactive current injected or absorbed.

9.5   CHAPTER SUMMARY

The smart grids, future power systems, are increasingly included and use more controllable power electronic devices and equipment to make the best use of existing circuits, maintain flexibility and optimum operation of the power system and to facilitate the connection of renewable energy resources at all voltage levels. Power electronics are enabling new dimensions in power conversion, control and uses, having a significant impact on the grid operation, management and structure. A selected set of power electronic converters and their most common applications are presented and discussed in this chapter. By definition, power electronics equipment and circuits are relating power semiconductor devices and circuitries, their design and role are including techniques of converting and processing higher levels of electrical energy. The end goals of a power electronic converter are to achieve high efficiency of conversion, minimize size and weight and achieve desired regulation of the output. Power electronics converters are critical elements of the renewable energy system applications, grid integration, maximum power tracking, operation, management and protection. Such power converter systems include DC-DC, inverters, voltage-source or current-source converters, each in various configurations and topologies. Protection of such systems and augmentation of reliability as well as stability highly depend on limiting the fault currents. Several fault current limiters (FCLs) have been applied in power systems as they provide rapid and efficient fault current limitation, improving the system stability and resilience. FCLs placement is important in limiting fault current and augmenting stability of power system. Shunt and series compensation are extensively used in power system for controlling reactive power flows and system voltages. Power flows in a transmission line depends largely on the line impedance and thermal line capacity. Apart from building new transmission lines, it may be quicker and cheaper to utilize the existing transmission system infrastructure by harnessing the dormant capacity of the underutilized lines, through the use of power electronic circuitry. Flexible AC transmission system devices are among the most promising power flow controllers used in smart grid power transmission over long distances, providing system stability, midpoint voltage support and reactive power control in grid interconnections. Reactive power compensation is a requirement of rid connection of distributed generation and renewable energy units, a specified by codes and standards and required by the utilities.

9.6   QUESTIONS AND PROBLEMS

1. What are the main reasons for increased uses and applications of power electronics in smart grids?
2. List few of the major applications of power electronics in smart grids.

3. Why some of the renewable energy conversion systems are using an AC-DC-AC conversion chain?

4. List the main power solid-state devices.

5. What are the major advantages of the power semiconductor devices that are used for power conversion and control?

6. List the major types of power converters.

7. What the major roles of power electronics converter?

8. Briefly describe the major types of DC-DC converters.

9. List the major types of power electronic converters used in smart grids.

10. What are the reasons that the use of power electronics is critical for smart grids?

11. A Buck converter of has the following parameters: $V_d = 24$ V, $D = 0.60$, $L = 25$ μH, $C = 24$ μF, and R $= 10$ Ω. The switching frequency is 80 kHz. Assuming ideal components, determine (a) the output voltage and (b) the output voltage ripple.

12. A Buck-Boost converter of has the following parameters: $V_d = 12$ V, D $= 0.6$, L $= 12$ μH, C $= 20$ μF, and R $= 10$ Ω. The switching frequency is 120 kHz. Assuming ideal components, determine (a) the output voltage and (b) the output voltage ripple.

13. A Boost converter of has the following parameters: $V_d = 30$ V, D $= 0.75$, L $= 20$ μH, C $= 60$ μF, and R $= 15$ Ω. The switching frequency is 100 kHz. Assuming ideal components, determine (a) the output voltage and (b) the output voltage ripple.

14. Repeat Example 9.3, but for an output power of 12 kW, output rated voltage of 100 V and the voltage range from 60 V to 140 V.

15. Why the power electronics is one of the key enabler of renewable energy and distributed generation?

16. Find the peak (amplitude) source voltage that a full-wave controlled rectifier is producing 100 V (DC) across a 25 Ω. What is the power absorbed by this load?

17. A single-phase inverter supplied by a 150 V DC source is delivering power to a resistive-inductive load, consisting of 15 Ω and 30 mH inductance (operating at frequency of 60 Hz), determine the amplitudes of the square wave load voltage, the amplitudes of the load current and the power absorbed by the load.

18. A single-phase full-wave bridge rectifier is connected to a 130 V (RMS), 60 Hz AC voltage source is supplying a resistive load of 10 Ω. Calculate the load current and voltage.

19. A single-phase square-wave inverter is connected to a 120 V DC voltage source. Calculate the magnitude of the fundamental and the first three harmonics produced by this inverter.

20. For a Buck converter with input voltage 12.5 V, output voltage 5 V and switching frequency 20 kHz, having and inductance of 1 mH and a capacitor of 470 μF, determine the ripple in output voltage.

21. A full-bridge square-wave inverter has the DC source voltage 105 V, and the output frequency of 60 Hz, and a resistive load of 15 Ω. Determine the average and RMS values of output voltage and current.

22. Compute the average value of the DC voltage of a CSC converter connected to a three-phase supply, having the line-to-line voltage, 460 V and a firing angle of 15°.

23. Briefly describe the CSC and VSC converters, and their main applications

24. Repeat problem 10, for a line-to-line voltage of 4.2 kV and a firing angle of 8°.

25. A PWM full-bridge inverter is producing a 60 Hz AC voltage across of a series R-L load of a 10 Ω resistance and a 25 mH inductance. The DC input to the bridge is 100 V, and the amplitude modulation ratio m_a is 0.8. Determine the amplitude of the 60 Hz component of the output voltage and load current, and the power absorbed by the load resistor. Hint: the power absorbed by the load is computed by using: $P_R = R \cdot I_{RMS}^2$

26. Why wind turbines or small hydropower systems need power electronics in order to properly operate?

27. What power converters are used in solar photovoltaic applications?

28. Repeat the Example 9.6, for an angle, α of 10° to 160°, and line-to-line voltage of 480 V. Plot V_d for angle increment of 10°.
29. Briefly describe the CSC operation.
30. What are the major advantages of using FACTS devices?
31. What are the main generator types used in wind energy, small hydropower and tidal energy?
32. Why the tidal stream energy systems require power electronics for opium energy generation?
33. Briefly describe the operation of fault limiting circuits (FLCs).
34. Why are FLC circuits needed in the smart grids?
35. What are the major VSC-ES device applications?
36. Briefly describe the STATCOM approach to control reactive power.
37. Briefly describe the power flow control in smart grids.
38. How an UPFC device works?
39. Describe the use of FACTS devices to enhance the existing transmission line power flow.
40. What are the reasons for reactive power compensation in modern power systems?

10 Smart Grid Energy Management and Design Tools

10.1 ENERGY MANAGEMENT PHILOSOPHY

Energy consumption is one of the most important factors that determines and limits future development and progress. Regardless of the economic, social and political contexts over long run, the energy uses tend to increase. Smart grid is envisioned to meet the 21st-century energy requirements and expectations in a sophisticated manner, by integrating the latest communications, sensing, monitoring and advanced, adaptive and smart control technologies to the existing power grids. Rapid advancements into control, information and communication technologies have allowed the conversion of traditional electricity grids into smart grids that ensure productive and complex interactions among energy providers (utilities), consumers and other stakeholders. Key components of the smart grids are smart meters, adaptive and intelligent; sensors; monitoring systems and data management systems that control the flow of power and information among various stakeholders, making the grids two-way communication and power networks. Other smart grid applications include Energy Management Systems (EMS), distributed generation (DG) and their reliable system integration, equipment diagnostics, control, overall optimized asset management, etc. Monitoring and optimizing the energy-economy balance became one of sustainable development crucial problem. Sustainable development represents a comprehensive and integrate framework for humankind future, attempting to plan future development based on past experience while predicting the future needs. Sustainable development concepts imply the understanding and reviewing their complex, multi-disciplinary, multi-dimensional and heterogeneous structure, including economic, technological, societal and environmental aspects. One of the most critical aspects of this concept is the energy sustainability, regarded as a comprehensive issue of energy generation and usage, with minimum environmental impacts, while providing the needs of the people and society. Energy sustainability implies conservation and optimum uses of energy sources, exploitation of renewable energy sources, energy efficiency and conservation and responsibility of all energy participants, utilities, energy traders, consumers, etc. Major energy issues are that most of the conventional energy sources are limited and are unevenly distributed, and the big energy users' concentration in a small number of countries.

The key questions for energy management are how to provide the best solutions for successful and optimum energy uses, control and practice. The EMS plan's purpose and standards are to provide an organizational framework for industries, regardless the activity domain to integrate energy control, optimum uses and efficiency into their management practices, including fine-tuning production processes and improving overall process efficiency. Businesses and organizations that are wasting energy are reducing profitability and causing avoidable pollution, primarily through increased carbon emissions, contributing to environment degradation and dwindling fuel reserves. Energy management seeks to apply to energy uses the same culture of continual improvement that has been successfully used by industrial and commercial organizations to improve quality and safety practices. Energy management guidelines recommend that companies track energy consumption, benchmark, set goals, create action plans, evaluate progress and performances and create energy awareness throughout the organization, enabling to integrate energy efficiency and conservation into management system for continuous improvements and to reach the requirements for high efficiency and ultimate to reduce costs and increase the revenue. An energy management standard is needed to set how energy is managed in a facility, system or process, thus realizing immediate

DOI: 10.1201/9780429174803-10

energy use reductions through operational practice, as well as creating a favorable environment for the adoption of capital-intensive energy-efficiency measures, practices and technologies. Efficient EMS requires the identification of where energy is used, wasted and where energy saving measures is effective. The EMS key successful feature is that it is owned and fully integrated within organization, as an embedded management process, its implications are considered at all implementation stages, being part of any process changes. Standards should lead to energy cost and pollutant emission reductions, providing a framework necessary to move beyond an energy saving approach to an energy efficiency approach that routinely and methodically seeks opportunities to increase energy efficiency and lower the energy use. Though one successful energy management program is implemented in a different way than the other, all share common elements. Making businesses more energy efficient is a largely untapped solution, addressing pollution and energy security. Energy management, an effective way of controlling energy uses, enabling overall cost reductions, is increasingly important due to the volatility energy cost and supply, or pollutant emissions and cost-related issues, e.g. carbon taxes, reducing the organization carbon footprint to promote a green, sustainable image, often a good public relation policy and the risks (the more energy is consumed, the greater the economic risks, due to the energy supply and price volatility). With energy management such risks are minimized by reducing the energy use, control, making it more predictable. In short, the energy management systems aim at reduced energy uses and costs, the EMS key elements. Energy management must be based on real-time information from process monitoring and control, and on production plans received from management, as detailed in Figure 10.1. Often total and comprehensive solutions include planning and scheduling tools to optimize energy uses and supply, energy balance management tools to support the real-time monitoring and control of energy balance, and reporting tools to evaluate and report energy consumption, costs, efficiency and any energy-related information. Opportunities for cost reduction are greatest when both electricity consumption and prices vary over time, which is common in process industries, and open electricity market environments.

In the context of power systems, the control centers, along with its energy management systems, the *electric grid brain*, are critical power system elements in operational and control reliability structure. Over the past few decades, power system EMS has evolved gradually from its early stages as digital computers to a more sophisticated form of hardware, smart sensing and monitoring, intelligent control and software applications, at the end of 20th century to essential supporting tools of the emerging smart grids. Efficient operation, advanced control and management lead to great energy saving potential, cost reduction and minimizing the environmental impacts without needs to change the structure or the parameters of the energy supply systems. Thus, a detailed investigation of the operational modes and measurements of electricity uses of every system component and elements can point out those devices and equipment that are not working and performing optimally. Such verifications and investigations can show the potential of efficiency increasing for each device and equipment as well as for the whole system. Another possibility of efficiency improvement for the whole system is the integration of additional capacities of renewable energy sources or energy storage units to ensure the needed system control and management flexibility. Renewable energy sources and energy storages systems can be used for load shifting during the peak load hours. Therefore, the usage of renewable energy sources by enterprises needs continual and reliable energy

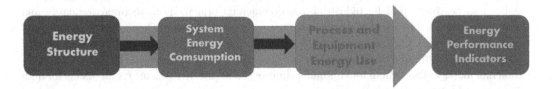

FIGURE 10.1   Energy management flow chart and structure.

management. The microgrid systems, local electric networks offer the possibility to improve the energy efficiency and to integrate smart elements, alternative energy sources and energy storage units into the buildings, facilities or processes. Standards define EMS entities as a combination of elements that an organization implements in agreement with its strategic energy policy objectives and specifics to improve energy uses, efficiency, energy conservation and supply security. In addition, monitoring and smart sensing actions and measures specifically adapted for the organization must be included. Utilities have been relying on EMS for over five decades; however, many EMS systems in use are outdated, under-maintained or underused compared with their potential that could be realized.

10.1.1 Energy Management in the Modern Power Grid Context

In the smart grid implementation framework, electric energy consumption, controllable loads, generation resources, energy storage, plug-in electric vehicles or smart monitoring and sensing should be managed and optimized in ways that are saving energy, improve efficiency, enhance reliability and service quality, maintain grid and supply security while meeting the increasing energy demand at minimum operating costs and environmental impacts. The expanding modern power grid requires system operation and control that is emerging at temporal and spatial scales, different from the ones traditionally considered by the EMS and control centers. New advances in the measurement, smart sensing, communication, computation and control are making possible an employable transition to the new EMS generations. Similarly new technologies are increasingly applied to power distribution, and the new generations of distribution management systems are now in the full deployment processes or uses. Power system state is ever changing, being a dynamic and random process since loads, and configurations of networks change, by making the power system operation difficult and complex and impossible to implement feed-forward control. In addition, the response of many power network components and devices is not instantaneous. Decisions must be made on the basis of predicted system future states, via processing, analyzing and visualizing huge amount of data and information. To automate the power system operation and control, the electric utilities rely on a highly sophisticated integrated computerized systems and software applications for monitoring, management and control. Operating electric grids close to normal frequency and voltage levels, in very narrow prescribed ranges, without causing any unexpected load or generation disconnections, knowing as maintain grid integrity and stability, is performed through control centers, using integrated software and hardware systems, the EMS having a multi-tier structure with several levels of elements and components. At the bottom is the high-reliability switchgear with facilities of remote control and monitoring, protective relays, automatic transformer tap-changers and other automatic equipment. At next tier are the tele-control cabinets at switchgears for actuator control, interlocking, frequency, voltage and current measurements, date concentrators, master remote terminal units, facilitating operator access to the date generated and collected by the lower tier equipment. At the top is supervisory control and data acquisition system. SCADA accepts telemetered data and displays them in meaningful ways to the operators. Other major SCADA component is the alarm management subsystem, sensing and monitoring all inputs and critical grid points, equipment and elements, informing and signaling to the operators of any abnormal operation conditions and events.

Effective energy management is essential in order to supply energy reliably, to minimize power system operating costs, to reduce emissions, to improve the performances and to extend the life of system equipment and elements (economizing). There are several EMS functions that are critical and essential to power system operation. Through *monitoring* function and application, the EMS must provide energy consumption information and data at various temporal granularities, such as 15 or 20 minutes, hourly, daily or weekly, the feedback is most successful when it is provided frequently and over a long period of time. To avoid consumers' misperception about the energy consumed by individual appliances or equipment, EMS needs to provide *disaggregated data* for different appliances and equipment. End-users can also benefit greatly from information about the real-time impact of specific appliances or equipment being powered ON or OFF. The disaggregated data also highlights the impact of long-term changes such as switching to an energy-efficient appliance. Many EM systems use indirect load sensing methods to provide disaggregated data and information based on specific current and voltage waveform *signatures* of individual appliances, installations or equipment. EMS must make the information and critical data *available and accessible* to the consumer at all times through an easy-to-use interface, either in the form of a physical device or through a web or mobile portal that also gives remote access to the information and data. EM systems may also use push technology to send urgent and critical notifications to consumers' smart phones, laptops or system screens. Besides current energy consumption, EMS systems must also integrate other types of information and data such as indoor and outdoor temperatures, humidity, acoustics, noise, cloudiness, light and consumers' historical data, usage data related to different appliances, equipment, as well as peers' consumption data, through the EMS *information and data integration*. Such data and information is collected at different timestamps, and needs processing and editing before being presented to the consumer. Semantic web technologies have been extensively used for such purposes. The system must be *affordable*, allowing easy installation without or minimum professional helps. Its configuration and maintenance must be simple, consuming minimal energy with a low running cost. These factors help reduce the barriers of the EMS system and facilitate widespread adoption and uses. The system must be able to provide remote, programmable and automatic *control* of devices and equipment. Traditionally, the end-users are expected to perform necessary control operations manually. However, a digital control option or automated actions are more effective, efficient and secure. The communication of data and control signals by EMS poses security challenges, and there are privacy issues related to disclosing personal consumption profiles. The system must authenticate all transactions to ensure that consumer's data and control operations are secure and not accessible to third parties without explicit consent and verifiable access. The *cyber-security* and *privacy* issues are critical for smart grid EMS. A desirable EMS feature is its *intelligence* or *smartness*. End-users often lack a basic electric system understanding and have limited time to make energy-related decisions. It is so desirable to have the system perform intelligent actions that balance energy consumption and consumer comfort. Such capabilities require techniques and methods from *machine learning*, *human-computer interaction* and *big data analytics* to discern usage patterns and predictive actions, reducing end-users' involvement to directly operate, control, and manipulate all appliances, installations and equipment all the time.

10.2 ENERGY MANAGEMENT IN SMART GRIDS

Electricity is an intermediate product, generated by using primary energy sources (oil, natural gas, coal, hydropower, nuclear energy, etc.) that are converted into electricity and transported, via transmission and distribution networks to consumers or end-users. Established electrical power systems, developed over the past 120 years, are feeding electricity from large generation stations through transformers to high-voltage transmission systems, through substations, sub-transmission and distribution networks to the consumers or end-users. Consumers are purchasing the electricity as an intermediate step toward some final non-electric product. The large-scale electricity usage is due to various factors, such as: easy to generate and transfer over long distances at high efficiency, easy to distribute to consumers any time, amount or almost everywhere to wide variety of uses (heating, light, electric motors, computers, air-conditioning, etc.). In order to meet the power operation criteria (safety, reliability, quality, economy, efficiency and supply security), within the power system operation the following tasks are performed: maintain the load-generation and reactive power balance, voltage profile control, an optimum generation schedule to control generation costs and environmental impacts and ensure the network security against contingencies, network protection against outages, equipment failure or unauthorized interventions. Major disruptive forces are reshaping the energy and electricity industries in the United States and abroad. Distributed renewable energy and storage, smart grid technologies, massively deployed sensors, evolving energy markets and security threats are drastically increasing the complexity of electric grid planning and operating processes. The rate of growth in the US electrical demand necessitates the development of new generation resources and the expansion of electrical infrastructure. With respect to this growing demand and the current trends of energy policy development, policymakers, corporations and private investors continue to push the development and integration of renewable energy generation, including nuclear, wind, and solar energy. The integration of novel and existing generation sources within the US electric grid presents new requirements and challenges for large-scale resource management and interconnection. The integration of conventional generation sources (coal, oil, and natural gas), green generation sources like nuclear energy and renewable energy sources presents technological obstacles to the current power system design and operating practices. The sheer magnitude of new generation being added to the grid, as well as the diversity of generation sources, drive expansion and investment within the transmission and distribution infrastructure sector and in enabling technology developments. For example, advanced metering alone is increasing the existing data acquisition by four orders of magnitude. Renewable energy sources are distributed and highly variable at on temporal scales that are different from those addressed by existing decision-support systems. Phasor measurement units (PMUs) sample at 30–60 times per second achieving wide-area synchronized monitoring, require adequate communication infrastructure, computing capabilities and database facilities. PMU concentrators integrate massive amounts of data to capture dynamic frequency oscillations and provide ultra-fast monitoring. These are examples of a broader trend of unprecedented availability of massive data acquisition and information and growing control complexity, suggesting that decision-making requirements are rapidly exceeding the human ability to comprehend and respond to emerging system conditions and events. Present monitoring and control processes through which the electricity infrastructure is operated can be compromised in a few years if the existing situational awareness tools are not replaced by drastically more powerful systems.

Electric grids are radically evolving and transforming into smart grid, characterized by new services and improved energy efficiency and manageability of available resources and assets. EMS was originally designed in the context of the conventional power grids, a vertically integrated system, having centralized communications and computing systems. With power industry deregulation and smart grid developments, the decision-making is becoming decentralized and the coordination between different stakeholder and actors in various markets are becoming important. A number of international and domestic standards are emerging, or are in place and use for the description of the systems and services and for system operation, management and functionalities. An advanced analytics and visualization framework (AAF) is required to present the power system operator with real-time conditions, data and information in a timely manner, having the ability to efficiently process, analyze and present the data for decision-making, including the ability to navigate and drill down to discover additional information, such as problem specific location and impact, together with ability to identify and implement corrective actions, in order to mitigate potential risks for grid operation. EMS can be a useful tool toward active demand side management, one of the major goals of the future smart grid. This functionality will have a significant impact on the consumer's behavior, remote controlling and/or rescheduling the appliances. This will require better understanding of energy use within homes and impacts on the overall energy consumption in the smart grid. Future EMS may have embedded intelligent functionalities that could automate decision-making and control household appliances in response to demand or price signals from the utility. Visualization of the real-time condition of the power grid is the best monitoring tool the operator has. Existing visualization tools have challenges in handling the growing data volumes, and it is becoming critical to develop advanced tools that can digest data while providing easy to use, actionable visualization and alarming. Visualization of the grid outside a control center boundary is currently unavailable or very limited, being a major drawback in the operation of large interconnected systems. Such limited ability of knowing what is going on in the neighboring systems has been a consistent element among the causes of large blackouts.

Innovative visualization methods, developed by the late 90s, have made their way into mainstream control center capabilities and attributes. However, the static 2D visualization may not be sufficient for the emerging smart grid requirements. In the aftermath of major blackouts, in the United States and other countries, it was clear that the control centers must be equipped with display and recording equipment and software applications to provide the most clear and accurate picture and meaningful information of the system conditions. Timely visualization of real-time grid conditions is essential and critical for successful electric grid operations. In particular, mechanisms for handling of emerging multi-dimensional and multi-scale data are needed. Planning, simulation, operation and control of power systems involves information and data that is naturally associated with spatial and temporal dimensions ranging from grid interconnections to homes and from years to microseconds, respectively. It also involves the analysis of a potentially large number of scenarios, dimensions and huge data amounts. Each one of the dimensions has specific scales, associated with contextually relevant behavior. Control centers are managing the transmission-generation subsystems, through the EMS, in existence since the 1960s. The EMS term has different meanings, depending on context, a common one being "the process of monitoring, controlling, and conserving energy in a power transmission-distribution system, organization, or a building". Over the past few decades, EMS systems have evolved gradually from early starts as digital computers to a slightly more sophisticated form, to major transformations to support emerging power system operations and smart grid objectives. The expanding power grid requires operation and control of system behavior at temporal and spatial scales, different from the ones usually considered by the EMS. Advances in measurement, sensing, communication, computation and control make such transition to a new EMS generation possible. Similar technologies are increasingly being applied to the distribution system, and new distribution management systems are now in full deployment process. Power system state is ever changing, being a dynamic process, loads and configurations of networks change, by making the power system operation difficult and complex and impossible to implement in a feed-forward control. In addition, the response of many power network components is not instantaneous. Decisions have to be making of the basis of predicted system future states. To automate the power system operation, electric utilities rely on a highly sophisticated integrated computerized system for monitoring and control. Operating grids close to normal frequency, in very narrow range, without causing any unexpected load or generation disconnections, knowing as maintain grid integrity, is performed through control centers, using the EMS systems having a multi-tier structure with a multitude of levels of elements. At the bottom are high-reliability switchgears with facilities of remote control and monitoring, protective relays, automatic transformer tap-changers and other automatic equipment. At next tier are the remote controls at switchgears for actuator control, interlocking, frequency, voltage and current measurements, data concentrators, master remote terminal units, facilitating access to the data generated and collected by the lower tier equipment. At the top is supervisory control and data acquisition system. SCADA accepts telemetered data and displays them in a meaningful way to the operators. Other major SCADA component is the alarm management subsystem, monitoring all inputs, informing the operators of any abnormal operation conditions.

The smart EMS was developed with ability to record, store and process energy use data of every major appliance, equipment and device in buildings, industrial and commercial facilities. Again, out of various management objectives in smart management system, such as improving energy efficiency, profiling demand, maximizing utility, reducing cost and controlling pollutant emissions, while integrating home and building energy management systems (HEMS, BEMS), distributed energy resources and energy storage with the microgrids. In addition to managing the existing energy resources, generating power effectively and intelligently is an equally important agenda. However, the biggest power system challenge, regardless of the use of renewable and non-renewable energy sources lies in preventing power grid imbalances. To limit power grid imbalances and other grid issues, over the years smarter energy management systems for modern power systems and smart grids were developed and employed. Supplementing the establishment of large power plants from conventional energy sources, there is also a need to focus on distributed small-scale generation of power particularly from renewable energy sources. Such SEMS entities were connected to building, industrial facility, process or home energy controlling and management systems, providing real-time, up-to-date energy information and data, such as current energy prices, energy supply and demand, grid status, weather forecasts, etc. They are also linked to various smart devices, equipment, distributed generation and energy storage units, such as solar PV panels, wind turbines, smart energy meters, smart sensors, electric car charging points or smart devices for appliances, equipment and installations. The two utility control centers are one for power generation-transmission system, the EMS and the other for the distribution system operation, the DMS. A simplified diagram of a power system, together with the energy management centers is shown in Figure 10.2, helping dispatchers and operators to properly monitor and control system operations. However, beyond control and operation of power systems, power system planning has numerous challenges associated with the limitations of the existing methods, tools and practices to manage a system that is dramatically changing with renewable energy generation, larger power transfers and distributed resources, including demand response resources. EMS is high-security computer system, receiving measurements and data from the grid and send out remote controls. EMS evolved over past six decades based on the centralized command and control paradigm into a quite large and complex system through computer automation. Major EMS objectives are energy use optimization, to minimize energy costs and environmental impacts without affecting production and quality. Major EMS functions are: *generation management* (managing energy demand and generation balance), *grid reliability analysis* (EMS must assure secure function of the entire grid and its components) and *grid operation* (EMS commands controllable equipment, circuit breakers, transformers, disconnecting switches and other assets in reconfiguration operations, grid maintenance or extension work to provide personnel safety working on them).

In order to aid transmission operators to monitor, control, operate and optimize the performances of the generation and transmission systems, a set of EMS software applications are used. EMS monitors and manages the power flows in the transmission networks. However, EMS is a generic term given to software applications used in a wide variety of applications to effectively monitor and control power generation, transmission, distribution or energy consumption. Early control centers were hardwired analog systems with meters, switches and thumbwheels used to change the power system operating points. Modern EMS functions were developed about six decades ago with the advent of modern computers. EMS essential functions are to increase energy efficiency and to optimally coordinate grid energy sources. As the energy demand continues to increase, the EMS importance to effectively manage energy, not only in power industry, but also in industrial, commercial and

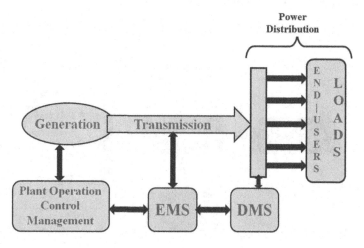

FIGURE 10.2 Simplified power system diagram.

residential sectors continues to grow. Over the years, EMS functions and capabilities have continually evolved. A large set of systems have been proposed, but not all have the same objectives or technologies. The increased distributed generation use, along with needs to increase energy efficiency and savings, are the main drivers for the implementation and utilization of modern energy management systems. EMS concept is commonly a computer system (hardware and software applications) used to monitor and control the high voltage power grid, including transmission and generation. Over the years, EMS evolved with the following objectives: *electrical network conditions' real-time monitoring, maintaining system frequency, reducing electricity production costs while following load changes, performing contingency analysis studies, optimizing grid operation conditions and assessing grid stability.* It is controlled by a SCADA system and several advanced applications that include forecasting and optimization, while providing electric grid control operation signals. In short, an EMS is the process of monitoring, coordinating and controlling the generation, transmission and distribution of electrical energy, as shown in the diagram of Figure 10.3. Generated electricity is fed through transformers to high-voltage transmission networks, interconnecting power plants and load centers. Transmission lines terminate at substations, which are performing voltage transformation, switching, measurement and control. Substations at load centers are designed to transfer electricity to the sub-transmission and distribution networks, typically operating radially, meaning no usually closed paths between substations through sub-transmission or distribution networks. Since transmission systems provide negligible energy storage, supply and demand must be balanced by either generation or load or both and this being the role of energy management system.

Historically, EMS have been in operation for quite a few decades, but until recently, it starts to require specialized instrumentation, smart sensing, advanced communication, computing, etc. EMS systems are real-time computer systems, introduced in early 1970s to provide power system operators with means to manage power grids in a reliable, safe, secure and efficient ways. Many utilities are still operating with EMS technology installed in 1990s, a three-decade old technology. Almost all EMS deployed in that period incorporated state-of-the-art electrical network models, e.g. state estimation, contingency analysis and operator study analytical tools, and training simulators. Their objectives are achieved by providing decision support and control systems for generation and transmission. Similar solutions have been developed for power distribution, via DMS, which evolves at much lower pace as EMS. Power distribution networks were managed using manual boards, containing static network diagrams. In the decades following the EMS introduction, software and hardware requirements presented major challenges. Notice that over last decades, due to advances in computer and information technology, communication, sensing and software technologies, there are significant DMS advances. Major power management challenges at the levels of architecture, data modeling, computation and visualization are:

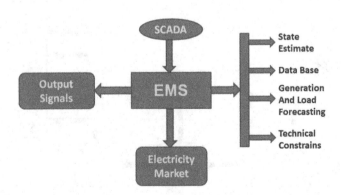

FIGURE 10.3   Diagram showing the EMS and SCADA connectivity.

1. Present communication infrastructure is not fully adequate for handling the increasing real-time data transfers imposed by transmission and new distribution measurements. These requirements are stretched by the data rates at the transmission level and by the data volumes at the distribution level.
2. The present proprietary data structures at each control center are a major impediment to data transfers between the distribution and transmission levels, as well as between EMS in the same interconnection. This issue impedes the ability to monitor, operate and control the interconnected power systems.
3. Models and computational frameworks used by EMS systems and planning tools are not unified and interoperable and are not easily compatible with high-performance computing. This makes integration of system components and functions difficult and, in many instances, not possible.
4. Power grid visualization is one of the most important tools available to operators for grid monitoring, but it is still limited to the individual control center jurisdiction. This inability to seamlessly move data means a system monitoring with insufficient information, one of the root cause of blackouts.

EMSs are high-security computer software systems designed to receive grid measurements, data and send out remote controls. Transmission represents one of the most critical power system processes. EMS enables the transmission operation processes by providing the operators with power system visibility, information and control. In recent years smart grid transformations has exposed it to new changes and technological advances. These systems are now very mature, costs being reduced and functions have been extended and standardized over past two decades. EMS systems have three main functions:

- *Generation management*: estimating, preparing and managing the energy demand and generation balance over timeframes ranging from few seconds to a year or more.
- *Grid reliability analysis*: in the plans drawn up by production management, the EMS must ensure the secure functioning of the entire power grid and each of its components; and
- *Grid operation*: the EMS commands controllable equipment, circuit breakers, transformers, disconnecting switches and other assets and installations in reconfiguration operations and grid maintenance or extension works. A primary concern being the safety of the people working on them.

10.2.1 DISTRIBUTION ENERGY MANAGEMENT

Power distribution networks, connecting the high-voltage transmission systems to end-users, are expensive and complex systems, difficult to monitor, control, manage, analyze and plan. Real-time monitoring and remote control are still quite limited in today's power distribution, so there are critical needs for interventions, especially during widespread faults and system emergencies. However, it is very difficult to deal with such a complex system through non-automatic procedures. Smart grids, a radical reappraisal of the function of power distribution networks, include uses and integration of distributed energy resources, load demand active control, and a more effective use of power distribution network assets. A DMS is a collection of applications and procedures used by distribution network operators (DNOs) to monitor, control, manage and optimize the power distribution performances, managing its complexity. It is an integrated decision support system of all the power distribution operation aspects, that it is making them visible and operable from a control center, being the power distribution equivalent of the EMS. However, one of the major differences of DMS functionality compared to the EMS is that the number of distribution equipment, devices and assets is huge compared to those in transmission systems. Factors and structures, contributing to the complexity of distribution systems, are related to both network structure and loads. Power distribution networks are often built as meshed circuits but operated radially, topology changes frequently during operation, due to faults and maintenance. Network structure changes as the distribution network expands, while the three-phase distributions are usually unbalanced. The time scales that have to be considered are ranging from milliseconds, e.g. protection operation to years, in the case of network expansion. The networks have strict performance objectives, and still there is limited communication between network elements and components, large part of control is local. Comprehensive monitoring and sensing of power distribution networks generate huge data amounts, requiring advanced technologies and data management. Composition, structure and diversity of the loads are complex, quite often not well known, patterns of distribution load consumption vary dynamically with time. Load variation trends are more difficult to predict than in the case of a large transmission network, being impossible to obtain simultaneous measurements of all loads. Load measurements usually are insufficient and may contain large errors and poor quality, while correlations between loads are not well understood. Advanced computing, communication, analysis and management methods and technologies are needed and must be used to optimize the DMS functionalities, in a real-time mode, in order to lead to better asset management and service quality, with the provision of new services and greater customer satisfaction.

From earlier DMS generations, integrating a number of simple applications into a computer system, and newer applications and functions were added, such as interactive graphical user interface, large relational databases, allowing the management of complex distribution networks and large data volumes. However, as more and more applications and functions were added, managing the information and data exchanges, processing and maintaining the DMS became a challenge. Standardized models such as the Common Information Model (CIM) were developed to support information management. For the smart grid, the DMS needs to use higher-performance ICT hardware, smart software applications and to be equipped with advanced intelligence and be deployed in a distributed, decentralized structure. DMS objectives implementation include, among others: enhancing safety by providing better visibility and control on system energization and de-energization, improving system reliability by reducing outage times, expanding power system assets life span by properly managing their use and operation, and enhancing system efficiency and optimizing the use of available resources. DMS functions are broadly divided in three main categories: (1) system monitoring, (2) decision support tools and (3) advanced control. Ultimate DMS goal is to enable a smart, self-healing power distribution system and to provide improvements in supply security, resilience, reliability, power quality, efficiency and effectiveness of system operation. A DMS must lead to better asset management and operation, able to provide new services, greater customer satisfaction and eventually to lower electricity cost. DMS consists of a collection of computer-based applications designed to monitor, operate and control the entire power distribution network efficiently and reliably, acting as a decision support system to assist the control center and field operating personnel with the monitoring, measurements and control of the electric distribution system. Improving the reliability and quality of service in terms of reducing outages, minimizing outage time, maintaining acceptable frequency and voltage levels are the key deliverables of a DMS. A DMS includes several applications used as modelling and analysis tools together with data sources and interfaces to external systems. The modeling and analysis tools are software applications and packages which are supporting one or more DMS applications. DMS accesses the real-time data and provides all information, measurements and data at the control centers in integrated manners and user-friendly ways. All these applications and functions are requiring modeling and analysis tools for which network parameters, customer and load information, and distribution network status data are used as inputs and constrains.

The development and implementation varied across different electrical distribution networks. In the United States, e.g. DMSs typically grew by taking outage management systems (OMS) to the next level of automation, digitalization and providing end-to-end, integrated view of the entire power distribution spectrum. In the United Kingdom, by contrast, the much denser and more meshed networks, combined with stronger health and safety regulations, had led to centralization of high-voltage switching operations, initially using paper records and schematic diagrams to show the current running states. DMSs originate from SCADA systems as these were expanded to allow digital centralized control and safety management procedures. SCADA provides real-time system information to the modelling and analysis tools, specifically for the DMS has the following attributes: data acquisitions, monitoring, event processing and alarms, adaptive and distributed control and data storage, processing, event location, analysis and reporting. These DMS systems are requiring even more detailed components and connectivity models than those needed by early EMS developments. Main domains offered by the DMS include distribution operation, engineering study tools, system operation planning, training simulators and quality assurance systems. Distribution operation environment provides visibility of the system, decision support and control functionality to manage power distribution operation. Operation planning study environment is used to conduct planning, e.g. outage impact analysis. Study systems are used to simulate possible system operation scenarios or alternatives to the ones generated by the production systems. Quality assurance system is designed for testing new applications or upgrades before implementing into production system. Control centers are large utility investments their major benefits are more reliable and safer power system operation, improved efficiency and usage of generation resources and better quality of services.

10.2.2 ENERGY MANAGEMENT FUNCTIONS AND COMPONENTS

Over decades, the utilities evolved beyond systems of few units supplying a group of loads, so effective power system operation requires that critical quantities be measured, monitored and transmitted to a central control center. EMS has evolved to schemes that can monitor voltages, currents, power flows, status of circuit breakers and switches in every transmission network substation and other critical grid components. In addition, other critical information, e.g. frequency, generator outputs and transformer tap positions are also transmitted. With such massive data and information simultaneously only a computer can handle, check and interpret in a reasonable time frame. For this reason, dedicated computers and software applications are installed in distribution control centers from which operators can monitor these data and information, in real time. In addition, software applications can check incoming information against pre-stored limits and alarm the operators in the events of overloads or out-of-limit voltages. State estimation is often used in such systems to combine telemetered system and sensing data with system models to produce the best estimates of the current power conditions. Such systems are usually combined with supervisory control systems, allowing the operators to control circuit breakers, disconnect switches and transformer taps remotely. Major actors involved in power system operation are: the operator, responsible to execute different power system functions; the process is providing a detailed description of how a function is or should be executed; and the technology that is enabling and facilitation processes. The power network state is ever changing because loads and network configuration changes is making the system operation very difficult. In addition, the response of many power network components and equipment is not instantaneous. Power supply quality is defined in terms of variables, such as frequency and voltage that must conform to certain standards to accommodate the requirements for proper operation of all loads connected. Supply reliability is not meaning a constant power supply, but rather it means that any break in the power supply is one that is agreed to and tolerated by both power supplier and consumer. Making the generation costs and losses at minimum motivates the efficiency and economy criteria while mitigating the adverse impacts of power system operation on the environment. Within an operating power system, the following tasks and functions are performed in control centers to meet the above criteria:

- Maintain load-generation balance;
- Maintain the reactive power balance in order to control the voltage levels and profiles;

- Maintain an optimum generation schedule to control costs and environmental impacts; and
- Ensure the network security against credible contingencies, meaning protecting the network against reasonable equipment failures or outages.

Power system security is divided into three major functions, carried out in operation centers: (1) system monitoring, (2) contingency analysis and (3) security-constrained optimal power flow. System monitoring provides power system operators with pertinent and essential up-to-date information on conditions on the power system, being the most important functions. System dispatchers and operators at the EMS are required to make short-term and long-term decision on operational and outage scheduling on daily basis. In addition, they have to be, at all times, prepared and alert to deal with contingencies that may occur in power systems. Several software and hardware functions are required and needed as operational tolls for the operators. These functions are classified as: (1) base functions, (2) generation functions and (3) network functions. Required EMS base functions are the ability of acquiring real time data for monitoring equipment throughout the power system and the capability to process raw data and to distribute the processed data and information with the central control system. Data acquisition function collects the data from remote terminal units (RTUs) installed throughout power systems, by using special hardware connected to data servers at the control centers. In addition, protection and operation of main circuit breakers, sole of the line insulators, transformer tap changers, alarms occurring at substations and other miscellaneous substation devices are also processed and transmitted by the data acquisition function.

Energy management, performed at system control centers by dedicated computer systems and software applications is providing the infrastructure to control and optimize energy consumption. Data acquisition and remote control is performed through dedicated computer applications by SCADA systems. An EMS typically includes a SCADA "frontend" to communicate with power plants, substations and other remote devices and equipment. SCADA system consists of a master station that communicates with remote terminal units, allowing observation, monitoring and control of physical plants, grid devices and equipment. Traditional SCADA system functions include: supervisory control, data acquisition and communication, tagging (identifying devices subject to specific operating restrictions), alarms (information of unplanned events and/or undesirable operating conditions), logging (logs all operator entry, alarms and selected information), load-shed and trending. Master station is critical to power system operations, its functions and capabilities are distributed among several computers depending on the specific design and system architecture. A dual computer system configured with one in primary mode and the other in standby mode is the most common. SCADA functions include: managing communication circuit configurations, check and correct message errors, detect status and measurement changes, monitor abnormal and out-of-limit conditions, detect malfunctions and announce alarms, respond to operator requests, such as: information display, data enter, execute control actions or acknowledge alarms, transmit control actions to RTUs, prevent unauthorized actions, maintain historical files, prepare reports, perform load shedding or down-line load RTU files. One of the most important EMS function is the automatic generation control (AGC), consisting of two major and several minor functions, operated online in real time, in order to adjust the generation in relation with loads and to minimize the costs. AGC's major functions are load frequency control (LFC) and economic dispatch (ED). Among minor functions are reserve monitoring (assuring enough system reserve), interchange scheduling (initiating and completing scheduled interchanges) and other monitoring, sensing and recording functions. Primary LFC functions are: (1) maintaining frequency at scheduled value; (2) maintaining net power interchanges with neighboring control areas at scheduled values; and (3) maintaining power unit allocations at economically desired and prescribed values. First two are achieved by monitoring and error signaling, through *area control error* (ACE), a combination of net interchange and frequency errors, representing generation and load imbalance at any time. ACE is filtered to prevent the translation of the excessive and random ACE changes into control actions. The control signal is divided among generating units through participating factors, so the units are loaded according with their cost, thus meeting the third LFC objective. However, the cost is not the only consideration because different units usually have different response rates, and maybe faster generators need to be set in order to obtain acceptable response. The control actions are repeated at prescribed rate (2–6 s). The LFC design philosophy is that each system follows its own load closely during normal operation, while in emergency each power system contribute according to its size in the interconnection, regardless of the emergency locality. In order to obtain a good control, the most important factor is the system inherent capability to follow its own load, which is guaranteed if the system has adequate regulation margin and response capability. However, systems with large portion of thermal generation have slow response and difficulties in keeping up with load changes. Controller design is an important factor and proper tuning of controller parameters is needed to obtain proper control.

All generating units that are connected to the electric grid have usually different generation costs, requiring finding the generation levels of each of them that meet the load at minimum cost. The process has to take into account that generation cost in one generator is not proportional with its generation level, being a nonlinear function of it. In addition, since the system is spread out geographically, the transmission losses depend on the generation pattern and must be taken into account to obtain the optimum. Other factor includes adequate reserve margin, which often is done by constraining generation level to a lower boundary than its generation capability. More difficult constrains are the transmission limits. Under certain real-time conditions, the most economic pattern may not be feasible due to unacceptable line power flows and voltage conditions. However, today ED algorithms cannot handle these constrains, and new approaches are proposed and tested. Minimum cost dispatch occurs when the incremental cost of all generators is equal. However, the generator cost functions are nonlinear and discontinuous. The demand and marginal costs are correlated, and algorithms for optimum cost are performed as the demand change. Power system losses are function of generation pattern, by multiplying the generator incremental costs by appropriate penalty factors, and by reflecting generator sensitivity to system losses. The ED algorithms apply to only thermal generation units that have cost characteristics, as one discussed here. Hydro units are dispatched with different characteristics, taking into account the amount of water available over a limited period and the displacement of fossil fuel by this available water, determining its worth. If the water usage limitation over a period is known, from previously computed hydro optimization, the water worth can be used to dispatch hydro-power units.

EMS periodically calculates the reserve requirements of the power system. Maintaining sufficient reserve capacity is a critical requirement in the case generation is lost. EMS methods have explicit formulas that are followed to determine spinning and ready reserve required, while taking into consideration operation circumstances, such as: largest unit on line, time frame required and needed to make changes, etc. to generate reports and alerts the operators if needed. Reserve availability can be assured by operator manually or through ED that can reduce the upper dispatchable limits of generators to keep the generation available. Another critical requirement is the contractual power exchange between utilities, taking into account by LFC and ED functions, by calculating the net interchange and adding it to LFC and ED generation needs. This is achieved automatically for the list of scheduled transactions and performed usually on hourly base. Modern electric grid, the smart-grid, is characterized by improved energy efficiency and manageability of available resources and assets. Newer EMS systems are often integrated with home automation systems, playing an important role in the control of home energy consumption and enable increased consumer or demand side participation. These systems are planned to provide consumers with information about their energy consumption patterns and help to adopt energy-efficient behavior. The newer EMSs leverage advanced analytics, information, computing and communication technologies and infrastructure to offer consumers actionable information and control features, while ensuring ease of use, availability, security and privacy. There is a growing interest in the smart grid evolution, supported by bi-directional communication between energy providers and consumers through smart metering, monitoring, and control. Key smart grid features are enhanced energy efficiency and

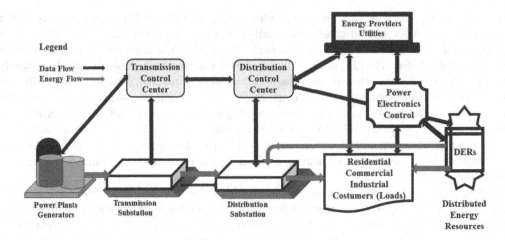

FIGURE 10.4 Typical smart-grid architecture.

manageability of available resources. These systems are providing infrastructure and means to the consumers to understand, control and optimize their energy uses. Widespread adoption of future EMSs by consumers will eventually lead to more efficient consumption behavior and will benefit the utility as well. Figure 10.4 is showing the EMS, DMS and communication infrastructure into the smart-grid structure.

10.3 LOAD AND ENERGY MANAGEMENT

EMS control centers, as grid nervous systems are critical components of the power system operational reliability. Although it is true that utilities have been relying on EMS for over four decades, it is also true that many of the systems in place today are outdated, under-maintained or underused compared with the total potential value that could be realized. SCADA can provide measurements and status for substation distribution feeders, via RTUs installed at distribution substations. Distribution automation equipment is now available to measure and control at several locations distributed all over the power distribution, helping system operators to identify potential problems, providing dynamic assessment program reports, in real-time, regarding the system equipment, reaching the rating threshold conditions. Examples include voltages approaching over limits, transmission lines approaching overloading, while considering thermal constrains and emergency ratings. This functionality is helping to identify potential problems before they happen and providing operating margins during emergency. EMS is also able of load shedding, during emergency, trip circuit breakers if the frequency declines. System operators can drop loads quickly and effectively or can coordinate rolling blackouts before the automatic load shedding relays start to operate. Static information about the systems lines, transformers, etc. enters in dedicated computer programs regularly, EMS computing and updating regularly power flow conditions. The software package can report the detailed power system information daily, weekly, monthly and yearly peak conditions. Power flow data is very useful to planning engineers in order to determine future power system additions, expansions or restructuring. EMS is an effective tool for planning power system needs, which include load forecast information, generation schedules, interchange or tie line exchange schedules, unit maintenance schedules, unit outage situations to determine and set the best overall generation implementation plan. Through AGC, the EMS effectively controls the generation dispatches, while system operations, area control errors and frequency are then monitored according to this schedule and generation plan to assure power system reliability and compliance. EMS is also able, up to 24 hours ahead to control reactive power resources for optimum power flow, based on economics, reliability and security.

Electric load (EL) or energy demand forecasting is a vital process in the electric industry planning, playing a crucial role in the operation and management of electric power systems. There are a huge number of loads in subtransmission and power distribution networks of various types, sizes and characteristics, while their energy consumption is usually quite small. When the balance between electrical supply and demand is broken, the resulting grid damage or supply outages may affect the quality of the service. Therefore, the accuracy of electric load forecasting has great importance for energy generating capacity scheduling and power system management, as these accurate load forecasts lead to substantial savings in operating and maintenance costs and correct decisions for future power system development. Furthermore, electric power load forecasting represents the initial step in developing and structuring future generation, transmission and power distribution infrastructure and facilities. Electricity load anticipation is therefore an important task for the transmission system operator of an electrical grid as it helps to reduce the risk of this happening; larger prediction horizons (greater than one year) help to anticipate the needs on production means and distribution, while shorter ones (hours, weeks) are employed to decide the production and distribution plans. In general, more accurate predictions result in lower production costs. Therefore, comprehensive real-time load measurements are too expensive, complex and impractical, and the load estimation and forecasting are used for both operation and planning of distribution networks. Load forecasting can be divided into three forecasting time horizons: *short-term load forecasting* which is usually from one hour to one week; *medium-term load forecasting* which is usually from one week up to one year; and *long-term forecasting* which is longer than one year. Load forecasting is also divided into two categories based on the forecasting scope: regional load forecasting, providing load forecasts and estimates for large geographical areas and busbar load forecasting, providing nodal load information for network control functions. For a number of DMS Applications, short-term load forecasting is of most significance. Load varies over the short term with time (e.g. weekday, weekend and holiday), weather (e.g. temperature and humidity) and the types of energy consumers (e.g. residential, commercial and industrial). Various methods and algorithms are available for short-term load forecasting, including regression, time-series methods, same day and persistence methods, computational intelligence-based methods, neural network forecasters and hybrid methods and algorithms.

Short-term forecasts, usually from one hour to one week, play an important role in the day-to-day utility operations, such as unit commitment, security analysis, economic dispatch, fuel scheduling, unit maintenance and load management. A short-term electricity demand forecast is commonly referred to as an hourly load forecast. Medium-term forecasts, usually from a few weeks up to a few months and even up to a few years are necessary in planning fuel procurement, scheduling unit maintenance, or in utility energy trading and revenue assessment. A medium-term forecast is commonly referred to as the monthly load forecast. Long-term electricity demand forecasting is a crucial part in the electric power system planning, tariff regulation or energy trading. A long-term forecast is required to be valid from horizons from 5 to 25 years. This forecast type is used to deciding on the system generation and transmission expansion planning. A long-term forecast is usually known as an annual peak load forecast. Electric load forecasting, the process used for forecasting future electric load, relays on historical load data, electrical demand patterns, economic and social information, weather and climatological data, current and forecasted weather information. The electricity demand patterns are usually affected by several factors including time, social, cultural and economic factors, as well as environmental factors by which the patterns will form various complex variations. Time is the most important factor in load forecasting since its impact on consumer load is the highest one. Social and cultural aspects, such as behavior and environmental concerns are large sources of randomness found on the load patterns. Diversity and complexity in demand patterns have been leading to developing complex load forecasting methods. The load shape may be different for different customer classes, electric utilities are serving customers of different types as residential consumer, commercial consumer and industrial consumer. The customer factors of electricity consumption are primarily the number, type and size of the electrical equipment and appliances of the customer. However, there are recognized types of customers which have similar properties. The residential load curve is somewhat different from commercial and industrial customers.

Various weather parameters, such as temperature, humidity, wind speed and cloud cover, have significant impact on the energy demand and must be considered in any load forecasting models. Most of the human activities or industrial processes involve electricity uses. For example, load and temperature are linked to some level through a positive correlation between temperature and electric load curve especially in summer season. This is because during summer changes in temperature affects the people's feeling of comfort level. During summer as temperature rises, there are increased usages of air conditioning and ventilation, increasing electricity consumption, whereas the temperature fall in the winter season, the more usage of heating appliances would increase the power consumption. Humidity affects short-term load forecasting since it increases the feeling of the temperature severity, especially during summer and rainy season. Thus, load consumption increases significantly during summer hot and humid days. Wind speeds also affect the electricity load consumption. During windy days, human body feels the temperature far below and heating is required thus increase the electricity uses. The effects of cloud cover on the electricity usage depend on the timings of usage. During daytime the cloud cover may disturb the sunlight, resulting in decrease of the temperature and hence lowering the electricity uses. However, among of the weather factors, temperature and humidity are the most commonly used load predictors to make better forecasts, minimizing the operational cost. The DER deployment programs are requiring, not only accurate weather information, but also very accurate energy demand (load) forecasting models. The traditional approach of obtaining the total energy uses and peak demands does not offer the required detailed information. In order to design appropriate demand response models and programs, accurate energy demand and power consumption models should be developed to simulate different scenarios for industrial, commercial and residential users. Conventional load forecasting models based on the previous consumption data in a top-down approach, or a bottom-up approach should be addressed to overcome the limitations of the former models, such as the inability to predict consumption pattern, load profile changes like the implementation of flexible demand, distributed generation or adoption of new technologies.

Load forecasting or energy demand, beside previous classification (short-, medium- and long-term forecasting methods) can also be classified into two major types based on the consumption of electricity, the spatial forecasting models and the temporal forecasting methods. Spatial forecasting refers to forecasts of the electricity load for the particular region or area, such as a whole country, a state, a province, an industrial area or a particular city or metropolitan area. Temporal forecasting means forecasts the electricity load for specific supplier or collection of consumers for future related particular or specific time, such as: hours, days, months, seasons, years and decades. Moreover, the temporal load forecasting is also classified into four categories, in addition to the previous three types, listed above, a very short-term load forecasting is sometimes considered, having a time horizons, from few minutes up to one hour. Some of the most common electricity demand forecasting techniques include: regression analysis, exponential smoothing, iterative re-weighted least-squares, adaptive load forecasting, stochastic time series, neural networks and artificial intelligence based models. Also, depending on the approach, the forecasting can be made from an aggregated level (e.g. from the electric utility side) in a top-down scheme, or from the user side, analyzing end-use activities, in a bottom-up scheme. The method to apply is chosen based on the nature of the available data and the desired nature and detail level of the forecasts. The power demand differs over the time of day. It depends on human activities and several other factors, such as the day time, season, weather conditions, days of the week, etc. A power load curve is a plot of variation values of the consumed power over time. The load curve reflects the behavior of consumers as they use different home appliances, house equipment, electronics, etc. A large variety of mathematical methods have been developed for load and energy demand forecasting. However, to design appropriate demand response programs and algorithms for accurate energy demand-consumption models should be developed to simulate different scenarios for industrial, commercial and residential users.

Load forecasting has always been important for planning and operational decision conducted by utilities. However, with the deregulation of the energy industries and smart grids, load forecasting is even more important. With supply and demand fluctuating and the changes of weather conditions and energy prices increasing by a factor of ten or more during peak situations, load forecasting is vitally important for utilities. The statistical models are embodying a set of statistical assumptions concerning the generation of sample data, representing the data-generating process with a considerably idealized form. Most forecasting methods, models and techniques are using statistical techniques or artificial intelligence algorithms such as regression models, artificial neural networks (ANNs), fuzzy logic and expert systems. Two of the methods, the so-called end-use and econometric approaches, are broadly used for medium- and long-term forecasting. A variety of methods, which include the so-called similar day approach, trends methods, various regression models, time series, neural networks, statistical learning algorithms, fuzzy logic and expert systems, have been developed for short-term forecasting. Trend method expresses the variable to be predicted as a function of time. It is a non-causal method, and therefore missed to explain the behavior of the trend line, making exclusively a projection based on the historical data. The main advantage of using this method is its simplicity and that only historic consumption data is required. The main limitation of trend method is that, since it does not include any type of demographic, socio-economic or end-use data, it cannot predict changes in the consumer behavior, adoption of new technologies or changes in policies for electricity use, required for infrastructure planning, policy changes and technology adoption. Similar day approach or method analyzes power load natural patterns and the forecasting day's weather features to define specific parameters that can be compared to previous days with similar characteristics. This information is used to create a training data bank to feed pattern recognition tools to emulate the non-linear relationships between load demand and the influencing factors. The most common pattern recognition tools used are artificial neural networks, expert systems, fuzzy logic and support vector machines (SVM). However, ANN is still the most used method for this approach due to its ability to learn complex and non-linear relationships, the availability of commercial tools for its implementation, the operational speed for pattern recognition once the network has been trained and the high accuracy level that can be obtained from this approach. Both, the trend method and the similar day approach, use a top-down approach because short-term forecasting is mainly used by utilities and to provide aggregate information from users' energy consumption. Statistical models are usually specified as a mathematical relationship between one or more random variables and other non-random variables. The accuracy of any load forecasting model depends not only on the technique itself, but also on the quality of the load data records or on the accuracy of forecasted weather parameters (topic beyond the scope of this book). Several statistical models have been developed, implemented and tested for making predictions and forecasting, according to some criteria of optimal fitting. In time series, forecasting seems to mean to estimate a future values given past values of a time series. The load prediction methods used in the energy industry in last decades are mainly based on AI theories because they allow the best way to deal with uncertainties, as well as non-linear functions. Similar-day approach is based on searching historical data for days within one, two or three years with similar characteristics to the forecast day. Similar characteristics include weather, day of the week and the date. The load of a similar day is considered as a forecast. Instead of a single similar day load, the forecast can be linear combination or regression procedure that can include several similar days. The trend coefficients can be used for similar days in the previous years. There are several scientific papers that prove the quality and robustness of the predictions based on these theories, specifically the ones using artificial neural networks. Regression methods are widely used to model the correlation between energy consumption and the impact factors such as time, day type, and customer class and weather conditions to predict future load at specific time. The forecasted load value can be computed by using this relationship:

$$F(t) = a + \sum_{i=1}^{N} b_1 \cdot x_i(t) \qquad (10.1)$$

where $F(t)$ is the forecast load value and $x_i(t)$ is the ith impact (influence) factor at time t, while a, b_i are regression factors, N is the number of influence factors considered in the regression function.

Exponential smoothing is the load forecasting method based on the previous data to predict the future load. Equation (10.2) shows the load $y(t)$, at time t, modeled using a fitting function, expressed as:

$$y(t) = b(t)^T f(t) + e(t) \qquad (10.2)$$

where $y(t)$ is the load at the future time t, $f(t)$ is the fitting function vector of the process, $b(t)$ is the coefficient vector, $e(t)$ is the white noise and T is the transpose operator. Iterative reweighted least-squares method uses the autocorrelation function and the partial autocorrelation function of the resulting differenced past load data in identifying a sub-optimal model of the load dynamics. Equation (10.3) presents the parameter estimation problem involving the linear measurement, expressed by:

$$Y = b \cdot X + e \qquad (10.3)$$

where Y is the vector of observations, X is the matrix of known coefficients based on previous load data, b is the unknown parameters and e is the vector of random errors. In the adaptive load forecasting, the model parameters of this method are automatically corrected to keep track of the changing load conditions. An adaptive load forecasting algorithm has the ability to predict load shapes in addition to daily peak loads. System operators can utilize the predicted load shapes even when the individual hourly errors are rather large. The total historical data set is analyzed to determine the state vector, not only the measured load but also the weather data. This mode of operation allows switching between multiple and adapted regression analysis. The model used is the same as the one utilized in the multiple regression as described by the Equation (10.1). In stochastic time series approach, the model is developed based on the previous data and then the future load is predicted based on this model. In the autoregressive (AR) model, the load is assumed to be a linear combination of previous loads (energy consumption data), then the AR model can be used to predict and to model the load profile. The main idea of this model is that the current value of the time series, can be expressed as a linear combination of previous (past) load values, then the AR model can forecast future load values by a relationship as one defined here:

$$\hat{L}_k = -\sum_{i=1}^{m} a_{ik} L_k + w_k, \quad k = 1, 2, \cdots \qquad (10.4)$$

where $\hat{L}_k$ is the predicted load at time k, a_{ik} is the unknown coefficients, L_k is the previous load and w_k is the random load disturbance. The moving average (MA) model is a linear regression model, relating the current values against the white noise of one or more past load values. Time series moving average is a model used to analyze a data set by creating a set of averages of different subsets of the full original data set. A MA algorithm is commonly used with time series to smooth out short-term fluctuations and to highlight the longer-term trends or cycles. Conventional statistical models are limited and often leading to unsatisfactory solutions. The reason is the large number of computational possibilities leading to large solution times and the complexity of certain non-linear data patterns. Hence, machine learning and artificial intelligence-based techniques provide a promising and attractive alternative. Artificial intelligence methods consist of artificial neural networks, fuzzy logic, expert systems (ES), machine learning and hybrid models. The time series methods, while being simple, have the limitation of being unable to handle nonlinear data, which is a common feature in load data. They are superior to the time series and regression models in terms of accuracy. Artificial neural networks can be classified into several categories based on supervised and unsupervised learning methods and feed-forward and feedback recall architectures. The most commonly used ANN model is multilayered perceptron (MLP) which has an input and output layer, and one or more hidden layers between these two. ANN consists of processing elements called neurons. Artificial neurons have more than one inputs and a single output, being connected with each other. The processing ability of the network is stored in the inter-node connection strengths, which are called weights. These weights are obtained by learning or adapting from a set of training patterns. A set of systematic steps called learning rules needs to be followed when developing an ANN. Further, the learning process requires learning data to discover the best operating point of the ANN. They can be used to learn an approximation function for observed data.

Long-term forecasting plays a very important role in policy formulation and supply capacity expansion. Since the impact of the adoption of new technologies and policies affects the demand itself, combined methods are usually employed to include as many relevant factors as possible, e.g. consumer behavior, equipment types and number, or technology adoption impact and simulated scenarios. End-use method analyses the impact of energy usage patterns of different devices, equipment or systems in the overall energy consumption in a disaggregated approach. End-use models focus on the various uses of electricity in the residential, commercial and industrial sector. For residential users, appliances, house sizes, equipment age, customer behavior and population dynamics are often included. End-use models are based on the principle that electricity demand is derived from user demand for individual requirements, e.g. lighting, cooling, etc.; therefore, these models are suitable for predicting demand changes with the adoption of new technologies, use of new policies or implementation of demand response programs. This demand prediction capability is necessary for long-term forecasting and helpful for the adoption of energy efficiency programs. To build an energy model using an end-use method, less historical data is usually required, compared to the trend method or the similar day approach. Ideally, this method is very accurate, but it is sensitive to the amount and quality of end-use data. However, this method requires more detailed information about the consumers. Econometric models combine economic theory and statistical analysis for forecasting electricity demand, by establishing the relationships between energy consumption and the influencing factors. These relationships are estimated by the least-squares method or time series methods. One of the options in this framework is to aggregate the econometric approach, when consumption in different sectors (residential, commercial, industrial, etc.) is calculated as a function of weather, economic and other variables, and then estimates are assembled using recent historical data. Integration of the econometric approach into the end-use approach introduces behavioral components into the end-use equations. When combined with end-use approach, the behavioral components are added to the end-use equations for more accurate forecasting and electricity use understanding. The most common econometric approach for end-use estimations is the conditional demand analysis (CDA). In a residential CDA, the total household consumption is the sum of consumption of various end-uses plus an error term or residual. Notice that a long-term forecasting model uses the data fed into the short-term forecasting model plus the econometric parameters obtained from the performed survey and the from the DR processor data.

The end-use and econometric methods are requiring a large amount of information and data relevant to appliances, equipment, customers, processes, economics, etc. Their application is complex, requiring human participation, while the needed information may not be available regarding particular customers and utilities keep and support a profile of an "average" customer or average customers for different type of customers. The problem arises if the utility wants to conduct next-year forecasts for sub-areas, which are often called load pockets. In this case, the amount of the work that should be performed increases proportionally with the number of load pockets. In addition, end-use profiles and econometric data for different load pockets are usually different. The characteristics for particular areas may be very different from the average characteristics for the utility and may not be available. In order to simplify the medium-term forecasts, make them more accurate and avoid the use of the unavailable information, a statistical model that learns the load model parameters from the historical data was developed. This methods can be applied to both medium-term and long-term load forecasting. The performance of any forecasting methodology can be evaluated using several statistical metrics. In general, the mean absolute percentage error (MAPE) and root mean square error (RMSE) are used when evaluating the forecasting accuracy of designed models for power load forecasting field as in the following equation. MAPE indicates the average relative error between the actual power load value and the predicted power load value. RMSE calculates the square root of the sum of the squared differences between the actual power load value and the predicted value from the model output. They are given by:

$$MAPE = \frac{1}{n}\sum_{i=1}^{n}\left(\frac{|y_i(t) - \hat{y}_i(t)|}{y_i(t)}\right) \tag{10.5a}$$

And

$$RMSE = \sqrt{\frac{1}{n}\sum_{i=1}^{n}(y_i(t) - \hat{y}_i(t))^2} \tag{10.5b}$$

where $y(t)$ and $\hat{y}(t)$ are the estimated load and actual load at time t, and n is the number of data used in the model. Lower MAPE means more accurate estimates. However, the performance metric suitability depends on the features of the load time series.

10.4   POWER SYSTEM MONITORING AND CONTROL

Among the most fundamental functions of power system operator are the system monitoring, sensing and measurement. These functionalities provide an accurate picture of the power system conditions and states. Energy management systems must provide energy consumption, generation and transfer information and data at various temporal scales such as 15 min, hourly, daily and weekly. Feedback is most successful when it is provided frequently over longer time periods, allowing consumers to directly relate near real-time information with their energy usage actions. Power system operators need to have access and be able to examine the prevailing system condition, via information and data to establish whether the system is operating within acceptable and prescribed limits and thresholds. SCADA portion of the EMS is obtaining information regarding the current power system operating conditions by continuously gathering data at key locations. The data are transmitted to control center through communication systems. Power system monitoring include substation voltages, transmission power flows, active and reactive power generation, total system load, frequency, interchange schedules or status of circuit breakers and switches. Operators need to monitor and measure these system parameters to be able to estimate the system conditions and establish if whether they need to take further actions to remove potential violations and operations outside the normal prescribed conditions. SCADA data acquisition, communication and telemetry activities are performed cyclically every few seconds, enabling control operators and supervisors to have as much instantaneous power system views as possible. Upon receiving the power system data ENS is processing it at control centers. Power system data and processing results of the analysis are made available to the control supervisors on various display devices.

DMS has a similar monitoring functionality, designed to provide an accurate picture of the distribution system conditions. DMS acquires a significant number of real-time or near real-time data and information about current status, performances, important parameters, loading of distribution system power devices. This real-time information and data would include analogue and status data reporting to DMS SCADA from RTUs in substations and feeders once every 4 seconds, while the near real-time information includes equipment measurement and status from sources, such as smart meters, advanced metering and monitoring infrastructure, updated between once every 5 minutes and once every 15 minutes. Example of monitored equipment and devices include distribution substation transformers, load tap changers, distribution substation switchgears, shunt capacitor and reactor banks, distribution substation bus meters, field reclosers, field voltage regulators, overhead and underground load-break switches with SCADA capabilities, field sensors and distributed generating resources. Information is usually conducted by using report-by-exception approach, meaning the values are transmitted to DMS when the value changes by user-definable dead band since the last value transmitted to DMS. Network management is a prerequisite for any smart grid of the future since these grids will have to incorporate and manage centralized and distributed power generation as well as intermittent sources of renewable energy like wind and solar and allow consumers to become producers and export their excess power. Enabling multi-directional power flow from many different sources and integrating real-time pricing together with load management data. In this regard, the communication systems and infrastructure will and are playing a critical role in the operation and management of large electrical power systems and future smart-grid. Like utility companies, data centers must insure continuous reliability. To maintain reliability, data centers are designed with intrinsic redundancies across numerous servers. Using the real-time data collected on energy use, the IT administrator could decrease the load of non-essential servers at specified times – maintaining necessary redundancy. Advanced data and communication networks are used for SCADA, power system protection, remote metering and monitoring, corporate data and voice communication. Modern equipment is used to provide communication services for costumer call centers, service center dispatch operation, corporate voice lines, system control communication lines, direct inter-agency communication lines and other services needed for power system operation. Video networks are used for surveillance, facility protection, video conferencing and enhanced training programs. These communication networks are designed, built, operated and maintained by the electric utility. Data, voice and video networks are usually made up of six communication technologies: fiber optics, microwave, radio, power line communication, leased telephone lines and satellite communication.

10.4.1 WIDE AREA MONITORING

Wide area monitoring systems (WAMS) are designed by the utilities for optimal transmission capacity and to prevent and mitigate the spread of disturbances. By providing real-time information and data on stability and operating safety margins, WAMS give early warnings of system disturbances for the prevention and mitigation of system blackouts. WAMS utilize sensors distributed throughout the network in conjunction with GPS signals for precise time-stamping of measurements in the transmission system. The integrated sensors will interface with the communication network. Phasor measurement units (PMUs) are a current smart grid technology, one of its critical components. Wide area monitoring (measurement) systems are being installed on modern transmission systems and smart grids to supplement data and information of the conventional SCADA, measuring and processing the magnitudes and phase angle of busbar voltages, and current flows through transmission circuits and eventually transferring such information and data to control centers. The transferred information and data to the control centers are employed for: power system state estimation, monitoring and warning and for power system event analysis. Because the phasor data are synchronized, the magnitudes and the phase angles of voltages at all grid busbars can be estimated using state estimation algorithms, and the estimates are then used to predict and estimate possible voltage and phase angle instabilities and to estimate power system damping and vulnerability to small-signal oscillations. The PMU-based wide area measurement system was instituted for providing the time-synchronized measurements pertaining to the health of the power systems. The timely detection of the faults and the subsequent contingency measures are not only dependent on the PMU but also on the underlying synchrophasor communication system (SPCS). SPCS is connecting the PMUs to the phasor data concentrator or to the control center resulting in the exchange of data and control commands. Hence, the SPCS is considered as the foundation of the WAMS infrastructure. Data from several such local PDCs may be passed on to a regional or super PDC which is on a higher level in the hierarchy. The PMUs, the PDCs and the SPCS together constitute the WAMS and a typical WAMS is illustrated in Figure 10.5. The collected phasor data allows the power system operating conditions to be monitored on a real-time basis, power system stability to be assessed and warnings issued. High-accuracy synchronized phasor data and information are also available before and after faults or other network incidents, enabling the study of the causes and effects of faults and to take countermeasures against subsequent events. Such applications are integrated into smart grid wide area monitoring, protection and control (WAMPAC) system. Examples of WAMPAC scheme functions include: actions to correct the system once a voltage, angle or oscillatory instability are predicted, e.g. switching of generators and controlling FACTS devices, power system stabilizers (PSS) and HVDC converters; and to generate emergency control signals to avoid a large-scale blackout (e.g. through selective load shedding or temporary splitting of the power network) in the event of a severe fault or disturbance.

Time-synchronized grid parameter measurements across dispersed locations and sites are a key and differentiating feature of a wide area monitoring, protection and control system. WAMS and WAMPAC systems are based on the synchronized sampling of power system currents and voltages across the grid using a GPS timing signal. The PMU or the synchrophasor is the WAMS and WAMOAC critical basic building block, sampling the power system signals from voltage and current sensors and converts them into phasors. The PMU measurements consist of bus voltage and branch current phasors, in addition to information such as locations and other network parameters. Phasor measurements are taken with high precision from different points of the power system at the same instant, allowing an operator to visualize the exact angular difference between different locations. A PMU measures the synchronously positive sequence voltages and currents, phase voltages and currents, local frequency,

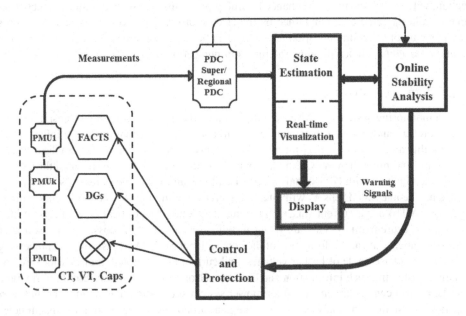

FIGURE 10.5   WAMS simplified diagram.

local rate of change of frequency and circuit breaker and switch status. These phasors are time tagged from a GPS timing pulse and then streamed communications network as fast as one phasor per cycle of the system frequency. Currently, the IEEE synchrophasor standard 37.118 defines the format by which the phasor data are transmitted from the PMU to control centers. The phasor angle information is referenced with the GPS timing pulse, and in order to have physical significance, is compared with and basically subtracted from other system phasor angle measurements. This phasor angle differences provide useful information concerning power system modes of oscillatory disturbances. In addition, PMU microprocessor-based instrumentation such as protection relays and disturbance fault recorders (DFRs) are incorporated into PMU module with other existing functionalities as an extended feature. The basic structure of PMU system consists of the synchronization unit, the measurement unit, the GPS circuit, and the data transmission unit. A PMU can be a dedicated device, or the PMU function can be incorporated into protective relays or other devices. Online applications of PMUs include monitoring of a large power system area for stability of frequency and current detection of any oscillation into the grid, while offline applications include post event analysis, validating model and data compilation. Post-event analysis recreates the events that could have happened after any power system disturbance by taking data from data recorders placed at various locations in the grid so as to take corrective measures for the future smart grid operations.

A PMU provides the critical synchronized time-lapsed information that enabled a clear understanding of events leading to a blackout or severe system instabilities. Modern PMUs have become more accurate and capable of measuring larger sets of phasors in an electric network. Offering wide area situational awareness, phasor measurement work to ease congestion, bottlenecks, and as mentioned above mitigate or even prevent blackouts. When integrated with smart grid communications technologies, taken measurements provide dynamic visibility into the power system. Addition of the real-time measurement in smart grids is enhancing every facet and aspect of the electric delivery system: generation, transmission, power distribution and usage, while increasing the possibilities of distributed generation. A phasor data concentrator (PDC) collects the phasor data from multiple PMUs or other PDCs, aligns the data by time tag, creating synchronized datasets, eventually passing the data on to applications' processors. PDCs send the data that is collected to a super (regional) PDC where there are applications for data visualization, storing the data in a central database and for integration with EMS, SCADA and WAMS, as shown in Figure 10.6. For applications that

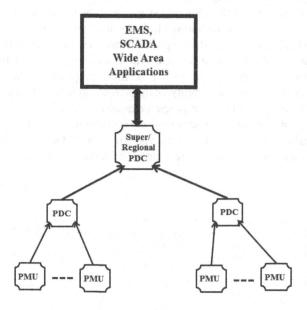

FIGURE 10.6   Typical PMU infrastructure and connections.

process PMU data from across the grid, all measurements are time-aligned based on their original time tag to create a system-wide, synchronized snapshot of grid conditions and states. To accommodate the varying latencies in data delivery from individual PMUs, and to take into account delayed data packets over the communication, PDCs typically buffer the input data streams and include a certain "wait time" before outputting the aggregated data stream. A PDC also performs data quality checks, validates the integrity or completeness of the data and flags all missing or problematic data. The functions of a PDC can vary depending on its role or its location between the source PMUs and the higher-level applications. There are three levels of PDCs: local or substation PDCs, control center PDCs and super or regional PDCs.

The power system centers supervise and control over the transmission networks, taking preventive actions to avoid any sort of system failure which can hamper electricity distribution. With ever-increasing size and complexity of the power system, the ability to detect any faults in the power system is heavily dependent on the real time information and data available. Traditionally, analog and digital information and data, e.g. circuit breaker status, power flow and frequency are measured at the substation level and transmitted to load dispatch center using SCADA or EMS. The major limitation of SCADA or EMS is the inability to accurately calculate the phase angle between a pair of substations. In SCADA or EMS, phase angle is either estimated from available data or is calculated offline. Phasor measurement units overcome the limitations of SCADA and EMS by accurately estimating and calculating the phase angle between a pair of grid points. PMUs measure the sinusoidal waveforms of voltages and currents, with a frequency of 50 Hz or 60 Hz at a high sampling rate, up to 1200 samples per second or even higher and with high accuracy. From these voltage and current samples, the magnitudes and phase angles of the voltage and current signals are calculated in the PMU phasor microprocessor, using the GPS clock signal to provide synchronized phase angle measurements at all their measurement and interest points, the measured phasors are usually referred to as synchrophasors. Figure 10.7 shows the voltage synchrophasors (bus voltages) at the two ends of an inductive transmission line.

Synchrophasors measured at different network points are transmitted to a PDC at a rate of 30–60 samples per second, and each PDC sends the collected data that is to a super (regional) PDC, where there are software applications for data recording and storing in a central database, data visualization, and for integration with EMS, SCADA and wide area application systems, for editing, later processing and analysis. The advantage of referring phase angle to a global reference time provided by the GPS is helpful in capturing the wide area snapshot of the power system. Effective utilization of this technology is very useful in mitigating blackouts and learning the real-time behavior of the power system. Since the bus voltage angle of a power system is very closely linked with the behavior of a network, its real-time measurement is a powerful tool for operating a network. The PMU and the synchrophasor measurements collected from the different points of the electric network and the state estimation is used for online stability analysis (see Figure 10.5 for details). When an event occurs, its location, time, magnitude (total generator capacity and/or transmission lines outage) and type (generator outage or transmission line outage) are first identified and then analyzed and eventually the control signals are generated. Real-time visualization of the event allows it to be replayed several seconds after it occurs, and the future system condition is then analyzed using the collected

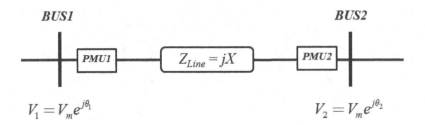

FIGURE 10.7   Voltage phasors at the ends of transmission line.

information and data. An on-line stability assessment algorithm continuously assesses the system status to check whether the system is still stable and how quickly the system would collapse if it became unstable. If instability is predicted, then corrective actions of the problem or to avert system collapse are taken. The voltage sinusoidal waveform is expressed as:

$$v_k(t) = V_{max,k} \sin(\omega t + \theta_k) \tag{10.6a}$$

Or in phasor notation as:

$$\hat{V}_k = V_{max,k} < \theta_k \tag{10.6b}$$

where k is the bus number at each end of the line (1 or 2), diagram of Figure 10.5, $V_{max,k}$ is the voltage peak value. Notice that IEEE is defining the PMU as "a device that produces synchronized phasor, frequency, and rate of change of frequency (ROCOF) estimates from voltage and/or current signals and a time synchronizing signal". The power flow on the transmission line, as one in Figure 10.5, is computed by using the relationship from the line ends voltages and line reactance (neglecting line resistance) as:

$$P = \frac{V_1 V_2}{X} \sin(\theta_1 - \theta_2) \tag{10.7}$$

For the network of Figure 10.6, power flow is estimated by using the phase difference between busbars 1 and 2, that is using the measured phase angle difference, $\theta_1 - \theta_2$, assuming the line-end voltages are known. If the measurement of the phase angle θ_1 has a time stamp error of $\Delta\theta$, and the phase angle, θ_2, zero error, the error of the estimated power is given by the following relationship:

$$\Delta P = \frac{V_1 V_2}{X} \sin(\theta_1 - \theta_2 + \Delta\theta) - \frac{V_1 V_2}{X} \sin(\theta_1 - \theta_2) = \frac{2V_1 V_2}{X} \sin\left(\theta_1 - \theta_2 + \frac{\Delta\theta}{2}\right) \sin\left(\frac{\Delta\theta}{2}\right) \tag{10.8}$$

Example 10.1: For the transmission line of Figure 10.6, the bus voltages are 1.5 p.u., and the power flow is 4.5 p.u. determine the error in the estimated power flow, if the measurement of the phase angle at bus 1 has a time-stamp error of 0.12 ms and zero error at the measured phase angle of bus 2, line impedance (inductive reactance) is 0.12 p.u. and line frequency is 60 Hz.

SOLUTION

The phase angle error $\Delta\theta$ is computed from the time-stamp error as:

$$\Delta\theta = 2\pi f \cdot \Delta t = 2\pi \cdot 60 \cdot 0.0012 = 0.045216 \ rad.$$

The phase angle difference $\theta_1 - \theta_2$ is calculated from Equation (10.2) as:
The estimated error of the power flow, computed by using Equation (10.8) is then:

$$\Delta P = \frac{2V_1 V_2}{X} \sin\left(\theta_1 - \theta_2 + \frac{\Delta\theta}{2}\right) \sin\left(\frac{\Delta\theta}{2}\right) \cong$$

$$\frac{2 \cdot 1.2 \cdot 1.2}{0.12} \cos\left(0.384397 + \frac{0.045216}{2}\right)\left(\frac{0.045216}{2}\right) = 0.498226 \ p.u.$$

10.4.2 POWER SYSTEM SECURITY CONTROL

Power systems are envisioned and designed to resist and survive to all probable contingencies. A contingency is defined as event causing that one or several important power system components (transmission lines, generators, transformers, etc.) are unexpectedly removed from service. Power system survival means the operation stabilization and continuation at acceptable of frequency and voltage levels without load losses. There are a huge number of possible operation conditions experienced by a power system, with large percentage of them, not anticipated or predicted, making impossible to predict all situations and scenarios. Usually, security control starts with a specific state, current state if executing real-time network sequence or a postulated state if executing a study sequence. Through a sequential execution of programs, determining the system state, based by current or postulated conditions, processing a list of contingencies and determining the consequences of each on the power systems, and finally determine the preventive or corrective actions for contingencies representing unacceptable risks. Security control requires topological processing for network models and use of large-scale network analysis to determine system conditions. Recent advances in synchronized and sub-second measurements, data acquisition and processing and advanced visualization capabilities dramatically improved the ability of managing grid operation in more effectively way, resulting in greater grid automation and helping grid operators in making better and more optimum decisions to maintain grid integrity. Grid automation is evolving toward more decentralized intelligent and local control, moving toward self-healing grid, one of the major attribute of the future smart grid. Self-healing power grid means to have capability to immediately identify and localize an intrusion, dealing with it locally, resolve it as fast as possible, while preserving the rest of the grid normal operation. Globally synchronized and in sub-second range measurements are currently used in control centers to facilitate earlier and faster detection of problems and to make timely and easier assessments of the conditions across the entire grid. This further improves the ability to maintain the power system integrity, and helping to faster identify contingencies, unplanned events and stability problems, at sub-second rates. The main objective is to provide the on-time and ahead information to the real-time power system conditions and potential problems or events. Speed and overall system condition information are critical to quickly navigate and identify the cause of a new problem. However, the data is growing dramatically, being critical to be able to concisely shown it on operator display screen in order that prompt decisions are made with confidence. Today, most of the control center operator decisions are reactive, based on some of the recent history, and assessment of current state and its vulnerability.

10.4.3 AVAILABILITY, ACCESSIBILITY AND AFFORDABILITY

Transmission system operation represents one of the most critical processes in utilities and power grids, EMS providing the operators with system visibility, information and control is enabling the optimum and efficient transmission operation process. Intelligent transmission systems and assets include a smart intelligent network, self-monitoring and self-healing, active and flexible dynamic power flow control and the adaptability and predictability of generation and demand robust enough to handle congestion, instability and reliability issues. Without EMS functionalities and control capabilities, operators can manage transmission operation for about 30 minutes, the timeframe in which the system load can change substantively. A substantive load change makes the predetermined remedial actions ineffective, while without system visibility and control removes the operator possibility of manually steering the system. EMS, being a mission-critical system, there are three criteria ensuring its availability and business continuity for prolonged outages. First is referring to restoration time following an EMS outage. Second criterion represents the system availability during a year. EMS availability is usually selected to be around 99.95%. Third is business continuity, ensuring that operators continue to manage the system operation in a conservative way, in the event of information unavailability. Future EMSs must have functionalities of making the information available to the consumers through easy-to-use interface, either as physical device, web or mobile portal that can give remote access to the information. The system should also allow installation and its configuration and maintenance must be simple one. However, EMS data and control signal communication poses security challenges and privacy issues. EMS must authenticate all transactions to ensure that user data and control operation are secure and not accessible to third parties without explicit approval. EMS may also push technology to send urgent notifications to consumers, via the newest or consumer's choice communication technology. EMS must also integrate other type of information useful to the users. Present power gird and future smart grid are considered critical infrastructures operating with specific and well-established functionalities. Situation awareness losses over the bulk electric transmission systems may and usually can lead to incorrect control actions, which furthermore may end-up in system damage or failure, and ultimately to blackouts. This may have severe effects on the local, regional and even country economy. The primary components ensuring control system availability are: system design and intrusion detection system (IDS). Availability losses can originate from denial of service (DOS) and command injection attacks, attempting to take control of the grid control system. Control injection attacks are detected with IDSs and prevented with authentication procedures. DOS attacks, consisting of flooding network with information at faster rates that it can process in order to deny network service can also be mitigate and detect by IDSs. Smart grids have to include IDS sensors to monitor network transactions at entry points to the control network or at points that can capture traffic from all control system network entry points. Entry points include modems, wireless terminals, local area network drops and connections to regional or independent system operators, connections to corporate LANs. System design is affecting control system availability, e.g. eliminating unneeded network services or including the requirements to disable all network ports and services that are not used for normal or emergency operations. Smart grid control systems must be designed to allocate a unique account ID and password for each user, allowing traceability and role-based access control. However, all cyber-security solutions must not harm the control system.

10.4.4 INFORMATION INTEGRATION AND SECURITY

The expanding size of electric power systems and the increasing complexity in operation have brought up a challenge to detect and mitigate abnormal events. Cascading detection and mitigation is an application that tries to detect cascading events at an early stage and prevent them from developing into large-scaled blackouts. In the meantime, new additions to the grid are taking place. The distributed generation blurred the separation between the generation, transmission and power distribution. Smart loads offered an opportunity to smooth out load curve by participating in demand side management (DSM); some loads such as electrical vehicle can also act as a resource to support part of the grid when power supply is interrupted by faults. Intermittent renewables such as wind generator brought in complexity to operation and control along with clean energy requirements. High penetration and integration of renewables introduces variability and uncertainty in power system security and operational challenges. How the EMS is evolving under such new developments in the electricity grid expansion still remains an important design requirement. Information and computing technology is one of the key elements of smart grids, enabling cooperation of distributed energy resources, local and adaptive control, smart sensing and monitoring and liberalized energy markets. Security assessment and control are important function to keep the power system in a secure state, considering power system operating under two constraint types: load constraints, load demand must be met; and operating constraints, maximum and minimum operating and stability limits are respected. In normal operation state, these constraints are respected. Security assessment and control applications include security monitoring and analysis, preventive and emergency controls, fault diagnosis and restorative control. When the system is insecure, the security analysis informs operators which contingency is causing the insecurity, as well as the anticipated emergency nature and severity. In the smart grid, the integrity cyber-security principle is intended to protect network traffic from unauthorized modification. Authentication responsibility is usually left to a high layer protocol. Most common insuring network traffic integrity method is through using digital signature algorithms (DSA) network traffic authentication. A key consideration in planning the encryption use in control systems are the latency added by it to the network traffic, and the hardware resources' availability.

The smart grid operation relies heavily on two-way communication for the exchange of data and real-time information that must flow continuously to and from the large central generation stations, substations, end-user loads and the distributed generators. Potentially sensitive personal data is transmitted and, in order to control costs, public ICT infrastructure such as the Internet is used. Obtaining information about customer loads could be of interest to unauthorized parties, infringing on the privacy of customers, opening up avenues for frauds. Smart grids require reliable and secure data and real-time information delivery. It not only needs *throughput*, the main criterion adopted to describe performance required for common internet traffic, but delays in the delivery of information accurately and safely is less tolerable in smart grids than for many commercial data transmission as the information is required for real-time or near real-time monitoring and control. Large part of sensing, monitoring and control information and data is repetitive, periodic, contributing to a regular traffic pattern in the smart grid communication networks, while during power system faults and contingencies there is a very large number of messages, information and data traffic. Any interruption resulting from security issues may have serious effects on the smart grid reliable, secure and safe operations. Information security measures ensure the following aspects:

1. Privacy that only the sender and intended receiver(s) can understand the message content.
2. Integrity that the message arrives in time at the receiver(s) in exactly the same mode it was sent.
3. Authentication ensures the receiver on sender identity, and messages do not come from an imposter.
4. Non-repudiation that a receiver is able to prove that a message came from a specific sender and the sender is unable to deny sending the message.

Providing information security is a common need of ICT systems since the Internet is one of the main modes of smart communication. In the smart grid contexts, the security must be: encompassing (i.e. identifying all smart grid cross-domain vulnerabilities is required because an exposure in any of these domains can, if not properly isolated lead to exploitation of smart grid network elements residing in other domains.), circulative (utilities must implement integrated security management and incident reaction controls) and aggressive (utilities must take the initiative in complying and exceeding security standards by implementing knowledge-based and artificial intelligence systems, able to detect network security events before occur). Security of smart grid network elements must also be pronominally current and up to date meaning that operating systems, applications, browsers, network interfaces, access control lists and other smart grid network elements must be constantly updated and protected. There are several standards which apply to the security of smart grid, substation equipment, advanced metering infrastructure, and several standards are under development or adaptation and upgrading. For overall security assessment, the standard ISO 27001 is widely used and specifies the assessment of risks for a system of any sort and the strategy for developing the security system to mitigate those risks. For example, the IEEE 1686 Standard is designed to provide guidelines for the substation intelligent electronic devices (IEDs) cyber-security capabilities. This standard is originated from an IED security effort of the NERC CIP (North America Electric Reliability Corporation – Critical Infrastructure Protection), being applicable to any IED where users require the *security, accountability and auditability* in the configuration and maintenance of the IED. This standard proposes different mechanisms to protect IEDs.

10.4.5 EMS Control and Monitoring Actions

Information and computing technology, advanced metering infrastructure, energy storage devise, renewable energy sources and distributed generation, among others, are elements that are becoming common in emerging electric grids. Massive deployment of sensors at all levels, renewable energy and storage and active loads result in highly complex data processing and MDMS problems. Many of these components did not exist a decade ago, including large-scale renewable wind and solar, advanced substation and feeder automation systems, utility scale storage and plug-in electric vehicles, rooftop solar installations, smart meters, AMI and two-way communication systems. While pursuing simultaneous objectives of efficiency, reliability and sustainability, the electricity industry faces numerous engineering challenges to ensure safe and integrated operation of all these components. In addition, the criticality of electrical energy delivery to the functioning of society, implies that control actions require the evaluation of "what if" events in real time. Thus, a very large number of scenarios must be tested to ensure that specified operating limits are not exceeded. Failure to do so may result in insecure systems conditions and potential cascading and broader failures. The resulting security dimension datasets contains the system conditions and operating regions that should be avoided. There are three types of control actions and functions implemented by the EMS: (1) direct closed control, (2) operator supervised control and (3) operator manual control. However, the system should be able to provide remote, programmable, and automatic control of power network devices and equipment. Direct control action is direct implemented by EMS, without operator intervention, such as automatic generation control, implanted automatically on frequency and schedule deviations. Another example of direct control action is the automatic remedial action schemes, taking place following specific power system conditions, such as major transmission line loss. Operator supervised control is implemented through EMS by operators, such as operator opening a SCADA switch. The last control action is implemented manually by field personnel after receiving orders from operators. An example is plant generation change as recommended by the EMS, in which the operator is calling plant operator requiring changing the generation because the operator does not have direct power plant control. The differences in objectives imply various technologies and implementations. Illustrated below is a focus on the next generation central EMS. From a technological point of view, we individuated three main concepts that are not commonly implemented in a classic central EMS:

1. The future power systems and smart grids are employing and are expected to integrate larger number of renewable energy sources and distributed generation units. Due to the intermittency of the sources themselves, stochastic and adaptive approaches to all the optimization applications (unit commitment, generation dispatch, etc.) are fundamental. For this reason, EMS applications should not optimize the resources operation based on deterministic conditions (fixed load or forecasted production from renewable energy generators), but finding solutions that optimizing the costs, minimizing the risk of load losses or over/under frequency operations under conditions, based on probability of each condition.
2. The other peculiarity of the new sources is the distribution on the territory and the size of each source. Compared with a classical power system with large power plants that supply electricity to the network, we are faced with a large number of small sources that have little importance individually, but become massive in total. To overcome this problem, hierarchical recursive monitoring and control system were proposed. Each resource (DG units or loads) belongs to a group controlled by a SCADA system, and the data collected by each SCADA are aggregated and posted to the upper level SCADA as a virtual resource. The upper level SCADA can monitor other real resources such as medium-size batteries or generators together with the virtual resources and report to another SCADA monitoring larger resources.

The system must provide energy consumption information at various temporal steps, such as 15 min, hourly, daily and weekly. Consumers can then directly relate near-real-time information with their energy use actions. The system should be able to provide remote, programmable and automatic control of devices. Generally, the consumer is expected to perform necessary control operations manually. However, a digital control option or automated actions are more effective. Telecommunication for utilities has a long history in the transmission level of the power grid. Nowadays almost all substations are monitored and controlled on-line by EMS. The main transmission lines are frequently equipped with fiber optic cables, mostly integrated in the ground wires and the substations are accessible via broadband communication systems. SDH (synchronous digital hierarchy) systems build the backbone of a utility communication hierarchy. Ethernet-based systems gradually complement these SDH systems and will replace them in the long run. SDH are used to connect substations in transmission and subtransmission networks and, depending on the market, also for medium-voltage substations. In the distribution networks, the situation is different from the transmission networks. Usually, the communication infrastructure at lower distribution levels is very poor. Most countries have less than 10% of transformer stations monitored and controlled remotely. Smart grid applications require a two-way communication network down to transformer stations and consumers. One of the difficulties that utilities face is that they may have multiple communications systems that have been implemented over a number of years, with changes in vendors. These changes present potential compatibility problems, even more when proprietary communications protocols are involved. The system should be able to provide remote, programmable and automatic control of devices. Generally, the consumer is expected to perform necessary control operations manually. However, a digital control option or automated actions are more effective. A desirable feature in new generation EM systems is that of intelligence. Consumers often lack a deep understanding of electrical systems and have limited time to make energy-related decisions. ISO 50001:2011 is the International Standard for Energy Management, released in June 2011 that replaces the British and European Standard BS EN 16001:2009. It provides a robust framework for optimizing energy efficiency in public and private sector organizations. Certification to this standard defines an organizations commitment to continual improvement in energy management. Implementation enables an organization to lead by example within their respective industries, ensuring that the related legislative and regulatory requirements are met. Significant financial savings is achieved through increased energy efficiency, considering that energy prices are forecasted to rise. It is desirable to have the system perform intelligent actions that balance energy consumption and consumer comfort, requiring techniques from machine learning, human-computer interaction and big data analytics to discern use patterns and predictive actions. This reduces the consumer to directly control and manipulate all appliances all the time.

10.5   SMART GRID DESIGN TOOLS

Smart grid as previously discussed is a cyber-physical system and also a critical infrastructure amenable for extensive modeling, simulation, analysis and design. A smart grid encompasses a large infrastructure through extensive use of information and communication technologies, smart sensing and monitoring, intelligent control or distributed and dispersed generation, extended use of energy storage or advanced communication infrastructure. Smart grid's intrinsic complex nature, distributed and dispersed generation, control and management, its increased component heterogeneity, bidirectional power and information flows, huge amount of measurements and increased service and power quality demands make classical power system optimization, analysis and design tools inadequate to handle the adaptability, complexity and stochasticity of smart grid functions and attributes. Thus, the computational tools and techniques required are defined as integrated and adaptive platforms for assessment, analysis, coordination, control, operation, design and planning of the smart grid under different uncertainties, requirements, expectations and operation conditions. As a consequence, the complex multidimensional nature (NP-hard type) [9] in development and planning of smart grid projects could not be addressed as a one-dimensional problem. Instead, these require the use of optimization techniques which consider the imbalance between stated objectives by different areas into the same energetic structure to decision-making. Modeling and simulation is a key technique for assessing the performance of virtually any system, including smart grids. Decision support tools combining game theory, decision support systems, risk analysis and analytical hierarchical processes (AHP) are used for computation of multi-objectives and risk assessment in smart grid planning, analysis and operations. Decision analysis is one of the powerful and critical tools that can make an uncertainty problem appear as a full rational decision that is based on numerical values for comparing and yielding fast results, based on the paradigm in which individuals or group decision-makers contemplate an action choice in an uncertain environment. Decision analysis is useful in handling multi-objective functions, or attaining several smart grid goals, where risk, choices and possibility are included.

One of the main challenges that energy fields confront during the deployment and design of a smart grid project is the availability of tools to support the decision-making processes, or designers and engineers in charge facing non-structured problems, bring decisions, evaluations and knowledge, based on the several and competing concerns, requirements and expectations. Besides the technical and engineering tools and methods, it is necessary to use qualitative approaches that allow becoming expert human judgment to mathematical representations as analytic hierarchy process or quality function deployment (QFD), which processes uncertain or diffuse information in the smart grid fields. The development of tools to support decision-making in smart grids projects requires solving problems of quantitative and qualitative nature at different levels of an organization. AHP is a decision-making approach and methods, providing alternatives, choices and criteria, while evaluating tradeoffs and performing synthesis and analysis to arrive at a final decision, being appropriate for cases involving both qualitative and quantitative analyses. Applications of AHP have been dominant in manufacturing, followed by the environmental management and agriculture field, power and energy industry, transportation industry, construction industry and healthcare. Other applications include education, logistics, e-business, IT, R&D, communication industry, finance and banking, urban management, defense industry, military, government, marketing, tourism, archaeology, auditing and the mining industry. In its general format, the AHP is a nonlinear framework, both deductive and inductive thinking, by considering several factors in the same time, allowing for dependence and feedback, and making numerical tradeoffs to arrive at a synthesis and conclusions. It is an approach well suited into smart grid planning, safety or efficiency analysis and studies.

Optimization techniques consist of static and dynamic methods and approaches for optimization, such as linear programming, mixed integer, dynamic programming, relaxation methods, etc. for development of smart grid optimization and planning activities and studies. Linear programming, nonlinear mixed-integer programming, dynamic programming and Lagrangian relaxation methods are extensively used for power system studies and operation issues, but they are limited for use in the smart grid due to fact that are suited for solving static networks. They work better when computed in conjunction with decision support tools and other computational tool techniques. Linear programming uses a mathematical model to describe the problem with linear objectives and constraints. A linear programming model consists of one objective which is a linear equation that must be maximized or minimized, and there are a number of linear inequalities or constraints. The general structure of problems solved by using linear programing methods is mathematically formulated in matrix format as:

$$Maximize\ C^T x$$
$$Subject\ to : Ax \le B, and\ x \ge 0$$

(10.9)

where C^T, A and B are constant matrixes, vector x are the variables (unknowns), all of them are real, continuous values. Note the default lower bounds of zero on all elements of vector x. The process to achieve the global optimum can use simplex techniques, variants of the interior point method or integer programming. These methods are applicable to the problems involving linear objective functions and linear constraints. Linear programming assumptions or approximations may also lead to appropriate problem representations over the range of decision variables being considered. In other instances, the nonlinearities in the form of nonlinear objective functions and/or nonlinear constrains are crucial for representing an application properly as a mathematical program. Nonlinear programming usually employs Lagrangian or Newtonian techniques for constrained and unconstrained optimization problems, assuming that all objective functions are modeled as continuous and smooth functions. For example, nonlinear programming with equality constraints that introduced slack variables can guarantee the robust of the algorithm and dispose the convergence problem. However, the power system mathematical responses are not usually complaining with such assumption in many cases and instances. The general structure of problems, including the one encountered in smart grids solved by nonlinear programming method is:

$$Maximize\ F(x_1, x_2, \ldots, x_N)$$
$$Subject\ to: A_i \leq F(x_1, x_2, \ldots, x_N) \leq B_i,\ i = 1, \cdots, N \tag{10.10}$$

Here, $F(x)$ is the objective functions subject to the two set of constraints, A_i and B_i. The two main issues of nonlinear programming method are its large computational burden and its limitation to static variables in the objectives and constraints. Integer programming is a special case of linear programming where all or some of the decision variables are restricted to discrete integer values, e.g. zero and one only, meaning YES or NO decisions, or binary decision variables. Pure integer or integer problems pose a huge computational challenge. While highly efficient linear programing techniques can enumerate the basic programing problem at each possible combination of the discrete variables or nodes, the problem lies in the huge number of combinations to be enumerated. Note that the direct use of these techniques for solving the smart grid optimization problem is quite limited, because they are generally static and are not designed for handling real-time and dynamic optimization problems. Dynamic programming approach is used to solve sequential, or multistage decision problems. Basically, it solves a multivariable problem by solving a series of single variable problems, achieved by tandem projection onto the space of each of the variables. Stochastic programming solves linear programming problems where the uncertainty assumption is so badly violated that the same parameters must be treated explicitly as random variables. The two ways to handle linear programing with variability are: stochastic programming (SP), and chance-constrained programming (CCP). SP requires all constraints to hold with probability whereas CCP permits a small probability of validating any functional constraint. However, to account for the smart grid predictive and stochastic nature of the smart grid involves the modeling of the components, accounting for the predictivity and stochasticity, and selecting new optimization methods.

10.5.1 Smart Grid Design Software Packages

The specific case of the power system analysis software that the power utilities currently use should evolve to cope with the fast-developing smart grid functionalities and applications. Several commercial and open-source tools are available for power system analysis, planning, studies and design, each of them has advantages and disadvantages. There are quite a large number of freely available software tools and packages for power system analysis, and quite a few of them have been created using the MATLAB®. However, MATLAB-based programs are requiring the MATLAB software license subscription to make use of these software tools, so are not quite free software. On the other hand, there are other power system analysis software packages, developed using Python programming language. Open-source software (OSS) use has allowed easy access to powerful power system analysis programs which are providing the opportunity of a flexible platform for development. A limited list of non-commercial power system analysis software is included: Dome (Python-based application), GridCal, GridLAB-D, MatDyn, MATPOWER, OpenDSS (distribution system simulator), Power Systems Analysis Toolbox (PSAT), etc. Commercial tools and software packages for power system analysis software has been available on the market for decades. Usually, the commercial power system tools have been closed, meaning limited access to change solution methods, models, database, functions, subroutines, etc. However, the software industry has progressively evolved to provide more and more flexibility and additional accesses to the user. The majority of the modern power system analysis software allows the users alternative ways to access the software, user-defined models, etc. The use of modern interfaces between software and hardware makes the power system analysis software even more flexible. However, the main drawback of the commercial tools is still the same – the cost, being quite expensive. A few of the power system analyses tools and software packages available on the market are: ASPEN, BCP Switzerland (NEPLAN), CYME, ETAP, IPSA Power, Power Analytics (EDSA), Siemens PTI (PSS/E, SINCAL), DIgSILENT PowerFactory. An important development aspect of smart grids is the interaction between a large number of grid components and active elements (i.e. components that actively affect the state of the power network by local or centralized algorithms), significantly increasing the complexity of the smart grids. The use of power system simulations and analysis tools is recognized as very well established and important method, used now for decades for the performance assessment of the power systems.

DIgSILENT is one of the most used power system analysis tools. It is used for modeling, analysis and simulation of the power system for more than 30 years, offering the most economical solution, as data handling, modeling capabilities and overall functionality replace a set of other power system analysis software. DIgSILENT PowerFactory can be used to calculate power flows, short circuits, harmonic content, stability simulations like transients or steady state simulations for balanced and unbalanced systems. Beyond the described classical functionalities of power system analysis, it offers a whole set of power system analysis functions dedicated to smart grids planning and operation, electric vehicles, renewable energy and distributed generation integration studies, e.g. load-flow with several optimization functions, optimal power flow (OPF), power distribution network optimization, open tie point placement, time sweep, state estimation, etc. But DIgSILENT PowerFactory also offers flexibility in modeling, simulation and communication. It also offers several alternatives and capabilities to be coupled with other power system models, design tools and software packages. The extensive integration of power system and ICT infrastructure mandates that the two systems must be studied as a single distributed cyber-physical system. DIgSILENT PowerFactory offers several alternatives to be coupled with other models include: MATLAB (PowerFactory built-in interface, DIgSILENT Simulation Language (DSL) for co-simulation, dynamic-link library (DLL, making possible the integration of an external event-driven C/C++ DLL (digexdyn.dll) is possible, OLE for process control (OPC), an asynchronous communication and data exchange mechanism used in process interaction (OPC client can be used with multi-agent systems and controller hardware-in-the-loop, OLE: object linking and embedding), remote procedure call (RPC), for constructing distributed client-server-based applications, and a wide set of interfaces, supported by the DGS, the PowerFactory's standard bidirectional interface. specifically designed for bulk data exchange with other applications such as GIS, SCADA and for exporting calculation results.

10.5.2　Green Energy Integration Technological and Infrastructure Challenges

The smart grid functionalities and applications require the implementation of devices or services with high computing performance into the grids and the development of fast communication networks between these devices and services. A summary of smart grid functions and attributes includes: fault current limiting, wide area monitoring, visualization, control dynamic capability rating; active and flexible dynamic power flow control, adaptive self-healing protections, automatic feeder and line switching (including optimizing performance or restoration), automating islanding operation and reconnection; diagnosis and notification of equipment condition, customer electricity use optimization. For example, there are several types of technical and infrastructural challenges associated with renewable energy sources, affecting planning and implementation of generation, transmission and distribution equipment, including aspects of power system dynamic performance. Smart grid design, analysis and study tools and EMS functions must include and handle such functionalities and attributes. Studies and actual operating experience indicate that it is easier to integrate PV solar and wind energy into a power system where other generators are available to provide balancing power, regulation and precise load-following capabilities. The greater the number of intermittent renewable generation is operating in a given area, the less their aggregate production is variable. High penetration of intermittent energy resources (larger than 20% of generation meeting load demands) affects the electric network in the several ways, as discussed here. Due to their highly intermittent nature of most of the renewable energy resources and systems, there are several challenges and issues related to dispatching them into grid, especially in distribution, such as:

1. Interconnected transmission and power distribution lines must not be overloaded, and the reactive power needed to be generated throughout the electric networks, not just at the interconnection point and needed to be compensated locally through the feeders. Due to the PV and wind power variations and required ramp rates larger than 1 MW/s, fast-acting reactive power sources needed to be employed throughout the feeders and electric networks to handle the reactive power flow requirements, so the smart grid EMS functions and design tools must be structured in suitable ways.
2. Voltage instability may occur due to the interconnection of the renewable energy resources to the grid. The output voltage of the generating utility and the grid operating voltage must be equal at the point of common coupling. Large differences between these voltages can lead to instability on the transmission grid. Voltage fluctuations can be caused by several factors, including changes in wind speeds, sunlight intensity, cloudiness or tidal heights, leading to potential system stability issues and problems.
3. Impact of additional generation power sources to short-circuit current ratings of existing electrical equipment on the network needed to be accurately determined. Notice e.g. that the PV inverters usually do not contribute the short-circuit duty and changes to the feeder networks.
4. Dynamic system behavior during contingencies, when sudden load changes and disturbance clouds can affect power system stability and power quality. Voltage and angular stability during such power system disturbances and production variance are very important. In most cases, fast-acting reactive-power compensation equipment, e.g. SVCs and distributed STATCOMs, is required for improving the network transient stability and power quality. Large PV plants or wind farms may require significant energy storage units to provide smoothing for the plant power output.
5. Ensure these fast operational switching transients have a detailed representation of the connected equipment, capacitor banks, their controls and protections, the converters and DC links. Due to PV power fluctuations, these network equipment and devices may switch much more than originally intended.
6. Voltage flicker issues, as the wind and solar power generators are intermittent, often resulting in fluctuations in output voltages that may cause an inconvenient noticeable voltage flicker. In the wind turbine cases, this is quite noticeable when wind turbines are switched on and off due to wind changes.
7. Wind turbine-related reactive power issue and similarly in the case renewable energy generators, utilizing rotating electric machines for electrical generation. Induction generators, e.g. wind turbine generators (WTGs) require substantial amounts of reactive power during their operation. This reactive power is absorbed from the grid and can cause depressed voltage conditions and voltage stability problems. This can be especially true for weak power systems that lack reactive power. In addition, this issue may conflict with the WTG

low voltage ride-through requirement, stating that a generator must be connected during grid faults and disturbances, and supply the grid with reactive power at low voltage. In the case WTGs, the reactive power usage increases with the wind speeds that may result in unacceptable low voltage levels at the receiving-end. Due to the load changes, renewable power systems need to adjust the reactive power consumption and supply to maintain the prescribed end-user voltage, in order to allow the utilities to maintain power quality and interconnection standards for renewable energy integration.

8. Unintentional islanding and reverse power flow may have large impacts on the network protection schemes, methods and settings. Large PV generation levels may reverse power flow during certain periods, and protection circuits need to be able to protect the distribution feeders under such conditions.

9. Due to the wind power natural variability, and the uncorrelated PV generation and loads, power generation has to be balanced with other very fast controllable generation systems and/or fast-acting energy storage units, to smooth out fluctuating power from wind and solar energy generators and increase the overall system reliability and efficiency. The costs associated with capital, operations, maintenance, and generator stop-start cycles have to be taken into account in any RES deployment.

10. Fluctuations in the PV and wind power generation and the transmission and power distribution network capacity and strength at interconnection points have direct impacts to the power quality. As a result, large voltage fluctuations may result in voltage variations outside the regulation limits, as well as violations on flicker and other power quality standards. Many renewable energy sources utilize rotating machinery, such as wind turbine-generator units. When such generation units start up, the electric machines act as large motors, until reaching grid voltage and synchronized with the grid, and then changing to the generating operation mode. During the motoring mode, a large reactive current (inrush current) resulting from the voltage difference between two systems, flows from the grid to the machines. This inrush current can be 2 to 5 times the rated current, and can cause large voltage sags into the network area around the wind turbines. Eventually, the voltage difference and inrush current tapers to zero after the two systems are synchronized. Wind turbine tripping has direct impacts on power system stability, e.g. transmission system faults may result in the power line losses, leading to WTG tripping. Such sudden losses in generation can be the start of a major cascading outage. Sub-synchronous resonance is a main contributor to wind turbine shaft damage, resulting from turbine tensional vibration, amplified by series capacitors installed to correct the line inductive reactance. This is especially problematic where renewable energy generation requires long distance transmission to the point of power system interconnection.

11. Several other DER and DG technologies are currently being integrated on the power distribution feeders as part of smart grids, e.g. PEVs, CHP generation and distributed energy storage units. The coordination of these DER devices is critical and essential to determine their combined impacts and effects on the power distribution feeders and networks, protection devices and power quality.

Power distribution systems have usually higher impedance on primary feeder lines, DER can be placed anywhere on such distribution lines. However, the DER and DG units have more impacts on the distribution systems than comparable DER and DG units placed at the substations. As previously stated, the conventional power distribution systems were designed for one-way power flow, from the power transformers to the end-users. DER placed on the feeder can cause reverse power flows, requiring different and additional protection and control equipment and devices. Usually, the security and safety of protective devices may be compromised if DER units cause fault levels larger than 5%. The main factor that affects the effects of DER on the power distribution is the network strength and stiffness. The closer the distance to the other large power sources, the stronger the system is. There are several stiffness ratios predicting the DER impacts on the system, with common one, specified by the IEEE P1547-D8 standard and employed when trying to evaluate the impact of DR on system fault levels, defined as:

$$\text{Stiffness Ratio} = \frac{\text{System Fault Current} - \text{Including DER}}{\text{DER Fault Current}} \qquad (10.11)$$

10.6   EMS IN BUILDINGS, INDUSTRIAL AND COMMERCIAL FACILITIES

Energy management system lies at the heart of all infrastructures from communications, economy and society's transportation to the society. This has made the system more complex and more inter-dependent. The increasing number of disturbances occurring in the system has raised the priority of energy management system infrastructure which has been improved with the aid of technology and investments. Modern industrial and commercial facilities operate complex and inter-related power systems. Energy conservation and facilities/equipment are only part of the approach to improve energy efficiency. Most energy efficiency in industry is achieved through changes in how energy is managed in a facility, rather than through installation of new technologies. Systematic management and the behavior approach have become the core efforts to improve energy efficiency today. An energy management system provides methods and procedures for integrating energy efficiency into existing management systems for continuous improvements in the energy uses. Energy management is often defined as a system, methods and procedures for an effective and optimum energy use in industrial processes and in operation of residential, commercial and industrial facilities to maximize profits and to enhance competitive positions through organizational measures and optimization of energy efficiency. Maximization of profits can be also achieved with reduction in energy costs during each productive and operational phase (in general the three most important operational costs are those for materials, labor and energy (fuels, electrical and thermal)). Moreover, the improvement of competitiveness is not limited to the reduction of sensible costs, but can be achieved also with an opportune management of energy costs which can increase the flexibility and compliance to the changes of market and international environmental regulations. Energy management is a well-structured process that is both technical and managerial in nature. In this chapter, we discuss the structure, methods and techniques used in energy management, as well as new approaches and developments in the field. A rich, comprehensive and up-to-date literature is also included in the chapter for professionals, engineers, students and interested readers in the energy management areas.

Traditionally, manufacturing, industrial and large commercial firms and facilities have lacked complex energy monitoring and control systems when compared to the ones frequently found in electrical generating plants. The reason for this is that the primary goal of any manufacturing firm is to cost effectively produce high-quality products for its customer and thus energy management often takes a somewhat secondary role to a plant's production objectives. However, with increasing energy prices and relentless competition in the marketplace, manufacturing firms are taking a closer look not only at energy saving opportunities but even the potential for onsite electrical generation from alternative energy sources. With these initiatives in mind, manufacturing firms are re-examining the functional requirements of their existing EMS and internal power networks to increase efficiency and better integrate with the future external power grid. Energy efficiency, energy conservation and energy cost have been top priorities all over the world, in particular for heavy energy consumers. Electrical energy costs are about 50% or so of total energy in many manufacturing industries, motivating companies to employ various strategies to control and reduce the rising energy costs. Certainly, many of our environmental problems today arise from the types of energy we use, and increased burning of fossil fuels will accelerate climate change. Energy conservation technology and facilities/equipment are only part of the approach to improve energy efficiency. Systematic management and the behavior approach have become the core efforts to improve energy efficiency today. Energy management represents a significant opportunity for organizations to reduce their energy use while maintaining or boosting productivity. Industrial and commercial sectors jointly account for approximately 60% of global energy use. Organizations in these sectors can reduce their energy use 10–40% by effectively implementing an energy management system. Fossil fuels are currently the major source of energy in the world. However, as the world is considering more economical and environmentally friendly alternative energy generation systems, the global energy mix is becoming more complex. Factors forcing these considerations are: (a) the increasing demand for electric power by both developed and developing countries; (b) many developing countries lacking the resources to build power plants and distribution networks; (c) some industrialized countries facing insufficient power generation; and (d) greenhouse gas emission and climate change concerns. Renewable energy sources such as wind turbines, photovoltaic solar systems, solar-thermal power, biomass power plants, fuel cells, gas micro-turbines, hydropower turbines, combined heat and power (CHP) micro-turbines and hybrid power systems will be part of future power generation systems.

The modern manufacturing plant has a large set of IT systems including EMS and manufacturing execution systems (MES) which are used to control processes and infrastructure. MES [2, 3] are single or even multiple pieces of software that together perform such roles as machine scheduling and monitoring, inventory and product tracking, quality monitoring, maintenance dispatching and operating allocation. These systems which all serve to manage some aspect of a manufacturing plant's operations are not always well connected nor integrated despite their common efficiency objectives which include improving energy efficiency. With the increased availability of all types of real-time information from plant-floor devices ranging from throughput, quality, reliability and energy consumption, there is growing interest from companies that want to take advantage of this valuable data to improve not only the productivity of their manufacturing operations but also simultaneously of their energy performance. However, due to the large amount of data and complex nature of most manufacturing systems, extracting meaning and understanding from this rich source of data is not always an easy task. To complicate matters further, manufacturing systems are highly dynamic environments where events like machine breakdowns, operator shortages and quality problems are just a few examples of situations that arise frequently. These events whether deterministic or stochastic, require an effective management response. Without a clear integrated view of an event, it is difficult at best to take appropriate actions to ensure efficient operations from both a productivity and energy perspective.

The key question for energy management practitioners is how to provide the best case for successful energy management within their organization, achieve the desired buy-in at top management level and implement a successful management system. The purpose of an energy management standard is to provide an organizational framework for industrial facilities to integrate energy efficiency into their management practices, including fine-tuning production processes and improving the energy efficiency of industrial systems. Energy management seeks to apply to energy use the same culture of continual improvement that has been successfully used by industrial firms to improve quality and safety practices. Energy management guidelines recommend that companies need to track energy consumption, benchmark, set goals, create an action plan, evaluate progress and performances, as well as to create energy awareness throughout the organization. These guidelines enable companies and organizations to integrate energy efficiency and conservation into their existing management system for continuous improvements and to reach the requirements for high efficient companies and ultimate to reduce the cost and increase the revenue. Among the requirements for high efficient companies are: efficiency is a core strategy, leadership and organizational support is real and sustained, company has energy efficiency goals, there is in place a robust tracking and measurement system, substantial resources are put for energy efficiency, and energy efficiency strategy is working and the results are communicated. An energy management standard is needed to guide how energy is managed in industrial facilities, thus realizing immediate energy use reductions through changes in operational practices, as well as creating a favorable environment and setting for the adoption of more capital-intensive energy-efficiency measures, practices and technologies. Efficient energy management requires the identification of where energy is used, where it is wasted and where any energy saving measures will have most effect. The key feature of a successful EMS is that it is owned and fully integrated as an embedded management process within an organization, energy management implications are considered at all stages of the development process of new projects, and that these implications are part of any change control process. Standards should lead to reductions in energy cost, greenhouse gas emissions and minimize the negative impact on the environment. A change in the organizational culture is needed in order to realize industrial energy efficiency potential. An EMS standard can provide a supportive organizational framework necessary to move beyond an energy saving project approach to an energy efficiency approach that routinely and methodically seeks out opportunities to increase energy efficiency, no matter how large or small.

Businesses that are wasting energy are reducing profitability and causing avoidable pollution, primarily through increased carbon emissions, which contributes to both climate change and dwindling fuel reserves. If comparisons were made between industries, corporations or specific manufacturing plants, one will find a wide variation in the structure of individual energy programs. Even with such variations in scope and implementation, there is a good chance of a successful energy management program. Comparing one industry or company with another would result in energy management program variations based on how energy is consumed. A company whose energy use is based mainly in lighting and heating would develop a different program and strategy than a cement manufacturer, with hundreds of electric motors in addition to a significant use of thermal energy. Though one successful energy management program may be implemented in a complete different way than another, they all are sharing common elements that are discussed later. A company with centralized management philosophy may dictate program structure across the company, while one with more decentralized management structure might set broad program goals, allowing individual units to develop programs according to local circumstances. Making businesses more energy efficient is seen as a largely untapped solution to addressing environmental pollution, energy security and fossil fuel depletion. As pressures mount on businesses to become more energy efficient, managing resources effectively is proving more essential than ever. In addition, customers are increasingly asking for assurance from organizations that they treat the environment responsibly and are able to demonstrate energy efficiency. Energy management for buildings, industrial facilities and processes is the means and an approach to control and to reduce the organization, building or process the energy consumption and uses. By controlling and reducing the organization, facility or process the energy consumption and use are very important, by enabling to:

1. Reduce costs, which becoming increasingly important as energy costs rise.
2. Reduce emissions and environmental damages that they cause, the cost-related implications of carbon taxes and the like, many organizations are keen to reduce their carbon footprint to promote a green, sustainable image. Not least because promoting such an image is often good for the bottom line.
3. Reduce risk – the more energy you consume, the greater the risk that energy price increases or supply shortages could seriously affect your profitability, or even make it impossible for your business/organization to continue. With energy management you can reduce this risk by reducing your demand for energy and by controlling it so as to make it more predictable.

10.6.1 Identification of Energy Usage Factors and Parameters

With energy costs and the importance attributed to climate change having risen over the past several years, energy efficiency has become paramount. Identifying key energy performance indicators is vital for the planning process, as it provides managers with a clear picture of how their company uses energy and can highlight ways to manage resources better. The energy manager's first step is to determine the energy consumption "structure" of his/her company; in other words, what energy resources does the company need to run its operations, be it gas, coal or electricity. A management process is required to proactively assess, manage and measure energy usage and conservation. However, many companies and organizations currently have limited levels of expertise necessary to achieve these reductions and so need guidance on how to do so – best practice, etc. Improvements in energy efficiency will require systems and processes necessary to improve energy performance. Companies and organizations will need to manage the way in which they use energy in order to reduce GHG emissions and other environmental impacts, as well as reductions in energy costs and wastage. Managers should set energy efficiency benchmarks that measure consumption and the number of energy consumption assets in operation and to set the opportunities and ways to reduce the energy usage. Such benchmarks allow companies to compare their energy management systems against best practices, both domestic and abroad. The benchmarks also allow companies to compare their energy efficiency results against national standards, such as the clean production standard or country's energy-saving technologies policy. Energy managers that are aware of their company's energy performance rating are in a better position to track energy and water consumption, and set targets and indicators. As a result, they can improve policies, identify areas that need improvement, and measure the investments required for facility upgrades. The energy management systems aim at reduced energy use and costs, are representing a key element in any company's energy management program. Energy management must be based on real time information obtained from process monitoring and control systems, and on production plans received from production planning systems. Often, total and comprehensive solutions include planning and scheduling tools to optimize energy use and supply, energy balance management tools to support the real-time monitoring and control of the energy balance, and reporting tools to evaluate and report energy consumption, costs, efficiency and other energy-related information. Opportunities for cost reduction are greater when both electricity uses and prices vary over time, which is common in process industries, and open electricity market environments.

Energy management concepts are built upon a plan-do-check approach (PDCA) cycle that it is used as a foundation for various management systems. Figure 10.8 illustrates how continued use of the PDCA process leads to continuous improvement. An EMS is a collection of procedures, methods and tools designed to engage staff at all levels within an organization in managing energy use on an ongoing basis. An EMS allows industrial plants; commercial, institutional and governmental facilities; and entire organizations to systematically track, analyze and plan their energy use, enabling greater control of and continual improvement in energy performance. Organizations may choose to pursue only certain components of energy management. While EMS implementation expertise is typically concentrated in technical and engineering positions within an organization, a variety of non-technical personnel also exert significant influence on energy decision-making (e.g. executive staff, accountants and financial managers). While these non-technical staff members need not be knowledgeable in all aspects of energy management, are included in this analysis since they can be critical to the success of an EMS within their organization. Many long-standing professional training or credentialing programs cover some skills or knowledge areas relevant to EMS implementation, but most falls short of providing the entire spectrum of skills and expertise needed to implement an EMS effectively. Workforce development programs that focus specifically on EMS implementation in response to the 2011 publication of ISO 50001 are still in the early stages of development or release, and most do not provide training for the broader range of personnel involved in instituting energy management at the organizational level.

The recommended knowledge and skills in this report may impart guidance to workforce programs under development, generate opportunities for collaboration among developing or expanding training programs, facilitate greater consistency among existing professional programs and increase awareness about the energy efficiency potential that can be achieved through skills programs.

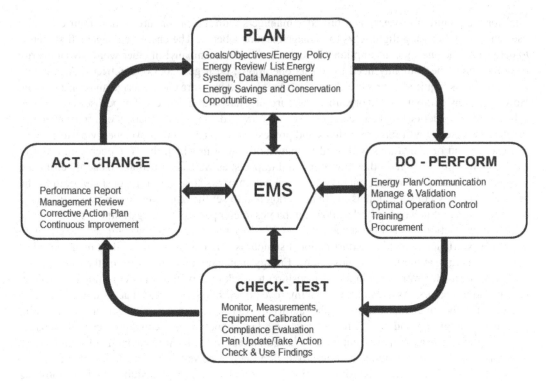

FIGURE 10.8 Energy management flow chart and structure.

Ultimately, building skills in the workforce will help countries achieve their national energy efficiency goals. Advancements in training and credentialing offer only part of the solution for improving effectiveness; supporting policies and national systems are also critical components. Although
this report does not analyze such policies, it acknowledges their vital role in improving workforce
programs and work quality. Steps in which are needed during implementation of an energy management system are listed below:

1. Initiating an energy management program: Understanding basic concepts and requirements; getting organization leadership commitment; establishing an energy team; developing an energy policy;
2. Conducting an energy review: Collecting energy data; analyzing energy consumption and
   costs; identifying major energy uses; conducting energy assessments; identifying potential
   opportunities;
3. Energy management planning: Setting a baseline; determining performance metrics; evaluating opportunities and selecting projects; developing action plans;
4. Implementing energy management: Obtaining resource commitments; providing training
   and raising awareness; communicating to all stakeholders; executing action plans;
5. Measurement and verification: Including the knowledge and skills required to monitor,
   measure, verify, track, and document energy use and savings; and
6. Management review: reviewing progress; modifying goals and action plans as needed.

These steps are usually embedded in a PDCA cycle, as represented in the diagram of Figure 10.8.
In addition to these common knowledge areas, the energy management plans identified the ancillary knowledge and skills that will enhance understanding of key energy management topics and
actions. To manage energy uses, a company or organization must gain a complete understanding
of its energy uses and demands. This understanding should be based on a comprehensive energy

review, consisting of analyzing all energy uses, determining where are significant energy uses and then identifying and prioritizing potential opportunities for improvements and conservation. An energy review is an essential element of energy management planning and should be performed by personnel with a broad range of knowledge and skills. An energy review requires the collection of energy consumption data from utility bills, energy meters and other sources. The data must be analyzed and interpreted within the context of the various sites, facilities, processes, business units and equipment in the organization. This analysis requires personnel that not only understand buildings, processes, energy-using equipment and other factors but also possess the knowledge and skills necessary to identify viable improvement opportunities. Some of the fundamental tasks involved in conducting an energy review include:

- Data logging and collection;
- Metering, monitoring, measurement and verification; and
- Facilitating and managing processes for identifying energy conservation and efficiency opportunities.

10.7 CHAPTER SUMMARY

There are few types of energy management systems that may be found in industrial, commercial and residential sectors, ranging from the most complex ones found in electricity industry down to the simplest ones utilized by the residential energy consumers. Regardless of the application magnitude and sophistication, each EMS type has its own unique requirements depending on the user's needs and requirements. The power system operation major objectives are safety, reliability, generation and load demand balance, supply security and efficiency. Power system operation was regarded from beginnings as a critical function because it can significantly change the utility bottom line, affecting user safety, impacting system reliability, having significant effects on operational costs associated with the transmission, power distribution and generation. While safety, service quality, reliability and supply security are primary objectives governing power system operation, efficiency, environmental impacts and economy are also very important, meaning using process optimization while complying with safety, supply security and reliability. However, efficiency applies to the operation of all power system segments: generation, transmission and power distribution. On the other hand, safety is the most important criteria, aiming to ensure personnel, environment and property safety in all aspects of power system operations. Power quality is defined in terms of variables, such as frequency, voltage, that must conform to specific standards, in order to accommodate the operation of all loads connected to the system. Two control centers are found in electric utilities, one for generation and transmission operation and management, and the other one dealing with power distribution. Former one is referred as energy management system, while the latter is referred to as the distribution management system. The two management systems are critical and essential for optimum power system monitoring, control, management and operation. Energy and distribution management systems must provide advanced and versatile functionalities while keeping the installation simple and running at lowest cost possible. These energy management systems should also be integrated with users' daily activities and expectations, while offering actionable feedback. The smart grid EMS field has been extensively researched in recent years, with many different approaches. Energy management system is one of the major software packages of any grid control center. As the power system central nervous system, the control center together with EMS is a critical component of power system operational entity. Due to the critical tasks performed by an EMS, there are strict system requirements, so a lot of attention is placed in choosing appropriate architectures to achieve the objectives. On the other hand, new technologies are required to achieve effective and efficient management, control and protection of smart grids, in the face of challenges presented by increased deployment of distributed generation and renewable generation resources, and needs for more efficient uses of the network infrastructure. Increased availability of information and communication technologies within electrical networks can be expected to facilitate the adoption of innovative protection techniques. There are also strong needs in developing in the electric power sector for improved integrated generation management with respect to the increase in green energy resource penetration. Many of the challenges faced by these new green energy resource portfolios that are emerging are within the power delivery sector, and the need for applying advanced transmission technologies to assure safe, reliable and efficient electricity delivery. The continuing changes in sensing, measurements, communications and computation have now reached a stage where the digital data acquisition architecture that has served the monitoring and control of the power grid for over half a century requires a fundamental change to fully utilize the new technologies and realize the promise of a new generation of applications. Smart EMS integrated with home automation systems are key smart grid components and are essential in bringing a multitude of benefits to the consumers. Load demand forecasting is an essential process in power system operation and planning. It involves the accurate prediction of both electric load magnitudes and geographical locations over the different planning horizons. Accurate load forecasting is critical for electric utilities in a competitive environment created by the electric industry deregulation and the smart grid advent.

10.8   QUESTIONS AND PROBLEMS

1. Briefly describes the overall scopes of the energy management.
2. What are the major attributes of any energy management regardless the application?
3. What are the main objectives of a power system energy management?
4. Briefly discuss the EMS main objectives in the context of modern power systems and smart grids.
5. What are the major attributes and characteristics of EM data and information?
6. List and discuss the key features of the modern EMS.
7. List the main EMS functions.
8. What are main SCADA attributes, related to the power system energy management?
9. Briefly describe the EMS evolution and changes.
10. What is the DMS main purpose and objective?
11. What is the main reason that a DMS is different from an EMS?
12. Briefly describe the distribution energy management structure.
13. What are the main power system security functions?
14. What are the primary load frequency control functions?
15. Why is load forecasting needed in power system management and planning?
16. List the most common load forecasting methods or algorithms.
17. What are the main factors influencing the electricity demand patterns?
18. List the weather factors affecting electricity consumption.
19. Repeat Example 10.1 for 50 Hz frequency and end-line voltages of 1.2 p.u.
20. Briefly describe the PMU structure.
21. Repeat problem 17, for a line impedance of 0.1 p.u., power flow of 5.6 p.u. and the rest of parameters remain the same.
22. What are primary components ensuring control system availability?
23. Why the load forecasting is needed in power system operation and management?
24. List the most common short-term load forecasting methods.
25. Briefly describe the PMU structure and operation.
26. List the main online and offline PMU applications.
27. What are the two most common medium- and long-term load forecasting methods?

28. A transmission line has the inductive reactance 0.15 p.u., power flow of 7.5 p.u., frequency of 50 Hz and the end (bus) voltages of 1.2 p.u. if the time stamp error at of first bus (end) is 0.1 ms and zero for the second bus, compute the power flow error.
29. What are the most common long-term forecasting methods?
30. What are the operating functions of short-term load forecasting?
31. List the smart grid information and data security measures.
32. Briefly describe the needs for wide area monitoring in smart grids.
33. List the open-source smart grid analysis tools.
34. List a few of the computational tools used in smart grid design or analysis.
35. Briefly describe the main capabilities of the DIgSILENT package.
36. List the major challenges of the grid integration of renewable energy sources.
37. Briefly describe the possible RES impacts on power quality.
38. What are factors affecting the DER impacts on power distribution?
39. What are the main phases of an energy management plan?

11 Microgrids in Smart Grid Architecture and Operations

11.1 MICROGRID CONCEPTS, TYPES AND ARCHITECTURE

Smart grids, envisioned to meet the 21st-century energy requirements and expectations in a sophisticated manner with real-time approach are integrating advanced communications, smart sensing and monitoring, adaptive and smart control technologies to the existing electric grids. Rapid advancements in control, information and communications technologies (ICTs) have allowed the conversion of conventional grids into smart grids that ensure productive and complex interactions among energy providers, utilities, end-users and other stakeholders. These multiple and enhanced interactions are helping to solve the issues raised in the existing power grids. For example, control and management systems are needed across broad temporal, geographical and industry scales, from devices, components to power system wide, from fuel supplies to consumers, or from utility pricing to demand response. With increased deployment of feedback and communication infrastructure, opportunities arise for reducing energy consumption, for better and efficient exploiting renewable energy sources, and for increasing the reliability and performances of the transmission and power distribution networks. Key components of the smart grids are smart metering, sensing and monitoring, adaptive controls, intelligent sensing and data management systems that control the flow of power, data and information among various stakeholders and subsystems, making the grids two-way communication and energy networks. Smart grid applications include Energy Management Systems (EMS), distributed generation (DG) and its reliable integration to the power systems, equipment diagnostics, control, overall optimized asset management etc. The concept of microgrids emerged as a way to handle the growing number of distributed generation units, such as dispatchable units using fossil fuels like fuel cells or combined heat and power (CHP) units, or renewable energy sources like photovoltaic panels, wind turbines, biomass or micro hydroelectricity. They usually are installed on low-voltage networks, closer to the customer loads, being able to provide power backup in case of blackouts. But the true potential of microgrids lies in their way of managing all-day power supply, through low-voltage and low-power units, integrated in the main power system. The idea is to create semi-autonomous sub-networks, composed of low-voltage equipment such as distributed generation, energy storage and loads that are considered as one single electric entity by the main networks. Microgrids, often described as a self-contained subset of local generation, distributed energy assets, protection and control capabilities and loads, are one of the smart grid major drivers. Microgrids may be operated in either utility connected or in grid isolated (islanded) mode, providing reliable and secure electricity supply, as well as a wide range of ancillary services, e.g. voltage support, frequency regulation, harmonic cancellation, power factor correction, spinning and non-spinning reserves. One of the key microgrid features in the smart grid consists of its ability to separate from mains during the unscheduled periods of interruption to continue feeding its own islanded load portion, at least supplying the most critical, essential and sensitive loads.

Microgrids are intrinsically distributive in their nature, by including several and diverse DG units, renewable and conventional energy sources, energy storage units, various protection systems, controllable and uncontrollable loads, power electronics, adaptive control and other components. Microgrids are essentially active power distribution networks, being a conglomerate of DG units, energy storage, local control and management and various types of loads (many controllable and smart loads) operating at power distribution voltage levels. In order to achieve a coordinated performance of a microgrid (or several microgrids) within the scope of a power distribution, it is required

DOI: 10.1201/9780429174803-11

to perform adaptive, distributed or cooperative control. *Agent* (a software and/or hardware entity) technology, exhibiting autonomy is one of the techniques for achieving the objectives of distributed microgrid control and management. An agent is a self-organized, decentralized and limited purview so as to progress the entire system toward a common goal such as in cooperative distributed application. Agents in microgrids are expected to perform sensory, communication, management and control tasks. In order to achieve reliable levels of distributed control of microgrids, it is imperative that the microgrid possesses a self-configurable sensor network that aids in communication among the constituent agents. A microgrid is often defined as a portion of the low-voltage power distribution network which includes loads and distributed generators. Loads can be residential or industrial entities, generators may be solar panels, wind turbines or other alternate power sources. The microgrid size ranges the whole power distribution network, or part of it, like a group of buildings (shopping centers, industrial parks, college or military campuses). A microgrid is connected to the power network in one point, point of common coupling (PCC), and it is managed autonomously from the network to achieve better service quality, improve efficiency and pursue specific economic interests. A smart microgrid can appear deeply different from traditional power distribution, whose unique task is to deliver energy power from the transmission grid to the loads. A smart microgrid may include a large set of intelligent entities, such as micro-generators, able to inject power instead of being supplied with only, electronic loads with their specific dynamic behaviors, *smart users* which can postpone their demand if financially rewarded, etc. All the micro-generators are connected to the microgrid via electronic interfaces, whose main task is to enable the generated power injection into the microgrid. However, these devices, if properly controlled and coordinated, can also perform other tasks needed to guarantee a desired service quality, and/or ancillary services, such as reactive power compensation, voltage support and regulation, harmonic compensation, reliability and robustness to faults, faster service restoration, etc.

A microgrid consists of a low- or medium-voltage electric network of small load clusters with DG sources, energy storage units, protection and local control. If a micro grid is connected to the grid, it is seen as a single aggregate entity. Microgrids are autonomous electricity networks, operating within a larger electric utility grid. The concept is not new, chemical plants, refineries, military installations and other large industrial facilities have had the ability to generate and manage their own electricity needs while, in addition, remaining connected to the grids for supplemental energy needs. Today, the promise of mass-produced electric vehicles and of affordable, locally produced, distributed energy resources is encouraging expansion of the microgrids. Such *micro-operation* within an electric utility has the potential to revolutionize electric grids, permitting local generation from distributed renewables (especially solar rooftops and residential wind generation), increasing the efficiency and environmental sustainability. One of the potential microgrid advantages is that it could provide a more reliable supply to customers by islanding from the system in the major disturbance event. If the microgrid concept was proposed as a way to better integrate renewable energy sources and distributed generation in general into the main power grid, consumers that are part of a microgrid would not just benefit from a more efficient energy system, but a more eco-friendly one. If distributed generators are placed close to the loads, there are reducing in the transmission losses, preventing network congestions, enhancing system reliability, stability or supply security. Microgrids offer various advantages to end-users, utilities and society, such as: improved energy efficiency, power quality, reliability, minimizing energy consumption, reduced pollutant emissions or cost efficient electricity infrastructure replacement. A microgrid links multiple distributed generation sources into a small network serving some or all of the energy needs of participating users, lowering energy costs, increasing efficiency and improving overall environmental performance and electric system reliability. Microgrids have the potential to better accommodate new demands for electricity by shifting the electricity paradigm away from an exclusive model of centralized, generation systems, such as distributed generation from local, clean resources supplementing centralized generation. Microgrids are small electrical networks embedded in an electrical utility service territory, such as neighborhoods, retail shopping areas, industrial or commercial parks. Microgrid elements include: (1) distributed energy resources, (2) energy storage and (3) flexible energy demand. Usually, a microgrid is interfaced to the power system by a fast semiconductor static switch, allowing islanding operation, while trying to use the available wasted energy. The approach relies on complex communication, control and protection techniques. It is essential to protect a microgrid in both the grid-connected or islanded operation modes against all faults, with major issues arising in island operation with inverter-based sources.

11.1.1 Microgrid Concepts, Drivers and Benefits

It has been proposed that one solution to the grid reliability, security and stability issues is to take advantage of microgrid technologies. *Microgrid* term is becoming a popular topic within the power and energy engineering community. However, the microgrid concept still remains not fully defined. A common approach is defining a microgrid as a *subsystem of distributed energy sources and associated loads, operating as a controllable entity*, allowing local control of the distributed generation and loads, reducing or even eliminating the needs of a central control and management. A microgrid can be a standalone system or it can be tied to a stiff AC grid that can be separated from it during disturbances and extreme events. It is offering three major advantages over a traditional electricity supply implying central generation stations, long distance transmission over a network of high voltage lines, then distribution through MV and LV networks: applications of combined heat and power (CHP), opportunities to tailor power quality to suit the requirements of end users and create a more favorable environment for energy efficiency and small-scale renewable generation investments. The wasted energy use, e.g. heat through combined heat and power processes implies an integrated energy system, delivering both electricity and useful heat from an energy source. A CHP process can convert as much as 90% of its fuel into useable energy. To maximize unit efficiency, energy sources need to be placed closer to the heat load rather than electrical loads since it is easier to transfer electricity over longer distances than heat. Small local energy sources can be sited optimally for heat utilization, and systems integrated with distributed generation are very pro-CHP, unexpansive and efficient. Power quality and reliability are often used in quantifying levels of electrical service. Both scheduled and unscheduled outages can affect the availability of certain services to the end-users, increasing dependence on on-site backup generation which can be costly. Power quality degradation has more subtle, important, effects as well. Voltage sags, harmonics and imbalances are triggered by switching events and/or faults. While power quality events do not lead to electrical losses, they can degrade the power in the end-use processes, affecting equipment performance and durability. DER units have the potential to increase system reliability and improve power quality due to the supply decentralization. Increase in reliability levels is achieved if DG units operate autonomously in transient conditions, mainly if there are outages or disturbances upstream in the electrical supply. Microgrids can also be inter-connected to make clusters of microgrids (multi-microgrid), or are integrated into a conventional grid. During emergencies, microgrids can be disconnected from the grids, operating in an islanded mode. In many cases, microgrids can provide higher service reliability, better power supply quality and higher efficiency. Unlike conventional power systems, microgrids possess the following properties: connecting lines have a higher resistive characteristic (higher R/X ratio), system capacity is limited by the capacity of the DER and RES units, the fault currents are low (i.e. low short-circuit ratio), loads can be highly unbalanced, system inertia is low and energy supply is prone to higher uncertainties, due to the intermittent nature of renewable energy sources, such as wind or solar energy. These features pose additional challenges for the operation of microgrids, and therefore represent active research areas.

The integration of renewable energy sources into the power system provides unique challenges to the designers of the electrical systems. Due to the intermittent nature of the sources, central generation is required to provide the base power supply as well as provide backup power when the sun is not shining or the wind is not blowing. Systems with intermittent sources can experience similar problems as systems with large, intermittent loads. Distributed generation can ease the burden of high penetration of renewable sources by filling in when intermittent generation is low and by smoothing the transmission system loading. Several studies of renewable energy sources tie to the electrical system through inverters, which lack the mechanical inertia that helps to bring grid stability. There are different inverter topologies that inherently have electrical storage on the output, such as voltage-sourced inverters, replicating the energy stored in a rotating machine, lending them well for microgrids. Much of the previous work on the RES integration into microgrids focuses on the inverter control that ties the source to the electrical system. The electrical energy generated by wind and solar energy conversion systems and even microgrid small local generators is reaching a considerable portion of the total produced energy in comparison to that of the previous decades. The presence of these new energy sources, distributed storage, power electronic devices and communication links make the power system control and protection more complicated than before, because they impose considerable and fundamental changes in the power system configuration, topology and power flow direction. Thus, in order to enhance the power system visibility and controllability, more data and communication links are needed throughout the entire power system; however, this huge data amount can cause heavy computational burden, as well as possibly negatively impacts the performance of protection schemes. Scale and location of the microgrid are important factors. Microgrids should be constructed at the LV or MV level. The key defining characteristics of a microgrid are: provides sufficient and continuous energy to a significant portion of the internal electricity demand, own internal control and optimization strategy, can be islanded and reconnected with minimal service disruption, can be used as a flexible controlled entity to provide services or optimization for the grid or the energy market, able to operate at various voltage levels (usually in the range of 1 kV to 20 kV) and has energy storage capacity.

There are several drivers pushing for microgrid deployment, development and research into modern power systems and smart grids. Among others, there are drivers and incentives such as environmental, economic, technological, power system operation, stability, improved resilience, supply security, grid integration of the renewable energy or power quality. Owners and operators of microgrids are benefiting in most countries from *governmental incentives* to help renewable energy deployments and developments, as well as with reduction of the environmental impacts of electricity generation, transmission and uses, making microgrids an efficient way to develop renewable energies and attain goals set by many countries in this and environment protection areas. Microgrids could constitute a relatively cheaper and efficient step toward rural electrification, meaning a *cost-effective electricity access* in many regions and countries. For many developing or undeveloped countries with low rural population density, for remote locations, farms, facilities, and islands, the high electrical infrastructure costs represents huge obstacle to completely electrify a territory and to provide electricity to such end-users. Microgrids could be more gradual solutions to solve such cases. Moreover, in regions where grid saturation and the transmission networks are operating near the capacity limits, where there are critical issues, such as frequent blackouts or brownouts, poor service quality, microgrids could offer solutions to alleviate pressure and stresses without heavy and large investments in large-scale power plants and high-voltage power lines. This, in turn, would give the end-users a better service and a more reliable and secure electrical supply. Power outages and electricity blackouts and brownouts are very costly for energy providers and end-users, being also very uncomfortable for the end-users. The *microgrid islanding capacity* represents one of the best ways to improve grid resilience in case of unforeseen difficulties, an important factor for several sensitive and critical loads and special end-users such as military bases, hospitals, first-response, fire and police stations, communication or server facilities. Therefore, it is advantageous for end-users and for energy suppliers to decrease the number and the duration of high demand periods: grid maintenance costs, energy losses and needs for stand-by generators are strongly reduced and a better quality and reliable service would be offered to end-users. DG and RES technologies have the benefit of reducing losses from long distance electrical transmission through closer to loads and end-users. Transmission losses currently vary from about 7%, in most of the advanced countries to 25% and even more in many developing and emerging countries. Microgrids could be *a strong facilitator towards renewable energy and distributed generation developments, deployments, research and grid integration*, and a way to minimizing losses, moving a portion of generation closer to loads and end-users, while improving power quality, resilience and supply security. Many countries have set ambitious goals in these matters, constituting microgrids is one way of achieving such goals. It is also a way to accelerate the development of smart grids with an easy integration. *Energy storage* is a vital part of the microgrids and smart grids, being a way to improve the operation, power flows, grid operation and management. Increased energy storage capacity has significant benefits for grid and utilities, reducing transmission stress, needs for larger peak generation, load shedding, etc. Also, the development of electrical vehicles can be seen as a plug-and-play storage capacity and for this reason, can have a large impact on microgrid development and research in the near future.

Microgrids can bring several benefits to the poser systems and smart grids, especially when distributed generation and renewable energy sources are used. Microgrids can reduce electrical losses, increase grid stability and security, and as a whole, reduce spending for consumers and power distribution systems. Indeed, microgrids may benefit at the same time DSOs, end-users and the micro-generation operators (DSO or associations of end-users). Microgrids can facilitate the use of micro-generation and DG units, providing direct benefits, by local generation and through governmental incentives to accelerate the implementation of renewable energy systems. Such schemes usually include a subsidized price and other incentives for the owners of renewable energy generation systems (PV arrays, wind turbines, small hydropower, biomass units. etc.) to sell back electricity to the electricity providers at higher than market price. Network spending reduction, in areas where the existing electricity infrastructure is under high demands or where there is no existing electrical infrastructure (e.g., rural areas in developing countries, islands or remote locations), microgrids could represent a much cheaper alternative to building expensive conventional generation and transmission infrastructures. Network spending is in that respect reduced or at the very least postponed such expenses. Microgrid intelligent automation systems can encompass relatively complex price setting mechanisms, through dynamic pricing software application to calculate in real-time the cheapest energy source of energy, main grid electricity or local generation sources (e.g., rooftop PV panels or wind farm integrated to the microgrid). Moreover, the dynamic pricing coupled with the local generation availability can be a powerful tool to shave or shift loads, lowering the peak load demands up to 10% and general electricity consumption up to 15% to their potential high-quality local automation capacities, microgrids should improve general grid stability and electricity reliability.

There are a certain number of factors that can slow down or even prevent microgrid developments and deployments that can be divided into two main categories: technical challenges and legal (regulatory) challenges. The first technical challenge concerns the power balance management between the load demands and generation. In order to improve grid efficiency, it is critical that the DG generation be scheduled and the load demand be controlled, while proper systems and securities to be implemented and to proper manage when to use the local generation and when to store the local generated electricity. All these actions imply also that the grid to have capabilities to forecast energy production and make *smart decisions* based on such forecasts. The microgrid must also be able to evaluate very accurate its reserves, to take optimum decisions at multiple horizons: *weeks, days, hours, 15 min or even few seconds. Protection, control and safety* are representing other important technical issues. Control and management programs must be designed to cope and handle how the grid reactions when a microgrid switches from a normal (grid-connected) operation mode to an islanded operation mode, and vice versa. It is also important to evaluate precisely the consequences of unplanned outages and the microgrid reconfiguration needed. A second concern refers to the *black start*, the process of restoring a power station to operation without relying on external energy sources. The final challenge concerns two-way electricity flows, e.g. how to regroup multiple microgrids, so they can feed each other with energy. For this system to function there are needs of two-way electricity flows at the transformer locations, by adapting the security rules and equipment capabilities. The third technical challenge concerns the issues and problems related to the microgrid and main grid interconnection. There are critical needs of the real-time monitoring and sensing of power flow, precise measurement tools and sensors to calculate and monitor key electrical characteristics, e.g. voltages, power flows, angles, Volt-VAR analysis, a controlled frequency and harmonics in islanded operation mode, and of course the optimum reconnection procedures after islanded periods. The final technical challenge concerns the information and communication parameters linked to the microgrid electrical aspects, requiring a data management system able of handling all the data generated by a microgrid, while keeping at minimum or even without redundant operations, adjusting the operation costs to the microgrid size. There several legal and regulatory challenges, such as: who operates the microgrid and what are the responsibilities? Can the microgrid stakeholders trade their electricity to other participants, other microgrids or to the transmission system operators, etc.? These legal and regulatory challenges also need to be addressed during microgrid deployment and design phases.

Microgrids are usually relying heavily on the local renewable energy sources and other DG units that can drastically reduce the electrical losses incurred on transmission lines and pollutant emissions. The possibility to obtain energy from the grid or locally can help to improve voltage quality, while the local generation reduces the need to transmit electricity on long distance, reducing the energy losses. Compared to conventional power grids, microgrids are covering much smaller geographical areas, in consequence are less vulnerable to the extreme events and natural disasters. Due to flexibility, reliability and islanding operation mode capabilities, microgrids are one of the major ways to improve grid resilience, service restoration, provide emergency power backup and help to mitigate power system outages. In summary, microgrids can provide improved electric service reliability, supply security, and better power quality to end-users and can also benefit local utilities by providing dispatchable loads for use during peak power conditions or allowing system repairs without effecting customer loads. Some less explored solutions to grid vulnerability that can be tackle by microgrids, such as: providing heterogeneous power quality and reliability, so that the quality provided better matches quality needed, hardening socio-technical systems to an inevitably imperfect grid, or providing secured power locally to sensitive loads. Microgrids enable small communities to take control of their energy generation and uses, while reducing pollution through an innovative way of generating and managing electricity. Considering alternative energy sources is vital, as technologically advanced economies struggle to meet inexorably growing electricity demands, pushing the limits of affordable power quality and grid expansion. While grid dependency has intensified, smaller generation using a generation mix has emerged as increasingly competitive with large remote generation stations. However, identification of microgrid benefits is a multi-objective and multi-party coordination process, strongly depending on business structure and models. However, microgrids have a lot to offer and several benefits to all stakeholders and entities of the electrical grids. Major MG advantages include: MG portfolio is tuned to local resources, enabling a hedge against fuel cost increase, larger RES penetration, emission reductions and finally the microgrids can actively control the network for improved reliability, resilience and service quality. A microgrid total value to a host include: significant power generation savings relative to the grid electricity generation paradigm, grid services (ancillary services, demand response, and sale of generated power excess), reliability, resilience and power distribution grid values, such as: improved system resilience, investment deferral or extended access to ancillary services.

11.1.2 MICROGRID STANDARDS AND TECHNOLOGIES

The IEEE 2030, addressing smart grid interoperability, and IEEE 1547TM, addressing distributed energy resources grid interconnection, have made substantial progresses since 2009 in terms of all electric grid interconnection and smart grid aspects. Small- and large-scale distributed energy resources are governed by the IEEE 1547 set of standards, including references to the UL1741 (Underwriters Laboratories) standard for the interconnecting on low-voltage networks. These standards were developed toward the end of the 1990s when DG, especially distributed PV and wind generation, was at very low penetration levels. IEEE 1547 describes the interconnection issues of DG resources in terms of voltage limits, anti-islanding, power factor and reactive power production mainly from a safety and utility operability point of view. The IEEE 2030TM and 1547 standards series focus on systems-level aspects and are also covering many of the technical integration issues involved in smart grids. The IEEE Standard 1547-2003 is the first in the 1547 series of interconnection standards and provides interconnection technical specifications and requirements as well as interconnection test specifications and requirements. The stated requirements of IEEE Standard 1547 are needed for interconnection of distributed energy resources, distributed generators and energy storage systems that involve synchronous machines, induction machines or power electronic converters. It provides requirements, sufficient for most installations, relevant to the interconnection performance, operation, testing, safety and maintenance. IEC has also published several standards for DC systems such as IEC 62040-5-3, IEC 61643-3 and IEC 61643-311 for existing DC applications. Recently, a number of activities in the area of low-voltage DC applications in Information and Communication Technologies, residential and commercial buildings etc. have led IEC to establish a new strategic group (SG) to study the standardization of DC distribution, in which SG4 has been approved for low-voltage DC distribution system up to 1500 V in relation to energy efficiency. In IEEE standard association, there also are a number of ongoing activities on the utilization of DC power distributions; some of them are listed here. WG 946 standard provides recommended practice for design of lead acid batteries based DC auxiliary power supply system. This standard covers the guidelines for selection of number of batteries, their capacity, voltage level and duty cycle. It also provides brief description about the effect of grounding on the operation of DC auxiliary systems. P2030.10 group is an ongoing work is mainly looking for the possibility to utilize DC microgrid concept to provide safe and economic electricity in remote areas where centralized utility system do not exist. This standard covers the design, operations and maintenance of a DC microgrid for rural or remote applications, providing also requirements for providing low voltage DC and AC power to off-grid loads.

Since 2008, equipment manufacturers, planning engineers, microgrid and system operators as well as all those linked with the microgrid and distributed generation technology have a guide, the IEEE Standard 1547.2TM-2008 covering various aspects of these technologies. This IEEE Standard for Interconnecting

Distributed Resources with Electric Power Systems (EPS) address aspects such as intentional islanding in systems containing DERs, connected with utility as well as the microgrid integration with the power system. In Annex A of this reference guide, for example, it can be found in aspects related to:

A.1.1 – Interconnection issues, which could be done, for example, through switchgears.
A.1.2 – Power source transfer, which can be automatic or manual. It also presents issues related to DER systems regarding the parallel operation.
A.1.3 – Metering and monitoring issues.
A.1.4 – Protection issues.
A.1.7 – DER dispatch, communication and control (e.g. energy management system/SCADA area).

Other issues discussed in the guide include: impact on voltage, frequency, stability issues, power quality, existence of single and multiple point of common coupling, identification of steady-state and transient conditions, understanding interactions between machines, reserve margins, load shedding, demand response and cold load pick-up. IEEE P1547.4 standard covers microgrids, including intentional islands that contain DER connected with local or area islanded electric power systems. It provides alternative approaches and good practices for the design, operation and integration of microgrids, covering also the ability to separate from and reconnect to an area EPS while providing power to the islanded local networks. The guide covers DER, interconnection and participating

EPSs. It is intended for use by EPS designers, operators, system integrators and equipment manufacturers. Its implementation expands the DER benefits by enabling improved power system reliability and building on the requirements of IEEE 1547-2003. Technical challenges include the design, acceptance and availability of MG technologies. Several such technologies were developed or under development, e.g. MG switches and advanced DER controls, allowing safe interconnection. IEEE 1547 has helped to modernize our electric power systems infrastructure by providing a foundation for integrating clean renewable energy technologies as well as other distributed generation and energy storage technologies. IEEE 1547 provides mandatory functional technical requirements and specifications, as well as flexibility and choices, about equipment and operating details that are in compliance with the standard. The full revision of IEEE 1547 is addressing DER interconnection and interoperability, including associated interfaces and per IEEE mandate must be completed by 2018. The full revision of 1547 development has substantial participation of individual utilities and participation from a number of transmission-level entities. Full revision of 1547 issues, concerns and updates are being coordinated with corresponding standards and codes such as the NEC and UL safety standards. The IEEE 1547 standard series of existing, and published standards that include:

1. IEEE Standard 1547-2003 (reaffirmed 2008), IEEE Standard for Interconnecting Distributed Resources with Electric Power Systems;
2. IEEE Standard 1547.1-2005, IEEE Standard Conformance Test Procedures for Equipment Interconnecting Distributed Resources with Electric Power Systems;
3. IEEE Standard 1547.2–2008, IEEE Application Guide for IEEE Standard 1547, IEEE Standard for Interconnecting Distributed Resources with Electric Power Systems;
4. IEEE Standard 1547.3–2007, IEEE Guide for Monitoring, Information Exchange and Control of Distributed Resources Interconnected with Electric Power Systems;
5. IEEE Standard 1547.4–2011, Guide for Design, Operation and Integration of Distributed Resource Island Systems with Electric Power Systems;
6. IEEE Standard 1547.6–2011, Recommended Practice for Interconnecting Distributed Resources with Electric Power Systems Distribution Secondary Network.

National Renewable Energy Laboratory (NREL) and its industry partners, with support from the US Department of Energy and California Energy Commission, are developing advanced technologies that allow for the use of microgrids. This research has focused on the development of smart (autonomous or semi-autonomous) switches, which are needed to evaluate an area's EPS conditions and coordinate operation modes, grid-connected and islanded. NREL provides a testbed to conduct technology characterization and examine issues related to the interconnection of multiple DER into a microgrid. This work is examining issues related to intentional islanding and will be a key in the development of the knowledge base for the IEEE P1547.4 standard. As part of this research, microgrid topologies and operational configurations have been defined and design criteria have been established for various possibilities of microgrid applications. NREL has been worked with ASCO Power Technologies, General Electric and Northern Power Systems to develop advanced interconnection technologies that allow distributed generators to parallel with the Area EPS for uninterrupted electrical service and sell electricity to the Area EPS. The objective is to consolidate the various power and switching functions (e.g., power switching, protective relaying, metering and communications) traditionally provided by relays, hardware and other components at the utility interface into a single system with a digital signal processor. These switch technologies are designed to meet IEEE 1547 and UL 1741 grid interconnection standards to minimize custom engineering and site-specific approval processes and lower cost. The ASCO 7000 soft-load transfer switch from ASCO Power Technologies is designed to provide a safe and controllable EPS interconnection. Such interconnection switch has been used for emergency power applications and is increasingly being used for grid-parallel operation. Avoiding significant voltage or frequency transients, the ASCO soft-load transfer switch can seamlessly transfer or share load between the generator and utility bus and has been designed to meet IEEE 1547 interconnection standard requirements. The General Electric Universal Interconnect is a prototype interconnection system, based on the General Electric G60 Universal Generator protection relay, a transfer switch type. The General Electric Universal Interconnect is intended to allow interconnection of any DG with the EPS and provide local load connections. It operates in a manner similar to that of the ASCO soft-load transfer switch, providing seamless transfer between the utility, generator and load. General Electric also developed a power electronic based interconnection switch that can be used with a variety of DER entities. Northern Power Systems and NREL developed a high-speed static switch for distributed energy resource and microgrid applications, the so-called the DER Switch. To maximize applicability and functionality, it was designed to be technology-neutral (i.e., the controls in the digital signal processor can be used with circuit breakers as well as with faster semiconductor switching technologies such as silicon-controlled rectifiers, integrated gate bipolar transistors, and integrated gate commutated thyristor) and applicable to DER assets with conventional generators or power converters. The generic control system design allows the use of faster power electronics. This power electronic interface has significant benefits compared with circuit breaker technology because the system can be designed for seamless transfer applications. This research promotes the development of new technologies, enabling faster switching, improved reliability and lower fault currents on electrical grids, providing fewer disruptions for customers while expanding capabilities.

11.2 MICROGRID ARCHITECTURE, STRUCTURE AND TYPES

A typical microgrid architecture is shown in Figure 11.1 with the microgrid connected to a larger power system, while a disconnect switch can "island" the distributed generation units, sensitive and critical loads. A major factor in microgrids is the disconnect switch, enabling the microgrid to compline with current standards such as IEEE 1547. Such a switch is necessary to realize the high reliability and power quality that microgrids offer. It has been found that in terms of energy security, multiple small generators are more efficient than relying on a single large generator. Small

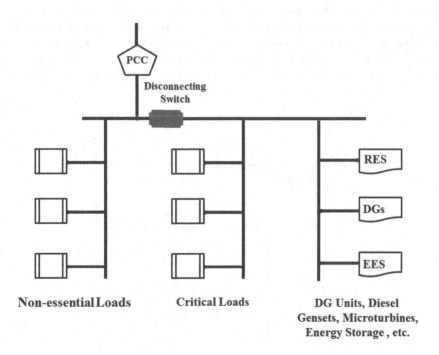

FIGURE 11.1 Typical microgrid architecture with critical, noncritical loads, DGs and disconnecting switch.

generators have a lower inertia and are better at automatic load following and help avoid large standby charges that occur when there is only a single large generator. Having multiple distributed generators available makes the chance of an all-out failure less likely, especially if there is backup generation capable of being quickly and easily connected to the system. A configuration of multiple generators creates a peer-to-peer network that insures that there is no master controller can be critical to the MG operation. Having a component such as a master controller creates a single failure point, not an ideal situation when the end -user demands high reliability in the electrical system. A peer-to-peer system implies that a microgrid can continue to operate with any generator lost. With the loss of one source, the grid can regain all its original functionality with addition of a new source, if available. This ability to interchange generators and create components with plug-and-play functionality is one requirement of microgrids. Plug-and-play elements imply not only that any unit is replaceable but that a unit can be placed at any point in the system without re-engineering the controls.

The MG functionality of plug-and-play gives the benefit of placing generation close to the load, further increasing efficiency by reducing transmission losses. The concept can be extended to allowing generators to sit idly on the system when there is more electrical capacity than necessary. As the load on the system increases, additional generators would come on-line at a pre-determined set-point necessary to maintain the correct power balance. The set-point could use voltage sag, frequency droop or any other factor that indicates the system is being stressed to signal an increase in generator capacity. Intelligent devices could sense when there is extra generation capacity on the system and would disconnect and turn off generators to save fuel and increase machine efficiencies. A scheme that would allow for generators to automatically drop off the system when there is no longer a high demand of power would assume a hierarchy of generators with different set-points so that not all the generators drop off at the same time. Figure 11.2 shows a microgrid schematic diagram. The microgrid encompasses a section of an electric power distribution system that is located downstream of the distribution substation, and it includes a variety of DER units, various end-users of electricity and/or heat. DER units include DGs and distributed energy storage (DES) units with different capacities and characteristics. MG electrical connection point to the utility, at the low-voltage bus of the substation transformer, constitutes point of common coupling (PCC). The microgrid serves a variety of customers, buildings, commercial entities and industrial parks.

The microgrid of Figure 11.2 normally operates in a grid-connected mode through substation transformer. However, it is expected to provide sufficient generation capacity, control and operational strategies to supply at least part of the loads (critical, sensitive and essential loads), when

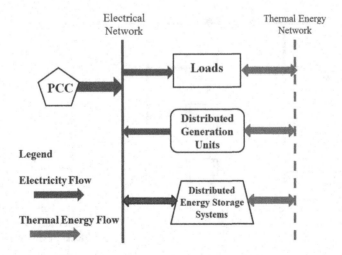

FIGURE 11.2   General microgrid representation, building blocks, electricity and thermal energy flows.

disconnected from the grid at the PCC and remain operational as an autonomous entity. The existing utility practice often does not permit microgrid accidental islanding and automatic resynchronization, primarily due to the safety concerns. However, the high amount of DER penetration, potentially necessitates provisions for both islanded and grid-connected operation modes and smooth transition between the two (i.e., islanding and synchronization transients) to enable the best utilization of the microgrid resources. DER units, in terms of their interface with a microgrid, are divided into two groups. The first one includes conventional generation units that are interfaced to the microgrid through protection devices and transformers. The second group consists of electronically coupled units that utilize power electronic converters to provide the coupling media with the host system. The control concepts, strategies and characteristics of power electronic converters, as the interface media for most types of DG and DS units, are significantly different than those of the conventional rotating machines. Therefore, the control strategies and dynamic behavior of a microgrid, particularly in an autonomous mode of operation, can be noticeably different than that of a conventional power system. Furthermore, in contrast to the well-established operational strategies and controls of an interconnected power system, the types of controls and power/energy management strategies envisioned for a microgrid are mainly determined based on the adopted DER technologies, load requirements and the expected operational scenarios. A typical microgrid includes loads, generation units, energy storage devices, electricity and thermal networks, and often two levels of controls.

11.2.1 Microgrid Types and Components

The microgrid concepts assume a cluster of loads, DG, micro-sources, control, energy management and protection systems and equipment operating as a single controllable entity that can provide power and heat to its area or facility. This concept provides a new paradigm for defining the operation of distributed generation. To the utility the microgrid can be thought of as a *controllable entity* of a power system. For example, such cell could be controlled as a single dispatchable load, which can respond in seconds to meet the needs of the transmission system. For customers the microgrid is designed to meet special needs, such as: enhance local service reliability, reduce feeder losses, support local voltages, provide improved efficiency and service quality through the use of waste heat, voltage sag correction or uninterruptible power supply functions, to name a few. The micro-sources of special interest for microgrids are small (<100 kW) units with power electronic interfaces. These energy sources, usually microturbines, PV panels and fuel cells are placed at or near customer sites. They are low-cost, low-voltage and have high reliability with lower pollutant emissions. Power electronics provide the control and flexibility required by the microgrid concept. Correctly designed power electronics and controls insure that the microgrids meet the customers and utilities' needs. Such characteristics are achieved by using system architecture with three critical components: DG controllers, system optimizer and distributed protection and control. One of the main technical issues to be addressed in the process of integration is the control and management of the non-dispatchable RES, like wind and solar energy, which feature somewhat unpredictable behavior. Another major concern is the assessment of the impact of such sources on the overall grid security and reliability. Also, there is a the need for re-engineering the protection schemes at the distribution level, to cope with the bi-directional power flows and take advantage of the fast response of power electronic devices commonly used for the grid connection of distributed energy sources.

Microgrid is not a specified term that is used similarly everywhere; however, common for all definitions is that a microgrid is usually a part of power distribution network, and it has island (standalone) operation capability. Microgrids can be: (1) a part of power distribution networks which has islanding capability, reducing the outages and provide emergency electricity supply; (2) an integration platform for supply-side (DERs and micro-generators) and demand-side resources (energy storage units and controllable loads) located in a local power distribution grid; and (3) an autonomous power system able to provide electricity and all necessary services, for extended periods, at least to essential and critical loads, islanded operation being only a special emergency case. Microgrids can be considered as a self-managed and self-serviced power system where islanded mode of operation is only a special case. Major microgrid types include institutional microgrids (hospitals, university or military campuses), commercial or industrial microgrids (factories, server farms, data centers, commercial malls or business towers) and finally community grids (multiple houses, apartment buildings, and commercial buildings). The latest type is not very common, but a huge increase is expected when regulatory and business barriers are lifted and mitigated. The autonomous, insulated microgrid segment consists of the areas not yet connected to the main grid, where no preexisting electric infrastructure exists. Another classification, found in the literature, separates microgrids into four categories: (1) separated island microgrid, (2) low-voltage customer microgrid, (3) low-voltage microgrid and (4) medium-voltage microgrid. The first category refers to standalone microgrids where DER provides electricity to one customer or small community outside the grid. Suitable places could be remote islands, isolated farms and small villages far away from the utility grid. Second class refers low-voltage customer microgrid. In this instance, a farm or detached house have own DER unit(s), providing needed electricity, operating usually parallel with grid. However, in case of fault in the utility grid, the microgrid operates in standalone mode. In this case of PV and wind power generation energy storage to offer constant electricity supply is needed. Third class, LV microgrid consists of a group of low-voltage customers it can include anything from few consumption points to whole LV network fed by a MV/LV transformers. Power production of this kind of microgrid can be based on several small-scale generation units of various types. The MV microgrids, where larger generation units are connected to MV network can consist of a part of HV/MV substation output or the whole output. It can also be possible to use all HV/MV substation outputs as a microgrid. MV microgrids offer an opportunity for wind parks and other large renewable energy systems to produce electricity during grid interruptions, another advantage for the energy producers. The basic principle is the same in all microgrid types. However, there are still many microgrid uncertainties, such as what are the roles of different stakeholders, how fast technology develop or implement, and whether microgrids are going to be cost effective way to limit outages. The microgrid future is dependent on development of smart grid features, such as fast, reliable, secure and inexpensive communication, cost effective energy storage, and DG, DER and RES implementation.

One of the main microgrid drivers is the possibility to supply good-quality electricity without spending on building and setting expensive transmission lines. In the areas where the main grid is saturated and hence there are problems on voltage stability and peak power demands, microgrids can be the key energy market players for this energy market segment. Microgrids can increase the stability and defer large-scale electric infrastructure expensive investments. On the energy security segment, the microgrids deal with all organizations where it is strategically important to get stable and good-quality electricity without any interruptions. Hospitals, military campuses, refineries, first-respond stations and the like can potentially be islanded for longer periods in the case of main grid outages. Within this category, it is possible to define subcategories depending on the critical-ity: typically hospitals need a higher service level (a few seconds of interruption can cost a life) than industries. Energy efficiency segment motivations are the environmental concerns and profits made by selling the electricity generated by renewable energy sources. This segment may include university campuses, office buildings, small communities, etc. From a technical point of view, dif-ferent categories of microgrids can also be defined along various level of voltages. Depending on microgrid architecture, level of voltages impacted, and components included (substations, feeders, generators, etc.), automation processes and communication networks may differ greatly. Based on such criteria, microgrid types include the following categories. LV microgrids, with power range up to 1 MW, include smaller individual facilities with multiple LV loads, such as small hospitals, schools, employing low-power solar, wind energy and energy storage units, containing also a switch for islanding the microgrid. MV microgrids, with power ranges up to 5 MW or so, include smaller to large conventional CHP units with a few MV range loads, mostly controllable loads, several solar, wind energy and energy storage units. It can also be a multi LV-type microgrid system. Such microgrids also have disconnecting switches for islanding mode, with the possibility of cutting-off the feeder. Feeder microgrids, with power range up to 20 MW, consist of small to large conventional CHP units, solar, wind energy, energy storage units, several low power or a few large MV loads, containing the entire feeder with switches at its ends for islanding when needed. Feeder microgrids may be also formed by several MV microgrids. Substation microgrids, sometimes referred as mini-grids, power range above 20 MW, consist of conventional CHP, biomass, large solar, wind energy, energy storage facilities, several MV loads, and all load types, up to several tens of MW for large industrial complex loads. Substation microgrids, which may be a combination of several feeded microgrids, are equipped with controllers of the sub-stations, and with HV and/or MV transformers.

A typical microgrid structure contains physical systems, control systems and interfaces with the other utility and main grid subsystems, devices and equipment. A microgrid is composed largely of off-the-shelf physical components, not necessarily specific to microgrid applications, and most of them are the same physical systems are used by the utilities. Microgrid building blocks include various types of generators and energy storage devices, power electronics, sensors, switches, protection devices and equipment and metering equipment. DG and DES units are usually connected at either medium- or low-voltage levels to the host microgrid. Generators can take a variety of types, with the most commonly are using diesel and natural gas combustion engines, being necessary even when renewable energy systems are available because they provide consistent energy that can be relied on with relatively high certainty. Power electronics allow for DC-AC or AC-DC conversions, or DER component voltage changes. This allows DER such as energy storage or microturbine generators to be grid-connected despite non grid-conforming generating modes. Energy storage helps smooth rapid changes due to external events or microgrid DER characteristics, while providing backup power. Intelligent (smart) switches are microgrid vital parts, allowing quick reconfiguration of the microgrid components and to electrically disconnect or reconnect with the grid, section-off areas of the circuit, or bring various DER components ON or OFF line. Sensors, and more generally information input, are required to determine whether criteria for islanding or reconnecting have been met, providing the microgrid operation and management needed data and information. Protection equipment is always needed, regardless of whether or not the DER units are configured in a microgrid. Nonetheless, there are special precautions that must be taken in a microgrid because of the added level of complexity involved with becoming a separate electrical entity and bi-directional power flows. Major issues that must be accounted for are protection when disconnecting and reconnecting with the grid and ensuring that there is an appropriate level of fault detection and protection in each part of the islanded microgrid. Control systems for protection equipment will also have to be modified to fit the operating paradigm of the microgrid. Advanced and/or smart metering must be in place at the substation level and preferably in the microgrid area so that power flow conditions are monitored in real-time. Figure 11.3 shows a DG unit comprising a primary energy source, an interface medium and switchgear at the unit point of connection (PC). In a conventional DG unit (e.g., a synchronous generator driven by a reciprocating engine or an induction generator driven by a wind turbine), the rotating machine: (1) converts the energy from the primary energy source into the electricity and (2) acts as the interface between the energy source and the microgrid. For an electronically coupled DG unit, the coupling converter: (1) can provide another layer of conversion and/or control, e.g., voltage and/or frequency control and (2) acts as the interface medium with the microgrid. The input power to the interface converter from the energy source side can be AC at fixed or variable frequency or DC. The converter microgrid-side is at standard frequency, 50 Hz or 60 Hz. Figure 11.3 provides a high-level representation of a DES unit for which the "primary energy source" is replaced by the *energy storage medium*.

Table 11.1 outlines typical interface configurations and methods for power flow control of DG and DS units for the widely used primary energy sources and storage media, respectively. It should be noted that in addition to the two basic types of DG and DES units, a DER unit can be of a hybrid type, i.e. a unit that includes both primary energy source and energy storage medium. A hybrid DER

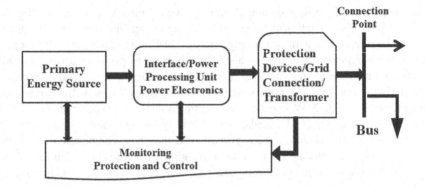

FIGURE 11.3   Block representation of a DG unit.

unit is often interfaced to the host microgrid through a power converter that includes bi-directional AC-DC and DC-DC converters. In terms of power flow control, a DG unit is either a dispatchable or a non-dispatchable unit. The dispatchable DG unit can be controlled through set points provided by a supervisory control system. A dispatchable DG unit is either a fast-acting or a slow-response unit. An example of a conventional dispatchable DG unit is a configuration utilizing a reciprocating engine as prime-mover. A reciprocating-engine-based DG unit is usually equipped with a governor for speed control and fuel in-flow adjustment. The automatic voltage regulator (AVR) controls the internal voltage of the synchronous generator. The governor and the AVR control DG unit real and reactive power outputs based on a dispatch strategy. In contrast, the output power of a non-dispatchable DG unit is normally controlled based on the optimal operating condition of its primary energy source. For example, a non-dispatchable wind unit is normally operated based on the maximum power-tracking concept to extract the maximum possible power from the wind regime. Thus, the

TABLE 11.1

Control Methods for Microgrid Distributed Generation Systems

DG Type	Primary Energy Source	Interface	Power Flow Control
Conventional DG Unit	Reciprocating Engines	Synchronous Generators	AVR and Governor
	Small Hydropower	AC or DC Generators	(+P, +/-Q)
	Microturbines	Permanent Magnet	DC-AC Converter
	Fixed-speed Wind Turbine	Generators	Stall and Pitch Control (+P, -Q)
		Induction Generators	
Nonconventional DG Unit	Variable-speed Wind Turbine	Induction Generators Power Electronics	Turbine and DC Link Voltage Control (+P, +/-Q)
	Solar PV Arrays	(AC-DC-DC Conversion)	MPPT and DC Link Voltage/
	Fuel Cell Stacks	Power Electronics (DC-DC-AC Conversion)	Frequency Control (+P, +/-Q)
Long-term Energy Storage (DES)	Battery Banks	Power Converter (DC-AC)	SOC and/or Output Voltage Control
	Compressed-Air Storage	Synchronous Generators	AVR and Governor
	Pumped Hydropower Storage		(+P, +/-Q)
Sort-term Energy Storage/Power Buffer	Supercapacitors	Power Electronics	State-of-Charge Control
	Flywheels	(DC-DC-AC Conversion)	(+/-P, +/-Q)
		Power Electronics (AC-DC-AC Conversion)	Speed Control (+/-P, +/-Q)

output power of the unit varies according to the wind conditions. The DG units that use renewable energy sources are often non-dispatchable units. To maximize output power of a renewable energy-based DG unit, normally a control strategy based upon maximum point of power tracking (MPPT) is used to deliver the maximum power under all viable conditions.

The lack of a master controller implies that autonomous controls are needed for both a peer-to-peer network and plug-and-play functionality. In addition to the increased network capabilities and functionality, if the distributed energy sources are allowed to operate autonomously in transient conditions, the system reliability increases. If the load demand suddenly increases or the system experiences generation losses, energy sources have the capability to get local on voltages, powers and frequency and to create a new operating set-point that ensure the system stability. Grid-tied microgrids can be operated in three different configurations. The first one is the unit power control configuration where each distributed energy source regulates the voltage magnitude at the connection point and the power that it is injecting. In this configuration, any load changes within the microgrid are provided power through the grid since every unit operates a constant output power. This configuration is ideal for CHP applications because power production also depends on heat demand. The second configuration is the feeder flow control configuration where each distributed energy generator regulates voltage magnitude at the connection point and the power flowing into microgrid. Extra load demands are picked up by the distributed energy sources and the microgrid looks like a constant load to the utility grid. With this configuration the microgrid becomes a dispatchable load as seen from the utility side, allowing for demand-side management. The third configuration is a hybrid of the previous two with some of the energy sources regulating their output power while others regulate the power into the microgrid. This configuration can have some units operating at peak efficiency utilizing waste heat, and other units ensuring that the power flow from the grid stays constant under changing load conditions.

11.2.2 Power Flow and Energy Management Systems for Microgrids

The microgrid concept is often defined as: *a small-scale power supply system, consisting of small electrical power and heat facilities, loads and their controller, and which manages them as a group and has one connection to a commercial power system.* Figures 11.1 and 11.2 are showing typical microgrid configurations, consisting of renewable energy generators, cogeneration facilities, electric storage units, thermal storage facilities, distribution network facilities, electrical thermal infrastructure, communication, control and protection devices, thermal and electric loads. The energy management system, which is a control device, plays an important role in a microgrid and has the following effects: (a) efficient operation of electric and thermal energy; (b) power flow control on the tie line (to protect power utilities from disturbances in the microgrid). Emerging technologies are making it possible to deploy advanced microgrids capable of integrating multiple DERs into a single system that can operate both independently from (i.e., in islanded mode) and seamlessly with the extant electric grid. By aggregating multiple loads and sharing supply resources, microgrids can take advantage of energy demand diversity, electric and thermal, in order to integrate DERs in a manner that may be more optimal than on a single-site basis alone. Due to their small scale and ability to coordinate and deliver both thermal and electric energy, microgrids may be viewed as demonstrations of the potential benefits of a smarter grid or as an alternative path to the aggrandized smart grid. Microgrids are customized solutions to the energy requirements of connected loads and as a result, it is unlikely that any two systems are using the same technologies or configuration. Important variables for determining microgrid design and technology include, but are not limited to: type, level and density of demand on-site for thermal energy, type and level of electric demand considered uninterruptible (i.e., affecting the amount of capacity that must be available at all times), local utility energy tariffs, electric grid interconnection and interaction requirements and local fuel supply.

Power grids, regardless their size, complexity, structure and the types of the generation units, must be designed, operated and controlled to provide rated voltages and required powers to the connected loads, requiring the calculations of the bus voltages from the scheduled transmission system, scheduled generation system and scheduled bus loads. A main objective of a power flow study or analysis, as discussed in Chapter 8, is to determine if a specific power system design, including a microgrid design, can produce bus voltages within required and needed limits. In a power flow problem, as discussed in Chapter 7, several bus types are defined; the three most important are a load bus, a generator bus and a swing bus. A power system bus has four variables: bus active power, bus reactive power, voltage magnitude and phase angle. For a load bus, the active and reactive power consumptions are given as a scheduled load for a given (specified) time. In the case of microgrid power flow studies, a PV (voltage controlled) bus modeling a generator bus is needed and adapted. For such bus, the injected power by the connected generator is given in addition to the magnitude of bus voltage. The reactive power injected into the network and phase angle must be computed from power flow analysis. However, the reactive power must be within the minimum and maximum limits of what the PV bus can provide. Let us consider a PV bus for microgrids with photovoltaic or wind generating systems. Such types of generating units are operated to produce active power, meaning that a PV or wind bus is operating at unity power factor, being modeled as given in Figure 11.4a. This diagram depicts the model of a PV or wind generating station connected to a bus when a microgrid is connected to the local grid. In this model, for the microgrid voltage analysis, bus active power generation is given and the reactive power generation is assumed to be zero, unity power factor. The bus voltage and phase angle is computed from the solution of the power flow problem subject to a minimum and maximum limitation as specified by the model of the PV generating station. Therefore, summarizing the PV generating model for bus k as P_{Gk} and $V_{min} < V_k < V_{max}$ as specified. However, the reactive power to be provided by a PV or wind generator must be within the minimum and maximum limits of the generating station. Considering the case when a microgrid is disconnected from the grid, the local microgrid must control its own frequency and bus voltages. When the microgrid of a PV and wind generating system is separated from the local grid, the PV or wind generating bus can be modeled as shown in Figure 11.4b. In this model, the bus voltage magnitude is specified with a minimum and maximum as defined by the power converter, active power generation and reactive powers are also specified, and the phase angle and voltage magnitude are to be computed from the power flow problem.

In a load flow study, all buses within the studied network have a designation, the load buses are modeled as a constant P and Q model where the active power P and reactive power Q are given and bus voltages are calculated, power is flowing toward loads, being represented as a negative injection into the power system network. The generator buses can be modeled as a constant PG and QG or as a PV bus, generators injecting positive active and reactive power into the network. For formulation of the power flow problem, the injected power into the power system network is of interest, and the internal impedance of generators is not included in the power system model. However, for the

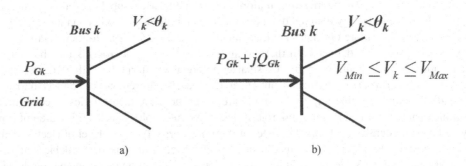

a)   b)

FIGURE 11.4   A photovoltaic or wind generating bus model: (a) grid-connected and (b) islanded microgrid.

short-circuit studies, internal impedance of the generators must be included into the system model. The internal impedance limits the fault current flow from the generators. In the model of Figure 11.4b, the bus voltage magnitude is specified with minimum and maximum voltages, as defined by the inverter modulation index, and the active power generation and reactive powers are also specified. The phase angle and voltage magnitude are to be computed from the power flow problem solution. However, a PV generating station without an energy storage system has very limited control over reactive power. To control an inverter power factor, the energy storage system is critical. To make an inverter with its supporting energy storage system must be able to provide active and reactive powers. In case of a wind energy conversion system, connected to the microgrid directly, the reactive power injection control is limited within the acceptable voltage range of the connected wind bus. For an isolated microgrid to operate at a stable frequency and voltage must balance its power loads and generation at all times. Because the load variations are continuous and the renewable energy sources are intermittent, it is essential that an energy storage system and/or a fast-acting generating unit such as high-speed microturbines, and/or a combined heat and power generating unit be part of the microgrid generation mix. For viable MG power flow, the balance of the system loads and generation must be maintained at all times, being expressed in terms of active and reactive powers as:

$$\sum_{k=1}^{N_G} P_{G,k} = \sum_{k=1}^{N_L} P_{L,k} + P_{Losses} \tag{11.1a}$$

And

$$\sum_{k=1}^{N_G} Q_{G,k} = \sum_{k=1}^{N_L} Q_{L,k} + Q_{Losses} \tag{11.1b}$$

where $P_{G,k}$ is the active power generated by generator k, $P_{L,k}$ is the active power consumed by the load k, N_G is the number of the system generators and N_L is the number of the system loads, $Q_{G,k}$ is the reactive power generated by generator bus k and $Q_{L,k}$ is the reactive power consumed by load k. To ensure the balance between load and generation, one must calculate the active and reactive power losses. However, to calculate power losses, the bus voltages are needed. The bus voltages are the unknown values to be calculated from the power flow analysis. Notice that commonly PV or wind energy sources modelled as PV buses have active power and voltage magnitude specified but in reality, renewable energy sources exhibit voltage variations over time, being more appropriate to model them as a PQ buses. Methods for solving power flow studies were fully discussed in the Chapter 8 of this book.

11.2.3 MICROGRID MODELING

The microgrid and its component models are important factors influencing and determining the power flow studies of microgrids. In grid-connected mode, DGs are usually modeled as PV or PQ buses. In islanded mode, since there is no slack bus, it is impossible to model all the DGs as PV or PQ buses. The overall stability and reliability of the MG are greatly influenced by the characteristics, types and operation of the loads. During proper load scheduling and management, the grid operating parameters widely influence the end-user energy usage, while during system failure or permanent blackout, proper control strategies must be taken by the grid operator and thereby the existing loads can be diminished to reduce the risk of overall power failure. In any studies of the power systems, load models are usually assumed to be voltage independent and, therefore, the demand of active and reactive power is assumed to be constant. Such assumptions, however, are not valid in many practical applications, particularly for microgrids where some loads are dependent to voltage or frequency. For voltage and frequency dependent loads, the active and reactive power load demand relationships are expressed as exponential equations:

$$P_{Lk} = P_{Lk0} \left(\frac{V_k}{V_0}\right)^{\alpha} \left[1 + K_P (\omega - \omega_0)\right] \qquad (11.2a)$$

$$Q_{LK} = Q_{Lk0} \left(\frac{V_k}{V_0}\right)^{\beta} \left[1 + K_Q (\omega - \omega_0)\right] \qquad (11.2b)$$

where V_0 and ω_0 are the nominal voltage magnitude and frequency, respectively, V_k is the voltage magnitude of bus k, ω is the frequency, P_{Lk0} and Q_{Lk0} are the active and reactive power of bus k corresponding to the nominal operating voltage, respectively, α and β are the active and reactive power exponents, $(\omega - \omega_0)$ is the deviation in the angular frequency, K_P and K_Q are the frequency sensitivity parameters. The values of coefficient K_P ranges from 0 to 3, and the values of the K_Q ranges from -2 to 0 depending on load type, geographic region and season. The exponent values α and β for different load types, as found in the literature, are summarized in Table 11.2. As the microgrid energy management system basis, reliable power flow analysis is critically important to unlock the potential of microgrids as primary resilience resources and enable situational awareness. For example, not only the special characteristics of the low-voltage grid pose significant challenges on the derivative-based methods, but also none of the existing algorithms is able to incorporate the hierarchical control effects in microgrids. In the case of a power system, consisting of N buses, the system admittance, Y_{Bus} is defined as a matrix representing the nodal admittances of all system buses. In case of droop control of DGs in an islanded microgrid, the system frequency cannot be considered as a fixed parameter. Since the frequency affects the line reactances, it should be taken into account for Y_{Bus} calculations. Therefore, for a system with N buses, Y_{Bus} is a function of the system (e.g. microgrid) frequency, if $Y_{ij}(\omega)$ is the admittance between bus i and j, as expressed by:

$$Y_{Bus} = \lfloor Y_{ij}(\omega) \rfloor, \ i = 1, 2, \cdots, N \ and \ j = 1, 2, \cdots, N \qquad (11.3)$$

In grid-connected mode, DGs are usually modeled as PV or PQ buses. In islanded mode, since there is no slack bus, it is not possible to model all the DGs as PV or PQ buses. So, DG units in islanded microgrids are modeled as droop buses. Most of the DG units in an islanded microgrid

TABLE 11.2
Load Types, Active and Reactive Exponent Values

Load Type	α	β
Constant Power	0.00	0.00
Constant Current	1.00	1.00
Constant Impedance	2.00	2.00
Residential Load	0.92	4.04
Commercial Load	1.31	3.40
Industrial Load	0.18	6.00
Typical Load	0.92	1.00

have as an interface a power inverter interface followed by a filter. Therefore, it is justified to assume the output impedance of the DG to be an inductive element. The complex power, $S = P + jQ$, represented by the active power and the reactive power delivered to the bus, respectively, are given by these relationships:

$$P = \frac{EV}{X} \sin(\phi) \tag{11.4a}$$

And

$$Q = \frac{EV \cos(\phi) - V^2}{X} \tag{11.4b}$$

Here, E is the inverter output voltage magnitude, ϕ is the power angle, V is the bus voltage magnitude and X is the output impedance of the DG unit. In the case of an inductive line ($X \gg R$ and $\phi \gg 1$), the relationships may be simplified to the Equations (11.5a) and (11.5b):

$$\phi \approx \frac{XP}{EV} \tag{11.5a}$$

And

$$E - V \approx \frac{XQ}{E} \tag{11.5b}$$

Alternatively, if the line resistance is much larger than is reactance ($R \gg X$ and $\phi \ll 1$, the relationships may be simplified to the Equations (11.6a) and (11.6b):

$$\phi \approx \frac{RQ}{EV} \tag{11.6a}$$

And

$$E - V \approx \frac{RP}{E} \tag{11.6b}$$

From these relationships, one may observe a direct relationship between voltage and active power and phase and reactive power. Conventional droop control is based on several assumptions such as the output impedance having negligible effect on the power characteristic (output impedance is small compared to line impedance) and there is no cross coupling in the $P - \omega$ or $Q - V$ relationships. However, these assumptions can make significantly impacts on the appropriate MG control approach and method. As discussed in previous chapters, the matrix bus admittance describing the flow of net injected current through the transmission systems, and the bus injected apparent power as given by these relationships:

$$I_{Bus} = Y_{Bus} \cdot V_{Bus} \tag{11.7}$$

And

$$S_k = V_k \cdot I_k^* \tag{11.8}$$

where S_k is the net injected complex (apparent) power at bus k, V_k is the complex voltage of bus k, I_k is the net injected current at the bus k, and the star represents the complex conjugate. Then, the bus net injected complex (apparent) power at bus from Equation (11.7) is substitute with (Equation 11.8) to obtain the residue form of the equation for each system bus k as:

$$S_k = V_k \sum_{j=1}^{N} Y_{kj}^* \cdot V_j^*, \quad k = 1, 2, \cdots, N \tag{11.9}$$

11.2.4 Energy Management in Microgrids

Unlike conventional power systems, microgrids often possess the following characteristics, properties and issues: connecting lines have a high resistive characteristic (higher ratio of R/X, often the resistance cannot be neglected), system capacity is limited by the capacity of DERs, fault currents are low (i.e. low, short circuit ratio), loads can be highly unbalanced, the system inertia is low, energy supply is prone to higher uncertainties, which is due to the intermittent and variable nature of RES. These features pose additional challenges for the operation, control and management of microgrids. For a microgrid consisting of more than two DERs, energy management system is needed to impose the power allocation among DERs, the cost of energy production, power exchange, load scheduling in the emergency case, and emissions. Optimal energy management for microgrids, including economic dispatch (ED), unit commitment (UC) and demand-side management (DSM), often without pursuing a robust formulation against RES uncertainty are used. The EMS in a microgrid is shown in the diagram of Figure 11.5. As it is shown in this figure, the forecast values of load demand, the distributed energy resources and the market electricity price in each hour on the next day are denoted as inputs. Furthermore, the operation objectives are considered to optimize the energy management, are given as follows: economic, technical, environmental and combined objective options. Based on Figure 11.5, the distributed microgrid control and computation architecture are based on the wind and solar power distributions and forecasting methods, market information, electricity cost and energy demand predictions and an ED problem is formulated to minimize the risk of overestimation and underestimation of available wind power. Several algorithms for the optimization of microgrid energy management are proposed and employed, including optimal energy management algorithms for island microgrids, e.g. ones using a rule-based management. The system operation depends on the developed rules; thus, the constraints are always satisfied, but the optimization is not global results. Fuzzy logic methods are often used to estimate the rule to improve the rule-based technique. The linear programming (LP) and mix integer linear programming (MILP) are used to find

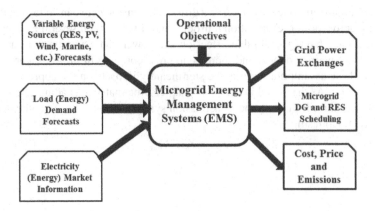

FIGURE 11.5 Microgrid energy management system's configuration.

the optimal energy management, giving good optimization results. However, the main limitation is known as the need of a specific and complex mathematical solver. In other applications, MG optimal energy management is solved by using game theory and multi-objective optimization, where the operating cost and the emission level are given as two objectives functions. Several implementations are based on artificial intelligence, dynamic programming or particle swarm optimization techniques to solve the energy management optimization problems and their inherent complexity.

For example, stochastic programming can also be used to cope with the RES variability. Single-period chance constrained ED problems for RES have been extensively studied, yielding probabilistic guarantees that the load will be served. Considering the uncertainties of demand profiles and PV generation, a stochastic program is formulated to minimize the overall cost of electricity and natural gas for a building. Without DSM, robust scheduling problems with penalty-based costs for uncertain supply and demand have also been investigated. Recent works explore energy scheduling with DSM and RES using only centralized algorithms. An energy source control and DS planning problem for a microgrid is formulated and solved using model of predictive control. Distributed algorithms are developed, but they only coordinate DERs to supply a given load without considering the stochastic nature of RES. In all the aforementioned works, however, robust formulations accounting for the RES randomness are not pursued. Often, a worst-case transaction cost based energy scheduling is proposed and employed to address the RES variabilities through robust optimization that can also afford distributed implementation. However, it considers only a single wind farm and no DS, and its approach cannot be readily extended to include multiple RESs and DES. The present paper deals with optimal energy management for both supply and demand of a grid-connected microgrid incorporating RES. The objective of minimizing the microgrid net cost accounts for conventional DG cost, utility of elastic loads, penalized cost of DES and a worst-case transaction cost. The latter issue stems from the ability of the microgrid to sell excess energy to the main grid or to import energy in case of shortage. A *robust* formulation accounting for the worst-case amount of harvested RES is developed. A novel model is introduced in order to maintain the supply-demand balance arising from the intermittent RES. Moreover, a transaction price-based condition is established to ensure convexity of the overall problem. The separable structure and strong duality of the resultant problem are leveraged to develop a low-overhead *distributed* algorithm based on dual decomposition, which is computationally efficient and resilient to communication outages or attacks. For faster convergence, the proximal bundle method is employed for the non-smooth sub-problem handled by the LC of RES.

One of the primary aims of introducing microgrids into a utility's service territory today is to promote the use of new technologies that are becoming available for both electrical based transportation and advanced electrical energy storage. Without a more localized focus, electric utilities will need to attempt to upgrade their transmission and distribution facilities (a very difficult and costly regulatory and legal proposition) in order to accommodate these new items. As anyone in electricity business knows, building new major transmission lines is next to impossible. Getting the necessary approvals for such things for such things as land siting/right away and environmental impacts involve a lengthy hearing process and numerous public meetings, requiring an extensive amount, while there is no guarantee of success. Due to that risk, utilities have increasingly shied away from new transmission line construction. Only a handful of new major construction projects have occurred over the past 40 years. The same is true for new major new generation projects. The trend in the past decades has been to build smaller steam generation plants fueled by natural gas. However, if the *intelligence* is provided in *a local microgrid* to manage the available local energy resources and serve the local demands, then utilities is able to meet the growing demands without building major new infrastructure. They will be able to provide the improved power quality needed for the new technologies as emerging. This requires utilities to plan today so that they can meet tomorrow's challenges. Figure 11.6 illustrates the basic microgrid architecture, assuming a radial electrical system with several feeders, a collection of electric and thermal loads, protection and energy management. The radial system is connected to the distribution system through a

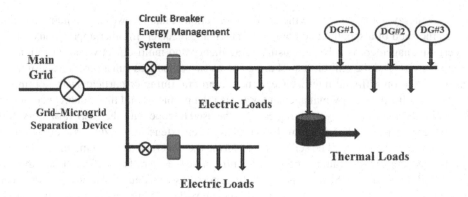

FIGURE 11.6   Microgrid architecture, control and energy flows.

separation device, usually a static switch. The feeder voltages at the load points are usually 480 V, 460 V or less, as required by the load characteristics. Upper feeder indicates the presents of several micro-sources, DG units with one providing both power and heat. Each feeder has circuit breakers and power flow controllers. Consider the power flow controller near the heat load in the upper feeder, regulating feeder power flow at a level prescribed by the local EMS unit. As loads downstream change the local micro-sources increase or decreases their generated power output to hold the power flow constant. EMS tries to provide power at least for critical loads, while non-critical loads may be disconnected in the emergency situations. Moreover, the microgrids have the potential to better accommodate new demands for electricity by shifting the electricity paradigm away from an exclusive model of centralized, remote generation to one in which DG and DER units from local, clean energy resources supplements centralized generation.

11.3   MICROGRID CONTROL, PROTECTION AND MANAGEMENT

Optimization, management, adaptive, *smart* control and protection are the core components of a smart grid EMS that is receiving data from SCADA system, performing complete data analyses and provides decision actions and commands. With smart grid advent, power electronic control devices and equipment are fully embedded into these new power systems. Conventionally, power system monitoring, control and decision-making are realized in a central control structure. A control center utilizes a centralized SCADA system to supervise and monitor the power system operation over a large geographical area and the EMS to analyze data, conduct needed optimal power flow studies and provide control actions. Such centralized control system plays a vital role in power systems, whose configuration is comparatively permanent and does not have high penetration of renewable energy and DG units. However, as part of smart grid and microgrid deployment, increased penetration of renewable energy and DG units are requiring more flexible and scalable control architectures. Moreover, the central control architecture suffers from higher cost and low reliability. Because the centralized control center requires expensive super-computing and data-storing centers, it is able to process and store massive data obtained by the meters spread over the huge system, analyze in real time and compute for possible control actions. Microgrid configurations and operating modes allow required dynamically changes. DG units may exhaust their fuel and shed, RES units may cut in or out with wind or solar energy variability, energy storage units may continue switching between discharging and charging modes, different loads may be connected or disconnected. With the participation of plug-in hybrid or electric vehicles, the extent of dynamic feature increases considerably. In microgrid context, conventional centralized structure restricts applications in terms of cost, flexibility and reliability. Distributed control is able to address these challenges and issues in a flexible and secure way by providing such merits: economic efficiency by integrating various applications, utilizing lower-cost devices, savings from the maintenance of the SCADA and EMS system, flexibility in terms of time-varying and adaptive configurations or functions and robustness by continue working in the presence of failures. Multi-agent system, smart management, adaptive and intelligent control are emerging as potential solutions that facilitates distributed control for microgrids and smart grids. However, all the microgrids in an electric network must work in a cooperative way to meet the overall load demands.

The energy management in microgrids consists of finding the optimal (or near optimal) unit commitment (UC) and dispatch of the available generators so certain selected objectives are achieved. A commonly pursued objective for stand-alone operation mode is to economically supply the local load, whereas under grid-connected mode of operation the maximization of profits is typically sought. Additional objectives such as the minimization of greenhouse gas emissions of the microgrid have also been proposed, in some implementations, applying heuristic and multi-objective optimization techniques. With regard to the architecture of the energy management system, two main approaches have been proposed to date in the technical literature: centralized EMS (CEMS) and distributed EMS (DEMS). The CEMS architecture consists of a central controller provided with the relevant information of every DER unit within the microgrid and the microgrid itself (e.g., cost functions, technical characteristics or limitations, network parameters and operation mode), as well as the information from forecasting systems (e.g., local loads, wind speed, solar radiation) to determine an appropriate UC and dispatch of the resources according to the selected objective. On the other hand, DEMS provides a market environment through the use of multi-agent systems where each microgrid agent sends buying and/or selling bids to a central microgrid operator (CMO) according to their specific needs and cost structures; the CMO then performs a binding process to determine the operation of the microgrid for the next period. In this case, a separated UC process must be realized to determine the agents that will operate in each particular period. In this paper, the characteristics of the existing EMS approaches are reviewed, to synthesize the conceptual design of a CEMS suitable for microgrids in stand-alone operation. A typical CEMS architecture is shown in Figure 11.7, where a central agent collects all the relevant information from the different microgrid actors to perform an optimization and determine the inputs of the control system for the next period. Depending on the particular resources present in the microgrid, the input variables of the CEMS can be:

- Forecasted power output of the non-dispatchable generators for the following N consecutive periods.
- Forecasted local load for the following N consecutive periods.
- State of charge of the ESS.
- Operational limits of dispatchable generators and ESS.
- Security and reliability constraints of the microgrid.
- Interconnection status.
- Main grid energy price forecasting.

Once all the CEMS input variables are gathered, a multi-stage optimization is performed to determine the optimal dispatch of the MG units according to a defined cost function, over a pre-specified timeframe. EMS output variables are the control system reference values (e.g., output power and terminal voltage) for each dispatchable DER, and binary decision variables for connecting or disconnecting loads for load shifting purpose. An additional output variable is the UC decision of the dispatchable generators, if it is required; however, this problem can be solved separately at a lower frequency than the dispatch, which is one of the main CEMS advantages. The most important issues

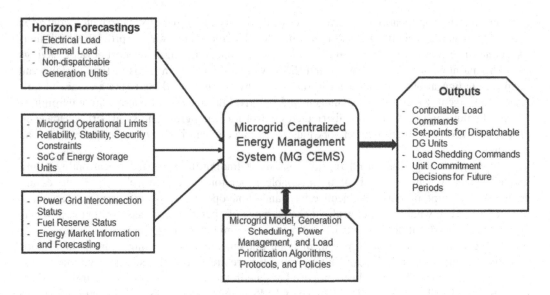

FIGURE 11.7 Typical microgrid centralized energy management system.

for DG and microgrids are the technical challenges related to control and protection of large number of micro-sources. This issue is very complex, requiring extensive development in fast sensors and complex control from a central point, which may be a potential for even greater problems. The main issue with complex control and protection systems is that the failure of a system component or a software error that can bring the entire system down. Moreover, one of the main challenges of microgrid protection system is that it must respond to both island mode and grid-connected operation mode faults. In the first case the protection system should isolate the smallest part of the microgrid when clears the fault. In the second case the protection system should isolate the microgrid from the main grid as rapidly as needed to protect the microgrid loads and equipment. DG and RES in a power distribution network or a microgrid need to be able to respond to events autonomously, using only local information and data. For any event (voltage drops, blackouts, faults, etc.) the generation units need to switch to island operation mode using local information, requiring immediate change in power output control of generators and micro-sources as they change from dispatched power operation mode to controlling voltage and frequency of the islanded network section along with loads.

A microgrid, regardless of the operation mode, is usually interfaced to the main grid by a fast semiconductor switch, the static switch, (SS). It is essential to protect a microgrid in both the grid-connected and the islanded modes of operation against all types of faults. The major issue arises in island operation with inverter-based sources. Inverter fault currents are limited by the ratings of the solid-state power devices to around two per-unit of rated current. Fault currents in islanded inverter based microgrids may not have adequate magnitudes to use conventional over-current protection techniques, requiring an expanded protection strategy. Modern microgrid are consisting of different distributed energy resources (solar PV, wind turbines, fuel cells, small-scale hydro, micro-turbines, combined heat power (CHP) systems, energy storage units, etc.), and these DER units are contributing to the system fault currents, depending on their type and characteristics. To ensure microgrid's safe operation, the protection equipment should be updated accordingly and when is needed. The microgrid dynamic structure and their operating conditions require development of adaptive protection strategies. The philosophy for protection is to have the similar protection strategies for both islanded and grid-connected operation modes. The static switch is designed to open for all faults.

With the static switch open, faults within the microgrid need to be cleared with techniques and methods that are not relying on higher fault currents. Figure 11.2 in previous chapter section is showing a simple microgrid architecture with PCC and disconnecting switch. The static switch can separate the microgrid from main grid, if needed. The microgrid energy sources needs to have adequate ratings to meet the load demands in island mode.

In order to ensure reliable operation of the LV DC microgrid, it is important to have a well-functioning protection system. As a starting point, knowledge from existing protection systems for high-power LV DC power systems, e.g. in generating stations and traction power systems can be used. However, these systems utilize grid-connected rectifiers with current-limiting capability during DC faults. In contrast, an LV DC microgrid must be connected to an AC grid through converters with bi-directional power flow and, therefore, a different protection system design is needed. Short-circuit current calculations for LV DC systems have been treated and fault detection. However, the protection devices have not been considered. So far, the influence of protection devices on the system performance has only been considered in studies of high-voltage DC applications, such as electric ships and HV DC transmission systems. An LV DC microgrid is well suited for naturally demarcated power systems, e.g. office buildings with sensitive computer loads or rural power systems, but also electric vehicles and ships. Since AC distribution is widespread and not all energy sources and loads benefit from having a connection to DC power systems, it is reasonable to consider a mixed, hybrid AC and DC microgrids. A mixed AC-DC microgrid can typically be used in systems up to a few megawatts, issue being how to interconnect different energy sources, loads and energy storage with the AC grid. However, the resulting design is often partially based on the assumption that no loads were connected directly to the DC bus. In summary, the main components used in an LV microgrid are: energy sources of various type and size, power converters, energy storage unit(s), and various loads. Photovoltaic arrays and fuel cells produce DC voltage and, therefore, are suitable to connect to a DC power system via DC-to-DC power converters. Microturbines are also preferably connected to a DC network due to their high-frequency output voltage, requiring conversion. Similarly, wind-turbine generators produce voltage with varying frequency, usually being connected through power electronic converters. Internal combustion engines or diesel engines, commonly used for standby-power generation are usually connected to AC bus. The control objectives of smart microgrids are divided in three main categories based on the reaction time of the controllers: primary, secondary and tertiary control. Primary controllers are usually fully distributed systems with a very fast response time, being responsible for the system stability. The secondary or tertiary controllers can operate with larger time steps, and with limited data-exchange due to lower sampling time.

Safety analysis and safety design concepts are very important issues in the microgrid protection and fault analysis. Proper safety models provide appropriate confidence level in protection systems and protection approaches. Intelligent control and monitoring need to meet the safety requirements, provided on basis of safety design criteria. A microgrid operational control and safety schematics is shown in the diagram of Figure 11.8. Central control and protection system should be designed to ensure required and proper personnel and equipment safety. Microgrid hazard analysis is important to design safety system and based on hazard level different safety threshold is settled. Usually, the microgrid master controller and its related field devices are communicating via either dedicated communication networks utilizing protocols such as the Internet Protocol (IP) and/or applicable wireless capabilities the needed control signals or commands. Additionally, the communication mechanisms between the microgrid controllers and the DMS, likely IP-based utilizing the existing utility communications infrastructure is used. So the dynamic microgrid structures and their various operating conditions require the development of adaptive protection strategies. One such strategy, as proposed in Figure 11.8, in which a central control unit communicates with all relays and distributed generators in the microgrid to record their status as ON/OFF, their rated current and their fault current contribution. Communicate with relay is required to update the operating current and to detect the direction of the

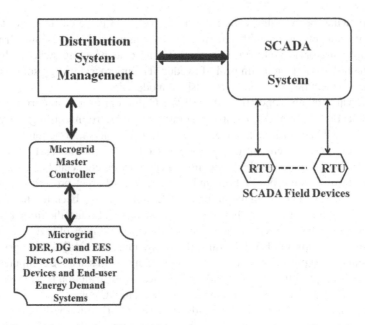

FIGURE 11.8 Microgrid master controller architecture.

fault currents and to mitigate the fault properly. The control unit also records the status of utility grid as connected or microgrid is islanded for adaptive and full protection.

Stability of microgrids is another critical topic in their operation and control. Microgrids have different stability-related problems in islanded operation mode than in grid-connected operation, because in grid-connected mode, the frequency and voltage are set by the grid, and stability issues can only be triggered by a failure of individual components. In islanded operation mode, additional stability issues can occur since the voltage and frequency regulation are performed by limited generation energy resources within microgrids. Stability issues in islanded microgrids can be categorized based on the origin, in two general groups: power-balance stability issues and control stability issues. Power-balance stability issues can happen when the power consumption and generation balance is lost, which is a critical issue in microgrids since higher shares of inter-mittent RES are connected to microgrids, and microgrids inertia is typically very low compared to conventional power systems. Therefore, in case of disturbances, e.g. loss of a generation unit or a load, large voltage and frequency oscillations can be observed, sensitive loads can perform vol-unteer tripping, further exacerbating the system power imbalance. Control stability issues refer to stability problems that can be addressed by modifying the control systems of the microgrid power electronic converters. These issues are also known as harmonic instability and can lead to sustained or increasing oscillations in a wide range of frequencies, from low sub-synchronous frequencies to the switching frequencies. Harmonic instability can happen due to various reasons. It has been shown that power electronic converters can interact with passive filters. Harmonic instability can also be triggered in parallel operation of power electronic converters with filters. The interactions happen in high frequency and often result from fast control loops of power elec-tronic converters. Moreover, the time delay induced by modulation and digital sampling process can also present negative damping in high frequency and consequently destabilizes the power electronics. Stability analysis of microgrids can be performed using two main approaches. The first one develops a microgrid state-space model in time-domain and then analyzes the stability based on the system's eigenvalues. This approach is not suitable for systems with a large num-ber of states, which makes it very difficult to trace a specific eigenvalue to a specific microgrid component and its controller parameters. The second approach is based on the impedance as seen from the component terminals, expressed in frequency-domain transfer functions and the stability is assessed by analyzing the system equivalent impedance. This approach has the benefit of scalability and reduced computation effort, and is widely used to analyze the stability of power-electronic-based systems.

11.3.1 GENERIC DESCRIPTION OF DROOP CONTROL

For power systems based on rotating generators, frequency and active power are closely interconnected. Any load increase implies that the load torque angle increases without a corresponding increase in the prime mover torque, which means that the rotational speed, and directly the frequency, decreases. The slowing of frequency with increased load is what a droop control is trying to achieve in a controlled and stable manner. Droop control methods are based on the behavior of synchronous generators in power systems. In this control technique, the active and reactive power sharing by the inverters are estimated by adjusting the output frequency and voltage amplitude, being able to avoid the needs for communication links. The main principle of droop control is to use the exchange of active and reactive power between a generator or storage unit and the grid to control the grid voltage magnitude and frequency. Grid impedance in transmission line is mainly inductive and thus grid voltage magnitude is influenced most by reactive power. The active power influences grid voltage phase angle or frequency. Droop control method is perhaps the most common distributed control method used in microgrids. Control strategy of micro-energy sources is the key to achieve stable operation of a microgrid. This microgrid control approach is usually realized through the control of power electronic inverters. This control strategy must ensure the voltage and frequency of the microgrid within the prescribed standard range and must maintain the power quality in both grid-connected and isolated operating modes. There are two effective kinds of basis control strategies of electronic inverters, constant power control (P/Q control) and voltage and frequency control (V/f control) based on droop characteristics. Basic function of the inverters in a microgrid system is to act as a connecting interface between all micro energy sources and the microgrid. Droop control is perhaps the most common distributed control method used in microgrids. Droop control can be considered as a "set-point control" where the voltage magnitude and frequency set-points change with the microgrid reactive and active powers. To understand this method, it is useful to initially approach the origin of the droop control, considering the problem of complex power transferred through a transmission line. The transmission line is modeled in Figure 11.9 as an R-L network with the voltages at the terminals of the line being held constant. Moreover, the applicability of droop control is readily apparent by considering the relationships that dictate here the power transfer. The power flowing into a power line at the terminal is described by the apparent power expressed by the following equation:

$$\bar{S} = P + jQ = \bar{V} \cdot \bar{I}^* = \frac{V_1}{Z} \exp(j\theta) - \frac{V_1 V_2}{Z} \exp(j\theta) \tag{11.10}$$

Typical transmission lines are modeled with the inductance being much greater than the resistance so the resistance is commonly neglected. The power flow equation, after some mathematical manipulation, can then be written by the well-known transmission line active and reactive power equations:

$$P = \frac{V_1 V_2}{X} \sin \delta \tag{11.11}$$

And

$$Q = \frac{V_1^2}{X} - \frac{V_1 V_2}{X} \cos \delta \tag{11.12}$$

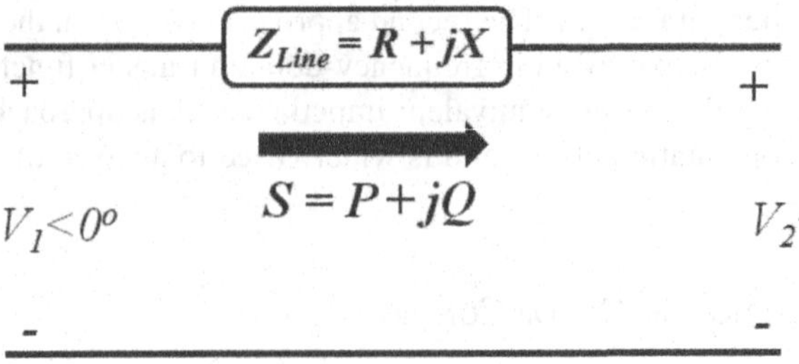

FIGURE 11.9 A transmission line parameters and power flow.

Here, X is the line inductive reactance, θ is the phase angle of the impedance and δ is the power angle. If the power angle is small, sine can be approximated by the angle in radians and the cosine by 1, the above equations can be simplified to show that the power angle depends heavily on the active (real) power and the input and the output voltage difference depends on the reactive power. Stated differently, if the real power can be controlled, so the power angle, and if the reactive power can be regulated, then the sending end voltage V_1 is controllable as well. From the power relationships, direct relationships between phase, active power, voltage and reactive power, assuming that the phase angle δ is small and the line inductive reactance is much larger than the line resistance ($X \gg R$), can be obtained as:

$$\delta \approx \frac{XP}{V_1 V_2} \tag{11.13a}$$

And

$$V_1 - V_2 \approx \frac{XQ}{V_1} \tag{11.13b}$$

The main principle of droop control is to use the exchange of active and reactive power between a generator or storage unit and the grid to control the grid voltage magnitude and frequency. So, the conventional droop control is based on several assumptions such as the output impedance having negligible effect on the power characteristic (output impedance is small compared to line impedance) and there is no cross coupling in the $P - \omega$ or $Q - V$ relationships. It is evident, however, that the assumptions made can significantly impact the appropriate control method and approach. Notice that the power transfer relationship is expressed in terms of phase and voltage differences, but droop method relies on frequency and voltage differences to transfer power. In the droop method, each unit uses the frequency, instead of the power angle or phase angle, to control the active power flows since the units do not know the initial phase values of the other units in the standalone system. Regulating system real and reactive power flows, the voltage and frequency can be determined, leading to the droop control equations:

$$f = f_0 - k_P \cdot (P - P_N) \tag{11.14a}$$

$$V_1 = V_0 - k_v (Q - Q_N)$$

$$V_1 = V_0 - k_V \cdot (Q - Q_N) \tag{11.14b}$$

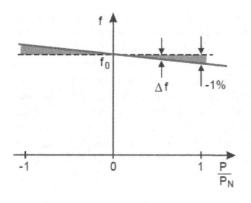 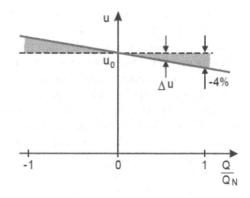

FIGURE 11.10   Drop control characteristic plots.

where f_0 and V_0 are the base frequency and voltage, respectively, and P_N and Q_N are the tempo-rary set points for the real and reactive power of the machine. In summary, a droop (proportional) control uses only local power to detect system changes and to adjust the operating points of genera-tors accordingly. The droop control uses the generator real power to calculate the ideal operating frequency. This relaxing of a stiff frequency allows the microgrid to dampen the fast effects of changing loads, increasing the system stability. A typical droop control characteristic diagram is shown in Figure 11.10.

From the droop equations and the graphical representations highlighted in Figure 11.10, as the load real power on the system increases, the droop control scheme is allowing the system frequency to decrease. However, the droop method has the inherent trade-off between the active power sharing and frequency accuracy, resulting in the frequency deviating from the nominal value. It is desirable to create a controller that is restoring frequency to its nominal value after a disturbance. Frequency restoration is not practical in systems with inverters due to the inaccuracies in inverter output fre-quency. These minor differences in inverter's frequency result in increasing circulating currents creating an unstable system. However, a system with rotating machines may lend itself to a droop control with frequency restoration. If an active power controller is built to include a frequency restoration loop, the controller is analogous to an engine or steam turbine governor. Engines are equipped with governors to limit the engine to a maximum safe speed when unloaded and to main-tain a relatively constant speed despite changes in loading. As the load varies, the speed may droop but over a usually short period of time returns to its nominal speed. Notice that droop control is basi-cally a proportional control therefore it is difficult to find a compromise between the errors needed for power sharing and stability analysis and management. The main advantages of the conventional droop method are avoid of needs of communications, great flexibility, high reliability, free lay-ing and different power ratings, while the drawbacks include trade-off between voltage regulation and load sharing, poor harmonic sharing, coupling inductances, subject to the influence of system impedance, slow dynamic response and poor integration of renewable energy systems.

11.3.2   Control Approaches and Types in Microgrids

Control strategy of micro-sources is the key to achieve stable microgrid operation, being often real-ized through the control of power electronic inverters and other power electronic convertor types. In fact, there are two effective kinds of basis control strategies of electronic inverters, constant power control (P/Q control) and voltage and frequency control (V/f control) based on droop characteristics. Basic function of the inverters in microgrids is to maintain an interface between a micro-source and a microgrid. Constant power control process is executed by controlling the active and reactive power output of the micro-sources which is equivalent to the current controlled voltage sources. By setting the reference value of active and reactive power, micro-sources can inject required power into the system. This method is generally used in grid operation mode, because the reference voltage and frequency are provided by the power grid. In this method voltage and frequency re-determined from the rate of droop characteristics and the power exchange between micro-source and load. The basic function of the voltage and frequency (V/f) control method is to sustain the voltage and frequency of the isolated microgrid, to ensure stability of the output voltage and frequency. By adjusting output power of the micro-sources, the frequency and voltage of the microgrid could be returned to the rated values, that is to say, the droop characteristic curves can be shifted left and right to maintain the frequency and voltage of the system, as detailed in the diagrams of Figure 11.10. One the other hand, the control strategies for microgrids can be based on centralized or decentralized paradigms. Microgrid control approaches are often divided into the coordinated control, supervisory control and energy management, and local or decentralized control. Coordinated control method allocates and optimizes DER power outputs, energy production costs and emissions. The forecast values of loads demand, weather, generation and the market electricity price in each hour on the next day are calculated to find the optimal DER power outputs, the consumption levels of utility grid, costs and emissions. In the second approach, the intelligent local controllers for DER units can enhance microgrid operations, by smart or optimum generation scheduling. In fact, the controllers participate to the frequency and voltage control in all microgrid operation modes, both islanded and grid connected modes. However, the basic microgrid control objective operating in island mode is to achieve accurate power sharing while maintaining close regulation of the microgrid voltage and frequency. To provide proper system operation is necessary to implement a comprehensive control strategy involving different time scales, usually referred as a hierarchical control. The hierarchical control structure spans local, primary, secondary and tertiary control, with time scales ranging from milliseconds to hours or a day. The hierarchical control scheme, having four coordinated control levels, each level is dependent on the lower control level. Zero (local) level control schemes are designed to properly regulate the currents and/or voltages of the DER units. Since the AC microgrids operate in both grid-connected and islanded modes, usually designed control scheme has two control loops, voltage- and current-controlled modes, when microgrids operate in islanded mode or in grid-connected mode, respectively. The local controllers are algorithms designed to assure the stability of lower level variables and counteract disturbances with fast response, good transient and steady-state performances. The local controllers are ensuring the system transient stability in range of milliseconds to seconds, currents and voltages references being given by higher-level controllers. The local control acts on the power electronic converters of the microgrid components. The primary control operates in a time range of few seconds, being responsible to adapt the grid operation points to a disturbance acting in the time interval that the secondary controllers need to calculate new optimal operation set-points. For smaller size microgrids, the primary control can be integrated into the local controller, in a master-slave topology. In this case, one converter is assumed to keep the grid stability, the master controller. The secondary level controller carries out the system power flow regulation taking into account the state of charge of energy storage units, and then the optimal power flow is set and controlled. Microgrid power flow is handled by sharing the load demand among the renewable energy generation units and the energy storage elements to allow for proper functioning and save battery lifetime. The secondary control provides a power reference to the grid assuring system power balance, being also related to power quality requirements and device operating limits, which constraints must be respected. The secondary and tertiary controls are typically implemented in centralized fashions, where a central entity communicates with converters through highly connected communication networks. Loss of any link in such topologies can lead to the failure of the corresponding unit, overstressing other units and potentially leading to system level instability and cascaded failures. Since future extensions add to the controller complexity, scalability is not straightforward. Distributed control has emerged as an attractive alternative as it offers improved reliability, simpler communication network and easier scalability. Notice that structurally it is desirable to extend the distributed control paradigm to the secondary or primary levels. Tertiary control deals with energy market, setting the energy dispatch scheduling, according to an economic viewpoint, taking into account negotiations between consumers (end-users) and producers. This level also deals with human-machine interaction and MG social aspects.

For improving the reliability of microgrids, it is desired to share and provide required power by the loads among different DER units. The power-sharing between DERs can differ and may be specified by different operation requirements. The power-sharing algorithms can be grouped into two categories, i.e. centralized and decentralized. Since decentralized approaches benefit from the plug-and-play feature as well as higher reliability, they are of great interest and often employed in practical applications. Centralized control relies on communications from and to each power electronic converter by a central controller that issues command signals to each power converter in a given microgrid. While a variety of algorithms and implementation strategies exist, a central issue with this approach is that it is not robust. If the centralized controller fails or any of the communications links fail, the entire system may fail. Conversely, the decentralized control approaches rely only on the local measurements, monitoring and sensing. The decentralized approach makes the system self-healing, this is to say that it can adapt to any number of configuration changes, and is therefore much more robust. The primary concerns with decentralized control are maintaining system stability over a wide operating range and limiting voltage and frequency excursions for perturbations in the system. The conventional droop algorithm is a decentralized power-sharing and control approach. This approach tends to emulate the behavior of synchronous generators by reducing the frequency in proportional to active power demand. However, it cannot be used as an effective method in low-voltage distribution systems. This is due to the fact that low-voltage distribution systems possess resistive characteristics rather than inductive characteristics, and unlike inductive grids, the active power distribution can be related to the magnitude of voltage. In this case, the inverse droop characteristic (approach) is used, which is formulated as:

$$f = f^{ref} + k_{Q,i} m_i \cdot Q_i \tag{11.15a}$$

And

$$V = V^{ref} - k_{P,i} n_i \cdot P_i \tag{11.15b}$$

where f^{ref} and V^{ref} are the microgrid nominal frequency and voltage magnitude, respectively; m_i and n_i are droop coefficients of the ith energy source defined by maximum permissible deviation from nominal parameters, $k_{Q,i}$ and $k_{P,i}$ are correction terms, often assumed equal to 1. The reactive and active power variables Q_i and P_i are obtained by passing the instantaneous reactive and active power of ith power source through a low pass filter. The purpose of the secondary control level of the hierarchical control is to compensate for the voltage drop and frequency deviation from their nominal values. This control level is typically slower than the primary control. Herein, the voltage magnitude set point V_0 and frequency set point f_0 of DERs are modified to account for the voltage and frequency deviations. This is equivalent to shifting up/down the droop characteristic lines. The tertiary control level is responsible for controlling the power exchange between the microgrid and the main grid. Hence, this control is generally operational only in the grid-connected mode of the microgrid. However, the role of the tertiary level can be modified to achieve the economical operation of microgrids in either grid-connected or islanded operation. The local and primary control generate the references for lower level power converters, while the secondary control deals with power flow regulation and the tertiary control covers the energy dispatch and energy market. With the advent of smart metering, development of smart grids with self-healing and high reliability, the real-time pricing becomes possible. With the real-time pricing feature, the grid operators can decide wisely about the economic operation of the power grid, which is conventionally known as the economic dispatch (ED) problem in power systems. With the real-time pricing help in microgrids, the

ED problem can be addressed for shorter time intervals, which in return leads to better economic operation. Moreover, the ED problem in microgrids results in less computational complexity, since in microgrids the numbers of resources are often limited, and some of the constraints (e.g., ramping rate, minimum up or down times) do not exist, meaning that the tertiary control level can be implemented easier.

In modern microgrids, there are used two other common control methods or strategies: master-slave control and peer-to-peer control. Master-slave control and peer-to-peer control each has advantages and disadvantages. They are suitable for different operation. There is a main control unit in master-slave control to maintain the constant voltage and frequency. The main control unit adopts V-f control while other distributed generations adopt P-Q control to output certain active and reactive power. Each unit is equal in peer-to-peer control. Peer-to-peer control is based on the method of external characteristics of declining. It associates frequency versus active power and voltage versus reactive power, respectively. Through a certain control algorithm, the voltage and frequency is adjusted automatically without the help of communication. The control method based on drooping characteristics is widely used in peer-to-peer control. As discussed before, one is f-P and V-Q droop control which produces reference active and reactive power of distributed generation units by measuring the system frequency and amplitude of output voltage of distributed generation systems. Conventional master-slave control can be considered a hybrid of centralized and decentralized, in which the master inverter utilizes droop control to set the voltage and frequency bus set-points. The master inverter communicates with the slave inverters via a current calculator which generates reference currents so that the master and slave inverters share the appropriate load currents. The centralized calculation of command signals and single loop control used in slave inverters simplify the microgrid control design. The disadvantage of the master-slave control is that it relies on calculating current references and communicating with all the slave inverters. This critical calculation and communication link make the master-slave approach less reliable than a system using inverters with droop control. There are a variety of the methods that are made to the conventional master-slave approach, allowing one of the slaves to become the master if the master or its controller fails. Such selection may be made according to a rotating priority or power capacity. However, these approaches suffer from greater complexity and lower redundancy when compared to a purely decentralized approach.

11.3.3 Microgrid Protection Technical Challenges and Issues

The most important issues for DGs and microgrids are the technical difficulties and challenges related to control and protection of a significant number of energy micro-sources and energy storage units, a very complex, requiring extensive development in fast sensors and complex control from a central point, which may be a potential for greater problems. The major issue with such complex control and protection systems is that the failure of a system component or a software error brings the entire system down. On the other hand, one of the major microgrid protection challenges is that it must respond to both island and grid connected faults. In the first case the protection system should isolate the smallest microgrid area when clears the fault. In the second case the protection system should isolate the microgrid from the main grid as rapidly as needed to protect the microgrid loads. DG and RES units in a microgrid need to be able to respond to the events autonomously by using only local information. For any event (voltage drops, blackouts, faults, etc.) the generation systems need to switch to island operation mode using local information, requiring immediate change in power output control of the generators and energy storage units as they change from dispatched power operation mode to controlling voltage and frequency of the islanded network section along with load followings. A microgrid is usually interfaced to the main power system by a fast semiconductor switch called static switch. It is essential to protect a microgrid in both the grid-connected and the islanded modes of operation against all types of faults. The characteristics of most protective devices used in microgrids are usually similar to those used in distribution networks, which are based on large fault currents. However, in island operation, the fault current is from the inverter-interfaced DERs, which generally can only provide very limited fault current, e.g., only about 20% above their rated current. As a result, conventional overcurrent protection schemes may be no longer applicable. One of the major issues arises in microgrid island operation with inverter-based energy sources. Inverter fault currents are limited by the ratings of the solid-state devices to around two per-unit of the rated current. Fault currents in islanded operation mode of the inverter based microgrids may not have adequate magnitudes to allow the use of the conventional over-current protection devices and technologies.

A general requirement of the DG units is to cease energizing the circuit to which they are connected before any reclosing attempt of that circuit. Automatic reclosing is widely used in power distribution networks to clear faults and to restore the supply with minimum operation impacts. Due to the common radial design in conventional feeders, where the main grid is the only power source, no voltage or synchronization check supervisions are needed to perform the reclosers. The presence of DG installations involves changes on such basic designs. The assumption that the power system is the only feeder energy source is invalidated by the presence of distributed generators, also energizing the network. In order to avoid significant damages, all DG protective devices are required to be carefully coordinated with the network recloser to which they are connected. If a temporary fault occurs in the feeder and the DG units do not trip off and extinguish the fault arc prior to the circuit reclosing attempt, the reclosing attempt is unsuccessful and network automatic restoration may be jeopardized. However, MG operating philosophy is that under normal conditions the microgrid operates in the grid-connected mode, while in case of a disturbance in the grid, a seamlessly grid disconnection at the PCC is performed and the microgrid operates as an isolated electric network. After fault clearing, the microgrid is switched back to grid-connected mode. It is essential to protect a microgrid in both operation modes and against all fault types. The microgrid protection philosophy is to assure safe and secure operation of its subsystems in both operation modes. However, the two operating modes pose challenges in protecting microgrids. Therefore, two sets of protection settings is the most probable solution to the dual operation modes. During grid-connected operation, mains can supply large fault currents, making possible the employment of the existing protection devices. However, the protection coordination may be compromised or even lost in some cases due to the DG presence. Moreover, large fault current contribution from microgrid cannot always be expected, especially when electronically coupled DG units dominate. Thus, conventional overcurrent protection use in the microgrid is no longer valid due to this low short-circuit currents from the micro-sources. On the other hand, the major issue arises in island operation with inverter-based sources. Inverter fault currents are limited by the ratings of the silicon devices to about twice of the rated current. Fault currents in islanded inverter based microgrids may not have adequate magnitudes to use traditional over-current protection techniques, requiring an expanded protection strategy.

Microgrid protection functions are expected to detect all fault types in a microgrid for both operation modes (islanded and grid-connected). A fast semiconductor switch is usually employed to connect the microgrid to the main grid, and the basic approach to protection is to disconnect the static switch for all types of fault, including main grid faults and microgrid faults. Most conventional feeder protections are based on short circuit current sensing. Over-current protection devices detect faults on the main power grid, but power electronic-controlled micro-sources cannot provide high enough levels of fault current. New algorithms are needed to detect microgrid faults. Furthermore, energy can flow in either direction through protection system sensing devices, making it more like a transmission line than a feeder. There are no bi-directional power flows on most of today's radial systems. Harmonics generated by power electronic devices and uncontrollable energy sources create further challenges. Relays should adapt to the intermittent energy source changes, and controls must limit harmonics. MG protection likely has more cross-content with grid control to achieve flexibility and reliability. It often depends on advanced communications technologies, following the development of microgrids, expecting reliable protection schemes. The microgrid concept faces several challenges in various areas, not only from the protection point of view, but also from the control and dispatch perspective. Nevertheless, due to their specific characteristics and operation, microgrid protection systems have to deal with technical challenges of diverse MV and LV generation systems, bi-directional power flows, grid-connected and islanded operation modes, LV network topological changes, due to connection or disconnection of generators, energy storage units and loads, generation intermittence of many of MG micro-sources (e.g. wind, PV solar), increasing uses of rotating machines, which may cause fault currents exceeding equipment ratings, insufficient short-circuit current level in islanding operation mode due to DG power electronic interfaces, reduction in permissible tripping times when faults occur in MV and LV networks in order to maintain the MG stability, and protection nuisance tripping due to faults on adjacent feeders. Many of the micro-sources are usually connected to the microgrid by a power electronic inverter, either because their output is not compatible with the grid voltage (PV panels, microturbines, etc.) or because of the flexibility provided by power electronics in the energy extraction management. Due to the low thermal inertia of semiconductor switches, inverters are actively current limited and, because of their low fault current contributions lead unavoidably to different problems that have to be considered by the protection system design:

1. Fault characteristics of the inverters may not be consistent with existing protection devices;
2. Throughout the whole microgrid, there may be different inverters with different characteristics;
3. Even in the individual inverter case, its characteristics may differ depending on design or application;
4. Difficulties in characterizing inverter behavior for short-circuit studies, set by control strategies; and
5. Significantly reduced fault current level when changed from grid-connected to island operation mode.

The characteristics of most protective devices used in microgrids are usually similar to those used in distribution networks and the protection of traditional distribution networks are based on large fault currents. However, under islanded operation, the grid cannot contribute to the fault and, therefore, its magnitude is limited to what the micro-sources provide. Consequently, conventional overcurrent protection schemes may be not applicable due to the inverter current limitations. Power flow in networks without DG unit is uni-directional, from meshed HV transmission networks down to radial LV distribution networks. However, the addition of large number of DG units, operating in parallel with existing grid, lead to a bi- or multi-directional power flow and a dynamic reconfiguration of short-circuit current sources, introducing nonlinear behaving power electronic interfaced generators, requiring new or redefined control and protection strategies. The MG control and protection system must be designed to safely operate the system in both operation modes. This system may be based on a central controller or embedded as autonomous parts of each DG unit. When grid is disconnected the system must control the local voltage and frequency, provide or sink instantaneous real power difference between generated and load consumed active power. There are special requirements for MG protection philosophy. First, the system is more likely be double end source systems, requiring fast fault clearance time; therefore, the one end single source time grading overcurrent protection cannot be applied. Second, the system is frequently switched between double and single end with possible much low fault currents, therefore protection adaptivity is required. Third, on one hand, a microgrid can operate in parallel with distribution network; on the other hand, it can operate as an autonomous entity. Therefore, IEEE Standard P1547 emphasizes the successive operation of the island in cases of power blackout. Whatever the case, the islanding must be detected quickly whether DGs are disconnected from microgrid or other control strategies are adopted. The most common technologies applied to microgrid protection, either in grid-connected or in island mode of operation, are adaptive protection schemes, voltage based methods, differential protection methods, distance protection methods and overcurrent protection techniques using symmetrical components.

The two common operation modes of microgrids may cause misoperation of conventional protection schemes in power distribution. The area, critically affected by DG penetration is the power distribution protection coordination. Traditional overcurrent protection schemes are designed for radial distribution system with unidirectional fault current flow. However, connection of DGs into distribution networks convert the singly fed radial networks into complicated ones with multiple sources. This changes the flows of fault currents from uni-directional to bi-directional. Furthermore, the use of traditional protection schemes to protect microgrids in islanded mode is also impractical. Islanded microgrid makes renewable energy generators keep working when there is a blackout. Although islanded microgrid has many advantages, it still has critical problems on their implementation. First, an effective method to detect faults occurring in power line is essential, so that the microgrid can enter the islanded mode in time. Second, since the renewable energy generation systems might not have enough power to support all the loads, a load shedding procedure to ensure the high-priority loads keep working so that the system can still operate is necessary. Third, when a MG operates at islanded mode, the fault current is much lower than that in grid-connected mode, so a protection system can detect faults rely on low fault currents is also required. Connecting generators to a distribution grid or microgrid changes its properties significantly. Voltage profiles and dynamic behavior are altered. The short-circuit power increases and current paths become complicated. Moreover, in a microgrid case, the short-circuit power may drop dramatically through the disconnection from a rather stiff grid. As a consequence, classical protection techniques may become inadequate. Problems concerned with selectivity losses, overcurrent protection level, earth-leakage protection, disconnection of generators, islanding and single-phase connections are occurring. The DG power output is often unpredictable, or even stochastic, e.g. wind turbines or photovoltaics, so the grid behavior when a fault occurs, changes constantly. Interaction with neighboring grid sections causes problems as well. Three types of generators have to be considered: synchronous, induction and inverter-interfaced generating units, each with very different properties. The protection has to depend on the local grid states, e.g. instantaneous production and consumption levels. Protection parameters have to be updated frequently. Protection of local grids during MG intentional islanding has to be investigated as well. The most important protection characteristics with a short description are listed below:

- *Selectivity:* System protection is selective if only the protection device closest to the fault is triggered to isolate the fault. If this takes too long, protection at a higher level takes over, restricting interruptions only to faulty components. Without DG units, power flows unidirectional, during normal operation or when faults occur, allowing creation of a selective system by applying time grading to overcurrent relays.
- *Overcurrent Protection and Earth-Fault Protection:* The presence of generators reduces the fault current or over-current detected at the beginning of a feeder or the fault current supplied by other local generators. Because of the generator's contribution to the short-circuit current, the voltage drop over the feeder section between the generator and the fault increases, which results in a lower fault current from the grid. If this reduction is sufficiently large, currents detected at various points are too low to trigger fast disconnection. This can result in prolonged over-currents or earth faults.
- *Protective Disconnection of Generators:* generators have to be protected against internal and external short-circuits, over- and under-voltages, unbalanced currents, abnormal frequencies, harmonic distortions and excessive torques. The grid itself and people have to be protected as well. Depending on the fault location, the protection mechanism of a generator should use a different time delay, ensuring selectivity. If there are one or more protection devices between a generator and a fault, these devices should be given the opportunity to disconnect the fault. If the fault is close to the generator, fast disconnection of this

generator is required. If generators can be disconnected quickly, classical selectivity can be restored.

- *Islanding – Microgrid Operation:* according to many grid codes, disconnection of generators is required to prevent unintentional islanding. The safety main concern refers to people and equipment. However, MG operation through intentional islanding for different periods has to be considered as an alternative. It can drastically increase the reliability of local grids because power delivery is independent of the grid state. However, the behavior of such a microgrid when a fault occurs, is completely different because of the changes in short-circuit power from the main grid, requiring adapted protective measures.
- *Single-Phase Connection:* certain generation units inject single-phase power into the distribution grid, e.g. small photovoltaic systems or Stirling engines. This affects the balance of the three-phase current, resulting in increased current in the neutral conductor and stray currents in the earth. This current should be limited to prevent overloading and to assure the safety of persons.

11.3.3.1   Existing Microgrid Protection Techniques

One of the major benefits of DG technology is the possibility of improving the reliability and continuity of energy supply by making possible that a network section operates autonomously, in stand-alone mode, during main grid power outages. For this purpose, there is updating in regulations to ensure that DG units supply adequate short-circuits current. Moreover, in order to improve system reliability and stability, fault-ride-through capability requirements for DG units connected in HV and MV electric networks have been introduced. For the LV level similar guidelines are also being considered. In designing a protection system of any power system, the protection requirements that must be considered are reliability, speed, selectivity and cost. The protection system main function is to quickly remove from service any of the system components that started to operate in abnormal manners. Other functions required are the safety and safeguard of the entire system, the supply continuity, the damage minimization and repair cost reductions. When a fault occurs, the protection system is required to disrupt fewest possible system sections. Protective devices' selective operations are ensuring the maximum service continuity with minimum system disconnections. Conventional power distribution systems are usually designed to operate radially, where only a single energy source is present with uni-directional current flows from the higher voltage levels at the substation, through the distribution feeders and laterals at lower voltage levels, to the users. The system relies on a simple and inexpensive protection schemes, consisting of fuses, reclosers, circuit breakers and overcurrent relays. An example of a conventional distribution system with protective devices is shown in Figure 11.11. Circuit breakers and reclosers are normally installed at the main feeder to allow clearance of temporary faults before lateral fuses blow. In normal condition, they are equipped with inverse time overcurrent relays which normally installed at the substation where the feeder originates. The protective devices are coordinated to operate according to criteria of selectivity based on current or time so as to ensure the device nearest to a fault will operate first. The basic criteria that should be employed when coordinating time or current devices in distribution systems are:

1. The main protection should clear a permanent or temporary fault before the backup protection operates, or continue to operate until the circuit is disconnected. However, if the main protection is a fuse and the back-up protection is a recloser, it is normally acceptable to coordinate the fast operating curve or curves of the recloser to operate first, followed by the fuse, if the fault is not cleared; and
2. Supply loss caused by faults is restricted to the smallest system part, for the shortest time possible.

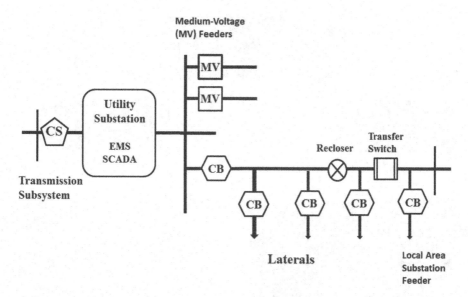

FIGURE 11.11 Typical distribution system with protective devices.

In a conventional power distribution, the protection schemes are designed assuming unidirectional power flow and are usually based on overcurrent relays with discriminating capabilities. Usually, for any fault situation, DG sources connected to the system are tripped off. In other words, islanded operation of DG sources is not allowed. When a microgrid is created in a distribution subsystem, the configuration becomes a complex multi-source and load power system. MG protection philosophy is to assure safe and secure operation of the subsystem in both operation modes. However, these operating modes pose challenges in MG protecting. Therefore, two sets of protection settings is the most probable solution to the dual operation modes. During grid-connected operation, the mains supply large fault currents, making it possible to employ existing protection devices. However, the protection coordination may be compromised or even entirely lost in some cases due to the DG presence. On the other hand, such large fault currents from the microgrid cannot usually be expected, especially from the electronically coupled DG units. Thus, the use of conventional overcurrent protection in the microgrid is not usable due to this low micro-source short-circuit current contributions. The MG protection must respond to both main network and microgrid faults. If a fault occurs on the main grid, the desired response is to isolate the microgrid from the main network as rapidly as necessary to protect the microgrid loads. This caused islanding of the microgrid operation. If a fault occurs within a microgrid, the protection system is required to isolate the smallest possible MG faulted section to eliminate the fault. Various possible microgrid protection schemes and coordination techniques that are available from the literature are summarized as shown in Figure 11.12. The protection schemes can be divided into overcurrent-based, voltage-based, current component-based, harmonic content-based, fault current limiter-based and current traveling wave-based. As for protection coordination techniques, time-current grading and optimization algorithms are used to ensure selectivity of the protective devices. The functionality and reliability of different protection schemes are required to fulfill different mode of operations, different types of DGs that exist in a microgrid, different issues that it has to tackle like power flow bi-directionality and DG locations.

11.3.3.2 Overcurrent, Modified Overcurrent and Adaptive Protection Schemes

For a fault on main distribution network, overcurrent relay protection and balanced earth fault protection are installed at the grid side of the network between the main and microgrid, with the capability of inter-tripping the microgrids. For a MG fault, overcurrent relay protection and residual

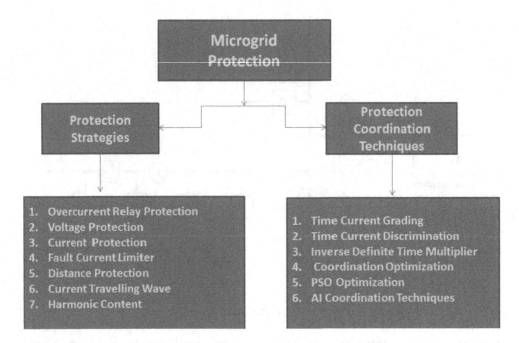

FIGURE 11.12   Different types of protection schemes for microgrid operation.

current device (RCD) are used to protect the feeder from such faults. After a fault occurs, the protection scheme disconnecting the feeder from microgrid, has also the capability of simultaneously inter-tripping all the micro-sources on the feeder. For a fault at the residential user, short-circuit protective devices (SCPDs), e.g. miniature circuit breakers, fuses and RCDs are installed at its grid side. The SCPD is used to protect the residential consumer against the phase-to-phase and phase-to-neutral faults while the RCD is used for phase-to-ground or earth fault. However, the use of conventional over-current protection scheme gives rise to several problems, such as less sensitivity, lack of operation and mal-operation. There are attempts to enhance the performance of conventional overcurrent protection schemes, by using to measurements and calculations with symmetrical components, as possible solutions for fault detection in islanded microgrids based on the measurements of current symmetrical components. For example, zero-sequence current detection in the event of upstream single line-to-ground (SLG) fault, coordinated with unbalanced loads and negative sequence current for line-to-line (LL) faults are often used. Conventional overcurrent method deficiencies to protect microgrids due to bi-directional current flows have led to the uses of directional elements in the overcurrent devices. The differential current components are used to detect fault, occurring up-stream the protection zone and symmetrical current components, zero- and negative-sequence current components to detect SLG fault in down-stream protection zones and LL faults in all protection zones. The directional over-current relay with communication for microgrid protection is also used. The relays are capable of detecting and isolating the MG internal and external faults. However, such protection schemes require relatively high investments compared to the conventional protection systems. While other methods use the directional comparison and current comparison pilot protection based on two terminals information to distinguish internal from external faults without coordination with other protection devices. A master-slave concept of the protection devices has been used to protect the microgrid through communication the fault direction, so its position could be identified.

A new approach for protecting LV microgrid by using programmable microprocessor-based relays with directional elements is often employed. This scheme does not rely on communication link and is fairly independent of fault current magnitudes and MG operation mode, being adequate for both MG operation modes. The use of symmetrical components for all types of faults (asymmetrical and symmetrical ones) in microgrids was also researched and implemented. However, the improved overcurrent method with directional element protection schemes still has limitations, being only suited the some of the protection requirements to some extent due to the large inclusion of DG and RES units and uncertainties of connecting and disconnecting of DG units from the main grid, power distribution network topology and protection setting values that change frequently. Communication-assisted protection selectivity strategies with three structural levels and with voltage-restrained directional overcurrent protection approach was also studied and applied. It is worth to be noted that the main problem in such protection schemes is related to the requirements of extensive communication system. For such protection systems, in the communication system failure event, the whole overcurrent protection and coordination may be jeopardized. Another method using a differential current protection scheme for inverter-based DG units is often used in some microgrids. Current sensors at both line ends can determine whether a relay must send or not a tripping signal to the circuit breakers. Schemes using differential protection and communication-assisted digital relays are also employed in microgrids. The synchronized phasor measurements and microprocessor-based relays were used to detect all types of fault conditions, including high impedance faults. Primary protection for each feeder relies on instantaneous differential protection. If absolute values of two samples are found to be above the tripping threshold, a tripping signal is sent to switching devices. Notice that connection of DG units to a distribution network increases the fault current closer to the ratings of the protection devices and can disturb the overcurrent protection coordination. The existing protective devices can be replaced with higher rating devices but such choice is quite expensive. One way to overcome overrating of protective devices is to limit the fault current to acceptable levels. The fault current limiter device is connected in series with power lines, limiting the fault current contributions from the DG units, while contributing small impedance additional power losses under normal operations. These approaches are mainly based on the use of adaptive relays, which can have their settings, characteristics or logic functions changed on-line, in a timely manner, by means of external signals or control actions. As a result of different microgrid protection approaches, the main adaptive protection approaches are:

1. The need for prior knowledge of all possible microgrid configurations;
2. Running extensive power flow and/or short-circuit calculations when a topology change is detected;
3. The need for communication infrastructure may be high;
4. The necessity to upgrade many protection devices, currently used in the existing power system.

Global positioning system (GPS) based adaptive protection schemes for power distribution networks with high penetration of DG units are often employed over last two decades. The power distribution network is divided into a number of zones by the circuit breakers according to a reasonable power balance of the DG units and local loads. The main relay at substation has the capability to store and analyze large amounts of data and to communicate with the zone circuit breakers and the DG units. The measurements of the GPS-based adaptive protection are synchronized current vectors of all three phases from every DG, main source and current directions of the zone breakers are collected and analyzed. Synchronized vectors are obtained using GPS-based phase measurement units. In normal operating conditions, the sum of all current phasors is equal to the total network load. In case of fault condition, the sum is significantly larger than the total network load. However, a novel adaptive protection scheme using digital relaying and advanced communication technique, based on a centralized architecture where protection settings are updated periodically by the microgrid central controller with regards to the microgrid operating states, by using numerical directional relays with directional interlock capability to selectively protect microgrids is often used. An adaptive fault current protection algorithm for inverter-interfaced DGs based microgrid was developed and implemented. By calculating the system impedance, the method adaptively changes the setting value of the protection to adapt to the grid-connected or islanded operation modes, as the operating modes affect the fault component. A comparison is made between the system impedance and the microgrid side impedance such that current instantaneous protection can automatically adjust settings. Recently, a novel adaptive protection scheme based on communication using IEDs is also employed. This method depends on the information available through measurement of current flow direction and voltage from multiple measurement locations, being implemented in two phases involving the detections of the fault condition, based on under-voltage and faulted zone based on current flow information from IEDs.

11.3.3.3 Voltage-Based and Differential Protection Methodologies

This approach mainly uses voltage measurements in order to provide an adequate protection system in microgrids, monitoring the micro-source output voltages, converting the measured signals from the a-b-c three-phase sequence operating frame to DC quantities to the d-q frame in two steps. To differentiate between in-zone and out-of-zone faults, a communication link is used between the relays. This technique also requires a decision-making procedure for the comparison of the average voltage values in each relay. They are also proposed to utilize the total harmonic distortion (THD) to improve the protection system in microgrids with inverter-interfaced micro-sources, for ground faults. After identifying the type of fault by monitoring the variation of the fundamental frequency (50 Hz or 60 Hz), voltage THD of different feeder relays is analyzed to determine the faulted zone. In order to avoid the difficulty of the previous methods associated with detecting the oscillation waveform of the voltage variation, instead of using the voltage magnitude, it was also proposed to use only its positive sequence. Besides, they conclude that, due to the different processes that have to be implemented, the total detection time might eventually be affected. In the same year, there was a claim that a distinction among the three fault types can be made only considering the direct and inverse voltage components, without using the homo-polar information. Additionally, the authors propose a microgrid protection strategy based on voltage and current measurements of the fault. A very similar approach consists of determining the fault occurrence and fault zone, based on a busbar voltage measurement and its transformation from a-b-c coordinates to d-q coordinates is also employed. However, the main problems that may be faced when dealing with voltage-based methods are:

1. Minor differences in voltage drop among the relays located at both ends of short lines lead to protection operation failures, due to reduction of the voltage gradient;
2. Relatively high calculation complexity when it comes to Park's transformation application;
3. Problems in detection of high impedance faults;
4. Problems with practical application of some of these methodologies, as well as with communication infrastructure, when high number of DG units are present;
5. Methods may be strongly dependent on the network architecture and on the definition of *protection zone* for the relay associated with each generator.

A protection scheme based on DG output voltage measurement can be used to detect and clear faults, in which the output voltages are monitored and transformed into DC quantities using the d-q reference frame. Any disturbance at the DG output due to a fault on the network is reflected as disturbances in the d-q values. Protection zones and communication link use between relays to aid the protection scheme in discriminating between in-zone and out-of-zone faults are also considered. Verification via simulation has been done for different fault locations and different types of faults. However, the proposed scheme is devised for islanded operation only without considering the grid-connected mode of operation and high impedance fault. The use of voltage based protection scheme for islanded microgrid dominated by inverter-based DG units is also used. In this scheme, a new fault estimation method based on detecting the positive sequence component of the fundamental voltage was used to judge the fault location and fault types in a microgrid. The waveforms of the three-phase voltages and the voltage magnitudes under symmetrical and asymmetrical fault conditions were transformed into the d-q reference frame and compared with the amplitude of the fundamental positive sequence voltages in the d-q coordinate system. The approaches outlined below base their performance on some kind of comparison between measurements in different parts of the microgrid. A combined methodology for microgrid protection based on differential protection and analysis of symmetrical components was also proposed, based on a network zoning approach, where the relay dividing each zone uses differential protection to detect single line-to-ground (SLG) faults that occur in its down-stream zone. To address the main challenges of microgrids, voltage and frequency control and protection are employed. Each line is equipped with two current transformers (at opposite ends of the line), and once a previously specified threshold is exceeded, the differential relays are designed to operate. The protection is additionally coordinated with the islanding detection algorithm, while a protection scheme for multi-phase faults in microgrids with inverter-interfaced generators, considering both amplitude and direction of the measured currents. Another protection scheme based on relays with a communication overlay, considering both radial and meshed microgrid architectures, also addressing the problem of high impedance faults was proposed. Inspired by the previous method, a PLC communication-based methodology aimed at meshed microgrids, whose protection is substantially different compared to the radial-configured ones. Apart from analyzing the differential protection in all the elements involved in the microgrid, the authors propose three different levels of protection: primary, secondary and tertiary or backup protection. A new differential approach to improve protection in meshed inverter interfaced microgrids, as well as those with radial configuration. Not only do they focus on feeder protection, but they also offer solutions to protect other subgroups (buses, DG units, etc.). Finally, a new approach to a differential protection for microgrids is proposed, based on a differential energy based protection scheme, which is less sensitive to synchronization errors than the conventional approach based on differential current. These methods have some implementation problems or difficulties as listed below. Need for communication infrastructure that may fail at some point, leaving the microgrid unprotected. For this reason, some authors provide different levels of backup protection. There are also needs for synchronized measurements and problems due to transients when connecting and disconnecting DG sources, unbalanced systems or loads, and due to relatively high cost.

Distance protection scheme is another solution for converter-controlled, e.g. having Mho characteristic with two protection zones is used. Zone settings are chosen such that Zone-1 covers 80% of the protected line and Zone-2 covers the whole protected line, plus 50% of the adjacent line. In this method, the fault currents in the faulted phases are limited by reducing power converter output voltage. Next, by analyzing fault characteristics, the sequence currents and voltages at the relay locations are calculated. Simulations were done for grid-connected and islanded modes of operation for different types of faults at different locations with changes in fault resistance and load conditions. However, the protection scheme effectiveness is still not fully proven. It is also suggested the use of distance relaying for multi-source systems so as to reduce complication in impedance-based setting of distance relays. Three zones of protection have been considered to protect transmission lines in this method. Other protection techniques are using the admittance or impedance measurements in order to detect the fault and set the trip. The most used method in this category is based on a new admittance relay with inverse time tripping characteristics and inverse time admittance (ITA), capable of detecting faults in both grid-connected and islanded operation modes. Apart from adding inverse time characteristics to each zone of protection, it also has the ability to isolate the faults occurring at either side of the protected circuit, since it can also operate for reverse faults. However, the reach settings should be different for forward and reverse faults. Nevertheless, there are a number of shortcomings when it comes to an accurate performance of the ITA relay are:

1. Limited fault resistance that can be reliably detected;
2. Errors in the measured admittance because of the fault resistance;
3. Increasing tripping time because of the downstream source in-feed;
4. Loss of accuracy due to problems in fundamental extraction, caused by harmonics, current transients and decaying DC magnitude and time constant.

11.3.3.4 Use of External Devices for Protection Improvement

In situations when fault current levels are drastically different between the grid-connected and the islanded operation modes (typically with inverter-interfaced DG units), the design of an adequate protection system, performing properly in both modes is a challenge. In this regard, there is a possibility of applying a different approach which actively modifies the fault current level when the microgrid changes from grid-connected to islanded operation and vice versa, by means of certain externally devices. Such devices can increase or decrease the fault level, as required. Here, the main options are:

1. To reduce the aggregated contribution of several distributed energy sources, which can alter the fault current levels enough to exceed the design limits of various equipment components, as well as to guarantee an adequate coordination despite the feeding effect of DG to fault current, fault current limiters (FCL) ae recommended and can be used. This effect is particularly evident with synchronous machine-based DG units.
2. To equalize the fault current level in both grid-connected and islanded operation, due to the reduced fault contribution by inverter-interfaced DG sources. This can be achieved in two different ways:
3. Incorporating energy storage units (flywheels, batteries, etc.) into the microgrid in order to increase the fault current level to a desired (required) one, allowing overcurrent protection devices and equipment to operate in conventional ways.
4. Installing protection devices between the grid and microgrid, to alleviate the grid fault current contribution.

The main problems associated with the use of these devices embedded in the microgrid are as follows:

1. Storage devices require large investment and need to match the main grid's short-circuit level so as to guarantee that faults are cleared in a timely manner.
2. The application of schemes based on a fault current location (FCL) technology is only possible up to certain amount of DGs connected. For very high levels of DGs, it can be difficult to determine the impedance value of the FCL due to the mutual influence of the DGs.
3. Sources with high short circuit capability (flywheels, etc.) require significant investments and their safe operation depends on the correct maintenance of the unit.
4. The methods based on an additional current source are highly dependent on the technology of islanding detection and the correct operation of the current source.

11.3.3.5 Protection Coordination Techniques for Microgrids

Protective device coordination or selectivity is the process of applying and setting the protective relays that overreach other relays such that they operate as fast as possible within their primary zone but have delayed operation in their blackout zone. The goal is to ensure maximum service continuity with minimum system disruption. Historically, protective device coordination was done on translucent log-log paper but modern methods normally include detailed computer based analysis and reporting. A properly coordinated protection maximizes power system selectivity by isolating faults to the nearest protective device, as well as avoiding nuisance operations that are due to transformer inrush or motor starting operations. System protection is said to be selective if only the protection device closest to the fault is triggered to remove or isolate a fault. If this takes too long, the protection at a higher level takes over. This rule of thumb allows disconnection of those components that are faulty. For a distribution system without DG, power flows in one direction during normal operation as well as when a fault occurs. This allows for the use of a selective system by applying time grading to overcurrent relays. When DG is installed, this system becomes inadequate. A possible scenario is disconnection of a healthy feeder by its own protective relay because DG contributes to short-circuit current flowing through a fault in a neighboring feeder. On the other hand, if a fault occurs on the connection between the supplying grid and a local network, disconnection of feeders or generators should take place. In an islanded microgrid, the voltage drop caused by a fault is almost the same across the entire network, due to limited geographical span. Therefore, it is almost impossible to coordinate protective devices based on voltage profiles. Based on the analysis of the wide range of publications, found in the literature the main conclusions and recommendations regarding the protection of microgrids can be summarized as follows:

1. There are still a relatively small number of references that address and describe the microgrid protection problem properly, let alone giving enough information to fully understand the approach.
2. Although the authors that are most related to protection manufacturing companies seem to favor adaptive protection systems, a general trend in this direction has not been observed.
3. It seems to be clear that there is an increasing need or even necessity to upgrade many protection devices (fuses, etc.) currently used in power systems, mainly in LV systems, in order to provide new capabilities for the application of new unconventional protection systems in microgrids.
4. Regardless of the protection methodology, it seems likely that some kind of communication is going to be necessary, either centrally operated or distributed.
5. In order to deal satisfactorily with the protection problems associated with a bi-directional power flow, the need for a directional feature is clear. This would imply different settings for each fault direction and, even in some cases, it may be necessary to use different methodologies for each fault direction.
6. Many references divide the microgrid into a number of smaller zones and subsequently apply the proposed methodology separately to each one of them: divide and conquer type approach.
7. Compared to other areas of power system protection, the number of references using artificial intelligence techniques in order to improve the microgrid protection is not very high.
8. There are some specific problems that are hardly analyzed in cited references, such as high impedance faults and protection of meshed networks.
9. Similar to other research areas in power system protection, it is likely that, in order to get an optimal protection system for microgrids, a combined action of different protection techniques is needed.

11.4 MICROGRIDS IN SMART GRIDS

Researchers and engineers working in power and energy industries are facing nowadays challenges that changed and are changing the ways that power system operated, and customers and utilities interact. These issues occur in large part due to the advent and development of smart grids and microgrids, enabling the possibility that end-users can be energy providers, not only consumers, depending on the power system conditions, management, structure and components. Microgrid is an important and necessary part of the smart grid development, being characterized as the *smart grid building block*. The microgrid concept was introduced as one of the main solutions for reliable interconnection of distributed energy systems and DG elements, i.e. dispatchable and renewable energy generating units, energy storage units or controllable loads. As an important smart grid driver and an entity used to implement renewable energy technologies combined with conventional generators and energy storage, microgrids are expected and required to optimally operate to minimize the use of fossil fuels, reduce pollutant emissions, provide better electric services, improved power quality, grid resilience and service restoration. In microgrid operation, as previously discussed in this chapter, a key point is the structure of the monitoring and control system, typically hierarchical and includes both distributed and centralized functions. The latter approach is assigned through the energy management systems, ability of supervising, operating and managing the microgrids in various configurations, structures and operation modes, as well as of cooperating and interacting with the host grid. The EMS pursues various objectives such as microgrid security operation, voltage profile and reactive power optimization, microgrid loss reduction, load balancing and minimization of costs in islanded operation, maximization of the economic profits from power exchange and ancillary services in grid-connected operation. The increasingly frequent natural disasters and extreme events highlighted the necessity of improving power system resilience, supply security and faster service restoration, which are also the smart grid expectations and requirements. On the other hand, the ever-increasing requirements for reliability and quality of supply suggest enabling the self-healing features of modern smart distribution networks. In such regards, the microgrids can act as emergency energy sources to serve critical and essential loads when utility power is unavailable.

11.4.1 Microgrid Applications for Service Restauration

The ability of the power distribution systems to efficiently withstand low-probability, high-impact events, extreme events or natural disasters while enabling a quick recovery and restoration to the normal states is interpreted as resiliency. In other words, the resilience is the ability of a power system to anticipate high-impact low-probability events, withstand against them, tackle their consequences in a proactive manner and rapidly recover from the degraded situation. Moreover, resilience goes beyond the impacts on customers (e.g., assessing the energy not supplied) and focuses on how rapidly and efficiently the infrastructure is restored to its pre-event operational state. Distribution system restoration (DSR) is aimed at restoring load after a fault by altering the topological structure of the distribution network while meeting electrical and operational constraints. Traditionally, electric service restoration is performed by first identifying alternative substations and possible routes, followed by network reconfiguration, so that the outage area can be re-energized via these substations. The ability of a microgrid to operate in islanded mode increases the reliability of the load points and the overall resilience of the host power distribution network. A power distribution system is considered to be resilient if it is able to anticipate, absorb, adapt to and/or rapidly recover from disruptive high-impact, low-probability events, such as natural hazards, extreme weather events and human induced disruptions. Microgrids with distributed generation, renewable energy and energy storage can provide resilient solutions in the case of major faults in power distribution systems due to natural disasters or other extreme events. Microgrids could be formed or are already in operation to supply the maximum lost loads after losing the upstream electric network supply. Each microgrid is expected to supply its local loads independently; however, it is probable that the formed or existing microgrids experience power deficiency (overloading) due to the intermittency of wind and solar DG units as well as load uncertainty. The restoration problem is a combinatorial and multi-disciplinary problem applying simultaneously knowledge from linear programing, graph theory and electrical power system discipline. Mathematical models are paned to be studied and developed for the optimal power scheduling of microgrids incorporating proper MG models, component representations, energy demand (load) and generation uncertainties, in two scenarios: the MG intentional (pro-active) islanding and the MG unintentional islanding operation modes. In order to take advantage of the microgrid capabilities to supply power at least to critical and essential loads, and for optimum power scheduling that accurate and simpler models and algorithms of microgrid, components (DG units and loads) and optimum power management to be available. Such models allow the microgrids to improve power supply availability and optimal generation scheduling in the event of natural and man-made disasters, through the use of distributed generators and local energy storage.

Common MG models to enhance distribution system restoration and improve resilience are usually considered: fixed boundary microgrids (already-installed microgrid), dynamic formation of microgrids (the microgrids is formed dynamically from section(s) of the power distribution network) and reconfigurable microgrids. However, most of the analyses are focusing on critical MG component models to improve and extend supply availability in the extreme events, through the optimized use of DER units. The EMS microgrid brain uses the historical data and other inputs (e.g., weather data, solar radiation, wind velocity) to forecast the DERs, the loads and the market and to optimize the generation scheduling. Notice that the accuracy of the forecast is crucial for the EMS to balance supply and demand in the microgrid. Forecast is challenging in a microgrid setting due to the inherent intermittency and variability of DERs (e.g., PVs and WTs) and the spatial and temporal uncertainty in controllable and even uncontrollable loads. The application of the defensive and optimized islanding solution is determined by a risk assessment, specifically, by the component weather-dependent failure probabilities by using existing unit fragility curves to identify the most vulnerable components, at higher risk of tripping due to the event. The risk assessment enables informed decisions regarding actions to mitigate risk. The defensive islanding algorithm is based on constrained clustering, using power flow analysis to split the system into islands and isolate the vulnerable components. The islanding operation mode can occur if there is a fault within the microgrid or in the upstream network to which microgrid is connected. In islanded operation mode, the energy supply to MG loads depends upon generated power by the microgrid DG units, ESS reserves, through the energy management system functions to curtail the non-sensitive and/or non-critical loads, based on load prioritization algorithms. The main aspects are: the scheduling of the MG dispatchable or non-dispatchable generators, generation control and MG management and operation systems able to monitor the variables that are relevant for operation of the various MG blocks (voltages, powers, weather parameters and load demand). Control systems are needed across broad temporal, spatial and industry scales, from devices, components to power system, from fuel sources to consumers, or from utility pricing to demand response. With increased deployment communication infrastructure, opportunities arise for reducing energy consumption, better and more efficient exploiting renewable energy sources, increasing reliability and performances of power distribution networks. Key components of the smart grids are smart meters, adaptive and intelligent control, smart sensors, monitoring and data management systems, controlling power and information flows among various stakeholders, making the grids two-way networks. Other smart grid applications include EMS, DG and its reliable integration to the system, equipment diagnostics, control, overall optimized asset management etc. Interconnection provides multiple benefits by exploiting the diversity that can exist between differing systems:

1. Diversity between times that peak loads occur, allowing the same generation to supply several loads.
2. Diversity of outages, allowing spare equipment to support more than one system.
3. Diversity of fuel sources, allowing the most economical fuel choice for any given situation.

However, new and emerging requirements find transmission and distribution in roles that they were not designed to perform. One example is the role of market channel, connecting buyers and sellers across very large geographic regions. Excessive transmission-use variability and far less predictability are the result. Further complicating factors come with the penetration of renewable generation such as wind and solar. In this vision, each smart transmission grid is regarded as an integrated power transmission system that functionally consists of three interactive, smart components, i.e., smart control centers, smart electricity transmission networks and smart substations. The features and functions of each of the three functional components as well as the enabling technologies to achieve these features and functions are discussed in detail in this chapter. Reduction of LV consumer's interruption time can be performed by allowing MG islanded operation, until MV network is available and by exploiting MG generation and control capabilities to provide fast black start at the LV level. Microgrids can act as emergency and backup energy sources to serve critical or essential loads when utility power is unavailable. As mentioned before, a microgrid includes various generation units, a distribution system, consumption (loads) and energy storage, and manages them with advanced monitoring, control, power processing and automation subsystems. A fully developed microgrid has the capability of automatically disconnecting and operating independently from the main grid, if needed or required. A major issue on utilization of microgrids for service restoration is that distributed generators within a microgrid have relatively small generation capacity. As a result, their ability to absorb shocks and maintain system stability is not as good as that of large electric generators in large-scale power systems. Due to such limited generation capacity of distributed generation units within microgrids, the dynamic performance, energy management and optimal scheduling of the DG units during the restoration processes are critical and essential. Moreover, the stability of microgrids, limits on frequency deviations and limits on transient voltages and currents of DG units are usually incorporated as constraints of the critical load restoration problem and optimal generation scheduling and load prioritization models. The limits on the amount of generation energy resources and energy storage units within microgrids are also critical important for restoration process.

11.4.2 PV and Wind Energy Microgrids

The main components of microgrids are distributed generation units (mostly renewable energy sources, PV arrays, wind turbines and energy storage systems), flexible, essential, critical and inflexible loads, control and communication subsystems and protection units. Solar and wind energy systems, specifically the development and design of PV and WTG systems with PV and WTG microgrid modeling are critical for microgrid applications. It is very important to be able to estimate the energy yield of a PV module, array or wind energy conversion system based on theory and input parameters developed in previous book chapters. On the other hand, the energy management problem deals with economic scheduling of various devices within a microgrid to supply its electrical demand in both grid-tied and grid-disconnected modes of operation. This problem could be challenging due to the issues associated with operation of different devices and microgrid elements. These issues include intermittent and uncontrollable nature of many types of renewable energy sources, load variations in time, different grid power tariffs, planned and unplanned outages, efficiencies of the distributed energy resources, DG unit characteristics, energy storage degradation, and availability of different options to balance the electric supply and demand. To operate a microgrid efficiently, there are two major tasks that should be accomplished by a management solution: optimizing the microgrids performance by defining long-term control strategies based on the application and operating and controlling the microgrid in real time and satisfying all operational constraints. Furthermore, it is important to model and limit degradation of the microgrid energy storage capacity. EMS and control schemes consist of the operation models for microgrid devices and components, e.g. DG units, energy storage devices, loads, including their cost model and operational constraints such as predicted generation of renewables and dynamic load prepared by forecasting module, dynamic tariff structure, deterministic and stochastic grid power outages, efficiency-based cost model for DERs or battery amp-hour based cost model and degradation model. In addition, the interaction between devices within the microgrid network is considered through the overall supply-demand balance model. The above-mentioned models are integrated into the framework of economic and generation scheduling problems or models. Optimization is the brain of the EMS, regulating microgrid power flows by adjusting the dispatchable DERs, the controllable loads, and the power imported/exported from/to the grid.

Photovoltaic systems are systems that convert solar energy into electrical energy that are typically either stored locally, supply electricity to local load(s) or fed back to the main power grid. PV systems have different characteristics that are non-linear, such as being constantly affected by changes in the temperature, solar radiation and weather that must be included in any model. In order to increase efficiency of PV systems and maintain stability of the power system without compromising the stability of the grid, several advanced control strategies have been developed, implemented and employed as discussed in one of the book chapter. *PV array models* are based on the estimations of the power generated by one PV panel, P_{Ipv} that is usually calculated according to this relationship:

$$P_{1pv} = f_V \cdot P_{STC} \frac{S_{G-t}}{S_{G-STC}} \left[1 + k_{temp} \left(T_{crt} - T_{STC} \right) \right] \qquad (11.16)$$

where f_V is a derating factor around 0.9 to account for the system losses, such as shading, aging or wire losses, P_{STC} is the rated power in standard test conditions (STC), S_{G-t} is the instantaneous solar radiation, S_{G-STC} is the solar irradiance for STC (1 kW/m²), k_{temp} is the temperature coefficient related to power (−0.43%/°C or −0.0043/°C), T_{crt} is the instantaneous temperature and T_{STC} is the temperature for Standard Test Conditions, usually 25°C. If the total number of photovoltaic panels is N-P, so the power output of each photovoltaic array is:

$$P_{1pv\text{-array}} = \frac{N_p \cdot P_{1pv}}{2} \qquad (11.17)$$

The PV system output power is constrained by the system safety, and constraint range is as follows:

$$P_{pv}^{min}(t) \leq P_{pv}(t) \leq P_{pv}^{max}(t) \qquad (11.18)$$

where $P_{PV}(t)$ is the output power of PV system at time t, usually expressed in kW, $P_{PV}^{min}(t)$ and $P_{PV}^{max}(t)$ are the minimum and maximum values of PV power output, usually expressed in kW. The output power of the photovoltaic power generation system is constrained by the safety of the system, and the PV power constraint ranges.

Wind turbine generators are modeled by the theory presented in previous book chapter. For a number of wind turbines taking into consideration of the generator efficiency, the total output power can be extracted as follows:

$$P_{WT-tot} = N_{WT} \times \eta_{WTG} \times P_{WT} \qquad (11.19)$$

where η_{wt} and N_{WTG} are the wind generator efficiency and number of wind generators, respectively. Wind turbine power is estimated by:

$$P_{WT}(v) = \begin{cases} 0, & v \leq V_{ci} \text{ or } V_{co} \leq v \\ 0.5 \cdot \rho \cdot A \cdot C_P (\lambda, \beta) v^3, & V_{ci} \leq v \leq V \\ P_{WT\text{-Rated}}, & V_r \leq v \leq V_{co} \end{cases} \qquad (11.20)$$

The wind generator power output is best estimated through interpolation of the values of the data provided by the manufacturers with wind velocity observations. As the power curves are quite smooth, they can be approximated using a cubic spline interpolation. The uncertainty of WT output is mainly originated from the inherent intermittency of wind speeds. The wind speeds follow at most of the sites, the Weibull distribution, given by the following equation:

$$f_{WB}(v) = \left(\frac{k}{c} \right) \left(\frac{v}{c} \right)^{k-1} \exp \left[-\left(\frac{v}{c} \right)^k \right] \qquad (11.21)$$

where v represents the actual wind speed; k is the shape factor (dimensionless), which describes the PDF shape of wind speeds; and c is the scale factor. The relationships between the WT power output P_{WT} and the actual wind speed v are described by a relationship as one given in Equation (11.22) or adaptions of it, and similar relationships, as one given here:

$$P_{WT}(v) = \begin{cases} 0, \text{if } v \langle v_{ci}, or\ v \rangle v_{co} \\ \dfrac{v - v_{ci}}{v_r - v} P_{WT-Rated}, when\ v_{ci} < v < v_r \\ P_{WT-Rated}, when\ v_r < v < v_{co} \end{cases} \tag{11.22}$$

According to Equation (11.20) and (11.22), the PDF of the WT output $f_o (P_{WT})$ is expressed as:

$$f_O (P_{WT}(v)) = \begin{cases} \dfrac{khv_{ci}}{P_{WT-Rated}} \left[\dfrac{(1+hP_{WT} / P_{WT-Rated}) v_{ci}}{c} \right]^{k-1} \\ \exp\left\{ -\left[\dfrac{(1+hP_{WT} / P_{WT-Rated}) v_{ci}}{c} \right]^{k} \right\}, when\ P_{WT} \in [0, P_{WT-Rated}] \\ 0, otherwise \end{cases} \tag{11.23}$$

where parameter h is expressed by: $h = (v_r/v_{ci}) - 1$.

Several types of *energy storage systems* are available in practice and being used in microgrids, such as super-capacitors, electrochemical batteries, super-conducting magnetic energy storage, compressed air energy storage and flywheel energy storage. Electrochemical batteries are selected in this study due to their popularity in storing electrical energy. The storage system can minimize the effects of the stochastic nature of RESs on microgrid to balance power generation and demand. The charging and discharging models of the battery is represented as follows:

$$E_{batt}(t) = E_{batt}(t-1) + \eta_{ch} \cdot P_{ch} \cdot \Delta t, \text{for charging phase}$$
$$E_{batt}(t) = E_{batt}(t-1) - \dfrac{P_{dh}}{\eta_{dh}} \cdot \Delta t, \text{for discharging phase} \tag{11.24}$$

These are subject to the following battery power and energy limits constraints:

$$0 < P_{ch}(t) < P_{ch-max}$$
$$And, \quad E_{batt-min} < E_{batt}(t) < E_{batt-max} \tag{11.25}$$
$$0 < P_{dh}(t) < P_{dh-max}$$

where $P_{ch}(t)$ and $P_{dh}(t)$ are charging and discharging powers of the battery at time, t, respectively, $E_{batt}(t)$ is the storage energy, η_{ch} and η_{dh} are the charging and discharging efficiency, respectively. The operating commands, $u(t)$, of the battery from the optimization scheduling algorithm determine the charging and discharging energy amount. The positive values of $u(t)$ indicate $P_{ch}(t)$ while negative values refer to $P_{dh}(t)$.

Good and simple *load models, types and load profiles* are critical for microgrid operation, management and control strategies. However, such load models and profiles are quite difficult to develop and characterize due to the load variability, often stochastic behavior and unpredictability.

Interruptible loads are disconnected during an outage or malfunction event. The realization of inter-ruptible loads is helping to save energy and to use power in a meaningful way, when sizing the local power generation capabilities. Examples of interruptible loads in the household are: loads in the kitchen, entertainment, washing machine, air conditioning or EV charging. Typical load profiles are determined by using recorded data from daily demand. In order to optimize the loads in an efficient manner, we are dividing into three different categories as: (a) *inflexible loads, sub-divided into: most critical loads, critical loads, essential loads and priority loads; and (b) flexible loads, sub-divided into: priority flexible loads, transferable and non-essential loads.* The essential loads are the first to be shed followed by the critical in the event of serious system contingencies, while the most critical loads should be maintained except otherwise, wherein the system is on the verge of collapsing. For all loads, we assume that the demand is bounded by:

$$P_{L-\min}(t) \le \sum_{k=1}^{T} P_L(t) \cdot \Delta t \le P_{L-\max}(t) \tag{11.26}$$

The core of the MG energy management system, in the extreme events, and grid interruptions is to provide power to essential and critical loads, though power management and allocation, and through the load optimization and prioritization. A widely used normal distribution model is usu-ally adopted for modeling load fluctuations. The probability distribution function is described by:

$$f_L(P_L) = \frac{1}{\sqrt{2\pi}\sigma_L} \exp\left(-\frac{(P_L - \mu_L)^2}{2\sigma_L^2}\right) \tag{11.27}$$

where P_L is the load active power, μ_L and σ_L are the mean and standard deviation of P_L. In order to facilitate the incorporation of multiple random variables, the power of an equivalent load model (P_{EL}) is defined as the difference of the load power and the joint power output of DG and RES units (e.g. wind turbine(s), PV system(s) and energy storage units), which is expressed as follows:

$$P_{EL} = P_L - \Sigma(P_{RES} + P_{DG}) \tag{11.28}$$

The low-voltage *electric grid* often presents a simple, radial topology. The full knowledge of the cable characteristics and lengths would allow a power flow calculation and imposing current and voltage limitations during the optimization. However, there is a practical drawback to this approach: the topology information at the LV level is not always available. To overcome this problem, the model uses simple power limits at the feeders and the secondary substation. The exchanged power at time t is denoted as $P_{grd}(t)$ in kW, with the following interpretation:

$$P_{grd}(t) > 0, - \text{power is taken from the grid}$$
$$P_{grd}(t) < 0, - \text{power is delivered to the grid} \tag{11.29}$$

For reliably grid operation, the power that is exchanged must be within the following limits:

$$P_{grd-\min} \le P_{grd}(t) \le P_{grd-\max} \tag{11.30}$$

where $P_{grd-min}$ and $P_{grd-max}$ are the grid minimum and the maximum power exchange limits.

Dispatchable distributed generators employed in microgrid applications include diesel engines, internal combustion engines, reciprocal combustion engines and microturbines, the most common being diesel engines. Various generators are employed in such systems. Diesel generators are used as backup power supply during power outages, or to generate power in off-grid applications such as remote locations or large ships. Diesel generators are also used to provide ancillary services such as voltage control, load regulation and frequency control. A diesel generator is often modeled as a synchronous generator with an excitation system driven by a diesel engine and speed governor. The fuel cost of diesel generators, natural gas generators or microturbines is expressed as a quadratic function of its active power as:

$$C_{DGen}(t) = \alpha + \beta \cdot P_{DG} + \gamma \cdot P_{DG}^2 \qquad (11.31)$$

where α, β and γ are the generator coefficients which can be obtained from the manufacturer. The DG output power should satisfy the power and ramp rate constraints at each time period. In practice the operation of a diesel generator in the range of 70–90% of full load is preferred for economic considerations and recommended by manufacturer. Diesel generators are, helpful in boosting power during peak energy demand hours as well as for energizing the energy storage units as and when required. Power limits for diesel generators can be expressed as:

$$k_{gen} \cdot P_{DG-rated} \leq P_{DG}(t) \leq P_{DG-rated} \qquad (11.32)$$

The value of k_{gen} is set to be in the range of 0.25–0.35 based on the specification of the manufacturers. The problem of electric service restoration in the presence of microgrids can essentially be modeled as an operation optimization problem which tries to minimize the operational costs of the network subject to technical and capacity constraints. Modeling DER units (DG, DS and DR) and considering their operational constraints make the problem nonlinear and add to its complexity level. The rigorous modeling of MG resources and options along with prevailing uncertainties are the key factor in achieving the desired MG operations in low-probability, high-impact events. To reduce the power system inherent losses and avert discontinuity of power supply during unfavorable weather events, blackouts or hazards the existing power systems must be re-engineered be resilient to these extreme events. The main aim of the microgrid is to optimally allocate the DER that minimizes costs and increase the performances of grid operations. Microgrid uncertainties are categorized into two groups: normal operation uncertainties and contingency-based uncertainties (including forced outages, unintentional islanding and resynchronization events). Therefore, random parameters are represented in the MG scheduling along with their associated probabilities. Such approaches can be easily extended to model any MG types. The framework minimizes the MG cost function, taking into account the operational constraints. Since the resiliency-oriented MG scheduling is computationally intensive, the scenarios associated with the uncertain parameters were usually reduced for computational tractability. The goal (cost) function (TF) which has to be minimized in order to obtain the minimum operation cost of the considered microgrid, is given by:

$$TF = \min\left\{ \Im\left(\sum P_{DSP}, CU_{DSP}, AV_{DSP}, P_{Non-DSP}, CU_{Non-DSP}, AV_{Non-DSP}, P_{ESS}, SOC, CU_{ESS}, \Delta t, \right) \right\} \quad (11.33)$$

In Equation (11.35), P_{DSP}, $P_{Non-DSP}$, P_{ESS} and SOC are the maximum available values of the active power that can be generated in the microgrid by the non-dispatchable units (e.g. photovoltaic panels, wind turbines), dispatchable units (e.g. by geothermal, biomass generator, microturbines. etc.) and energy storage devices. $CU_{DSP}(t)$, $CU_{Non-DSP}(t)$ and $CU_{ESS}(t)$ are the usage coefficients in relation to the maximum values of the active powers (rated power for generators and maximum capacity for energy storage systems), in their variation domain being [0, 1]. $AV_{Gen}(t)$ represents the *availability function* of RES units, given by input data, e.g. the solar irradiance level at a certain moment of the day, related to the maximum value of the solar irradiance (e.g. implying the possibility to generate P_{PVmax} in the system) or wind availability, by using the weather forecasts or prediction models. The loads are specified based on load profiles for critical, essential, non-critical, priority or transferable types. The decision variables include each of the source-generated power and stored energy,

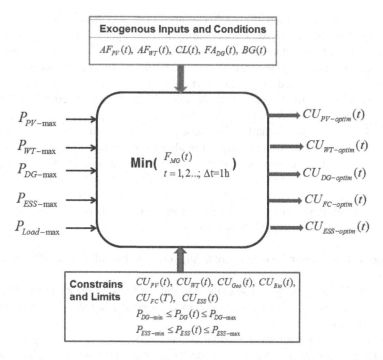

FIGURE 11.13 Minimization process diagram.

computed by multiplying its estimated usage coefficient with its rated (maximum) power at each time step. The optimization energy management and optimal generation scheduling are constrained by the generation and stored energy to meet the specified loads, by solving the MG energy balance equations with specified constrains. The minimization procedures are applied at each time step, over a specific time interval, with input data: weather forecasts, prediction models, generation, load profiles (CL(t) and time reserve estimates, represented as a black-box of the minimization process, as is shown in the diagram of Figure 11.13).

The energy and power balance equations for the microgrids and hybrid power systems are based on the load forecast and/or load profiles, wind velocity and solar radiation data and forecasts, generation predictions, energy storage data, dispatchable and distributed generation units, expressed as:

$$P_{Grid(In-Out)} = \sum_{i \leq N_{Load}} P_L + \sum_{k \leq N_{WT}} P_{WT} + \sum_{k \leq N_{PV}} P_{PV} + \sum_{k \leq N_{DDG}} P_{DDG} \pm \sum_{l \leq N_{EES}} P_{EES} + P_{ELZ} \qquad (11.34)$$

In above equation, the additional term, P_{ELZ} refers to the power used by the electrolyzers if there are such systems included into the microgrid configuration. The energy balance equation is obtained, by integration of Equation (11.34), over a specific time interval:

$$E_{Grid(In-Out)} = \sum_{i \leq N_{Load}} E_L + \sum_{k \leq N_{WT}} E_{WT} + \sum_{k \leq N_{PV}} E_{PV} + \sum_{k \leq N_{DDG}} E_{DDG} \pm \sum_{l \leq N_{EES}} E_{EES} + E_{ELZ} \qquad (11.35)$$

11.4.3 DC Microgrids and Nanogrids

In the microgrid context, DC configurations and networks have major advantages, since distributed generation systems (PV, fuel cells, batteries, supercapacitors, many wind energy conversion systems, etc.), electronic loads or electric vehicles are DC technologies. If such systems are connected through a DC grid, they require a fewer number of converters, and those converters are simpler than if they are connected through an AC grid, resulting in less expensive materials, higher efficiency and lower losses. Direct current technologies can be more efficient due to their simpler topologies, absence of controlling reactive power and frequency, while the harmonic distortions are no longer a problem, and there is no need of synchronization with the network. Moreover, simpler control structures based on the interaction of currents between the power converters, the DC bus voltage is the main control priority, the voltage being the natural indicator of power balance conditions. At the same time, the DC microgrid is a challenge because the structure of the power grid, power supplies, transformers, cables and protection is designed in alternating current. For such reason, sometimes hybrid AC-DC microgrids are seen as a compromise between AC and DC to allow for better integration of these DC devices, equipment and the classical electric grid. A general DC microgrid configuration consists of various renewable energy systems, energy storage units, several loads and grid-connection point. Regarding the renewable energy sources, there are a few advantages worth mentioning when using DC microgrid technologies:

- Reduced system power losses, by reducing the number of AC to DC or vice versa conversions;
- Loads, supplied with energy through the distribution lines, in the event of a blackout, the local power sources and/or local energy storage, especially in redundant architectures can still provide power;
- There is no need to synchronize distributed generators;
- Fluctuations of generated power and the loads can be compensated through energy storage units; and
- Such systems do not require long and high-capacity transmission lines.

DC microgrids can have various configurations with different renewable energy sources that affect the system in certain ways, being desirable to design control strategies that are applicable to any microgrid configuration, with minimal changes, generic multi-terminal configurations are often employed. All components of such configurations are usually interconnected through power electronic converters. Focusing on the DC side of the microgrid, power converters are responsible for maintaining the grid voltage between required limits. Among other benefits, this concept has some difficulties. A DC system does not experience harmonic issues, its fundamental frequency is 0 Hz and multiple frequencies of 0 Hz do not exist. However, in reality voltage oscillations could come from grid resonances, controller interactions, making the issue of harmonics to DC system relevant. In DC microgrids several power converters are connected to a DC bus and nonlinear effects of these converters can cause oscillations, damaging resonance or unacceptable electromagnetic interferences (EMI). Special filters must be included to mitigate and filter such as *harmonic currents*. Another critical issue in DC microgrids is the impact of fault currents that can be drawn only through power electronic converters, the current being limited by the converter ratings. The low fault currents in a DC system can create voltage disturbances in other system points, making difficult to select the proper protection settings that accurately differentiate between fault currents and heavy load conditions. In addition the DC networks also suffer from the absence of periodic zero crossings, especially problematic for arc faults. Grounding is another problem, closely related with fault issues. Grounding configuration has impact on system power quality and safety in fault conditions. In applications, where AC appliances are powered with DC sources as a way to reduce electricity consumption and harmful emissions, similar issues may occur. On the other hand, internally in most of computers, laptop, TV, radio and electronic systems are DC systems. Most of the appliances are able to work with DC power supply, if are equipped with a transformer and a rectifier to obtain the proper DC operating voltage. When such appliances are in stand-by mode, the transformer absorbs small open-circuit current, creating stand-by losses. The US total domestic power consumption of consumer equipment in stand-by mode has been estimated to over 40 TWh/year and it is predicted to increase. A DC network can eliminate transformer-rectifier stages, eliminating stand-by losses, improving efficiency and losses due to the absence of reactive power. By considering a DC residence, the conversion stages can be removed, increasing efficiency and decreasing the number of components needed for each appliance. Furthermore, the absence of reactive power lowers the current needed to transfer the same amount of power. Though some appliances might need a DC-DC converter to properly adjust the DC voltage needed for the appliance, harmonics and power factor issues are eliminated. Furthermore, with advances in power electronics technology, motor driver devices are being considered to replace actual motors in home appliances.

For improved uses of distributed energy sources and reducing the impact on loads, new microgrid types based on *DC energy pool* are proposed, as the configuration shown in Figure 11.14. All distributed energy sources, whether DC sources such as photovoltaic panels or fuel cells or AC sources, such as wind turbine, are converted into DC energy and eventually fed into a DC-energy-supplying network. DC power customers take energy directly from the DC electricity network, while AC power customers take energy by inverting DC power into AC power with low-cost inverters. In case that the total generation of distributed energy sources is more than the total loads, the energy excess may be stored in the energy storage units, and in case that the total generation is less than the total loads, the energy lack may be supplied by the energy storage units or from other energy sources, if available. A DC energy pool is usually formed by the DC-power-supplying network, connecting distributed energy sources, end-users (loads) and several energy storage devices. The energy pool is not an energy storage concept like an impounded hydropower reservoir, it just provides a power (energy) exchanging platform, helping to circulate energy among all the components connected with the pool (platform). Through coactions and interactions of all these components (elements), the energy generated by the distributed energy sources can be utilized sufficiently and optimally, reducing the energy needed to be stored, so there may not be needs for large-capacity energy storage devices and higher costs. Notice that commercially, most appliances are readily available with an input voltage of 12 V, or 24V, either some appliances are available at input voltages of 48V. However, low-voltage DC appliances demand higher currents to comply with power requirements, remaking feeder losses considerable, and the overall appliance efficiency is practically low. In a near future, the market for DC appliances will be so high that appliances will be manufactured at higher voltage levels. Considering this, along with the fact that at higher DC voltages feeder losses decrease substantially, the chosen appliance voltage for a DC residence is 48V. In an AC microgrid

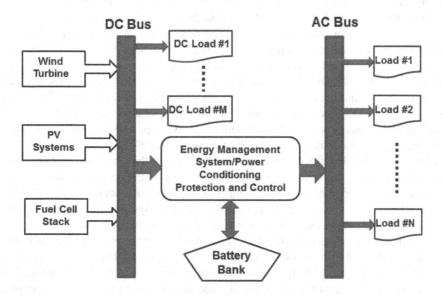

FIGURE 11.14 A typical hybrid AC-DC microgrid configuration.

system, control operations require frequency control (P/f) and voltage control, where two control units are required which receives information from a hierarchical entity or supervisory control entity based on which control algorithm is implemented to maintain system stability, reliability and operability. Whereas in a DC microgrid, control aspect refers to a single entity, usually voltage rather than two different parameters, frequency and voltage as in AC microgrid control. This makes simpler the implementation of control algorithms in DC microgrids as compared with control of AC microgrids.

11.4.4   DC Microgrid Power Quality and Protection

The DC networks, including DC microgrids can operate in an islanded mode, but usually have an interface (connection point) with an AC grid system in order to exchange power during normal operation. Therefore, the power quality issues in the DC microgrid can arise either internally or from the AC grid side. The most common power quality issues in DC microgrid system are: transient voltage from AC grid, harmonics due to resonances and power electronic converters, electromagnetic interference compatibility (EMC) issue, communication failures, inrush currents, DC bus faults, voltage unbalance in bipolar DC bus and circulating currents, arising if there is a mismatch in the converters output voltages. Voltage transients are frequently encountered in an AC grid system mainly due to capacitor bank switching, load changes and power fluctuations in grid connected renewable energy systems. The DC bus voltage variation is generated due to abruptly change in input powers and voltage or current feedback error. The circulating current among source converters is raised by the unbalance cable line impedances between the DC bus and power source converters. Recent data center studies have showed that the voltage transient could be a critical problem for the DC grid system. It has been found that if a transient occurs in a DC network, the transient overvoltage not only reaches 194% of the operating (rated) voltage but also stabilize at new voltage level of 111% of the nominal value that can be very dangerous for the equipment sharing the same DC bus. The knowledge, identification and analysis of fault currents are an essential part of designing an appropriate protection schemes in any electric network. Moreover, coordination of protection devices (PDs) such as relays or circuit breakers requires knowing the fault current characteristics. Fault currents can damage microgrid elements, while microgrid power converters can lose the voltage and current control, so power converters need higher power ratings that increase the overall cost, and may initiate cascade tripping in the other protection zones. In the DC microgrids, faults occur in two different ways: the pole-to-pole (PP) and pole-to-ground (PG) fault types. In the PG faults, one or both of the conductors are connected to the ground, the PG faults are high-impedance faults. In the PP faults, the conductors are directly connected to each other. PP faults are low-impedance faults, and such faults are easily detectable but more dangerous. All faulted systems are usually modeled by using Thevenin DC and AC equivalent circuits. However, the response characteristics of faults in the DC systems are divided into two cases: transient and steady-state. The transient part of fault injected from DC-link capacitors and cable discharge of converters, while the steady-state part is injected from energy sources. In addition, the transient part of the fault currents is divided into three types: slow, medium and fast front transient. Voltage-dependent loads, converter control and batteries cause slow front transients. Surge current in the filter capacitors usually causes medium front transient, while the transient recovery voltage at the opening of the PDs cause fast front transient. In addition, there are no reactances or the values of reactances are negligible in DC networks, the DC fault peak value is higher than the one in AC systems, giving also higher rates of change in the DC microgrids. Due to this high fault raising rate, faults in DC microgrids are faster than in AC microgrids. Moreover, DC microgrids typically are using voltage source converters (VSCs), whose withstand rating is lower than devices in AC systems. Therefore, the protection systems for DC microgrids must be faster than AC systems to prevent damage to converters. Moreover, the upstream grid can have an impact on the DC network protection system. For example, a fault occurs in a load of a grid-connected DC microgrid, consisting of a PV array, a wind turbine and a battery. In a passive system, the relay must clear the main grid injected fault current. However, in DC microgrids, the fault may also be injected from PV array, WT and battery. The relay settings must be modified based on the characteristics of all faults occurring in DC microgrids. Hence, when the number and size of DG units increase, the main grid fault contribution decreases, but the fault sources increase. If a fault occurs at the main grid, the protection must island the DC microgrid, but if the fault originates from microgrid, the relays must disconnect part of it, so DC microgrid relays must operate bi-directional.

Faults in the DC systems have unique challenges and short-circuit currents on the DC bus because it can increase faster to a high level. Moreover, due to the differences in the characteristics of conventional and DC microgrids, the protection schemes are designed for traditional power systems and DC microgrids have fundamental characteristics. Conventional power systems are usually radial, and because of the unidirectional current nature are protected by current-based relays. Hence, the protection schemes of such power systems are designed for uni-directional fault currents. Conversely, due to the connection of energy sources in different locations, fault currents in the microgrids are bi-directional, and the conventional protection schemes are not usable in microgrids. In other words, in the ring systems, all DER or DG units can contribute to faults, and change the fault direction. Therefore, the non-directional relays cannot protect the microgrids, reducing their reliability. On the other hand, the topology of DC microgrids may often change, as well as the direction of fault changes, needing adaptive protection methods to provide the protection scheme for the new topology. Due to the low value of line resistance and the high value of fault rising rate in the DC microgrids, coordination of current-based relays is also a challenge. Therefore, all series current-based relays sense high-value fault currents, causing a lack of coordination between relays. Moreover, the faults in DC microgrids are of PP and PG types, meaning the relays must coordinate for both faults. However, because of the differences of PP and PG fault currents, if the relay settings set for one of them, in another case a dis-coordination can happen. Due to the lack of selectivity, typical overcurrent relays cannot be implemented in the DC microgrids. Thus, several protection methods for current-based relays are proposed and employed in DC microgrids. Circuit breakers interrupt fault currents in power systems by a specific mechanism, and the AC CBs interrupt the fault current in the cross zero point at every half period. Due to the lack of cross zero point in DC systems, conventional CBs cannot be used in these systems. In addition, DC microgrids require faster fault current interrupters to prevent damages to the VSCs. On the other hand, DC CBs can be made by magnetic arc blowers and arc chutes to force the fault current to zero. This method is expensive, slow, and not suitable for all DC microgrids. Therefore, the best solution to this problem is using power electronic devices such as Insulated gate bipolar transistor (IGBT) and integrated gate-commutated thyristor (IGCT).

In the AC power systems, SC faults are typically ten times greater than the rated current of the system, and this greater fault current assists protection schemes to easily detect faults. Nevertheless, fault currents in DC microgrids may be limited by power converters, and the currents may be smaller than the threshold of fault detection schemes, making fault detection more difficult. On the other hand, the operation modes of DC microgrids impact the short-current level and direction. In the grid connected mode, grid and DERs contribute to the fault current, up to 50 times larger than the nominal rated current, while in the islanded mode, only DERs generate fault currents, about 5 times larger than rated current. Therefore, the fault level in grid-connected mode is higher than in islanded mode, and the fault current direction may be different in each mode. To overcome this problem, the settings of relays must set according to the various fault current level, or an adaptive protection scheme must be implemented in the DC microgrids. The fault current capacity of inverters inside the microgrids is usually less than design fault current half. Hence, during the faults, if the DER penetration into DC microgrid is low, the operation can be changed to islanded mode to reduce the fault current level. Consequently, the coordination and sensitivity of current based relays is affected, and it may induce a delay or dis-coordination between relays. In addition, the main inverter between the grid and DC microgrid, having a limiting role has the most fault current, needing most fault tolerance, being able to limit the fault with a high flexibility. One of the main settings of current-based relay is fault current amplitude, and for PG faults, which is affected by the grounding system. On the other hand, the grounding system helps detection of PGs by providing a fault current path. Hence, designing a proper grounding system is vital for reliable protection of DC microgrids.

In the conventional power systems, the most common topology of the system is radial, and the protection of these systems designed based on unidirectional power flows. On the other hand, increasing the DG penetration makes the system protection more challenging. DG units cause the increase of fault current levels, changing also the system power flow direction. Therefore, it can affect the PDs coordination. In addition to the common issues related to the both AC and DC protection, due to the nature of DC power systems, such as large DC capacitors, low impedance of DC cables, high transient current and voltage, several challenges are only related to DC protection. Circuit breakers interrupt the fault current in the AC networks using zero crossing. However, due to lack of zero crossing in DC systems, conventional CBs cannot be used. On the other hand, DC systems need a faster protection scheme, because of the prevention of any damages to the voltage-source inverters. Also, grounding in the DC microgrids must be designed properly to detect the faults. Hence, a grounding system must minimize the DC stray current and common mode voltage. In recent years, several protection methods have been proposed and used in DC microgrids. The most common protection methods, used in DC microgrids include: current- and voltage-based methods, impedance-based methods, communication-based methods and local protection methods. In the AC systems, the distance protection uses the analysis of the symmetrical component to avoid the impact of fault resistance on the protection method. However, in the DC systems, these are not possible. The impedance and travelling wave methods have been accepted as an industry standard for AC power systems; however, it is difficult to directly implement these methods to the DC microgrids due to the lack of phasor parameters. In AC microgrids, by changing the voltage angles, the magnitude of the travelling waves is changing. Thus, this is a problem for traveling wave protection methods. On the other hand, there are no such problems for DC microgrids. High transient faults cause a voltage collapse within 5 ms during the fault. Therefore, the protection methods required for the DC microgrids must be much faster than the AC power systems. Hence, in summary, the protection scheme must include: detect high resistance faults, fast protection method, isolate fault section, cost-effective, adaptive in changing of topology, considering the dynamics of renewable resources.

11.5   CHAPTER SUMMARY

A microgrid is a smart integration of generation and energy storage units with a cluster of loads, some of them controllable, which can work as an isolated local controllable power system by disconnecting from the utility at the point of common coupling. The microgrids include a set of microsources such as microturbines, fuel cells, photovoltaic arrays and wind turbines, energy storage units, such as flywheels, supercapacitors and batteries, controllable and uncontrollable loads. It can be connected to the grid (grid-connected operation mode) or operated independently when isolated from the grid (island operation mode) during faults, other external disturbances or by choice, thus increasing the supply quality and security, efficiency, providing cheaper and cleaner energy. These abilities of microgrids are key features; however, microgrids may enhance local reliability, provide lower investment costs, reduce emissions, improve power quality and reduce the power losses of power distribution network. A microgrid is an integrated energy system consisting of interconnected loads and distributed energy resources, operating as a controllable entity. Its objectives are to ensure better energy reliability, security and efficiency, providing improved electric services and better power quality to the end-users. Some microgrid implementations have been deployed to improve energy reliability, resilience and efficiency in industrial plants, commercial buildings, military or university campuses. Microgrids are usually installed on low-voltage networks, close to customer loads, and can also provide power backup in case of blackouts. However, the true potential of microgrids lies in the way they are managed as all-day power supplies. Indeed, such low-voltage, low-power units cannot be directly integrated into the main power systems. The idea is to create semiautonomous networks, the microgrids, composed of LV equipment that is considered as one single electric entity by the main network. Microgrid structures are a reality for utilities wishing to integrate local generation or implementing grid relief solutions in areas that are poorly served by the transmission subsystems. Microgrids can also benefit local utilities by providing dispatchable load for use during peak power conditions and alleviating or postponing distribution and transmission system upgrades. Microgrids may be a quick alternative to the building or reinforcement of power transmission lines. Main challenges for utilities are to guarantee grid reliability, stability and security and also to optimize energy efficiency. Uniform adoption of the IEEE P1547.4 guidelines for DR island systems can lead to the elimination of technical barriers posed by project-specific operational requirements. Renewable energy sources, energy storage systems and loads are the basics components of DC microgrids, as for AC structures. However, these components can be better integrated thanks to their DC nature, resulting in simpler power converter topologies, as well as the control strategy required for such applications. Low-voltage DC microgrids can be used to supply sensitive electronic loads, by combining the advantages of using a DC supply for electronic loads, and local generation to better supply sensitive loads. With a DC power system, AC-to-DC conversion within these loads is avoided, and losses reduced. Moreover, many of DG technologies, e.g. PV and fuel cells are DC making DC networks a better choice.

11.6 QUESTIONS

1. In your own words, briefly describe the microgrid concept.
2. What are the microgrid main operation modes?
3. List the major microgrid advantages.
4. Is microgrid a new concept?
5. What is the microgrid usefulness for the smart grids or modern power systems?
6. List the main microgrid specific features, unlike as the one of conventional power systems.
7. What are some of the ancillary services that a microgrid can provide?
8. List the major microgrid drivers.
9. What are the main microgrid types?
10. Why a microgrid is one of the smart grid drivers?
11. List and briefly describe the main microgrid benefits.
12. What is the power ranges of a feeded microgrid?
13. List the two most common microgrid classifications.
14. Briefly describe the microgrid major classes.
15. Briefly describe the major microgrid types.
16. What are the most important challenges for the microgrid development and deployment?
17. Describe the total values that a microgrid can bring to its host.
18. List and describe the major microgrid building blocks and components.
19. Why accurate load models are important for microgrid analysis?
20. Describe the microgrid energy management operation, parameters and input variables.
21. What is microgrid centralized energy management?
22. Briefly summarize the microgrid droop control.
23. What are the advantages of MG decentralized control?
24. What the major issues of microgrid protection?
25. Briefly describe the microgrid stability issues.
26. Describe the microgrid hierarchical control methods.
27. In your own words, describe the microgrid power scheduling models.
28. What is the main issue of microgrid decentralized control?

29. Briefly describe the most common microgrid protection issues.
30. What are the most important protection characteristics?
31. List the main convention protection deficiencies when micro-sources are present.
32. Briefly describe the microgrid droop control method.
33. What are the common microgrid protection approaches?
34. List the main advantages and drawbacks of conventional droop control.
35. Why microgrids are a way to enhance power system resilience?
36. List the most important microgrid protection schemes and methods.
37. What are the main advantages of DC microgrids and nanogrids?
38. List and briefly describe the major DC microgrid challenges.
39. What are the protection issues of the DC microgrids?

References and Future Readings

1. N. Cohn, *Control of Generation and Power Flow on Interconnected Systems*, Wiley, New York, 1971.
2. O. Elegerd, *Basic Electric Power Engineering*, Addison-Wesley, 1977.
3. R.D. Schultz, and R. A. Smith, *Introduction to Electric Power Engineering*, John Wiley & Sons, New York, 1988.
4. D.J. Glover et al., *Power System Analysis and Design* (5th ed.), Cengage Learning. 2012.
5. F. Sioshansi (Ed.), *Smart Grid: Integrating Renewable, Distributed and Efficient Energy* (1st ed.), Academic Press, 2011.
6. Stuart Borlase (Ed.), *Smart Grids: Infrastructure, Technology, and Solutions*, CRC Press, 2013.
7. A B M Shawkat Ali, *Smart Grids - Opportunities, Developments, and Trends*, Springer, 2013.
8. C. W. Gellings, *Smart Grid Planning and Implementation*, River Publishers, 2015.
9. S. Borlase (ed.), *Smart Grids - Advanced Technologies and Solutions*, CRC Press, 2018.
10. Chen-Ching Liu (Editor-in-Chief), Stephen McArthur (Editor-in-Chief), Seung-Jae Lee (Editor-in-Chief), *Smart Grid Handbook, 3 Volume Set*, Wiley, 2015.
11. B. K. Bose, Power electronics and motor drives – recent progress and perspective, *IEEE Transactions on Industrial Electronics*, Vol. 56(2), pp. 581–588, 2009.
12. F. Bouhafs, M. Mackay, and M. Merabti, Links to the future: communication requirements and challenges in the smart grid, *IEEE Power and Energy Magazine*, Vol. 10(1), pp. 24–32, 2012.
13. M. Liserre, T. Sauter, and J. Y. Hung, Future energy systems: integrating renewable energy sources into the smart power grid through industrial electronics. *IEEE Industrial Electronics Magazine*, Vol. 4(1), pp. 18–37, 2010.
14. R. A. Huggins, *Energy Storage Fundamentals, Materials and Applications* (2nd ed.), Springer, 2016.
15. J. R. Cogdell, *Foundations of Electric Power*, Prentice-Hall, Upper-Saddle River, NJ, 1999.
16. P. Schavemaker and L. van der Sluis, *Electrical Power System Essentials*, John Wiley and Sons, 2008.
17. M. E. El-Hawary, *Introduction to Power Systems*, Wiley – IEEE Press, 2008.
18. M. A. El-Sharkawi, *Electric Energy: An Introduction* (2ed ed.), CRC Press, 2009.
19. J.L. Kirtley, *Electric Power Principles: Sources, Conversion, Distribution and Use*, Wiley, 2010.
20. F.M. Vanek. L.D. Albright, and L.T. Angenent, *Energy Systems Engineering*, McGraw Hill, 2012.
21. M. Amin, B. Wollenberg, Towards a smart grid, *IEEE Power and Energy Magazine*, Vol. 3(5), pp. 34–38, September-October 2005.
22. G. W. Gellings, *The Smart Grid: Enabling Energy Efficiency and Demand Response*, The Fairmont Press, Lilburn, 2009.
23. The smart grid – an introduction, (US) Department of Energy, 2008; http://www.oe.energy.gov/SmartGridIntroduction.htm.
24. U. S. DOE, "Smart Grid: an introduction," Tech. Rep., US Department of Energy, 2010, http://energy.gov/sites/prod/files/oeprod/DocumentsandMedia/DOE SG Book Single Pages(1).pdf.
25. M. McGranaghan, D. Von Dollen, P. Myrda and E. Gunther, "Utility experience with developing a smart grid roadmap," 2008 *IEEE Power and Energy Society General Meeting - Conversion and Delivery of Electrical Energy in the 21st Century*, 2008, pp. 1–5, doi: 10.1109/PES.2008.4596927.
26. R. L. King, "Information services for smart grids," *2008 IEEE Power and Energy Society General Meeting - Conversion and Delivery of Electrical Energy in the 21st Century*, 2008, pp. 1–5, doi: 10.1109/PES.2008.4596956.
27. Amin, S. M., and A. M. Giacomoni. Smart grid: Safe, secure, self-healing, *IEEE Power & Energy Magazine*, Vol. 10(1), pp. 33–40, January–February 2012.
28. J. Taft, *The Intelligent Power Grid*, IBM Glaonal Services, 2006.
29. H. Akagi, E. H., Watanabe, and M. Aredes, *Instantaneous Power Theory and Applications to Power Conditioning*, Wiley, 2006.
30. R. W. Uluski, "VVC in the Smart Grid era," *IEEE PES General Meeting*, 2010, pp. 1–7, doi: 10.1109/PES.2010.5589850.
31. H. Johal, W. Ren, Y. Pan and M. Krok, "An integrated approach for controlling and optimizing the operation of a power distribution system," 2010 *IEEE PES Innovative Smart Grid Technologies Conference Europe (ISGT Europe)*, 2010, pp. 1–7, doi: 10.1109/ISGTEUROPE.2010.5638859.

32. R. F. Arritt, and R. C. Dugan, Distribution system analysis and the future smart grid, *IEEE Trans. Ind. Appl.*, Vol. 47(6), pp. 2343–2350, 2011.

33. M. Daoud, and X. Fernando, On the communication requirements for the smart grid, *Energy and Power Engineering*, Vol. 3, 2011, pp. 53–60; doi:10.4236/epe.2011.31008.

34. W. Wang, Y. Xu, and M. Khanna, A survey on the communication architectures in smart grid," *Computer Networks*, Vol. 55(15), pp. 3604–3629, 2011.

35. S. Amin and A. Giacomoni, Smart grid: Safe, secure, self-healing, *IEEE Power and Energy Magazine*, Vol. 10(1), pp. 33–40, Jan 2012.

36. L. Peretto, The role of measurements in the smart grid era, *IEEE Instrumentation Measurement Magazine*, Vol. 13(3), pp. 22–25, 2010.

37. D. Manz, R. Walling, N. Miller et al. The grid of the future: Ten trends that will shape the grid over the next decade, *IEEE Power and Energy Magazine*, Vol. 12(3), pp. 26–36, 2014.

38. M. L. Tuballa, and M.L. Abundo, A review of the development of smart grid technologies. *Renew. Sustain. Energy Rev.* Vol. 59, pp. 710–725, 2016.

39. I. Alotaibi et al., A comprehensive review of recent advances in smart grids: A sustainable future with renewable energy resources, *Energies*, Vol. 13, p. 6269, 2020; doi:10.3390/en13236269.

40. L. T. Bergers and K. Iniewski, *Smart Grid Applications, Communication and Security*, Wiley, 2012.

41. J. Momoh, *Smart Grid – Fundamentals of Design and Analysis*, Wiley – IEEE Press, 2012.

42. R. Fehr, *Industrial Power Distribution* (2nd ed.), Wiley, 2015, ISBN: 978-1-119-06334-6.

43. EPRI 1016097, *Distribution System Losses Evaluation*, Electric Power Research Institute, Palo Alto, CA, 2008.

44. E.E. Michaelides, *Alternative Energy Sources*, Springer, 2012.

45. N. Hatziargyriou (Ed.), *Microgrids: Architectures and Control*, Wiley-IEEE Press, 2014.

46. H. Bevrani, M. Watanabe, and Y. Mitani, *Power System Monitoring and Control*, Wiley-IEEE Press, 2014.

47. IEC 61970-n, Energy management system application program interface (EMS-API); Part 1: Guidelines and general requirements; Part 2: Glossary; Part 301: Common information model (CIM) base; Part 302: Common information model (CIM) extensions; Part 401: Component interface specification (CIS) framework; Part 402: Common services; Part 403: Generic data access; Part 404: High Speed Data Access (HSDA); Part 405: Generic Eventing and Subscription (GES); Part 407: Time Series Data Access (TSDA); Part 452: CIM Network Applications Model Exchange Specification; Part 453: CIM based graphics exchange; Part 501: Common Information Model Resource Description Framework schema

48. IEC/TS 62351-n, Power systems management and associated information exchange – Data and communications security. Part 1: Communication network and system security - Introduction to security issues; Part 2: Glossary of terms; Part 3: Communication network and system security - Profiles including TCP/IP; Part 4: Profiles including MMS; Part 5: Security for IEC 60870-5 and derivatives; Part 6: Security for IEC 61850; Part 7: Network and system management - data object models; Part 8: Role-based access control; Part 9: Cyber security for key management for power system equipment; Part 10: Security architecture

49. IEEE standard for interconnecting distributed resources with electric power systems, New York, NY, 2003.

50. IEEE 1547-2003 standard for interconnecting distributed resources with electric power systems; Amendment 1, New York, 2014.

51. IEEE Recommended Practice for Utility Interface of Photovoltaic (PV) Systems, IEEE Std 929-2000, pp. 4–6, Jan 30, 2000.

52. J. W. Smith, W. Sunderman, R. Dugan and B. Seal, "Smart inverter volt/var control functions for high penetration of PV on distribution systems," 2011 *IEEE/PES Power Systems Conference and Exposition*, 2011, pp. 1–6, doi: 10.1109/PSCE.2011.5772598.

53. M. Peskin, P. W. Powell, and E. J. Hall et al., Conservation voltage reduction with feedback from advanced metering infrastructure, *IEEE PES T&D Conference and Exposition*, Orlando, FL, 2012.

54. W. Sunderman, R. C. Dugan, and J. Smith, Open source modeling of advanced inverter functions for solar photovoltaic installations, *IEEE PES T&D Conf. and Exposition*, Chicago, IL, USA, 2014.

55. A. Short (Ed.), *Handbook of Power Distribution*, CRC Press, 2014.

56. R. Belu, *Industrial Power Systems with Distributed and Embedded Generation*, The IET Press, 2018.

57. M. W. Maier, Architecting principles for systems-of-systems, *Systems Engineering*, John Wiley & Sons, Inc., Vol. 1(4), pp. 267–284, 1998.

58. S. M. Amin, and B. F. Wollenberg, Toward a smart grid: Power delivery for the 21st century, *IEEE Power Energy Magazine*, Vol. 3(5), pp. 34–41, Sep.-Oct. 2005.

59. P. Schavemaker, and L. van der Sluis, *Electrical Power System Essentials*, John Wiley and Sons, 2008.
60. European SmartGrids Technology Platform, European Commission, 2006 [Online]. Available: http://ec.europa.eu/research/energy/pdf/smartgrids_en.pdf
61. C. W. Gellings, *The Smart Grid: Enabling Energy Efficiency and Demand Response*, CRC Press, Aug, 2009.
62. A. Yazdani, and R. Iravani, *Voltage-Sourced Converters in Power Systems: Modeling, Control, and Applications*, New Jersey: Wiley-IEEE Press, 2010.
63. X. Zhou, Y. Ma, Z. Gao and H. Wang, "Summary of smart metering and smart grid communication," 2017 *IEEE International Conference on Mechatronics and Automation (ICMA)*, 2017, pp. 300–304, doi: 10.1109/ICMA.2017.8015832.
64. A. Carvallo, *The Advanced Smart Grid: Edge Power Driving Sustainability*, Artech House, June, 2011.
65. L. T. Berger, and K. Iniewski (Eds.), *Smart Grid, Applications, Communications, and Security*, Wiley, 2012.
66. S. F. Bush, *Smart Grid, Communication-enabled Intelligence for the Electric Power Grid*, Wiley-IEEE Press, 2014.
67. E. Kabalci, and Y. Kabalci (Eds.), *Smart Grids and Their Communication Systems*, Springer, 2019.
68. Q. Huang, S. Jing, J. Yi, and W. Zhen, *Innovative Testing and Measurement Solutions for Smart Grid*, John Wiley & Sons Singapore Pte. Ltd., 2015.
69. C. Cecati, G., Mokryani, A., Piccolo, and P. Siano, An overview on the smart grid concept, Proceedings of 36th Annual Conference IEEE IECON, pp. 3322–3327, 2010.
70. P. Zhang, F., Li, and N. Bhatt, Next-generation monitoring, analysis, and control for the future smart control center, *IEEE Transactions on Smart Grid*, Vol. 1(2), pp. 186–192, 2010.
71. U. S. DOE, Communications requirements of Smart Grid technologies, Tech. Rep., US Department of Energy, 2010. Available: http://energy.gov/gc/downloads/communications-requirements-smart-grid technologies.
72. CEA, Thesmart grid: a pragmatic approach, Tech. Rep., Canadian Electricity Association, 2010, http://www.electricity.ca/media/SmartGrid/SmartGridpaperEN.pdf.
73. V. C. Gungor, and D. Sahin et al., A survey on smart grid potential applications and communication requirements, *IEEE Transactions on Industrial Informatics*, Vol. 9(1), pp. 28–42, 2013.
74. B.V. Mathiesen, H. Lund, and D. Connolly et al., Smart energy systems for coherent 100% renewable energy and transport solutions, *Applied Energy*, Vol. 145, pp. 139–154, 2015.
75. N. Cohn, *Control of Generation and Power Flow on Interconnected Systems*, New York: Wiley. 1971.
76. P. Palensky, and D. Dietrich, Demand side management: Demand response, intelligent energy systems, and smart loads, *IEEE Trans. Ind. Inf.*, Vol. 7(3), pp. 381–388, Aug. 2011.
77. F. Bouhafs, M. Mackay, and M. Merabti, Links to the future: Communication requirements and challenges in the smart grid, *IEEE Power Energy Mag.*, Vol. 10(1), pp. 24–32, Jan.–Feb. 2012.
78. V. Cagri Gungor et al., A survey on smart grid potential applications and communication requirements, *IEEE Trans. Ind. Informatics*, Vol. 9(1), pp. 28–42, 2013.
79. EPRI 1000419, Engineering Guide for Integration of Distributed Generation and Storage into Power Distribution Systems, Electric Power Research Institute, Palo Alto, CA, 2000.
80. EPRI 1016097, Distribution System Losses Evaluation, Electric Power Research Institute, Palo Alto, CA, 2008.
81. EPRI 1023518, Green Circuits: Distribution Efficiency Case Studies, Electric Power Research Institute, Palo Alto, CA, 2011.
82. EPRI 1024101, Understanding the Grid Impacts of Plug-In Electric Vehicles (PEV): Phase 1 Study—Distribution Impact Case Studies, Electric Power Research Institute, Palo Alto, CA, 2012.
83. IEEE P2030 Guide for Smart Grid Interoperability of Energy Technology and Information Technology Operation with the Electric Power System (EPS), End-Use Applications, and Load, [Online]. Available: http://standards.ieee.org/findstds/standard/2030-2011.html.
84. IEC Smart Grid Standardization Roadmap, SMB Smart Grid Strategic Group (SG3), 2010 [Online]. Available: http://www.iec.ch/zone/smartgrid/pdf/sg3_roadmap.pdf, 1.0.
85. Standards Identified for Inclusion in the Smart Grid Interoperability Standards Framework, Nat. Inst. Standards and Technol., Release1.0, 2009 [Online]. Available: http://www.nist.gov/smartgrid/standards.html.
86. Farret, F.A., and M.G. Simões, *Integration of Alternative Sources of Energy*, New York: Wiley-Interscience, 2006.
87. R. O'Hayre et al., *Fuel Cell Fundamentals*, Wiley, 2006.

88. H. Chen, T. N. Cong, W. Yang, C. Tan, Y. Li, and Y. Ding, Progress in electrical energy storage system: A critical review, *Progress in Natural Science*, Vol. 19, pp. 291–312, 2009.

89. O S. Vazquez, S. M. Lukic, E. Galvan, L.G. Franquelo, and J.M. Carrasco, Energy storage systems for transport and grid applications, *IEEE Trans Ind. Electron*, Vol. 57, pp. 3881–3895, 2010.

90. J. Eyer, and G. Corey, Energy Storage for the Electricity Grid: Benefits and Market Potential Assessment Guide, Sandia National Laboratory, Report No.SAND2010- 0815, February 2010.

91. NERC, Report, 1996. Available Transfer Capability Definitions and Determination, Princeton, New Jersey, North American Electric Reliability Council. [Accessed 21-02-2017].

92. F. S. Barnes, and J. G. Levine (Eds.), *Large Energy Storage Systems Handbook*, CRC Press, 2011.

93. R. Belu, Renewable energy: Energy storage systems, *Encyclopedia of Energy Engineering & Technology (Online)*, (Eds.: Sohail Anwar, R. Belu et al.), CRC Press/Taylor and Francis, Vol. 2, 2014.

94. B. M. Buchholz, and Z. Styczynski, *Smart Grids – Fundamentals and Technologies in Electricity Networks*, Springer, 2014.

95. J. Stoustrup, A., Annaswamy, A., Chakrabortty, Z., Qu (Eds.), *Smart Grid Control - Overview and Research Opportunities*, Springer, 2019.

96. Recommended Practice for Signal Treatment Applied to Smart Transducers, IEEE Standard 21451–001-2017, Mar. 2017. [Online]. Available: http://standards.ieee.org/findstds/standard/21451-001-2017.html.

97. A. G. Phadke, "Synchronized phasor measurements-a historical overview", In: *Transmission and Distribution Conference and Exhibition 2002: Asia Pacific*, Vol. 1. IEEE/PES, pp 476–479, 2002.

98. IEEE Standard C37.118.1-2011 – IEEE standard for synchrophasor measurements for power systems. IEEE, Piscataway, NJ, 2011.

99. A. G. Phadke, and J. S. Thorp, Synchronized phasor measurements and their applications, *Power Electronics and Power System Series*, Springer, New York, 2008.

100. IEC 61850-5 communication networks and systems in substations, part 5: Communication requirements for functions and device models, 2003.

101. The precision time protocol (ptp), IEEE 1588-2008, IEEE standard for a precision clock synchronization protocol for networked measurement and control systems, July 2008, 2003.

102. D. Haughton D, and G. T. Heydt, A linear state estimation formulation for smart distribution systems, *IEEE Trans Power Syst*. Vol. 28(2), pp. 1187–1195, 2013.

103. D. Hart, *Power Electronics*, McGraw-Hill, 2010.

104. P.T. Krein, *Elements of Power Electronics*, Oxford University Press, 2012.

105. R.G. Belu, Power electronics and controls for photovoltaic systems, *Handbook of Research on Solar Energy Systems and Technologies* (Eds.: Dr. Sohail Anwar et al.), IGI, Global, pp. 68–125, 2012; DOI: 10.4018/978-1-4666-1996-8.ch004.

106. F. Blaabjerg, *Control of Power Electronic Converters and Systems* (1st ed.), Academic Press, 2018.

107. E. W. Erickson, and D. Maksimovic, *Fundamentals of Power Electronics* (3rd ed.), Springer, 2020.

108. R. Galvin, K., Yeager, and J. Stuller, *Perfect Power: How the Microgrid Revolution Will Unleash Cleaner, Greener, More Abundant Energy*, McGraw Hill, New York, 2009.

109. EMS History (last accessed 04/04/2012), Available: www.bacnet.org/Bibliography/STB-1988.pdf

110. A. Handschin, and E. Petroianu, *Energy Management Systems, Operations and Control for Electric Energy Transmission Systems*, Springer-Veralg, Berlin, 1991.

111. E. Vaahedi, *Practical Power System Operation*, Wiley-IEEE Press, 2014.

112. J, Heydt, *Computer Analysis Methods for Power Systems*, John Wiley & Sons Inc., New York, 1984.

113. A. S. Debs, *Modern Power Systems Control and Operation*, Kluwer Academic Publishers, Boston, MA, 1988.

114. J. Arrilaga, Y. H. Liu, and N.R. Watson, *Flexible Power Transmission: The HVDC Options*, John Wiley & Sons, Ltd., Chichester, 2007.

115. J. W. Evans, Energy Management Systems, Survey of Architecture, *IEEE Computer Applications in Power*, Vol. 2, pp. 11–16, 1989.

116. A. Vaféas, S. Galant, and T. Pagano, *Final WP1 Report on Cost/Benefit Analysis of Innovative Technologies and Grid Technologies Roadmap Report Validated by External Partners*, REALISEGRID Deliverable D1.4.2 (June 6, 2011), http://realisegrid.rse-web.it/content/files/File/Publications%20 and%20results/Deliverable_REALISEGRID_1.4.2.pdf.

117. R. Podmore, Criteria for evaluating open energy management systems, power systems, *IEEE Transactions on Power Systems*, Vol. 8(2), pp. 466–471, 1993.

118. L. Murphy and F. Wu, An open design approach for distributed energy management systems, *IEEE Transactions on Power Systems*, Vol. 8(3), pp. 1172–1179, 1993.

119. EMS Systems (last accessed 07/04/2012), Available: http://en.wikipedia.org/wiki/Energy_management_system

120. EMS Solutions (last accessed 05/04/2012), Available: http://www.noveda.com/solutions/energymanagement/energyflowmonitor?gclid=COzKidW-j68CFcYOfAodAQ8tzg; Siemens EMS (last accessed 05/04/2012), Available: http://www.energy.siemens.com/hq/en/energy-topics/smartgrid/? stc=wweccl20054

121. Cisco quits EMS Market (last accessed 05/04/2012), Available: http://www.forbes.com/sites/williampentland/2011/08/14/cisco-exitsenergy- management-software-market/

122. T. Gonen, *Electric Power Distribution Engineering*, CRC Press, 2014.

123. T. E. Dy-Liacco, Modern Computer Control Centers and Computer Networking, *IEEE Computer Applications in Power*, Vol. 15, pp. 17–22, 1994.

124. H.L. Willis, *Spatial Electric Load Forecasting*. Marcel Dekker, New York, 1996.

125. C.W. Gellings. *Demand Forecasting for Electric Utilities*. The Fairmont Press, Lilburn, GA, 1996.

126. S. C. Sciacca, and W. R. Block, Advanced SCADA concepts, *IEEE Computer Application in Power*, Vol. 8(1), pp. 23–28, 1995.

127. Advancing Smart Grid Interoperability and Implementing NIST's Interoperability Roadmap NREL CP-550-47000 Report, Nov. 2009.

128. A Moser, G. Ejebe, J. Frame, *Network and Power Applications for EMS within a Competitive Market*, Proceedings of the IEEE Power Engineering Society Transmission and Distribution Conference, pp. 280–285, 1999.

129. D. Marihart, Communications Technology Guidelines for EMS/SCADA Systems, *IEEE Transactions on Power Delivery*, Vol. 16(2), pp. 181–188, 2001.

130. A. G. Phadke, J. S. Thorp, and M. G. Adamiak, A new measurement technique for tracking voltage phasors, local system frequency, and rate of change of frequency, *IEEE Trans. Power App. Syst.*, Vol. PAS-102(5), pp. 1025–1038, 1983.

131. T. Lobos and J. Rezmer, Real-time determination of power system frequency, *IEEE Trans. Instrum. Meas.*, Vol. 46(4), pp. 877–881, 1997.

132. D. Fan and V. Centeno, Phasor-based synchronized frequency measurement in power systems, *IEEE Trans. Power Del.*, Vol. 22(4), pp. 2010–2016, 2007.

133. IEEE Standard for Synchrophasor Measurements for Power Systems, IEEE Std. C37.118.1-2011. [Online]. Available: http://standards.ieee.org/findstds/standard/C37.118.1-2011.html

134. IEEE Standard for Synchrophasor Data Transfer for Power Systems, IEEE Std. C37.118.2-2011. [Online]. Available: http://standards.ieee.org/findstds/standard/C37.118.2-2011.html

135. S. Bricker, L. Rubin, and T. Gonen, Substation automation and advantages, *IEEE Computer Application in Power*, Vol. 14(3), pp. 31–37, 2001.

136. B. Milosevic, and M. Begovic, Voltage-stability protection and control using a wide-area network of phasor measurements, *IEEE Transactions on Power Systems*, Vol. 18, pp. 121–127, 2003.

137. A.G. Phadhke, Synchronized phasor measurements in power systems, *IEEE Computer Applications in Power*, Vol. 6, pp. 10–15, 1993.

138. J. E. Tate, and T. J. Overbye, Line outage detection using phasor angle measurements, *IEEE Transactions on Power Systems*, Vol. 23, pp. 1644–1652, 2008.

139. C. Dufour, and J. Bélanger, On the use of real-time simulation technology in smart grid research and development, *IEEE Trans Ind. Appl.*, Vol. 50(6), pp. 3963–3971, 2014.

140. N. Hatziargyriou, M. M. G. Contaxis, J. A. P. Lopes et al., Energy management and control of island power systems with increased penetration from renewable sources, " 2002 *IEEE Power Engineering Society Winter Meeting*. Conference Proceedings (Cat. No.02CH37309), 2002, pp. 335–339, Vol.1, doi: 10.1109/PESW.2002.985008.

141. F. Moghsoodlou, R. Masiello, and T. Ray, Energy management systems, *IEEE Power and Energy Magazine*, Vol. 2(5), pp. 49–57, 2004.

142. M. Shafie-khah, *Blockchain-Based Smart Grids* (1st ed.), Academic Press, 2020.

143. IEEE Standard 1547.2-2008, IEEE Application Guide for IEEE Std. 1547, IEEE Standard for Interconnecting Distributed Resources with Electric Power Systems, 2009. pp. 1–207.

144. IEEE Standard 1547-2003, IEEE Standard for Interconnecting Distributed Resources With Electric Power Systems, 2003. pp. 1–16.

145. R. H. Lasseter, MicroGrids, *2002 IEEE Power Engineering Society Winter Meeting Proceedings*, Vol. 1, pp. 305–308, 2002.

146. A. C. Zambroni de Souza, and M. Castilla (Eds.), *Microgrids Design and Implementation*, Springer, 2019.

147. F. Katiraei, and M. Iravani, Power management strategies for a microgrid with multiple distributed generation units, *IEEE Transactions on Power Systems*, Vol. 21, pp. 1821–1831, 2006.

148. H. T. Haider, O. H. See, and W. Elmenreich, A review of residential demand response of smart grid, *Renewable and Sustainable Energy Reviews*, Vol. 59, pp. 166–178, 2016.

149. A. Abur, and A. Gómez Expósito, *Power System State Estimation*, Marcel Dekker, 2004.

150. A. G. Phadke, and J. S. Thorp, History and applications of phasor measurements, *Proc. IEEE PES PSCE*, pp. 331–335, 2006.

151. J. A. P. Lopes, C. L. Moreira, and A. G. Madureira, Defining control strategies for MicroGrids islanded operation, *IEEE Trans. Power Syst.*, Vol. 21(2), pp. 916–924, May 2006.

152. A. Engler, Applicability of droops in low voltage grids, *International Journal of Distributed Energy Resources*, Vol. 1(1), pp. 3–15, 2005.

153. F. Katiraei, R. Iravani, N. Hatziargyriou, and A. Dimeas, Microgrids management, *IEEE Power Energy Mag.*, Vol. 6(3), pp. 54–65, 2008.

154. J. C. Vasquez, J. M. Guerrero, A. Luna, P. Rodriquez, and R. Te, Adaptive droop control applied to voltage-source inverters operating in grid-connected and islanded modes, *IEEE Transactions on Industrial Electronics*, Vol. 56(10), pp. 4088–4096, 2009.

155. J. P. Lopes, C. L., Moreira, and F. O. Resende, Control strategies for microgrids black start and islanded operation, *International Journal of Distributed Energy Resources*, Vol. 1(3), pp. 241–261, 2005.

156. A. Kwasinski, and C, N. Onwuchekwa, Dynamic behavior and stabilization of DC microgrids with instantaneous constant-power loads, *IEEE Trans. Power Electronics*, Vol. 26, pp. 822–834, 2011.

157. J. M. Guerrero, J. C. Vasquez, J. Matas, L. G. de Vicuna, and M. Castilla, Hierarchical control of droop-controlled AC and DC microgrids - A general approach toward standardization, *IEEE Trans. Ind. Electron.*, Vol. 58(1), pp. 158-172, 2011.

158. E. Planas, A. Gil-de Muro, J. Andreu et al., General aspects, hierarchical controls and droop methods in microgrids: A review, *Renewable and Sustainable Energy Reviews*, Vol. 17, pp. 147–159, 2013.

159. E.-K. Lee, R. Gadh, and M. Gerla, Energy service interface: Accessing to customer energy resources for smart grid interoperation, *IEEE J. Sel. Areas Commun.*, Vol. 31(7), pp. 1195–1204, 2013.

160. V. A. Boicea, Energy storage technologies: The past and the present, *Proceedings of the IEEE*, Vol. 102(11), pp. 1777–1794, 2014.

161. L. E. Zubieta, Are microgrids the future of energy? DC microgrids from concept to demonstration to deployment, *IEEE Electrification Magazine*, Vol. 4(2), pp. 37–44, June 2016.

162. R. H. Lasseter, Smart distribution: Coupled microgrids, *Proceedings of the IEEE*, Vol. 99(6), pp. 1074-1082, 2011.

163. N. Hatziargyriou, *Microgrids Architectures and Control*, New York: John Wiley & Sons, 2014.

164. C. Cecati, C. Citro, and P. Siano, Combined operations of renewable energy systems and responsive demand in a smart grid, *IEEE Trans. Sustain. Energy*, Vol. 2(4), pp. 468–476, 2011.

165. J. J. Justoa, F. Mwasilua, J. Lee and J-W. Jung, AC microgrids versus DC microgrids with distributed energy resources: A review, *Renewable and Sustainable Energy Reviews*, Vol. 24, pp. 387–405, 2013.

166. N. W. A. Lidula, and A. D. Rajapakse, Microgrids research: A review of experimental microgrids and test systems, *Renewable and Sustainable Energy Reviews*, Vol. 15(1), 186–202, 2011.

167. J. J. Justo, F. Mwasilu, J. Lee, and J.-W. Jung, AC microgrids versus DC microgrids with distributed energy resources: A review, *Renewable and Sustainable Energy Reviews*, Vol. 24, pp. 387–405, 2013.

168. J. G. de Matos, F. S. F. de Silva, and L. A. S. Ribeiro, Power control in AC isolated microgrids with renewable energy sources and energy storage systems, *IEEE Transactions on Industrial Electronics*, Vol. 62(6), pp. 3490–3498, 2015.

169. E. Unamuno and J. A. Barrena, Hybrid AC/DC microgrids: Part I: Review and classification of topologies, *Renewable and Sustainable Energy Reviews*, Vol. 52, pp. 1251–1259, 2015.

170. E. Planas, et al., AC and DC technology in microgrids: A review, *Renewable and Sustainable Energy Reviews*, Vol. 43, pp. 726–749, 2015.

171. E. Unamuno and J. A. Barrena, Hybrid AC/DC microgrids: Part II: Review and classification of control strategies, *Renewable and Sustainable Energy Reviews*, Vol. 52, pp. 1123–1134, 2015.

172. T. Dragicevic, X. Lu, J. C. Vasquez, and J. M. Guerrero, DC microgrids, Part I: A review of control strategies and stabilization techniques, *IEEE Transactions on Power Electronics*, Vol. 31(7), pp. 4876–4891, 2016.

173. T. Dragicevic, X. Lu, J. C. Vasquez, and J. M. Guerrero, DC microgrids, Part II: A review of power architectures, applications, and standardization issues, *IEEE Transactions on Power Electronics*, Vol. 31(5), pp. 3528–3549, 2016.

174. H. J. Laaksonen, Protection principles for future microgrids, *IEEE Trans. Power Electron.*, Vol. 25, pp. 2910–2918, 2010.

175. P. Karlsson, and J. Svensson, DC bus voltage control for a distributed power system, *IEEE Transactions on Power Electronics*, Vol. 18(6), pp. 1405–1412, 2003.

176. L. E. Zubieta, Are microgrids the future of energy? DC microgrids from concept to demonstration to deployment, *IEEE Electrication Magazine*, Vol. 4(2), pp. 37–44, 2016.

177. C. L. Moreira, F. O. Resende, and J. A. Pecas Lopes, Using low voltage microgrids for service restoration, *IEEE Trans. Power Syst.*, Vol. 22(1), pp. 395–403, Feb. 2007.

178. A. Selakov, D., Bekut, and A. T. Sarić, A novel agent-based microgrid optimal control for grid-connected, planned island and emergency island operations. *Int. Trans Electric Energy Syst.* Vol. 26(9), pp. 1999–2022, 2016.

179. T. Kato, H. Takahashi et al., Priority-based hierarchical operational management for multiagent-based microgrids, *Energies*, Vol. 7, pp. 2051–2078, 2014; doi:10.3390/en7042051

Index

Note: Locators in *italics* represent figures and **bold** indicate tables in the text.

Printed in the United States
by Baker & Taylor Publisher Services